Silicon Surfaces and Formation of Interfaces

Basic Science in the Industrial World

Silicon Surfaces and Formation of Interfaces

Basic Science in the Industrial World

JAREK DABROWSKI | HANS-JOACHIM MÜSSIG
Institute for Semiconductor Physics, Frankfurt (Oder)

World Scientific
Singapore • New Jersey • London • Hong Kong

Published by

World Scientific Publishing Co. Pte. Ltd.
5 Toh Tuck Link, Singapore 596224
USA office: 27 Warren Street, Suite 401-402, Hackensack, NJ 07601
UK office: 57 Shelton Street, Covent Garden, London WC2H 9HE

British Library Cataloguing-in-Publication Data
A catalogue record for this book is available from the British Library.

SILICON SURFACES AND FORMATION OF INTERFACES
Basic Science in the Industrial World

ISBN-13 978-981-02-3286-3
ISBN-10 981-02-3286-1

To my Parents, and to Ewa and Wojtek,
for their understanding, patience, and love.

Moim Rodzicom oraz Ewie i Wojtkowi,
za ich zrozumienie, cierpliwość i miłość.

– J. D.

Preface

Silicon is the material of choice of the multibillion semiconductor industry. Ongoing miniaturization of microelectronic devices drives critical regions of integrated circuits ever closer to the place where the silicon ends and another material begins, to the interfacial region. Performance and reliability of electronic microelements depend more and more on the microscopic quality of the interfaces and on the microscopic-scale quality of the original Si surfaces. Suffice to say that as soon as in the first decade of the coming millenium the customers will be offered microelectronic chips with minimum features about hundred atoms long and a couple of atoms thick! Small wonder that atomic-scale knowledge is gaining a commercial value.

Determination of the atomic structure and the chemical bonding at semiconductor surfaces and interfaces is a demanding task. Advances in the experimental and computational techniques make this task increasingly managable. The work of physicists is assisted by dramatic improvements in the performance of computers. Fruitful interaction between predictable theories and reliable experiments has greatly influenced our knowledge on the intricated relationship between the behavior of individual atoms and the behavior of mesoscopic and macroscopic objects. We are approaching the ultimate practical goal of modern semiconductor surface research: efficient usage of microscopic information in the control of semiconductor structure and chemistry and, consequently, efficient control of technological processes and design of optimal circuits and systems.

The main objective of this book is to give a comprehensive overview of silicon surfaces and interfaces: what is known and unknown about them from scientific point of view, what is their industrial role, what are the relations between applications and basic research, what are the most probable directions of future developments. Since the scope is broad and the volume small, I took care to avoid the danger that this presentation evolves into an unreadable encyclopaedia accessible only to experts. In fact, there is no particular reason to compile all the data; first, this has been done many times before, and second, this would be a task for a large team. So I decided to focus not on numbers and curves, but on problems. A reader interested in more details is often asked to consult an easily available book or a journal article.

Each technical chapter begins with a general overview, intended to be understandable to any person with some basic knowledge on solid state physics. Individual Sections begin with overviews as well, but these have a more specialized character. Subsections deal with their subjects much more thoroughly and they do not avoid controversial or speculative issues. We have used the advantage of the storyteller and limited the scope of material covered in detail to the topics we found either most interesting or of a major practical importance.

Practical importance is likely to invoke scientific interest. But many subjects which are scientifically interesting have no technological relevance. When combined with lame budgets of many govermnents, an unfortunate but far-reaching consequence of this trivial fact may be that science will be commercialized. Let us hope that scientists will eventually contain this danger by recognizing the problem, by some self-control, and by occasional refreshing exercises in broader, even philosophically oriented thinking.

Acknowledgments. This book would not appear if we did not receive a lot of support from many persons and institutions. I am particularly thankful to Abbas Ourmazd who invoked our interest in technological aspects of scientific research, to Sylwia Hinrich, Roland Nuffer, and Günther Wolff who participated in some of our experiments described here, to Michael Duane, Scott T. Dunham, Ron Goossens, Eugenia Kaczmarek, Tadeusz Figielski, Armin Fischer, Andrzej Fleszar, Kurt W. Kołasiński, Olaf Krüger, Pierre Masri, Mathias Scheffler, and Tadeusz Wosiński, for encouragement, comments, and discussions, and to Piotr Bogusławski, Marilia J. Caldas, Emily A. Carter, Anne Chaka, Ewa Dąbrowska, Steven C. Ervin, Thomas Fauster, Kristin Fichthorn, Peter Gaworzewski, Frank Herzel, Dietmar Krüger, Max G. Lagally, Juin J. Liou, John E. Northrup, Eckhard Pehlke, Wolfgang Ranke, Catherine Stampfl, Brian S. Swartzentruber, Roger I. G. Uhrberg, Vladimir A. Ukraintsev, Andrea Vittadini, John H. Weaver, Don R. Wolters, and Harold J. W. Zandvliet who gave us additional support by reading various portions of the manuscript. Their remarks have greatly helped us to improve the presentation of the material; any imperfections and faults that remained are solely our responsibility. I am also grateful to Acta Chemica Scandinavica, The American Geophysical Union (AGU), THe American Institute of Physics (AIP), The American Physical Society (APS), CRC Press, The Electrochemical Society, Elsevier Science, Gordon and Breach Publishers, The Institute of Electrical and Electronics Engineers (IEEE), The Institute of Physics Publishing (IoP), International Union of Crystallography (IUC), Lattice Press, The Physial Society of Japan, Vacuum Solutions, and Vieweg for permission to use illustrations from their publications.

Part of this work was financially supported by the German Research Society (Deutsche Forschungsgemeinschaft, DFG) Research Projects DA 308/5-1 and MU 1187/4-1 and by the Institute for Semiconductor Physics (IHP) Research Project BAP P902. Our numerical results described in this book were partially obtained on computers of the Max Planck Society in Garching and Berlin, Germany, and of the John von Neumann Institute for Computing (NIC) in Jülich, Germany.

Jarek Dąbrowski
Frankfurt (Oder), 9 May 1999

Contents

2. The Famous Reconstruction of Si(001) 82

How the reconstruction of the clean Si(001) was unraveled and what can one learn from this story.

List of Figures

List of Tables

In chaos Nature's laws hid veiled in night.
God said, "Let Crystal be," and all was light.
It did not last; the devil howling "Ho!
Let Surface be!" restored the status quo.

– anonymous, after Alexander Pope

CHAPTER 1

Fundamental Concepts and Methods

J. Dąbrowski and H.-J. Müssig

Silicon as a technological material. Basic concepts. How to determine the atomic structure of a surface and why this determination is needed. Silicon surfaces on the (001) to (110) azimuth.

1.1. Introduction

Why should one read a book on atomic details of silicon surfaces? Why should one *write* such a book at all? Why should one write or read a scientific book in a world in which the main value is market value and a search for knowledge for its own sake is viewed as a costly extravagance[1]?

The answer may be clear. Silicon surfaces are one of the few places where big industry and truly basic science can meet and profit substantially from their encounter. An industrial manager can convert a share in this profit into money. An academic scientist can win even more: the security of his or her own intellectual independence. Win – or lose it.

The danger is that the progressing commercialization of science, driven by towering costs of scientific research, would turn physicists into a production-line working class. Blue-collar workers at contemporary factory lines work with their hands. The physicists at future lines may have to rely on brains. But they would be similarly exploited: all their skills would be used to assembling the final product as dictated by the policy of the enterprise. As artisans of old left their art at the factory gates, so would scientists leave their science. The freedom of research would be gone.

The master is he who is free to define the problem. He, whose freedom is limited to a search for solutions, is the servant. The precious independence a scientist needs is the freedom of a fashion designer compared to that of a tailor in a factory. A fashion designer uses professional skills and knowledge of the market to dictate what people want to wear next summer. Likewise, a researcher should use professional skills and knowledge of up-to-date technologies and/or current trends in psychology, philosophy, and even religion to dictate what problems the industry or ordinary taxpayers want to have addressed next. That one must look for tasks in a given environment does not impair the creative freedom. Research subjects are otherwise picked up habitually or even at random. The problem is in communication: we must learn the dragon's language lest he devours us!

The main purpose of this book is therefore to portray the physics of silicon surfaces and interfaces from two points of view: scientific and technological. These themes are often interwoven. The scientific thread dominates the first Chapters (2 through 4), which describe atomically clean surfaces. The technological tone gains volume in the Chapter 5 ("Adsorption on Silicon Surfaces") and becomes quite loud in the Chapter 6 ("On the Road to Devices: SiO_2/Si(001)").

A special role is reserved for Chapter 2 ("The Famous Reconstruction of Si(001)") and for the Afterword. The former is a historical narrative illustrating certain aspects of scientific studies: how what we call "scientific knowledge" is shaped by interactions between facts and interpretation, between understanding and assumptions, between experiment and theory. Attention is paid to the interplay between the evolution of surface science methods and the sophistication and accuracy of atomic models. We use a conceptually simple example, the fundamental atomic structure of Si(001), and we assume that the reader has some knowledge about surface physics. The Afterword, a summary and an outlook, begins with a philosophically oriented essay on science in the industrialized world.

Above all, this is however a book on silicon surfaces and interfaces, on a thoroughly studied system with two great advantages: it allows for conceptually transparent experimental studies on the atomic-scale level, and is of growing interest to a multibillion-dollar industry. Though an easy experimental access and technological relevance are often mutually exclusive, analogies can be pursued. Knowledge acquired from an "easy but technologically unimportant" system can be applied to analyse a "difficult but technologically important" system.

This technological bias has certainly affected the contents. We put emphasis on the Si(001) surface and on interfaces between Si(001) and silicon dioxide. When non-silicon atoms are considered, the species are chiefly those which are likely to be interacting with the silicon surface in a commercial device: hydrogen, halogens, oxygen, nitrogen, dopant elements, transition metals. We hope that the contents of Chapters 5 and 6 will help the reader to focus research on systems promising for microchip technologies and will provide the stimulation and knowledge needed to initiate a search for technologically relevant and scientifically interesting topics.

In brief, our choice of subject was motivated by the observation that "silicon is the driving force behind the information superhighway, multimedia, and all the other much-hyped trends that have suddenly been noticed by the media in the past two years" (Dave J. Eaglesham,[2] in *Physics World*, November 1995), and that "semiconductor surfaces are the launching platform for the solid state electronic evolution" (Harry C. Gatos,[3] in a jubilee issue of *Surface Science*, July 1994).

This Chapter begins with an introduction to silicon. We describe the properties of Si as well as the basic features, trends, and problems of the mainstream microelectronic technology, Complementary Metal Oxide Semiconductor (CMOS). Section 1.3 collects key information about the theoretical and experimental approaches instrumental for our discussion. Section 1.4 overviews the fundamental concepts and issues pertinent to this monograph. It uses other Chapters and Sections for illustration and can be therefore viewed as a kind of extended summary. Section 1.5 goes partially beyond the main topic of the book, which is Si(001), giving an account of primary surfaces of silicon and their vicinals.

1.2. The Silicon Age

Silicon has always been one of man's most useful elements.[4] For example, sand and clay are key components of brick (known from possibly over ten thousand years ago) and concrete. Sand is a principal ingredient of glass. Silicates find their use in pottery. Adding 0.5-5% silicon to steel makes a silicon steel, a highly magnetic material. Silicon carbide is one of the most important abrasives. It has been also used in lasers to produce light of 4560 Å and is a promising semiconductor for hard-duty microelectronics working at high temperatures and at high radiation levels. But most importantly, silicon rendered the informatic revolution possible by providing the basic material for inexpensive microelectronic devices. These microchips can be combined with silicon-made, micrometer-sized mechanical devices,[5,6] such as micromotors[7] or accelerometers and gyroscopes.[8]

1.2.1. *The omnipresent silicon*

The name "silicon" comes from the Latin "silex", "silicis", meaning "flint". Si is the second most abundant element on Earth, surpassed only by oxygen. It makes 25.7% of Earth's crust weight. Yet elemental Si does not occur in Nature. It is bound in compounds. Si compounds exist in all natural waters, in many plants, and in the skeletons and tissues of some animals, including the human skeleton. Si is present in the Sun and stars and is the main component of meteorites known as aerolites. **Silica** (SiO_2) is the basis of flint, quarz, sand, opal, agate, amethyst, jasper, or silica glass. **Silicates** (silicon-oxygen compounds of metals) make feldspar, clay, mica, talc, beryl, tourmaline, garnet, olivine, topaz, granite, or asbestos. A very useful class of materials akin to SiO_2 and silicates is **zeolite**[9] (hydrated microporous aluminosilicate containing exchangable Na, K, Mg, or Ca cations). Zeolites are used, for example, in detergents and as a catalyst in the synthesis of petrochemicals. Another important class of chemical compounds manufactered from Si is **silicone**. Silicones are polymers; their backbone structure is a polysiloxane chain, in which Si and O atoms alternate. Silicones may be prepared by hydrolyzing a silicon organic chloride, such as dimethyl silicon chloride. A very great number of polymeric products can be obtained, ranging from liquids through resins and rubbers to hard, glasslike solids.

Though silicon is not poisonous, inorganic dust containing Si attacks the air passageways in lungs. Miners, stonecutters, potters, and others exposed to siliceous dust may develop a serious, sometimes fatal, lung disease known as silicosis. Inhaled asbestos dust may cause cysts and lung cancer.

The scientific history of silicon begins in the year 1800, when Sir Humphry Davy suggested that silica is a compound. The isolation of elemental Si is usually credited to the Swedish chemist Jöns Jacob Berzelius, who obtained in 1824 amorphous Si by heating potassium with silicon tetrafluoride. Impure amorphous Si was probably prepared in a similar way by Joseph Gay-Lussac and Louis Thenard as early as in 1811, but Berzelius

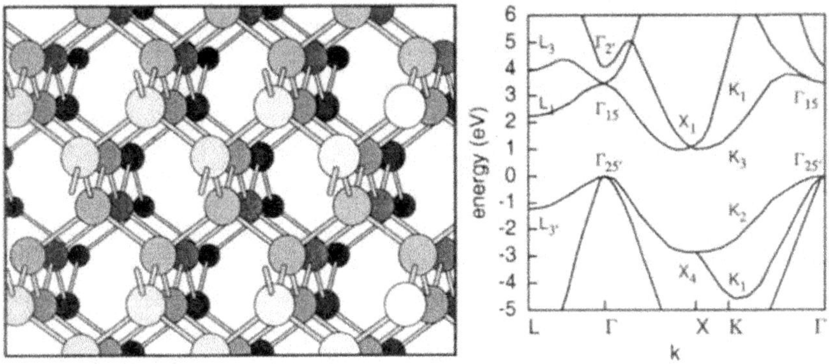

Fig. 1.1. Ball-and-stick model (left) and a schematic band structure (right, adapted from Ref. 10) of crystalline Si. Atoms closer to the observer are brighter and bigger.

succeeded to removed fluorosilicates from the material. Crystalline silicon was first obtained in 1854 by Henri Sainte-Claire Deville.

The major commercial method for preparing **elemental silicon** is to reduce SiO_2 (remove oxygen from it) by heating it with carbon in a furnace. Amorphous Si often has the form of a brown powder which melts easily. Crystalline Si is hard, dark gray, and has a metallic luster. Single Si crystals for the semiconductor industry are grown by the Czochralski process. In this method, a seed is introduced into a melt and then slowly withdrawn. The industry uses now (001)-oriented wafers cut from single crystals of 200 mm diameter. The first 300 mm technology facility is working in Dresden, Germany; 400 mm wafers will be used by the end of the next decade. Hyperpure Si can be obtained by the thermal decomposition of ultra-pure $SiHCl_3$ in a hydrogen atmosphere, and by a vacuum float zone process. While regular grade Si (99%) costs around \$0.50/kg, silicon 99.9% pure costs about \$100/kg; hyperpure Si may cost as much as \$3000/kg. Hydrogenated amorphous Si can be used to produce economical solar cells.

Crystalline Si (Fig. 1.1, left) is a group-IV semiconductor. It crystallizes in the **diamond structure**, with the lattice constant $a_b = 5.43$ Å. This corresponds to the density of $5.0 \cdot 10^{22}$ atoms/cm^3. The *fcc* unit cell contains two atoms, located at (0,0,0) and $\frac{a_b}{4}(1,1,1)$, and the primitive translation vectors are $\frac{a_b}{2}(1,1,0)$, $\frac{a_b}{2}(1,0,1)$, and $\frac{a_b}{2}(0,1,1)$. Each Si atom has four neighbors. The bonds run along $\{111\}$ crystallographic directions, the bond length is $b = a_b\sqrt{3}/4 = 2.35$ Å, and the bond angles are tetrahedral (109.47°). Each $a_b \times a_b \times a_b$ cube contains four fcc cells, that is, eight Si atoms. The volume Ω per a bulk Si atom is thus $\Omega = 20$ Å^3 = 12.1 cm^3/mole. The **bulk modulus**[11] of Si is 98.8 GPa, that is, hydrostatic pressure of approximately one thousand atmospheres is needed to compress the volume of a Si crystal by 0.1%. The **electronic configuration** of an isolated Si atom is $1s^2 2s^2 2p^6 3s^2 3p^2$. The valence and conduction bands of the crystal (Fig. 1.1, right) are formed by bonding and antibonding combinations of Si $3s3p^3$ hybrids of the third shell. The contribution of core electrons of Si ($1s^2 2s^2 2p^6$) to the bonding is negligible. The **cohesive energy** of silicon is E_c = 4.63 eV (page 49), that is, the energy of an Si–Si bond in bulk silicon is 2.32 eV (Table 1.4, page 57).

The electronic **band structure** of bulk Si (Fig. 1.1) has been thoroughly studied, experimenatlly as well as theoretically.[12,10,13] Interestingly, not all details of this structure have been clarified completely. For instance, recent theoretical work[14,15,16] indicates that the Si valence band may be 11.5 eV wide, which is approximately 10% less than the generally accepted value of 12.5 eV. Such a narrower band appears to be consistent with newer experimental data,[17] acquired after the width of 12.5 eV had been established. It is noteworthy that these new measurements yielded a well-resolved band structure of the bulk because the states associated with the crystal surface were passivated (by adsorbed Sb atoms). The energy **band gap** of crystalline silicon is 1.2 eV at 0 K and 1.1 eV at room temperature.[13] Although this corresponds to near-infrared light, the gap is indirect, so silicon crystals are not good for optoelectronics.

Silicon is **doped** by elements from groups III (shallow acceptors) and V (shallow donors). But group-III Al is not a useful dopant as it diffuses quickly in bulk silicon. Group-III Ga is not a good choice because it diffuses fast in SiO_2. Silicon dioxide remains in contact with doped Si in devices manufactured by mainstream microelectronic technologies; see Sect. 1.2.2.

Polycrystalline Si (**polysilicon**) is widely used in the semiconductor industry for gate electrodes (Sect. 1.2.2), interconnections between elements, and for resistors. It is usually deposited from low-pressure silane (SiH_4) vapor at substrate temperatures $600 < T_s < 650$°C. Deposition below $T \simeq 575$°C leads to amorphous films with some dangling bonds saturated by hydrogen; the H content increases with decreasing T_s. Deposition above 625°C produces a film with columnar grains. Their sizes depend on the film thickness. In 100-500 nm films, the grains are 10-50 nm in diameter. Another popular precursor is dichlorosilane, SiH_2Cl_2.

Si is relatively inert chemically at room temperature. It resists most acids (except HF), but is attacked by OH^- ions in basic solutions and by atomic F. Other elements only attach themselves to surface Si atoms. When *cooled*, silicon can be etched by atomic H; this produces volatile silane. When heated to a few hundred °C, Si combines vigorously with halogens (F, Cl, Br, I) to form volatile halides and with most metals to form silicide films (notable exceptions are Au and Ag). **Silicidation** is the standard industrial way to make electric contacts to silicon. At red heat, Si is efficiently **oxidized** by water steam or by oxygen gas, and a surface layer of SiO_2 is formed. Oxidation is a key step in contemporary Si microelectronic technologies because SiO_2 is used in integrated circuits as a versatile isolator. In particular, gate dielectrics in Si MOS (Metal-Oxide-Semiconductor) transistors are made of SiO_2. In future, gate dielectrics for highly miniaturized transistors may be prepared by **nitridation** or oxynitridation. A promising gate nitridation scheme has been recently proposed.[18,19]

Si crystals **melt** at 1414°C. The melting begins at the surface around 1300°C, whereby atoms in the topmost few layers disorder progressively.[20]

1.2.2. *The MOS technology*

Were it not for a fortituous coincidence between the properties of silicon and silicon dioxide, the role of silicon in the semiconductor industry might have been marginal and the information revolution might have been delayed. The Metal-Oxide-Semiconductor (MOS) technology, which has enabled and still maintains the mass-scale production of ever cheaper and faster microprocessors and memories, builds on this chemical relationship between a good and inexpensive semiconductor and an inexpensive and stable insulator, its oxide.

Originally, MOS processes employed SiO_2 layers sandwiched between metal and silicon, hence the name. The metal strip is now substituted by a heavily doped polycrystalline Si strip, but the name has been retained.

The workhorse of MOS is a field effect transistor,[21] or MOSFET (Figure 1.2). The flow of electric current in a "channel" between the source and the drain of a MOSFET is controlled by the potential of the transistor gate. The gate is isolated from this channel by a thin oxide layer, called the "gate oxide".[a] The gate is made of heavily doped polysilicon with a transition

[a] According to the roadmap[22] set by the Semiconductor Industry Association, already by 2001 the gate oxides will be approximately 30 Å (3 nm) thick, with rms (root-mean-

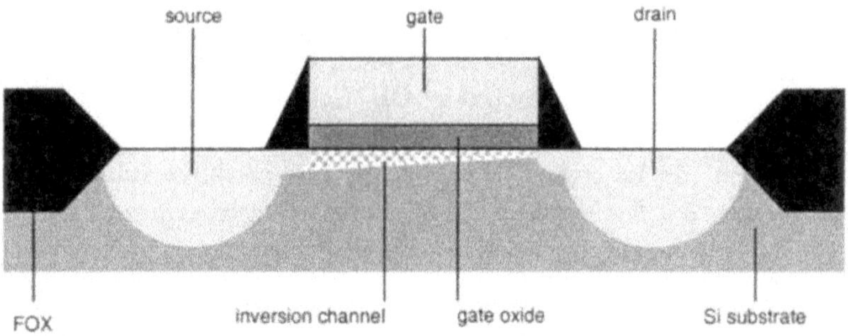

Fig. 1.2. A schematic cross-section of Metal-Oxide-Semiconductor Field-Effect Transistor (MOSFET). The gate oxide (SiO_2) is shown in dark gray. Black areas on the sides of the picture correspond to the isolation oxide ("field oxide" or FOX). FOX surrounds each device, separating it electrically from the rest of the circuit. Semi-triangular terminations of the FOX are "bird's beaks", side-effects of LOCOS isolation (page 377). Also in black is shown the gate spacer (SiO_2 or Si_3N_4) which defines edges of the conducting gate. The gate electrode is a heavily doped polysilicon strip. The other two electrodes, source and drain, are also highly doped. They are separated by the substrate, which is lightly doped with a dopant of the opposite type. The electrostatic potential in the silicon below the gate oxide is controlled by the gate. When the gate voltage V_G reaches a threshold value (V_T), an "inversion channel" opens between the source and the drain.

metal silicide contact on top; contemporary technologies use $TiSi_2$.

In our discussion we will assume, for simplicity, that the source terminal and the back side of the wafer are *grounded*, that is, that their potential is 0. In reality, the electrical characteristics of the transitor depend also on the voltage applied to the back side, the so-called "substrate voltage".

The **working principle** of a MOSFET is simple: depending on the potential of the gate, electric current either flows or does not flow between the source (S) and the drain (D). The heavily doped source and drain (S/D) areas are electrically separated from each other by the source-substrate and substrate-drain *pn* junctions. Carriers on the way between the source and the drain must thus go uphill on a potential curve in order to reach conducting electronic states of the substrate (conduction band for *n*-type S/D, valence band for *p*-type S/D) which is doped oppositely to S/D. When no potential is applied to the gate, no significant electric current can flow between the source and the drain electrodes: the transistor is of the "normally off" type. Naturally, some leakage ("standby") current cannot be avoided.

The gate voltage V_G affects the energy bands of the silicon below

---

square) fluctuations barely exceeding a silicon monatomic step height of 1.38 Å.

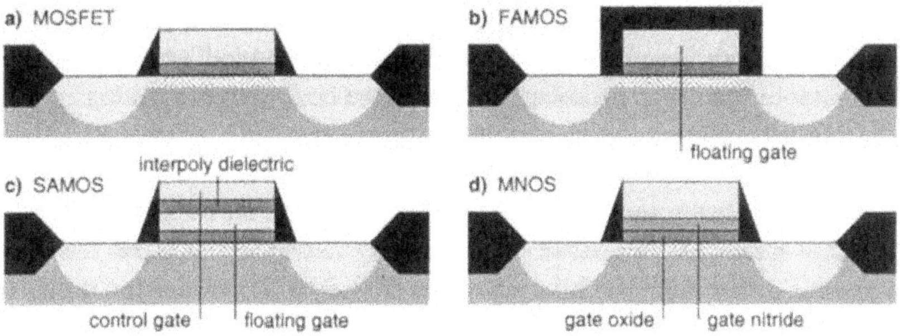

Fig. 1.3. MOS transistors for memory cells. **(a)** MOSFET (Fig. 1.2) stores the charge in an additional capacitor of a dynamic memory. The other three structures are employed in nonvolatile memories. **(b)** Floating-gate Avalanche-injection MOS (FAMOS) stores the charge in the floating gate. **(c)** Stacked-gate Avalanche-injection MOS (SAMOS) stores the charge in the floating gate and erases it through the control gate. **(d)** Metal Nitride Oxide Semiconductor (MNOS), stores the charge at the SiO_2/Si_3N_4interface.

the gate oxide. For example, they can be shifted towards the energy of the carrier states in the source and the drain. The substrate below the channel eventually becomes available to carriers from S/D and a conducting "inversion channel" opens below the gate oxide between S and D.

The inversion takes place at the so-called **threshold voltage** V_T. But the S/D voltage V_D offsets some of the influence of the gate voltage V_G: the potential difference between the channel and the gate is larger near to the source than near to the drain. As a consequence, the channel is wider at the source than at the drain (Fig. 1.2) and its width decreases with increasing V_D. At a certain V_D the channel becomes shorter than the S–D distance. This happens at the **pinch-off** voltage, $V_D = V_{Dsat} \simeq V_G - V_T$. In an ideal MOSFET, the S–D current does not increase for voltages higher than V_{Dsat}, but remains constant at its **saturation** value I_{Dsat}.

In **digital systems** logical data is represented by high (logical 1) and low (logical 0) potentials. MOSFETs act as switches controlling the flow of current between MOS capacitors of other MOSFETs, loading or unloading them and setting them ON or OFF. In this way the data is processed.

Analog systems have stringent requirements on the electric characteristics of transistors. Since analog circuits process information contained in the frequency and amplitude spectra of the signal, their design must assure that uncontrolled nonlinearities do not deteriorate this information. Simple analog amplifiers are needed even in digital systems. Namely, the difference between high and low signal tends to decrease as the data makes its way through the circuit. The "low" potentials become higher and the "high" potentials become lower. At a certain point, a small analog amplifier must restore the order necessary to guarantee error-free processing.

MOS transistors are used also in **memory cells** (Fig. 1.3). Random Access Memories (RAM) employ typical MOSFETs coupled to vertical MOS capacitors assembled in trenches etched in the substrate. Such a capacitor stores one bit. Read Only Memory (ROM) circuits are designed somewhat differently: the information is stored by the transistor itself. In the simplest structure, Floating-gate Avalanche-injection MOS (FAMOS, Fig. 1.3b) the charge is injected from the substrate across the gate oxide and trapped in a fully isolated gate (floating gate). The stored information can be erased only by UV light. Stacked-gate Avalanche-injection MOS (SAMOS, Fig. 1.3c) adds to this structure a control gate which can be biased such that the charge escapes from the floating gate through the interpoly dielectric ("poly" is a technical term for "polysilicon"). The advantage is that in this way the information can be erased *electrically*. Finally, Metal Nitride Oxide Semiconductor (MNOS) transistors (Fig. 1.3d) use a gate dielectric made of two layers: SiO_2 and Si_3N_4. The charge injected from the silicon is trapped at the SiO_2/Si_3N_4 interface. SAMOS designs may adopt similar multilayers in interpoly dielectrics.[23]

MOSFETs are **named** after the carriers flowing between the source and the drain. An n-channel MOSFET (nMOSFET) is built on p-doped material, with n^+-doped source and drain (the index "+" in n^+ and p^+ indicates heavy doping). A pMOSFET is built on n-doped substrate, with p^+-doped S/D. Since the carriers which flow between the source and the drain are of a single type, MOS transistors are termed **unipolar**. A typical *npn* (*pnp*) transistor in which p-type (n-type) base remains in electrical contact with n-type (p-type) emitter and collector is known as **bipolar** because carriers of both signs contribute to its current. A **heterobipolar** transistor (HBT) is a bipolar device which uses a modified material (usually SiGe) in the base region. This SiGe layer is grown epitaxially. Hole mobility in strained Si/SiGe/Si layers is higher than in bulk silicon; this effect has been used to produce fast pMOSFETs with SiGe channels.[24] Recently, a pMOSFET of this type has been obtained by Ge implantation.[25] Hybrid technologies are being developed which combine HBT and MOSFET devices on the same wafer. Technologies of this class are known as **BiCMOS**.

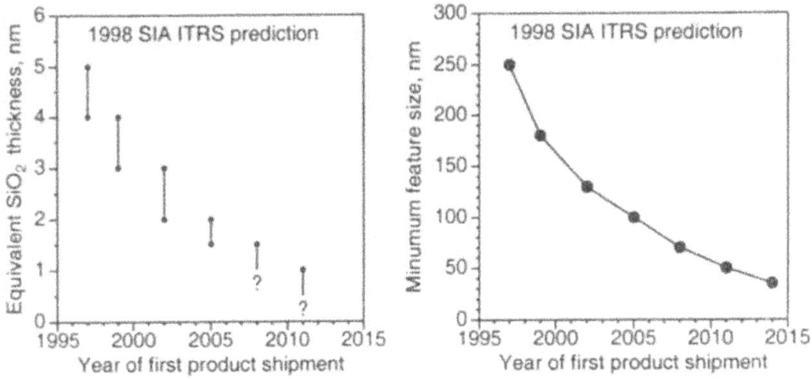

Fig. 1.4. Trends in the miniaturization of MOSFET devices, as predicted by the 1997 version of the Semiconductor Industry Association National Technology Roadmap for Semiconductors.[22] **Left**: Equivalent SiO_2 thickness t_{ox} of the gate MOSFET dielectric in microprocessors. The question marks indicate that the lower limit is uncertain. **Right**: Minimum feature size (the width of the gate oxide strip, or "linewidth").

a. *Miniaturization*

In not so distant past, the race to the outer space and to the Moon stimulated the miniaturization of electronic systems. The background of these space projects, apparently scientific, was in fact military. Today, the miniaturization of microchips[26,27,22] (Fig. 1.4) is driven not by military demands and not by science, but by economy.[28] The *cost* per bit must drop, implying that interconnects between ever more circuits are brought on a single chip. This enforces making smaller, more densly packed devices with shorter channels, shallower junctions, and thinner oxides. A chip can then deal with more data and works faster.

What oxide is a "thin" oxide? Usually, one speaks of a thin oxide film when its thickness t_{ox} is around 50 nm or less. MOSFET gate oxides are *much* thinner than 50 nm. A typical gate t_{ox} was 7 to 8 nm (70 to 80 Å) in the year 1997, but state-of-the-art 0.25 μm technologies already used gate oxides with t_{ox} = 4-5 nm. Calculations published in 1997 indicated[29] that this can be scaled down to about 2 nm (20 Å). Later in the same year, a working MOSFET with only 1.2 nm of gate oxide and 0.06 μm gate length was demonstrated.[30] Transistors with such ultrathin oxides work fast but have high power losses due to tunneling through the oxide. This is more pronounced for nMOSFETs than for pMOSFETs, because electrons tunnel more easily than holes. The high tunneling currents leaking through ultrathin gate oxides cause (controversial) reliability concerns (Sect. 6.6.2a).

Cutting-edge microchip technologies often contribute only a small fraction of the total revenue, particularly in microprocessors; scaling down memory chips pays off better. At the time these words are written, "cutting edge" is 0.18 μm linewidth (the length of the gate oxide strip, somewhat exceeding the channel length). Maintaining the increasing performance of transistors is very expensive.[31] Costs of new facilities and production costs increase dramatically from technology generation to technology generation. Around a billion dollars must be invested in order to build a modern factory! A great majority (say, 95%) of the products shipped today by a prospering company may thus represent "old" 0.5, 0.8, and even 1 or 2μm technologies. The message is: *though the future has its challenges, innovations which optimize yesterday's technologies may be very welcome today.*[28]

b. *Technological MOS processes*

In order to produce a typical integrated circut, several hundred fabrication steps must be performed. They include such processes as lithography, cleaning, etching, oxidation, implantation, deposition, and polishing.

The wafer must be divided into electrically **isolated** areas which will be occupied by individual circuits and, within each circuit, by individual devices. This is done conventionally by LOCal Oxidation of Silicon (LOCOS; state-of-the-art technologies use Shallow Trench Isolation (STI, Sect. 6.3.7). The substrate must be structured electrically, as well. The dopants are **implanted**[32,33,34] and subsequently activated by **annealing**. Activation of boron may take approximately one minute at 1000°C, or three hours at 800°C. Boron is the standard acceptor; it is implanted usually as BF_2.[b] Indium is a candidate for the future. Phosphorus and arsenic are the standard donors. In future designs, antimony donors may be used more and more frequently.

A factory purchases *p*-type wafers doped with boron. In the Complementary MOS (CMOS) technology, which dominates today, nMOS and pMOS devices are grown on the same wafer. The *n*-type substrate "tubs" for pMOS, all sources and drains, and dopants adjusting the profiles of channel regions are implanted through a screen oxide film, sometimes cov-

[b] BF_2 is used due to its size: boron ions are small and do not stop easily enough in the silicon lattice when implanted alone. The dopant is however the usual substitutional boron. The molecules break down from the impact and the fluorine mostly escapes.

ered by a Si_3N_4 layer. These films not only protect the substrate but also help to reduce channeling of the implants through the "holes" in the crystalline lattice of Si. Channeled ions are troublesome because they may travel much deeper and in a different direction than was planned. But channeling becomes of little concern when ultra-shallow junctions (such as 50 nm for 0.18μm CMOS or 30 nm for 0.1μm CMOS) are implanted.[33]

As next, **gate oxides** must be prepared and covered by gates (but see page 6.3.7 for a new technology, EXTIGATE, in which the gate oxide and the gate polysilicon are prepared on a blank wafer *before* any structuring). The pad oxide is thus etched away and high-temperature oxidation in dry oxygen ambient is done (Section 6.3). This is followed by deposition of **polysilicon** gates. The deposited silicon may be doped, but often at least some of the gates must be implanted, as the device parameters are optimal when the gate and the channel have the same work function (are both n- or both p-type).

Subsequently, **contacts** to sources, drains, and gates are prepared. A transition metal layer (usually Ti, in the future possibly Co) is deposited and allowed to react with silicon to form a silicide. Titanium and cobalt react with Si but not with SiO_2, making silicidation easy: the unreacted metal is washed away, and the contacts are ready. This convenient procedure is known as self-aligned silicidation, or **sallcidation**.

When the active devices and the contacts are made, the whole circuit is wired. This is currently done with horizontal Al lines running in a few planes (**metallization** levels) and joined by vertical W pins ("plugs"), but copper emerges as the future material of choice. At the end of 1997, IBM announced[35,36] the first technology (0.2 μm) using Cu instead of Al.

The wires are separated by a dielectric. Glasses such as boronphosphosilicate glass (BPSG) are used, but their relatively high dielectric constants hinder propagation of electric signals in the wires (RC time constant) and facilitate interference (cross-talk). New dielectrics will thus be needed soon, and even Cu may ultimately have to be replaced by another material.

Creating a network of wires between elements of sub-micrometer size is not easy. Most of these several hundred fabrication steps are needed to connect the devices into a complete circuit. Such processes are known under the common name of **Back End** Of the Line (BEOL). The steps needed to make active devices belong to the **Front End** Of the Line (FEOL).

Some BEOL processes, like electrochemical etching and plasma deposition, involve charging of the wafer. This is harmful to the gate oxides and their interfaces with silicon.[37,38] Also stray electrons from processes like electron beam lithography can deteriorate the oxide. Electron beam lithographies are not used in production lines, mainly because exposure times are too long. But SiO_2 lenses go opaque for light wavelengths shorter than about 0.2 μm. Methods to overcome this problem have been proposed;[39] for instance, a 0.2 μm lithography can be followed by chemical shrink (an additional film is grown on the photoresist, so that the opening between the masking strips is reduced). Recent advances in optical materials (CaF_2 optics) and lasers (F_2 157 nm laser) suggest that it will be possible to produce features greater than ~60 nm by optical lithography.[40] However, deep submicron technologies (below 0.06 μm) may eventually have to employ electron beams, ultraviolet light, or soft X-rays.[41] The first integrated circuits with channel lengths of this magnitude are expected[22] to appear on the market around the year 2010 (Fig. 1.4, page 11).

1.3. Methods of Modern Surface Science

The interest in the science of surfaces and interfaces is in the studies of their geometric, chemical, electronic, and dynamic properties. These goals are pursued experimentally as well as theoretically. The degree of coupling between experiment and theory depends on the problem and on the applied technique. Sometimes, as for AES,[c] not much more than standard mathematical tools such as deconvolution or Fourier transform is needed to extract a lot of information. In some cases, for example, when atomic positions are inferred from LEED, advanced computer simulations are necessary. In other cases, as for STM, a visual inspection gives many hints about the actual atomic structure, but a reliable analysis requires a thoroughful optimization of the energy of supposed reconstructions by a reliable theoretical method as well as calculation of their electronic structures. Design of new technological processes is often supported by simulation runs which usually interpolate or extrapolate experimental data on the basis of models with widely varying degrees of complexity.

Be it simple rule-of-thumb estimates, or modeling by analytical formulas and integrable equation sets, or sophisticated numerical simulations, cal-

[c] For acronyms and references see Table 1.2 on page 29.

culations are indispensable in science and technology. Since mathematical constructs represent formalized patterns of logical thinking while physical processes correspond to the data on which our thoughts feed, this "trivial" correlation between physics (the observable) and mathematics (the logical) may have deep epistemological roots (Chapter 7).

Homo sapiens tries to generalize and extrapolate observations, changing them into predictions and designs. Thus, theoretical methods evolve in a natural way along with their experimental counterparts. As illustrated in Chapter 2 on the simple example of the clean Si(001) surface, theory and experiment have been continuously stimulating each other, even when (or especially when) their conclusions seemed contradictory. Reliable atomistic calculations and reliable measurements of atomic positions on the surface became possible approximately at the same time. Today, the accuracy with which such a tiny structural detail as the buckling angle of Si(001) surface dimers can be measured is the same as the accuracy with which it can be computed ($\sim 5°$, or $\sim 25\%$). On the other extreme, Nature is sometimes so friendly that relatively superficial understanding may be good enough for astonishingly complex applications. Vide the unpreceded success of the CMOS technology. But as the size of commercial CMOS devices shrinks rapidly towards the atomic scale (Fig. 2.9, page 123), the ability to predict the behavior of individual atoms is gaining practical importance.[28]

1.3.1. *Theoretical techniques*

Below we briefly describe the salient features of the theoretical methods mentioned in the coming Chapters. We will focus on those approaches which are instrumental in getting a reliable atomic-scale picture of silicon surfaces. These are the *ab initio* techniques, free from adjustable parameters and constructed from the first principles of quantum mechanics. We will also consider less accurate and less expensive atomistic tools which are needed for exploratory purposes or to extrapolate more exact results to systems too big for *ab initio* treatment. We will devote a few words to continuous models, that is, to models which ignore the discrete atomic structure. These models make the interface between the world of atoms and "our" world. In all cases we will refrain from presenting technical details, so that the physics rather than the formalism is emphasized. Table 1.1 lists some technical references and review articles.

Advanced solid state physics uses **quantum mechanics** (QM) to describe electrons. Of interest to us are the descriptions providing direct

Ref year	Subject relevant to theoretical methods of surface science
42 1964	Fundamentals for density functional theory
43 1965	Fundamentals for density functional theory, LDA and GGA introduced
44 1965	Quasiparticle theory: GW approximation introduced
45 1966	The popular valence force model introduced by Keating
46 1974	Harrison's bond-orbital model introduced
47 1979	Chadi's tigh-binding scheme introduced
48 1985	Classical Stillinger–Weber (SW) potential introduced
49 1986	Classical Tersoff potential introduced
50 1986	Stillinger-Weber potential and nonideal atomic configurations
51 1987	Stillinger–Weber potential improved
52 1987	Dodson potential: a many-body Tersoff-type potential for silicon
53 1987	SW, Tersoff, and Dodson potentials applied to simulate Si(001)
54 1987	A GW approach for semiconductor surfaces; As/Ge(111)
55 1988	A Green's function approach for surfaces; Si(001) 2×1
56 1988	Interaction between surface steps; elasticity theory
57 1989	Sankey's DFT-based tight-binding scheme introduced
58 1989	**Review**: Density functional theory
59 1990	Configuration interaction; numerical efficiency
60 1990	Modelling of silicon surfaces: a comparative study
61 1991	Kinetic Monte Carlo: basic implementation
62 1991	Monte Carlo: Poisson distribution, kinetic pathways, detailed balance
63 1991	Keating model modified by Tersoff for surfaces
64 1991	**Review**: Generalized gradient approximation (GGA)
65 1992	GGA and pseudopotential calculations for semiconductors
66 1992	**Review**:Total energy calculations in surface studies
67 1992	Analysis of the role of GGA in surface calculations
68 1992	**Book article**: DFT, quasiparticles, band-gap problem
69 1993	Density functionals analysed by Monte Carlo calculations
70 1993	Density functionals analysed by analytical models
71 1995	**Review**: Methods for total energy calculations in surface studies
72 1995	A GW scheme efficient for surfaces
73 1995	Anharmonic Keating model for Si, Ge, C, and their alloys
74 1996	A modified Stillinger–Weber potential for Si surfaces
75 1996	Potential of a Stillinger–Weber type succesfully fit to HF
76 1997	Equilibrium shapes of epitaxial islands; elasticity theory
77 1997	Why does LDA work well for silicon; quantum Monte Carlo
78 1997	Screened-exchange LDA justified theoretically
79 1997	**Review**: Ab initio dynamics by quantum chemistry methods
80 1998	**Review**: Ab initio dynamics by DFT methods
81 1998	GGA: core-valence exchange correlation and binding energies
82 1998	Beyond the Born-Oppenheimer approximation
83 1998	**Review**: quasiparticles, GW approximations

Table 1.1. Theoretical methods of surface science: selected references.

links between atomic and electronic structures. Our sole purpose is to introduce the basic concepts needed to comprehend the role and reliability of the approaches which simulate atomistic processes on Si surfaces.

Such methods can implement QM equations on various levels. Different approximations are made which reduce the complexity of the problem to a managable size, but in all practical cases the calculations are done on a computer. There are two major divisions of QM computer codes: (1) semi-empirical, employing hamiltonians with parameters which must be fitted to reproduce certain experimental data, usually electronic structures, and (2) *ab initio*, with hamiltonians generated without empirical input.

The distinction between these two divisions may seem fuzzy. On one hand, the parameters in a nominally semi-empirical code can be fitted also to *ab initio* data. On the other hand, a nominally *ab initio* calculation may describe the interactions between atoms and electrons through parametrized potentials, which are generated by another *ab initio* method and verified by comparison of the results with experiment. In practice, one speaks of an *ab initio* calculation when the "parametrization" is transferable between a reasonably wide range of atomic systems. A truely *ab initio* input is merely a table which lists atoms (their species and initial positions); the physical rules responsible for interactions between the atoms and the electrons are contained in the numerical algorithm. In other words, all silicon atoms in an *ab initio* input deck are safely given the same properties, while a semi-empirical code should in principle adopt a scheme in which silicon atoms acquiring different bonding configurations are given different parameters. Clearly, *ab initio* descriptions are much more flexible and their results are less likely to be biased by the expectations of the researcher.[d]

Ab initio calculations are done **self-consistently**, that is, the simulation is done iteratively and stops only when a measure of the difference between the potentials generated in consecutive iteration steps is sufficiently small. This measure can be, for example, the change in the computed total energy. Certain QM semi-empirical schemes may also adopt self-consistency, but this is an exception rather than the rule. On the other hand, in rare cases approximate *ab initio* calculations are done without completion of the self-consistent loop.

The size of an atomic system which can be treated *ab initio* is nowadays limited to around thousand atoms. Semi-empirical quantum mechanical schemes can increase this size significantly, but a real breakthrough comes with **classical methods**. They ignore the quantum behavior of

[d]This bias is otherwise difficult to avoid because it is interwoven with the most useful tool of a physicist: the physical intuition. See Chapter 2 for profound examples.

electrons and describe the energy and motion of atoms in a model interatomic potential. The most popular model functions are potentials of the Stillinger-Weber and Tersoff type, and even simpler valence-force schemes based on the Keating model. These are beautiful and versatile schemes, extremely useful as means to get an educated guess as well as means to explore systems containing hundreds of thousands of atoms. This may be enough to directly address certain problems of technological relevance.[28]

However, any direct link to the electronic structure is lost in a classical treatment. Examples of unphysical behavior caused by this oversimplification are known particularily for structures in which dangling bonds or unusually coordinated atoms play a role (for example, see pages 210 and 227). Critical configurations must thus be cross-checked against lower-level methods. In any case, dynamical calculations are limited even with classical potentials because the time scale in all strictly atomistic approaches is determined by the atomic vibrational frequency (about 10^{-13} s). There are thirteen or more orders of magnitude from the atomic to the technological time scale of seconds, minutes, and hours!

Monte Carlo methods (see Sect. 4.3.1a for an example of a Monte Carlo approach to a surface phase transition) get around this problem by describing atomic dynamics in terms of energy barriers and hopping frequencies. This increases the time scale by two to six orders, to the hopping frequencies. If deposition and/or surface diffusion should be simulated, the density of the mobile species drops from the bulk atom density ($10^{23}/\text{cm}^3$) to the density of the mobile surface atoms (below $10^{15}/\text{cm}^2$). Monte Carlo simulations are even closer to technological applications than classical potentials. Surface studies may help to develop schemes for rigorous transition of parameters from *ab initio* through model potentials to Monte Carlo.

The approaches which used to be most efficient for systems of real-life time and size scale are those which ignore the discrete structure of matter. Most interestingly, a recent implementation[84] of **Kinetic Monte Carlo**[61,62] (KMC) for simulation of technological processes[28] presents a serious challenge to continuous schemes. KMC can simulate dopant profiles in 0.18 μm CMOS devices as fast as continuous models used in commercial simulation tools can do it. Since KMC uses physically transparent parameters such as activation energies and hopping frequencies, it provides quite a straightforward link between atomistic *ab initio* or experimental data and technological process database. The key difference between Kinetic Monte

Carlo and the "classical" Metropolis algorithm of Monte Carlo techniques is that individual KMC events are assigned thermally activated jump rates in KMC instead of thermally activated jump probalilities. As a consequence, every KMC event is guaranteed to take place at every iteration.

Continuous models are formulated in terms of continuous differential equations which can be integrated efficiently by standard numerical algorithms. Their typical application is in simulation of dopant diffusion in the semiconducting substrate.[85] Since all commercial tools for simulations of technological processes still rely on continuous models, experience in passing well-defined parameters from atomistic studies to continuum equations might turn out to be of real practical importance. An example of an approach which closely links atomic and continuous description is given in Section 4.3.2 in the context of interaction between steps. A somewhat related issue is the problem of island formation during deposition.[76]

a. *Approximations in* ab initio *studies*

We will now browse through the various types of *ab initio* studies, emphasizing the approximations made and the practical consequences of these approximations. We will focus on the methods of treating many-body wave functions and on the common tricks that allow one to reduce the number of atoms and electrons for which the calculation has to be done.

All practical QM codes use the **Born-Oppenheimer** approximation which separates the motion of electrons from the motion of atomic nuclei. The nuclei are then treated classically. This separation and the classical description of the ions is possible because nuclei are much heavier than electrons. Even the nucleus of hydrogen, a single proton, weighs ~2000 times more than an electron. Ideas for calculations going beyond the Born-Oppenheimer approximation have been recently discussed in Ref. 82.

The electrons, even assumed to adiabatically adopt to the positions of the nuclei, are still a very complex system. Their motion is mutually corelated, and there are around 10^{22} of them in each cm^3 of a solid. In general, they are described by the time-dependent many-electron wave function Ψ which becomes time-independent only when the ground state is considered.

Since many-electron functions are unwieldy, the problem is usually recast in a single-particle framework. A physically transparent single-electron basis set $\{\phi\}$ is chosen. It may contain, for example, atomic orbitals, plane waves, or gaussians. Auxillary wave functions $\{\psi\}$ of individual electrons are expressed as linear combinations of the basis functions $\{\phi\}$. Finally,

Ψ is approximated by a combination of the functions from $\{\psi\}$, an effective hamiltonian is constructed, and the resulting Schrödinger equation is solved. The reliability of the solution depends on the completeness of the basis sets, on the approximations made to Ψ and to the hamiltonian, and on the numerical accuracy with which the Schrödinger equation is integrated.

There are three different ways to construct and solve this equation. The first one is the Hartree-Fock (HF) approximation which can be refined by including Configuration Interactions (CI). This method is popular among chemists[79] and hence sometimes called "quantum chemistry". The second one is the Density Functional Theory (DFT). It is often applied together with the Local Density Approximation (LDA) which can be refined by General Gradient Approximations (GGA); this method is popular among physicists.[80] The third one is the quantum Monte Carlo approach[86] (see also Ref. 87 and references therein), the most exact but also the most elaborate way of treating many-electron systems; this is the method of choice for hard-core theorists. Monte Carlo results for homogeneous electron gas can be used as a firm basis to construct LDA or GGA potentials.[88,89,67] We will now look closer at HF- and DFT-based methods.

In the **Hartree-Fock** approximation, the many-body function Ψ is assumed to have the form of a Slater determinant built from the single-electron functions $\{\psi\}$ so that the Pauli exclusion principle is obeyed. In other words, the *exchange* energy of the many-body system of electrons is properly described.[e] However, a single Slater determinant is not flexible enough to represent a true many-body state. The effects not captured by the HF approximation are termed *correlation* effects. **Configuration Interaction** accounts for the correlation of electron motion by constructing Ψ from a linear combination of many Slater determinants. Since the HF wave function is a reasonable approximation, one starts from the HF "ground state" solution Ψ_0 and adds to it weighted Slater determinants obtained from Ψ_0 by exchanging some of its ground-state single-particle elements by single-particle functions corresponding to "excited" states (that is, those unoccupied in a single-determinant calculation). The weights are then varied to minimize the total energy. As expected, the weights tend to be smaller for higher excited Slater determinants: the expansion converges.

The **Density Functional Theory**[43,58,90,91] (DFT), which is in fact

[e]The exchange energy is in fact just a part of the real electrostatic energy sorted out formally by the introduction of Slater determinants.

a general approach to inhomogenous many-particle systems,[92,93] provides an alternative to Hartree-Fock calculations. Walter Kohn won the 1998 Nobel Prize in chemistry "for his development of the density-functional theory" (together with John A. Pople, who received it "for his development of computational methods in quantum chemistry").

DFT builds on the Kohn-Sham theorem.[43] This theorem assures that the electrostatic potential $V_{nuc}(\vec{r})$ of the nuclei in the crystal defines a functional $F[\rho]$ of the real-space electron density $\rho(\vec{r})$ such that the minumum of F is at the density of the electrons in the many-body ground state. The minimum value of F corresponds to the total energy E_{tot} of the many-electron quantum system. When ρ is formally written down as a sum of contributions from single-particle states $\{\psi\}$, that is, $\rho(\vec{r}) = \sum_i f_i \psi_i^*(\vec{r})\psi_i(\vec{r})$, the variational principle yields Schrödinger equations from which ψ_i functions of $\{\psi\}$ can be computed:

$$H_{eff}\psi_i(\vec{r}) = T_{eff}[\rho(\vec{r})]\psi_i(\vec{r}) + V_{eff}(\rho(\vec{r}))\psi_i(\vec{r}) = E_i\psi_i(\vec{r}). \qquad (1.1)$$

These equations contain an effective kinetic energy functional T_{eff} and an effective potential V_{eff}. In principle, Eq. 1.1 allows one to determine the electron density and the total energy of any system of atoms. This formalism can be extended to finite temperatures of the electrons.[94]

The density $\rho(\vec{r})$ obtained from the solutions of Eq. 1.1 could be exact if the functional F and, consequently, the effective hamiltonian were known exactly and could be written down without approximations. Unfortunately, this is not the case. The kinetic energy is in practice approximated by a single-electron expression and the dependence of the exchange and correlation potential on ρ is taken after *ab initio* data for homogeneous electron gas. The **Local Density Approximation** does it by assuming that the exchange and correlation energy E_{xc} can be approximated through:

$$E_{xc} = \int \mathcal{E}_{xc}(\rho(n(\vec{r})))d^3\vec{r}, \qquad (1.2)$$

where $\mathcal{E}_{xc}(\rho(n(\vec{r})))$ is the energy density of homogeneous electron gas of density n. This works astonishingly well,[77] even though one of the consequences is a clear tendency to overbind atoms (LDA bonds are usually ~1% shorter than real bonds). When total energies of atomic structures with very different degrees of electron localization are compared, LDA must be augmented by approaches like the **Generalized Gradient**

Approximation[64,67,69,81] (GGA). A gradient expansion of LDA must be done with care, so that certain sum rules are not violated; a GGA result is not automatically guaranteed to be better than an LDA result. Various GGA schemes are currently under intensive investigation (for example, see Refs. 95, 96, or 81).

By construction, DFT is a method to obtain the electron density and the total energy of a many-body system in its ground state. The single-particle eigenfunctions ψ_i and their eigenvalues E_i are only technical by-products of the calculation. However, since the effective DFT potential V_{eff} resembles the potential seen by real electrons, ψ_i and E_i are often interpreted as "true" wave functions and their energies. This works quite well for occupied states, but is generally less reliable for empty states. A troublesome consequence is that LDA band gaps come out much smaller than in expermient. Part of the problem can be blamed on the way the exchange and correlation is treated in standard DFT codes. The so-called **screened exchange** LDA[97,98,78] removes some of these shortcomings and improves LDA energy gaps and total energy differences.

The correct computation of experimentally accessible electronic energies requires, however, that a theory specifically designed for many-body *excitations* is used. Indeed, the energies accessed in various measurements of the electronic structure are obtained through excitation of the sea of electrons. Though the excited object often does resemble a singled-out electron, all electrons remain in fact identical. The excitation may be viewed as a **quasiparticle** of a limited lifetime and with a wave function resembling that of the DFT eigenfunction.

The most popular method to compute quasiparticle energies, the **GW** approach of Hedin,[44] is as old as the DFT. In the name "GW", G stands for Green's function and W the for screened electrostatic potential. This is the characteristic term in the expression for the self-energy operator Σ used to iteratively generate quasiparticle functions and energies. Good first guesses are LDA wave functions and eigenvalues.

An important feature which exemplifies the difference between a quasiparticle energy and a DFT single-particle eigenvalue is that Σ depends explicitely on the *wave function* of the quasiparticle, while the DFT hamiltonian depends only on the electron *density*. This distinction is essential: DFT electronic structures of several important systems, such as the buckled Si(001) 2×1 surface or Ge bulk, are metallic, while their quasiparticle structures are semiconducting.[f] The difference between DFT eigenvalues and GW excitation energies depends on the wave function, that is, also on the $\vec{k}$ vector from the Brillouin zone. The GW band gaps are close (usually within 0.1 eV) to those observed in experiment.

Rigorous quasiparticle calculations are time consuming, chiefly because screening plays a major role. The dielectric response of the system has to be calculated, which is a very elaborate process. Use of model dielectric functions greatly speeds up the calculations and only moderately deteriorates the accuracy of the obtained electronic structure.[72]

No matter how the Schrödinger equation is written, it still contains nearly 10^{24} electrons for each cubic cm of silicon. The size of the problem can be reduced by treating only a limited number of atoms explicitly. This means that the rest of the crystal must be represented by properly chosen boundary conditions. One popular choice is to cut a **cluster** of atoms from the crystal and passivate all cut bonds by hydrogen atoms; a more sophisticated method would use so-called siligens, that is, fictitious hydrogen-like atoms with the potential tailored such that it reproduces the electronegativity of silicon, and with it the properties of a true Si–Si bond. However, the boundaries of clusters break the continuity of wave functions, imposing artificial localization on electron states and deteriorating electronic structures. A proper inclusion of lateral relaxation of the stress imposed by the surface reconstruction is also problematic.

The boundaries become less critical when the cluster is periodically repeted as a "**supercell**". This method, while producing in a natural way the infinite bulk of a perfect crystal, requires that the crystal with a surface is represented by a **slab** of a limited thickness. The slab is then periodically copied in the direction normal to the surface. Periodicity allows one to expand the wave functions into plane waves. The advantages of plane wave basis set include easy convergence tests (Sect. 1.3.1b) and easy implementation of efficient numerical algorighms, such as the Car-Parrinello scheme.[99,100,101] But care must be taken that the surfaces of the slabs are decoupled, that is, that the vacuum region separating the slabs is thick enough and that the slabs are sufficiently thick. The slab separation and thickness must be checked for convergence; the accuracy of a calculation which does not refer to corresponding tests may be difficult to judge.

---

[f] These are not failures of DFT, but simply the prominent cases for which the interpretation of DFT eigenvalues as excitation energies works worse than naively expected.

The third method to accomodate boundaries is to use the **Green's function** formalism. The advantage is that a single surface can be computed, with vacuum on one side and the bulk on the other. The Dyson equation allows one to limit the actual calculation to a surface region of a limited thickness and keep the result matched to the solution obtained previously for the perfect crystal. The surface may have an imposed lateral periodicity, but when the Green's function of the periodic surface has been tabulated, a similar formalism can be used to treat a localized defect.

The size of the system which needs to be computed may often be further reduced, by taking advantage of the very small contribution of core electrons to the interatomic bonding. The only electrons retained in such a simplified calculation are those of the valence atomic shells. The core electrons are assumed to be frozen, that is, to occupy the same states as in free atoms. The common influence of the bare nuclei and the frozen cores on the valence electrons is described by a ficticious potential, a **pseudopotential**. This potential is often nonlocal, that is, it depends on the angular momentum of the electron it acts upon. The reason for this non-locality is simple: valence states with different angular momenta are orthogonal to different core states, which is reflected in the construction of the pseudopotential.

The earliest pseudopotentials were empirical, fitted to measured electronic energies. Modern pseudopotentials are *ab initio*-derived, constructed to reproduce the valence eigenvalues obtained in all-electron *ab initio* calculations for isolated atoms. Also the shape of the atomic valence wave functions beyond a certain cutoff radius r_{cut} and the total valence electron density inside r_{cut} are retained (norm conservation). Transferability of a pseudopotential to different chemical environments is guaranteed by the properly reproduced dependence of the eigenvalues on the externally applied potential (logarithmic derivatives). Since the logarithmic derivatives can be matched only in a limited range of energies, the high-energy scattering properties of a pseudopotential may be unphysical. This is, however, not a problem in the cases we will consider.

A pseudopotential is not uniquely defined by the atomic potential. Many pseudopotentials can be constructed for the same species. They are tailored for different purposes, but when carefully used, they should yield basically the same results within their common range of applicability. The reader should not feel confused by the various schemes found in the literature. The essential point is, however, that the combination of certain species, like Ga and N, requires that some *core* electrons must be excluded from the frozen core and treated together with valence electrons. This happens when the energy of these core electrons falls close to the energy of the valence electrons of the other species. Moreover, the electron density of the core electrons must be sometimes[102,103] accounted for in the DFT exchange-correlation energy (the so-called **nonlinear core correction**[104]).

b. *Convergency issues*

One of the key criterions which determine the practical value of an *ab initio* result is its numerical convergency and an estimate of the error bar. Though a first-principles calculation is free from empirical parameters, it does employ parameters which control its numerical performance. Resources such as computing time, memory, and disk space are limited and their usage must be traded against accuracy. It may be the responsibility of the researcher to perform and discuss convergency tests, yet for many reasons this is not always respected. We will therefore mention a few key points which should help the reader to judge whether a particular work may be viewed as reliable or should be approached with care.

Basis sets are obvious candidates for testing. Functions F such as orbitals, electron densities, or potentials are represented in the calculation as linear combinations $F_N = \sum_i^N c_i B_i$ of N functions B_i from the basis set: F is written as $F = F_N + R(N)$ and the remnant $R(N)$ is rejected because it cannot be represented by B_i. Naturally, the magnitude of $R(N)$ depends on the quality of the basis. The calculation becomes more accurate when more basis functions are used and/or when the shape of the basis functions is more similar to the shape of the actual solution.

Localized sets, such as gaussians or atomic orbitals, are efficient in basis set for species with highly localized valence orbitals, that is, when the most important parts of the functions F occupy a small fraction of the whole volume of interest. At the same time, they are difficult to improve systematically and convergency tests are not trivial. Systematic tests are easy for plane wave basis sets. In this case, the set is defined by the cutoff energy E_{cut}, which corresponds to the energy of a plane wave describing an electron in vacuum. All plane waves up to E_{cut} are included in the set. Yet plane waves, being delocalized, are troublesome when the expanded functions F are very localized. The required number of plane waves grows with increasing localization, sometimes leading to inadequate convergence. Energy differences tend to converge faster than absolute energies.

Of particular interest is here the case of CI calculations. While simple molecules can be treated by this method with a high accuracy, the application of CI to surface studies requires heavy computing. Therefore, small clusters[g] and/or small basis sets are used. This may make the results unrealistic. While DFT-LDA and DFT-GGA treat exchange and correlation approximately but with the contributions from exchange (E_x) and from correlation (E_c) on roughly the same level, CI contains all of E_x and attempts to compute E_c as accurately as possible by introducing as many excited configurations as practicable. Since E_x and E_c are both large and partially cancel one another, it is difficult for CI to converge. A suspected problem of this type is mentioned on page 274.

An issue specific to slab DFT calculations is the **Brillouin zone sampling** ($\vec{k}$-point sets). The need to compute the energy at many different $\vec{k}$-points is the price paid for selecting only a piece of the crystal for the calculation. If all atoms of an infinite, perfect crystal were included in a hypothetical calculation, then all states which build the electron density would be contained at $\vec{k} = 0$, that is, at Γ. On the other extreme, with atoms only from the unit cell included, all states of the crystal can be recovered by taking into account all states in the first Brillouin Zone (BZ). Naturally, taking all $\vec{k}$-points is of little use: the number of states in the first BZ is by construction the same as the number of states in the whole crystal. The calculation is therefore done at a limited number of $\vec{k}$-points.

This is greatly facilitated in semiconductors. If energy bands are either completely occupied or empty, already a few carefully selected $\vec{k}$-points can give a very accurate electron density.[105,106,107] Interestingly, the Γ point, (0,0,0), is not among these so-called special $\vec{k}$-points, although it may be used successfully for big supercells.[108] The surface supercell dimensions must be at least a few 1×1 translation vectors if Γ is used.

However, metallic systems require a brute-force treatment. No theorems for special points in metals exist and densely spaced sampling must be employed. Thus, a comparison of total energies computed for reconstructions with both metallic and semiconducting densities of states must be done with care. This includes the calculation of the energy gained by surface reconstruction. Reconstruction may open a surface band gap, but truncated-bulk surfaces of silicon are metallic.

The **number of atoms** N treated explicitly in the calculation is

[g]HF-type calculations are much easier to implement for clusters than for slabs.

a critical issue because the numerical effort scales usually as N^3 or even N^4. Although promising *ab initio* $O(N)$ algorithms, that is, the algorithms scaling as N, have been designed,[109,110,111,112,113,114] their convergency for large systems is still not tested well.

Slab techniques (which model the crystal with a surface by a periodically repeated supercell) suffer from size effects when relaxation of many subsurface layers plays a role (as it is the case for step formation energies), when disordered structures are addressed, and when isolated surface defects or adsorbates are considered. Finite size effects become even more troublesome when the surface is represented by a single cluster. Tests of the effects of the cluster size and shape,[115,116,117] often indicate serious convergency problems. Embedded clusters at least as big as $Si_{15}H_{16}$ are needed to reproduce buckling of Si(001) dimers.[116] Clusters smaller than $Si_{21}H_{20}$ are insufficient for calculations of interactions between H_2 molecules and Si(001) surfaces.[117] Sometimes it is possible to overcome the difficulties related to the cluster boundaries by fitting a model potential to *ab initio* data obtained for a cluster of a managable size. A succesfull approach of this type was reported[118] for the case of H diffusion on Si(001).

Unconverged results can too often be easily accepted as correct if the outcome agrees with the physical intuition (see Chapter 2). A similar danger, much more difficult to avoid, appears when the atomic structure of a newly investigated object is discussed. Even a perfectly converged *ab initio* calculation done with no approximation would contain a biased input: the atomic configurations before the energy optimization. The number of **structural degrees of freedom** is much greater on the surface than in the bulk. It is not only that the atoms may be interconnected in various ways; also the *number* of surface atoms may differ between physically plausible structures! It is therefore imperative that as many structures as practicable are treated and compared to as many experimental results as possible, before a particular model is accepted as reliable.

c. *Tight-binding methods*

Structures computed *ab initio* can be used to calibrate approximate QM calculations. Probably the most natural implementations are within Tight-Binding (TB) schemes. The idea behind the popular TB methods is to treat the matrix elements of the hamiltonian as adjustable parameters. The fit, although usually done to experimental data, can also readily utilize *ab initio* data. Some specialized approaches have been designed, for example, on the basis of DFT[57,119] or HF calculations.[120] The advantage is that much larger systems can be treated than allowed for by *ab initio* studies. In particular, TB methods are well suited for $O(N)$ algorithms.[109,121]

A well-known TB approach is due to Harrison.[122] It is highly simplified but can give one a feeling about the major chemical trends in the system of interest. In general, properly designed and parametrized semi-empirical calculations can be both efficient and quite reliable. For instance, the TB scheme designed by Chadi[47] has been used succesfully for many systems. One of its most spectacular successes was the first correct prediction of the structure of the clean Si(001) surface (page 100). Tests have shown[115] that semi-empirical calculations for Si adsorption on Si(001) give results intermediate in quality between those obtained with classical potentials and various *ab initio* schemes. A recent tight-binding $O(N)$ implementation gives good results for Si(001) energetics, for instance, for step energies.[123]

A difficult problem in parametrized QM calculations is to obtain a correct description of the matrix elements for the interaction of atoms of different species, as in the case of foreign atoms adsorbed on a surface. In general, handling of surface atoms may be problematic. Also the choice of the basis set is not always obvious.[66]

1.3.2. *Experimental techniques*

Excellent surveys of surface and interface sensitive methods can be found, for example, in Refs. 124 and 125. Ref. 126 discusses spatial resolution of numerous surface techniques. Table 1.2 collects acronyms of selected approaches and gives a single technical reference for each entry; for obvious reasons the choice of bibliography is very arbitrary.

Many popular experimental techniques are based on the interaction with photons, electrons, ions, or neutral particles. After interaction, the flux arriving at the detector imparts structural information. We focus on scanning probe microscopies, which are among the newest and most widely applied of the direct methods for the determination of surface structures. We discuss as well the key techniques that use low energy electrons ($E \leq 1000$ eV). The mean free path of such electrons is only a few atomic layers, making them an ideal tool for surface research.[h] Also X-rays, which since their discovery in 1895 by Röntgen have enjoyed an unprecendented career,[127] are valuable in surface science; we describe the ideas behind sev-

[h] However, it should be noted that irraditation with electrons of the sort used in low-energy electron diffraction and Auger-electron spectroscopy creates surface defects.[128]

Ref year	Reviews on experimental techniques of surface science
129 1994	AES; Auger Electron Spectroscopy
130 1988	ARPES; Angle-Resolved Photoemission Spectroscopy
131 1994	ARUPS; Angle-Resolved Ultraviolet Photoelectron Spectroscopy
132 1982	EELS; Electron Energy Loss Spectroscopy
133 1992	IPE; Inverse Photoemission
134 1990	LEED; Low Energy Electron Diffraction
135 1994	LEEM; Low Energy Electron Microscopy
136 1994	LEISS; Low Energy Ion Scattering Spectroscopy
136 1994	MEISS; Middle Energy Ion Scattering Spectroscopy
137 1994	NRA; Nuclear Reaction Analysis
137 1994	RBS; Rutherford Backscattering
138 1995	REM; Reflection Electron Microscopy
134 1990	RHEED; Reflection High-Energy Electron Diffraction
139 1994	SCLS; Surface Core Level Spectroscopy
140 1994	SEXAFS; Surface Extended X-ray Absorption Fine Structure Spectroscopy
141 1996	SHG; Second-Harmonics Generation
142 1994	SIMS; Secondary Ion Mass Spectroscopy
143 1996	STM; Scanning Tunneling Microscopy
144 1996	STS; Scanning Tunneling Spectroscopy
145 1994	XPS; X-ray Photoelectron Spectroscopy

Table 1.2. Selected reviews on experimental techniques of surface science. Entries ordered alphabetically after the acronym.

eral related techniques, including photoelectron spectroscopies, absorption, and diffraction. Finally, we consider the various ion spectroscopies.

In all scanned probe microscopies, a sharp tip is brought to within about a nanometer from the surface. In the case of Scanning Tunneling Microscope (STM), the detected signal is related to the tunneling current flowing between the tip and the substrate, while for the Atomic Force Microscope (AFM) it is linked to the force acting on the tip. The major advantage of these direct methods over diffraction or spectroscopic techniques is that they provide real space images of non-periodic structures. In plain words, in these topographic images one can *see* atomic-sized features. Thus, methods based on STM and AFM are ideal for characterization of surface imperfections such as point defects, impurities, steps, or grain boundaries. Though the suggestive appearance of STM micrographs can (and does) easily lead to misinterpretation, the danger of drawing entirely false conclusions is in fact much lower than in the case of indirect approaches. The human mind is, after all, visually oriented.

The routine availability of ulta-high vacuum (UHV), that is, pressures in the range of 10^{-10} mbar[i] or lower, is instrumental to obtaining reproducible results in fundamental studies on surfaces. Under high-vacuum conditions ($\sim 10^{-6}$ mbar), a clean surface would become rapidly (within $\sim$1 s) covered with such components of the residual gas as water (Sect. 5.5).

a. *Scanning Tunneling Microscopy (STM) and Spectroscopy (STS)*

Probing of surfaces on the atomic scale in real space became possible with the invention of the STM by Binnig and Rohrer[146] in 1982 (for a review on STM studies of semiconductor surfaces see Ref. 147). The invention of STM won Binning and Rohrer the 1986 Nobel Prize for Physics. An STM study can associate local details of the electronic structure to local atomic geometries virtually atom by atom. This provides a unique chance to make direct connection between intuitively transparent experimental data and state-of-the-art atomistic calculations.

An STM consists of a tunneling junction where one electrode is an atomically sharp tip and the other is the studied surface. The tip is brought within a few Ångstrøms of the sample, so that there is a significant overlap of their electronic wavefunctions. When a small voltage is applied, a tunneling current can flow from the occupied states near the Fermi level of one electrode into the empty states of the other electrode. The voltage between the sample and the tip is chosen usually between 1 mV and 4 V, depending on the electronic structure and stability of the investigated system; typical tunneling currents are between 0.1 nA and 10 nA.

The tunneling current depends exponentially on the tip-sample distance. This is of key importance for the atomic resolution, though nonlinearities in the response of the tunnel junction can affect the stability of the servo system, especially at low bias voltages. Under favourable conditions, a vertical resolution of hundredths of an Ångstrøm and a lateral resolution of about 1 Å is reached. In most cases the sample is fixed and the tip is scanned by using a piezoelectric drive system.

STM has two basic imaging modes: the constant current mode and the constant height mode. The **constant current mode** is the most popular one. A feedback system controls the vertical position of the tip such that a preset value of the current is maintained. While scanning the sample, the tip follows thus the surface of a constant tunneling current. Since in many cases this surface can be approximated by the surface of the constant

[i] The unit of 1 Torr, often used in older publications, equals 1.33 mbar or 133 Pa.

density of electronic states,[148] it may be considered a topographic profile of the surface electronic states ("sample profile"). The STM electronics records the voltage applied to the tip piezo, that is, the voltage controlling the vertical position of the tip. This voltage is a proportional representation of the sample profile.

In order to increase the scan speed, the **constant heigh mode** of operation has been introduced. The movement of the tip is now confined to the lateral directions. In place of the piezo voltage, the tunneling current is recorded. The contour line "heights" on the imaged map of the current are not proportional to the contour line heights of the sample profile, because the current-distance dependence is exponential. The advantage of the constant height mode is that relatively high scan frequencies (1 kHz) can be used. This makes a real-time video recording of surface dynamics possible. Unfortunately, for a simple implementation of this technique to be successful, the surface must be atomically flat and perfectly aligned with the plane in which the tip moves.

Since most commercial STMs are programmable and many STM devices are home-made, the actual measurement may go far beyond these two basic modes. For example, the electronics may be set up such that the tip follows an atomic-sized structural defect diffusing on the surface.[149]

Further information on the local density of states as a function of energy can be obtained from STM by varying the bias voltage. This is the **Scanning Tunneling Spectroscopy** (STS), earlier known as the Current Imaging Tunneling Spectroscopy (CITS). The I–V characteristics of the tunneling tip–sample diode are recorded at each pixel as the topography is scanned. To analyse the energy dependence of the surface Local Density Of States (LDOS), the normalized differential tunnel conductance ($d\ln I/d\ln V$) is usually plotted as a function of the tip–sample voltage V. This representation allows in many (but not all[150]) cases for elimination of the dependence of the I–V characteristics on the tip–sample distance.[151]

The atomic resolution acquired in the topographic image associated with the STS data may not necessarily be reflected in the STS characteristics. The spurious current collected simultaneously from many places of the surface by a blunt tip is likely to average out in a topograph, producing a harmless background. In a spectrogram, the voltage dependence of this current may overshadow local variations of the I–V curve.

A simple and practical implementation of STS is to record two data points for each pixel: one for a negative and one for a positive bias. In this way, occupied and unoccupied states are simultaneously imaged.

Since the STM measures the tunneling current, the outcome of the experiment depends not only on the local electronic states of the sample but also on the electronic states of the tip, on their relative distance in energy, and on the energy barrier which separates them. For example, local changes in the work function caused by electric fields in the sample are immediately reflected in the scanned image. The interpretation an STM image, though sometimes obvious, in a general case requires a comparison with an atomistic calculation.[152,153] Different methods of numerical simulation are used. The most widely used is the simplified but very efficient perturbation theory of Bardeen[154] and Tersoff and Hamann[148] in which no current flows, the electronic states of the sample are decoupled from the states of the tip, and the electronic structure of the tip is approximated by a single orbital. These approximations are often sufficient for a qualitatively correct simulation. Extensions aimed to include the electronic structure of the tip and the influence of the sample-bias voltage in a more realistic way have been formulated as well.[155,156] Finally, non-perturbative treatments which take into account the process of tunelling may be needed in some cases.[153] Since several physically plausible structures may lead to remarkably similar STM images,[157] a reliable total-energy optimization of the atomic position and analysis of measurements done at several sample-tip biases are instrumental for the reliability of the conclusions.

b. *Atomic Force Microscope (AFM)*

The Atomic Force Microscope (AFM) is a close cousin of the STM. It was invented in 1986 by Binnig, Quate, and Gerber[158] to cope with the major limitation of the STM that both the sample and the tip must be conducting. A highly competitive variety of AFMs is now commercially available to cover applications under UHV (where atomic resolution is possible) and in air. AFM can image the surface topography, measure surface forces and adhesion, and determine mechanical properties. This is the reason why AFMs have entered not only laboratories but also industrial settings. Instruments adapted to such diverse problems as fixed disk manufacturing or biological applications are on the market. A number of theoretical models of AFM imaging, including *ab initio*-based schemes, have been developed.[159,160,161] In the following we will examine the use of AFM in UHV environment.

For **contact imaging** the sample is scanned under an atomically sharp tip mounted on a catilever. The deflection of the cantilever due to the force exerted by the sample on the tip is monitored, usually optically. Typical forces are some tenths of nN and are limited by adhesion. They are small enough for reproducible, non-destructive imaging of most surfaces. It is practical to use a feedback system which maintains a constant deflection of the cantilever by adjusting the z-piezo voltage. A plot of $z(x, y)$ represents then a surface of the constant force, approximating the total charge density. The corrugation can be modeled by hard spheres in contact.

UHV AFM in the contact mode easily achieves resolution in the nanometer range. It is an ideal tool for the characterization of insulating films and epitaxial semiconductor layers with low dopant concentrations. True atomic resolution is rare and depends on the stability of the structures.

The AFM can also measure long-range interactions (**non-contact mode**). The advantage is that the tip remains now at a distance of some nanometers from the sample so that there is no possibility of damaging either the tip nor the surface. The tip is oscillated at a frequency slightly above its natural resonance frequency. If the distance to the surface is 1–5 nm, the resonance frequency of the vibrating tip is changed noticeably by an amount proportional to the surface force gradient. A feedback system adjusts the tip-sample distance to keep the frequency shift constant at ~10–20 Hz. In this mode, however, the tip is not only influenced by van der Waals forces (which dominate in the contact mode), but also by charges on the surface and by capacitive forces. In order to be sensitive to surface morphology, the contact potential has to be compensated accurately, whereby the capacitive forces are minimized. Atomic resolution may be then achieved (Fig. 3.3, page 135).

An interesting version of AFM which uses the capacitive forces is the Scanning Capacitance Microscopy[162] (SCM). The usefulness of this method is in its capability of detecting the dopant profile and in particular, the position of the *pn* junction, even under an oxide layer. The lateral resolution limited in principle mostly by the Debye length and (in the case of *pn* junction detection) by the width of the depletion region. SCM is therefore a promising method[28] which can evolve into an analytical tool for measuring two-dimensional[163] (2D) or even 3D[164] junction profiles. A higher degree of confidence may be provided by 2D junction delineation by the so-called Scanning Capacitance Spectroscopy[165] (SCS), a new approach which allows one to record C–V curves at each pixel of the 2D scan.

c. *Low Energy Electron Diffraction (LEED)*

The most important distinction between the real space methods and diffraction (reciprocal space) techniques such as Low Energy Electron Diffraction (LEED) is that the former provide information about the atomic positions of periodic and disordered structures while the latter rely on periodicity and measure the Fourier transform of the atom-atom correlation function.

LEED dates back to 1927, when Davisson and Germer performed experiments of electron scattering from nickel.[166] LEED is the most common technique in studies of single crystal surfaces. Electrons backscattered elastically from a crystalline surface bombarded by primary electrons with energies in the range of 20–500 eV are detected by a Retarding Field Analyzer (RFA). According to the famous de Broglie relation, which links a flux of particles with velocity v and mass m to a plane wave with length $\lambda = h/(mv)$, where h is the Planck constant, the electron wavelengths are then in the Ångstrøm range, that is, of the same order as the interatomic distances. It follows that these waves interfere when scattered from the periodic lattice of the crystal. LEED is used to determine:

- periodicities of reconstructions, directly from the diffraction patterns;
- atomic positions within the unit cell of a periodic structure, whereby intensities of Bragg diffraction spots measured as a function of the primary electron energy (I–V spectra) are compared to computed curves;
- partial disorder on the surface (for example, identification of defects, arrangement and size distribution of domains and terraces). This is done by **Spot Profile Analysis** (SPA), that is, by examination of the broadening or splitting of the diffraction spots.

Its high incident currents and large elastic scattering cross sections make LEED sensitive to adsorbates with very low Z, which are difficult to detect in X-ray measurements. The high surface sensitivity of LEED (only several atomic layers are probed) is caused by the short mean free path of slow electrons in the solid due to their strong interaction with the conduction electrons. Yet this strong interaction makes the application of the dynamic theory[167,168] necessary for a meaningful structural analysis. As the paramount example of the clean Si(001) surface shows (Chapter 2), this analysis is sufficiently complex that a good knowledge about the investigated structure is needed in the first place. Otherwise, the parameter space becomes so vast that the "optimized" structure is likely to represent a local, unphysical minimum of the fit function.

An obvious remedy would be *ab initio* verification of each of the structures deduced from LEED. Unfortunately, since LEED vendors rarely offer experimental setups equipped with *ab initio* experts, such verification is usually (if at all) done *post mortem* by other teams. Seriously speaking, analysis of LEED I–V spectra is another example (the first one was STM) where *ab initio* support would be very helpful.

d. *Auger Electron Spectroscopy (AES)*

Auger Electron Spectroscopy (AES) is one of the most popular surface analytical tools. The first application of AES to surface analysis was reported by Lander[169] in 1953. Today, there are numerous books on AES, for example, Refs. 170 and 171. AES can access the chemical composition and provide quantitative data on the elements in the topmost atomic layers. Furthermore, information on the electronic structure and bonding is obtained by comparing experimental and theoretical line shapes. A quantitative analysis, however, requires a precise knowledge of the experimental parameters or comparison with reference samples and some knowledge about the sample structure in the analyzed near-surface region.

AES is a three electron process. The sample is bombarded with primary electrons of high energy[j] E. They ionize atoms in the near-surface region, whereby secondary electrons are emitted from the core states of energy E_1. Electrons from higher levels (energy E_2) fall then to the level E_1 and release the energy $E_2 - E_1$. This may be achieved by the radiationless Auger process or by emission of a photon. In the former case, the excess energy is transfered to the third electron (Auger electron). This electron is emitted from the energy level E_3 and escapes with the kinetic energy $E_A = E_2 - E_1 + E_3$. Since the energetics of the Auger process reflects the energy structure of the emitter, it is characteristic to the chemical composition of the material. For transitions of K-type (that is, E_1 is in the first electronic shell), the Auger electron yield is larger than the X-ray yield up to atomic number Z=32; for L-transitions (E_1 in the second shell) it is larger than 90% up to Z=50. This means[172] that the emission of X-rays is negligible especially for lighter elements and energies below ~500 eV and becomes comparable to the Auger yield only in the range of about 2000 eV.

For a qualitative chemical analysis, the energies of the measured peaks can be compared with tabulated spectra. An extensive database exists.[173]

[j] Usually 3–5 keV; about 20 keV is needed for small spot sizes (below 100 nm).

With the exception of H and He, most of the elements can be identified on the surface. In order to quantify the chemical composition of the surface, one must know the Auger electron current and take into account all factors which contribute to the emitted Auger electron current.

The intensity I_{SXY} of a single atom Auger transition SXY without interactions with other atoms is given by

$$I_{SXY} = I_0 \sigma_S(E_0) P_{SXY} T(E_A) D(E_A). \tag{1.3}$$

I_0 represents the intensity of the beam of primary electrons with energies E_0, $\sigma_S(E_0)$ stands for the ionization cross section of the inner shell S (that of energy E_1), P_{SXY} denotes the probability for the excited atom to decay via the Auger process SXY, $T(E_A)$ characterizes the analyzer transmission at the Auger electron energy E_A, and $D(E_A)$ describes the detector efficiency. If the atom is embedded in a solid, there are additional complications which must be taken into account:

- The energy and intensity of the primary electrons is modified by inelastic and elastic scattering;
- The interaction of the emitted Auger electron with the solid limits the range of low energy electrons to several inter-atomic distances;
- Valence band electrons can take part in the Auger process.

If valence band (V) electrons are involved in the Auger transition, as in KVV or LVV, the shape of the Auger peaks carries information on the chemical state of the surface, orbital hybridization, electron delocalization and correlation, screening effects, charge transfer, bonding, and covalency. But in order to retrieve this data, one must have reliable theoretical models for all the factors contributing to the line shape. The "true" Auger line shape extracted from the measured spectra can be viewed as a self-convolution of the Local Density of States (LDOS). In other words, deconvolution of the "true" signal allows one to determine the DOS localized at atoms which take part at the Auger transition. In this way, one can get information about the electronic structure and the bonding at the surface.

A typical AES setup consists of an electron gun which produces the primary electron beam with energy of 2000 to 5000 eV and an energy dispersive detector. A popular analyzer, frequently used in combination with LEED, is the retarding field 4-grid system (RFA). Also common are the Cylindrical Mirror Analyzer (CMA) consisting of two coaxial cylinders and the 127°-analyzer employed in the pioneering work of Harris.[174,175]

e. *X-Ray Photoelectron Spectroscopy (XPS)*

The roots of X-Ray Photoelectron Spectroscopy (XPS) can be traced back to the discovery of the photoelectric effect by Hertz, the description of its energetics by Einstein, and to the publication of XP spectra from several metals by Robinson and Rawlinson.[176,177] Siegbahn *et al.*[178] managed to develop XPS into a sensitive, high-resolution method for the determination of binding energies of electrons. Today, XPS is applied to a wide range of materials and provides information about their composition and, to a lesser extent, bonding and structure.

In an XPS experiment electrons are photoionized from the core levels by X-rays. The surface is irradiated with photons from a soft X-ray source, most commonly Al K_α (1486.6 eV) but also Mg K_α (1253.6 eV). Such beams have narrow line widths; moreover, due to their high heat conductivity, Al and Mg are ideal materials for anodes in the X-ray lamps.[170]

The cornerstone of XPS is the measurement of the kinetic energy E_K of the outgoing electron. This energy is related to the electron binding energy E_B and to some external parameters:

$$E_K = h\nu - E_B - E_R - \Phi - \delta E. \tag{1.4}$$

Here, $h\nu$ is the photon energy of the incident X-rays, E_R is the recoil energy (this is a minor component, 0.1–0.01 eV, and is generally ignored), Φ is the spectrometer work function, and δE describes the electrostatic charging of the specimen which slightly (by up to 5 eV) reduces the energy of the photoelectrons. E_K is determined experimentally; $h\nu$ and Φ are known.

Contemporary XPS provides information about the environment of surface atoms.[179,139] Though core electrons are very localized and their wavefunctions are hardly affected by the surrounding, the binding energy of a core electron depends on the charge of the atom as well as on the background electrostatic potential in the core region. The matter is somewhat complicated by the dependence of the binding energy on the efficiency of dielectric screening. These factors (background potential, charge, screening) are different on the surface than in the bulk. One speaks of **Surface Core Level Shifts**, or Spectroscopy (SCLS). The sign and the magnitude of the shift, which may range from a few tenths of eV to a few tens of eV, are signatures of the local environment of a surface atom.

Screening enters into the description of core level shifts because the electrons ejected from the solid leave a core hole.[180,181,182] The energy of this hole depends on how efficiently its charge is screened by the electrons of the crystal. One may ask whether the quantity measured in SCLS, the kinetic energy of photoelectrons, reflects the energy of the core electron before it is ejected (initial state appoximation) or rather the energy of the core hole (final state approximation). The answer depends on how efficiently the crystal electrons can screen the hole within the time needed for the photoemission process. If the electron sea reacts extremely slowly, the ejected electron escapes as in the initial state picture. If the electron sea reacts extremely fast, the ejected electron takes with it the whole screening energy and the final state picture with static screening of the hole is valid.

Generally, the photoelectron can take only a part of the screening energy. One should thus consider dynamical screening effects; the energy of the photoelectron is expected to be somewhere between the limits set by the initial state and final state approximations. In any case, screening of the hole causes the photoelectron to appear at a higher kinetic energy, reducing its measured binding energy.

Core level spectroscopy uses the sharpest photoemission peaks. In the case of silicon surfaces this means electrons from the Si($2p$) cores,[k] with binding energy of about 100 eV. Tunable **synchrotron radiation** sources allow one to achieve excellent surface sensitivity. With photon energies in the range 130-150 eV, the escape depth of photoelectrons becomes as short as two silicon layers. This short escape depth has also another adventage: it makes the analysis virtually insensitive to any band bending under the surface. The conventional X-ray spectroscopy suffers from relatively large escape depths of more than ten silicon layers.

Since the cross-section for ionization of a core does not depend on the state of the valence electrons, the intensity of the emitted photoelectrons is proportional to the number of atoms in the detected volume. This is the basis of a quantitative XPS analysis.[183] After a proper calibration, the chemical composition of the investigated region can be determined.

The very low binding energy region of the spectrum (0–35 eV) contains information about the valence band structure. This information is, however, difficult to access because the light from conventional X-ray sources is not sufficiently monochromatic (the line width is about 1 eV for Mg K_α radiation) and because the photoionization cross-section of valence electrons

[k] The Si(2p) level is split by spin-orbit interaction into two states, $2p_{1/2}$ (more bound) and $2p_{3/2}$ (less bound). See Fig. 3.10, page 152.

is small in the X-ray energy range. Valence band spectra are therefore better investigated using the light from synchrotron sources. Particularly suited for valence band studies is ultraviolet light (Ultraviolet Photoelectron Spectroscopy, Section 1.3.2f).

A standard XPS setup consists of a primary source of X-rays and a detector with an electron analyzer. This analyzer is the heart of the system. Two commercial types of analyzers are available for XPS: the Concentric Hemispherical Sector Analyzer (HSA or CHA) and the double pass Cylindrical Mirror Analyzer (CMA). Because of the paramount importance of the spectral resolution, an electron analyzer for XPS will usually be operated in the constant energy mode.

f. *Ultraviolet Photoelectron Spectroscopy (UPS)*

Ultraviolet Photoelectron Spectroscopy (UPS) has established itself since the mid fifties as a standard technique for the investigation of solids, surfaces, and interfaces.[184,185,186] The application of UPS in surface science focuses either on the intrinsic properties of clean surfaces or on the properties of adsorbed species, such as atoms or molecules.

To achieve high surface sensitivity, the spectroscopy of valence electrons is generally done with photon energies between 10 and 100 eV. The low kinetic energy (below 40 eV) of the excited electrons results in a short mean free path of the photoelectron; the escape depth is thus very small, about 0.5 nm. UPS is therefore much more surface sensitive than XPS.

UPS investigations with He(II) excitation are widely used. Due to the narrow distribution of the primary energy (~80 meV) in comparison with the XPS , He(II) provides a high energy resolution for weakly bonded electrons. But the key factor which made UPS and XPS powerful is the easy access to **synchrotron radiation** of tunable wavelength, high intensity, and well defined polarization.

The photoelectron emission is described in Section 1.3.2e. In UPS, electrons bound in the valence shell of surface atoms or molecules are emitted. This transfers the system from the initial state of energy E_i (neutral atom) to the final state of energy E_f (ionized atom + photoemitted electron). The photoelectron spectrum correlates with the energy distribution of the occupied states. The kinetic energy E_K of the photoelectron is:

$$E_K = h\nu - E_B - \Delta E_{vib} - \Delta E_{rot}. \tag{1.5}$$

ΔE_{vib} and ΔE_{rot} account for vibrational and rotational excitations in molecules upon ionization, $h\nu$ is the energy of the incident monochromatic radiation, and E_B can be interpreted as the binding energy of the electron in a particular orbital of the studied object. While the gross features of UP spectra can in most cases be understood as arising from such single-particle orbitals, dynamic screening effects (Section 1.3.2e) may also play a role.

According to Eq. 1.5, each orbital of an atom or molecule is – within the single-particle scheme – imaged as a characteristic peak in the angle-integrated photoelectron spectrum. This means that densities of occupied electronic states are measured. Since the electronic valence structure is a more or less unique fingerprint of a particular species, this principally allows for a chemical analysis of the adsorbates. In particular, one should be able to distinguish between molecularly and dissociatively adsorbed molecules. In reality, this task may be not so trivial (Chapter 5).

In order to investigate the dispersion of the surface bands, the electron wave vector must be determined. Besides the kinetic energy one must thus know the momentum transfer, that is, the direction of the emission. This is provided by **Angle-Resolved PhotoEmission Spectroscopy** (ARPES). This generic name conveys both UP and XP spectroscopies; the UPS version is often described as Angle-Resolved UPS, or ARUPS.

The UPS setup is in principle the same as that for XPS. It consists of the source of light and a spectrometer. Two types of light sources are common: noble gas discharges which emit resonance radiation and synchrotrons which emit in a broad spectral range. The popular spectrometers are Hemispherical and Cylindrical Mirror Analyzers (HSA and CMA) which may be used simultaneously with XPS and/or AES.

g. *Absorption and Diffraction of X-Rays*

The **Surface X-Ray Diffraction** (SXRD) is an old and well developed method for structural analysis.[187] Its advantage is that the bulk diffraction effects can be reduced by using a high intensity, weakly interacting probe, for example, X-rays in an appropriate geometry. The experiment can be designed such that a simple kinematic analysis can be used to determine the atomic structure of the unit cell.

X-ray Standing Waves[188,189] (XSW) are generated through dynamical interference between the incident and diffracted X-rays during Bragg diffraction from nearly perfect single crystals. Surface films and the positions of incorporated atoms with respect to the crystal lattice can be determined by recording the reflected and fluorescent radiation emitted from the crystal at the same time. This method provides information about the long range correlations for specific atoms within crystalline solids.

Surface Extended X-Ray Absorption Fine Structure (SEXAFS) is a powerful technique for determining the adsorption site and bond length of atoms and molecules.[190,191]. In comparison with diffraction techniques, SEXAFS has the advantage that it does not require long-range order. SEXAFS is an indirect surface-sensitive measurement of the X-ray absorption in the energy range up to about 10 keV. In other words, it is EXAFS modified to overcome the low surface sensitivity of the latter.

In EXAFS, the X-ray absorption cross-section of a particular element in the solid state is measured as a function of increasing photon energy. The observed absorption edges reflect the onset of transitions of electrons from core states of the atom to unoccupied states above the Fermi level of the material. Additionally, a fine structure is visible caused by the modification of the final state wave function of the emitted electrons by the matter surrounding the absorbing atom. For higher photon energies, the fine structure in this region is due to scattering of the photoelectron wave function by neighboring atoms. This means that this fine structure is sensitive to the structure of the material in the short range order. To get high surface sensitivity one records the photoemitted Auger electrons, which are a deexcitation product of the photoionization process. The surface sensitivity is particularly high if the escape depth of the Auger electrons depending on their energy is only a few Ångstroms.

The experimental requirements for SEXAFS studies consist of a source of tuneable, monochromatic X-rays (synchrotron radiation source) and a CMA such as used in AES (Section 1.3.2d).

h. *Ion Spectroscopies*

Secondary Ion Mass Spectroscopy[192,142] (SIMS) is probably the most popular of the ion-based techniques. A beam of ions or neutrals removes atoms from the substrate. The ionized fraction of the removed material is then analyzed in a mass spectrometer. Secondary Neutral Mass Spectroscopy (SNMS) has the advantage that secondary neutrals can be ionized in a controlled way using electron impact, plasmas, or lasers.

The standard, dynamic SIMS is not a technique for surface analysis, though it is sometimes applied to investigate the chemical composition of interfaces. Dynamic SIMS employs ion currents of high densities. The reactive ions erode successive surface layers so that a depth profile is obtained.

Since the detection limits are in the ppb range, this technique is used extensively for the characterization of dopant profiles in semiconductors.

The version designed for surface studies is known as static SIMS (SSIMS). The surface is now accessed with a low intensity beam which is intended to sample less than an atomic layer. This provides chemical information about the outermost atomic layer.[193]

In **Rutherford Backscattering**[137] (RBS), a MeV beam of light ions is used as the primary ion source. Measurements of the energy of the scattered particles lead to a quantitative composition and depth analysis without the need for reference standards. While this technique has poor surface sensitivity (~20 nm), it is indispensable in studies of buried interaces.

Ion Scattering Spectroscopy (ISS) uses Low Energy (LEISS) and Medium Energy (MEISS) ions, that is, particles with energies below ~200 keV. The development of ISS for surface crystallography is reviewed in Ref. 194. The energy spectrum and angular distribution of the scattered ions contain a wealth of information on the structure and composition of near-surface regions of the target. Ion scattering may be used for a wide variety of measurements, from compositional analysis to the structural analysis of adsorbates on highly ordered surfaces. However, ISS has never acquired the popularity of LEED, AES, STM, or AFM.

At LEISS or MEISS energies an electrostatic analyzer can be used for accurate resolution of the energies of the scattered ions. This allows for very sensitive analysis of the surface composition. A typical ISS instrument can access the top 0.3 to 0.5 nm (3 to 5 Å) of the specimen. If the energy of the incident beam is low (less than 0.5 keV), the ions have a high probability of neutralization. Virtually all particles scattered on deeper layers lose their charge,[195] so that the only signal detected is that of ions scattered from the few topmost atomic layers. The adventage is that the surface sensivity becomes independent of the energy resolution of the system.

Basically, ISS involves an elastic two-body collision between a noble gas ion beam and the sample surface. For sufficiently high energy ions the process may be considered as a Rutherford scattering event between the bare nuclei. The energy of the incoming ions is selected and the energy loss after scattering at a chosen angle is measured. The varying energy transfer to target atoms of different masses gives rise to the elemental sensitivity of ISS. The biggest difference is for target atoms with masses similar to that of the incident ion. But the main advantage of ISS, especially at low energy, is in the detailed structural analysis of surfaces.

The energy of the scattered ion may be approximated by a simple function of the scattered angle and the mass of the incident ion and target atom.[196] No account is taken of the binding energy of the atom within the target. The energy E_s of a scattered ion is indicative of the mass M_T of the target atom and is given by the relation for inelastic scattering:

$$\frac{E_s}{E_i} = \left(\frac{M_T \cos\Theta + \sqrt{M_T^2 - M_i^2 \sin^2\Theta}}{M_T + M_i}\right)^2, \tag{1.6}$$

where E_i and M_i are the energy and the mass of the incident ion, respectively, and Θ is the scattering angle.

The basic requirements for an ISS setup consist of an ion source and analyzer. The ion source must produce an incident ion beam of low energy spread and low angular divergence. To study the angular distribution of the scattered ion flux, a precise angular motion of the sample with respect to the beam is needed. The detection of low-energy ions is usually performed by similar techniques as those used for AES and XPS, at a constant scattering angle, that is, at a constant energy for each target element (Eq. 1.6). Since electrostatic analyzers have high energy resolution but are unable to detect neutral particles, modern systems use time-of-flight methods that have lower energy resolution but are able to detect all back-scattered particles, ions as well as neutrals.

i. *High Resolution Electron Energy Loss Spectroscopy (HREELS)*

High Resolution Electron Energy Loss Spectroscopy (HREELS) is widely used in studies of vibrations of adsorbed atoms or molecules. The purpose of such investigations is to identify the adsorbed species and to get information about adsorption sites and bonding geometries.[132] Since the vibrational energies in molecules are some 10 meV up to some 100 meV and line widths are in the order of 1 meV, HREELS needs highly sophisticated monochromators and analyzers. The primary energies are typically 2 to 5 eV, so that the main interaction is surface scattering.

Consider a molecule adsorbed at a surface of a dielectric medium with the dielectric constant ϵ. Because of their charge, electrons are particularly able to excite surface vibrations if a **dipole moment** is associated with the oscillation. A dipole field created by a vibration of an adsorbate is screened by a factor of $1/\epsilon$ if the dipole is oriented parallel to the surface. Only the dipole moments normal to the surface give rise to significant dielectric scattering. The moments parallel to the surface can only be detected in off-specular geometry. In other words, chemical bonds oriented parallel to the surface cannot cause a strong dipole scattering from their stretching mode in the specular direction. The application of these selection rules allows one to draw conclusions about the adsorption geometry of a molecule.

In addition, there is a short-range interaction between the incoming electron and the adsorbed molecule. The range is limited to atomic dimensions (0.1 nm) so that the scattering also shows a broad angle distribution. In this case, the selection rules of dipole scattering (dipole moment perpendicular to the surface, scattering into the direction of elastic diffraction) are not valid and measurements outside the specular beam are needed. The mechanism is called **impact scattering** and gives rise to completely different characteristics as compared with the dipole scattering. The concept of impact scattering is necessary to explain angular profiles of loss intensities which do not exhibit the sharply peaked structure typical of dipole scattering. The theoretical description of impact scattering requires that the electron-molecule scattering potential similar to that in LEED is known.

j. *Other Surface Science Techniques*

Finally, we briefly discuss some of the experimental techniques of surface science which are less frequently addressed in the coming chapters.

Low Energy Electron Microscope (LEEM) is an emission electron microscope in which low energy electrons (2-100 eV) are used to illuminate the surface of single crystal specimens. Electrons diffracted and/or reflected from the surface are accelerated into an immersion objective by a voltage of several kV applied between the specimen and the first objective lens. Electron optics forms both a diffraction pattern and a real space image. The contrast in LEEM depends on the diffraction and the electron reflectivity of the specimen surface at very low electron energies. A narrow energy spread in the illuminating beam and an optimum aperture would allow for a resolution in the 5 nm range and better.

This extremely sophisticated instrument was invented by Bauer.[197] At present, there are less than a dozen LEEM instruments in the world, about half of them actively producing data. To date, the main work of LEEM was to study adsorption phenomena on semiconductor and metal surfaces. Above that, real-time in-situ investigations were made for understanding chemical reactions and surface morphology evolution.

The **Transmission Electron Microscope** (TEM) and the **Reflection Electron Microscope** (REM) at glancing angles are widely used to study surfaces and interfaces. The imaging in an aberration-free electron-optical system consists of a sequence of two Fourier transforms that combine the Fourier analysis of the object wavefunction into the diffraction pattern and the Fourier synthesis of the diffraction pattern into the image. In a TEM, the electrons are scattered after transmission through the object. In a REM, the reflected electrons are altered by interaction with the potential of the object or object surface. Deviations from the ideal imaging conditions are due to such factors as spherical aberration, chromatic aberration, defocus value, finite aperture size, and specimen and illumination characteristics, but atomic resolution in TEM is now relatively common.[198,199] An important, emerging technological application of TEM is electron holography, which allows one to access the distribution of dopants in CMOS transistors in two dimensions with nearly arbitrary lateral resolution.[200]

Reflection High-Energy Electron Diffraction (RHEED) works with primary electron energies between 10 and 100 keV and grazing incident angles (3–5°). Since the sample is far from the electron gun and the screen, the application of RHEED is not limited by the temperature of the sample. It can be also implemened for in-situ monitoring of the epitaxial growth process. The surface sensitivity is similar to that of LEED (Sect. 1.3.2c).

Ellipsometry is an optical method popular in academic as well as industrial environments as nondestructive means of accessing thin films. A beam of light sent through the film reflects from the interface; the polarization of the reflected light is then analysed. In particular, the Variable Angle Spectroscopic Ellipsometry (VASE) has been shown to be a powerful technique for the post-deposition characterization of multilayer heterostructures. It can be employed to determine film thickness of films, alloy composition, number of superlattice periods, and homogeneity.

1.4. Silicon Surfaces and Interfaces

"Surface" and "interface" are ambiguous notions. We interpret surface as atoms terminating a bulk, and interface as a contact region between two media. A free, clean Si surface is thus an interface between silicon and vacuum. Its oxidation is the first step in making a SiO_2/Si interface.

Atomically clean surfaces belong to the domain of basic science. Applied exercises begin when one considers foreign atoms coming into contact with them. Growth of films, etching, and implantation are needed to build systems sufficiently complex to wink and blink, count, play music, talk, control an experiment, drive a car or a spaceship. Much of the activity needed to perform these tasks has its roots in (or is affected by) processes which take place near or at interfaces, notably with SiO_2 and with silicides. Interfaces often bear a resemblance to free surfaces (interfaces with vacuum). After all, atoms know only a limited number of tricks to play when their neighborhood is perturbed, be it by vacuum or by other species.

We first survey the fundamentals: geometries and electronic pecularities of ideal and realistic clean surfaces (Section 1.4.1). As next we consider chemical activity of free surfaces (Section 1.4.2), including methods by which clean surfaces for scientific and industrial usage are prepared. Together with notes on buried interfaces (Section 1.4.3), this introduces and overviews the subjects to be considered later.

1.4.1. *Fundamental concepts*

It is said that God created the bulk and Satan added the surface. If so, this must have happened when Satan was still the wise Lucipher, the Son of Light, readily helping the Creator. A design so perfect as a crystal surface can hardly be born of a mind darkened by evil. Or maybe God used his lost friend as an unaware assistant? And Satan added a final touch to the creation, thinking in his selfish madness that he opposed his abandoned Master but was in fact guided by His unspoken wish.

If perfectness is measured not by simplicity but by the ability to make a variety of structures from simple elements, then Si surfaces are more perfect than Si bulk. Breaking an infinite crystal may harm its symmetry. Yet the wound – the truncated-bulk surface – heals. It reconstructs, creating a world in which the nature of the crystal can be expressed in ways forbidden by the rigid, unanimated order of the endless array of atoms.

Reconstruction is likely to leave an easily detectable signature: periodicity changes are reflected in diffraction patterns produced by scattered particles. This provides convenient **classification** means. If translation vectors $\vec{t}_i$ of a reconstructed surface are represented in the basis of the primitive translations $\vec{a}_j$ of the truncated-bulk surface: $\vec{t}_i = \sum_j T_{ij}\vec{a}_j$, the periodicity is uniquely described by the transformation matrix $\mathbf{T} = (T_{ij})$. Thus, if the translation vectors of the reconstructed surface are $\vec{t}_1 = k\vec{a}_1 + l\vec{a}_2$

and $\vec{t}_2 = m\vec{a}_1 + n\vec{a}_2$, the surface periodicity is unequivocately denoted as $\binom{k\ l}{m\ n}$. This is the Park-Madden matrix notation,[201] very useful for surfaces which, as Si(110), support reconstructions with complicated translations.

Sometimes a convenient cell described by a diagonal matrix can be chosen (Fig. 1.6 on page 70, Fig. 1.8 on page 75, Fig. 2.2 on page 89). This is typical for Si(001) and clean Si(111). If vectors $m\vec{a}_1$ and $n\vec{a}_2$ span this cell, the structure is named $p(m\times n)$ if the cell is primitive or $c(m\times n)$ if the cell is centered. The designator p is omitted unless needed for clarity; $\binom{m\ 0}{0\ n}$ becomes thus $m\times n$. The $m\times n$ notation is a special case of Wood's method,[202] in which a cell spanned by vectors $\vec{t}_i$ with lengths $|\vec{t}_1| = x\vec{a}_1$ and $|\vec{t}_2| = y\vec{a}_2$ is denoted as $x\times y\text{-}R\phi^\circ$. Here, ϕ is the angle by which the vectors $\vec{t}_i$ are rotated with respect to the primitive translations $\vec{a}_j$. The full Wood's notation is useful for adsorbate-induced reconstructions of Si(111).

Sometimes the angle between $\vec{t}_1$ and $\vec{a}_1$ is not the same as that between $\vec{t}_2$ and $\vec{a}_2$. While the matrix T_{ij} provides a unique notation, diffraction patterns may suggest a simplified method which refers to the number of primitive unit cells in the reconstructed unit and gives a feeling about its aspect ratio (Fig. 1.10, page 80). A cell with mn primitive cells is then written as "$m\times n$". The shorthand is indicated by the quotation marks. Finally, scattering from a *disordered* surface may be dominated by diffraction on the ordered substrate: such surfaces have an apparent 1×1 periodicity.

While the local structure (reconstruction) of the surface is approximately characterized by its periodicity, its global structure (morphology) can be approximately described by the degree of **roughness**. Roughness can be observed convenently by scanning or transmission electron microscope (SEM or TEM), by AFM, or even by STM.[l] Its degree can be quantified by the random mean square (**rms**) deviation from the flat plane. Note that the measured rms value may depend on the size of the scanned region.

a. *Ideal truncated bulk and surface energy*

In order to create a Si(klm) surface, take a ball-and-stick model of the perfect crystal and truncate it along a $\{klm\}$ crystallographic plane, removing or breaking the sticks. Figure 1.5 illustrates several such ideal 1×1 surfaces. It is clear that all atoms with **broken bonds** on Si(111) 1×1 are equivalent. Each of them has a single broken bond. Each atom on Si(110) 1×1 has one broken bond as well, but these atoms are of two classes, belonging to two

[l] See Sect. 1.3.2 for experimental techniques, Table 1.2 on page 29 for their acronyms.

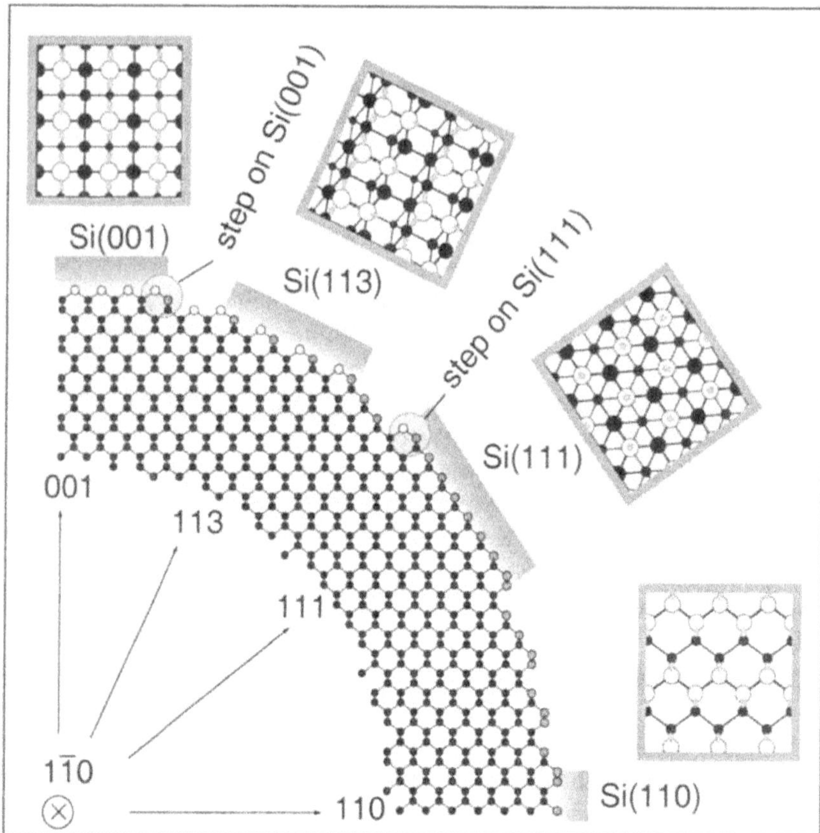

Fig. 1.5. Ball-and-stick models of several bulk-terminated surfaces in the (001) to (110) azimuth. The surface atoms are represented as open circles, the bulk atoms are shown in black. Bottom left: a view along the [1$\bar{1}$0] direction on a section of a cylinder cut from Si bulk. From the left to the right: top views of Si(001), Si(113), Si(111), and Si(110) surfaces. Dangling bonds on surfaces are shown as gray clubs.

sublattices of the diamond structure. All atoms on Si(001) 1×1 are from the same sublattice, but each of the surface atoms has two broken bonds. With its two clases of atoms, Si(113) 1×1 is more complex. An atom of one class has a single broken bond, an atom of the other class has two.

The first question which springs into mind is: how much energy is needed to make such a unit area of surface in a perfect, infinite silicon crystal? What is the **surface energy** γ^0_{klm} of Si(klm)?

It is natural to expect that the energy γ^0_{klm} required to produce a unit

Ideal surface	Broken bonds (atoms) per cell N (M)	Cell area A, Å^2	Atoms per area 10^{14}cm^{-2}	Broken bonds/area N/A, Å^{-2}	Energy per area $1.1N/A$, eV/Å^2
Si(111)	1 (2 atoms)	12.5	16.0	0.080	0.09
Si(110)	2 (2 atoms)	20.8	9.6	0.096	0.11
Si(113)	3 (2 atoms)	24.0	8.3	0.125	0.14
Si(001)	2 (1 atom)	14.4	6.9	0.139	0.15

Table 1.3. Densities of broken bonds and atoms and estimated surface energies of truncated-bulk Si surfaces.

area of such ideal Si(klm) 1×1 can be measured by the density of broken bonds. The energy E needed to break a Si-Si bond when a surface is created can be estimated from the **cohesive energy** of silicon,[203,204] $E_c = 4.63$ eV. Cohesive energy is the energy gained when a perfect elemental crystal is build from free atoms. Each Si atom makes $n_b = 4$ bonds, and to each bond $n_a = 2$ atoms contribute. The energy released by forming a bond is thus $E = n_a E_c / n_b$, or ~2.3 eV. The same energy is needed to break it.

The formation energy is then estimated by multiplying the bond energy E by the density of broken bonds. The latter is the number N of broken bonds per total surface area $2A$ (two surfaces are created), which gives $\gamma^0_{klm} \approx EN/2A$. The estimates for the surfaces depicted in Fig. 1.5 are listed in Table 1.3. The result suggests that Si(111) 1×1 can be made most easily. Indeed, {111} is the best cleavage plane of silicon.

What happens when the surface normal is not oriented exactly along one of the directions listed in Table 1.3? Figure 1.5 indicates that the directions of Table 1.3 correspond to those truncation planes which cut through as few bonds as possible. The (001), (113), (111), and (110) surfaces of silicon are thus expected to be **primary**, that is, small deviations from these orientations (by a few degrees) should increase the surface energy. Surfaces in vicinity of primary orientations, or **vicinal** surfaces, consist largely of flat primary terraces separated by steps. Atoms on such steps have usually more broken bonds than atoms on a primary terrace.

The surface of a cylindrical crystal will thus tend to develop flat faces in of the primary orientations. But a crystal terminated only by primary surfaces, though geometrically possible, is not optimal energetically because the total area of its surfaces is much larger than the surface area of a rounded crystal. The **equilibrium shape** of silicon is the result of a compromise between step energies and energies of primary surfaces. It can be obtained by means of the Wulff construction.[205,206]

In any case, the energy γ needed to create an ideal Si surface is of the order of 0.1 eV/Å^2, or ~3 eV per area of $a_b \times a_b$, where a_b is the natural microscopic length, the bulk lattice constant. This is roughly equal to the energy needed to produce in the bulk a typical defect, such as a self-interstitial.[m] It seems that the Nature dislikes ideal surfaces as much as it dislikes other imperfections in this material. Hence it heals them. The satanic malice is then defeated by the insight of God: surfaces reconstruct.

b. *Realistic clean surfaces*

The atoms exposed by truncation of the bulk may adapt their positions to the new situation. In metals, this effects is usually weak and limited to surface **relaxation**, a minor correction leaving the surface periodicity unaffected. But bonds in a semiconductor are strongly directional. Atoms on a semiconductor surface search for new configurations in which their requirements for the number and position of neighbors are better satisfied. Thus, a semiconductor surface often **reconstructs**, that is, the topology of bonds changes. Superstructures, sometimes really beautiful (Fig. 1.7, page 72), may be formed. As usual, a moderate dissatisfaction drives progress!

How do the atoms rearrange on silicon surfaces? This question is as fundamental as the question about the crystallography of the bulk. Since the atomic structure is the key factor behind all microscopic processes, it also determines macroscopic material properties. For example, it controls diffusion of atoms and defects. On the macroscopic scale this is reflected by such bulk properties as mechanical strength of the material. Similarly, such macroscopic properties as surface morphology or sticking coefficients of adsorbates have their origin in the surface atomic structure.

It is also important, though sometimes forgotten, that surfaces are just a class of defects, whose features are determined by the same principles which determine behavior of bulk defects.[207] Since surfaces are more accessible to structural measurements than bulk defects, surface studies can stimulate development of theoretical methods and make them more credible in analyses of problems relevant to microchip technologies, such as implantation and diffusion of dopants or growth of thin films.

A few notions are frequently invoked in the context of surface reconstructions. First of all, there is a tendency to reduce the number of

[m] Which is natural. In all these cases broken bonds play the decisive role.

broken bonds. This was the idea behind the first model of the clean Si(001) surface (Section 2.2.1), and this model immediately captured the most characteristic feature of Si(001), that is, pairing of the surface atoms.

Saturation of broken bonds is most efficient in σ bonds, when the orbitals point to each other: $\rightarrow\leftarrow$. Also π bonds, between parallel orbitals ($\uparrow\uparrow$), are important. The π-bonds in Si(111) 2×1 chains (Section 1.5.2) are strong enough to make some of the subsurface bonds reorganize. The weak π bonds of Si(001) dimers (page 134) determine the electronic structure of the surface near the Fermi level (Section 3.2.3c). In addition, rather complex bonds of molecular type may be responsible for the most characteristic features of a surface, as in the bonding between the pentamers and the interstitials of the clean Si(113) surface (Section 1.5.4).

The **electronic structure** is important as well. A reconstruction which produces a metallic surface, that is, that with fractional occupation of electronic states, is likely to be unstable.[n] The instability can be removed by making the surface semiconducting (with an energy gap between occupied and empty states) through relaxation of Jahn-Teller type (Fig. 3.4, page 136). This requires some relatively small atomic displacements within the surface cell, such as buckling of the Si(001) dimers (Sections 3.2.2b and 3.2.3). Alternatively, additional atoms, such as the interstitials in the case of the Si(113) surface, can be acquired in order to adjust the number of surface states to that needed to make the surface semiconducting.

The formation of a surface, the resulting reconstruction, and particularly the distortions needed to change a metallic surface into a semiconducting one are likely to cause a redistribution of charge among the affected atoms. This results in the appearance of a **dipole moment** oriented parallel to the surface normal. This dipole moment affects the **work function**, that is, the energy needed to remove an electron from the Fermi level to the vacuum infinitely far away from the crystal and from other charges. Indeed, electrons on one side of the surface are repelled, and on the other side of the surface are attracted by the surface dipoles.

Another effect of surface states is the **band bending**. If the reconstruction places empty surface states below the bulk Fermi energy E_F or occupied states above E_F, electrons flow from or to the bulk until the electric field that builds up becomes strong enough to stop them, that is, until the bulk and surface Fermi energies equilibrate. This retarding electric field,

[n] A well-know exception is Si(111) 7×7.

which corresponds to bent electronic bands of the bulk, is screened in the crystal; the screening length depends on the doping level N_D as $1/\sqrt{N_D}$.

Surface **passivation** flattens the bands. It can be achieved, for example, by saturating the dangling bands with hydrogen atoms. Bonding states move then into the valence band and empty states into the conduction band. The charge returns to the bulk.

The distortions needed to build the structure with the minimal density of dangling bonds may generate a considerable **surface stress** (Sect. 3.2.2d). Since stress and broken bonds cost energy, the most stable structure is not necessarily the one which minimizes the number of broken bonds. The examples are the Si(111) 7×7 and Si(113) 3×2 reconstructions. The latter case is somewhat extreme, in the sense that the stress produced by rebonding of surface atoms is used to stabilize in the subsurface region a high concentration of rather unusually bonded, ordered interstitial atoms.

The broken bonds are sometimes efficiently saturated after mere displacements of the surface atoms, as on Si(001). In other cases it may be necessary to remove some atoms in order to form objects such as corner holes of Si(111) 7×7, or extra atoms must be added, for example, the adatoms of Si(111) 7×7. When the number of atoms in a "$m\times n$" cell differs from that in mn primitive cells of the truncated bulk, the reconstruction may be **kinetically limited** by the transport of material from or to surface steps. In consequence, metastable reconstructions may appear, such as the 5×5 structure of Si(111). Also deposition of atoms on the surface or extraction of atoms from it may produce kinetically limited, metastable reconstructions, such as the $c(4\times4)$ class of structures on the clean Si(001) surface (Section 3.4.2). Kinetic limitations may also influence macroscopic objects, for example, the observed equilibrium shape of the crystal.[208,209]

If no material is being deposited, **surface diffusion** plays a role only when temperature is high enough for atoms to detach from weakly bound sites. These are usually surface steps (Sect. 3.3.2), though atoms can also escape from exit points of dislocations (Sect. 4.4), whereby characteristic thermal etch pits are formed. At somewhat lower temperatures atoms can still travel along step edges. In this manner steps fluctuate in a wavy motion without changing their average position (Sect. 4.2.3b), and small **islands** (objects surrounded by a closed step line) can migrate.[210,211]

Diffusion across the surface allows neighboring steps to exchange atoms. The steps can rearrange in this way and build patterns depending on the vicinality, temperature, external strain, and on the anisotropy of the surface stress (Fig. 4.9, page 255). The resulting **phase diagram** may be rich (Section 4.3.2), with steps of varying heights and shapes.

Steps-step interaction is usually repulsive. When it is attractive, as around the (113) orientation of silicon,[212] a **faceted surface** may form. The steps coalesce into high walls which are in fact surfaces with another orientation. The terraces and the walls alternate in a sawtooth pattern of up and down slopes running across the faceted plane: . Clean Si(112) is an interesting example of such a faceted system.[213]

These rather straightforward mechanisms become somewhat complicated when atoms of species different than the substrate are deposited on the surface. Section 1.4.2b summarizes these chemical interactions.

1.4.2. *Free surfaces*

By "free" we understand here surfaces which make interfaces with the vacuum. They may be either chemically clean or covered with a small amount of adsorbates. "Small amount" means that the coverage does not exceed a monolayer, that is, that the areal density of the adsorbates does not exceed that of surface atoms on the corresponding truncated bulk.

The purpose of this Section is to introduce and overview the key issues and concepts related to surface defects and to chemical activity of surfaces. In brief, though defects observed on clean surfaces under Ultra High Vacuum (UHV) have little technological relevance,[o] they are much more abundant than bulk defects. Since the UHV conditions are instrumental in fundamental surface science, one must understand defects in order to put one's knowledge of surface processes on solid footing. Some limited technological interest may begin when interactions of the clean surface with molecules and atoms arriving from the gas ambient (adsorption, etching, formation of surface compounds) are considered. Though it is a long way from there to a commercial innovation, physical intuition gained in such studies may result in practical ideas (see Section 6.5.5d for an example) and may facilitate practical simulation of technological processes (page 282). Probably the most applied field here is cleaning (Sect. 1.4.2c).

a. *Defects on the surface and in the bulk*

Neither real surfaces nor real interfaces are free from faults. Indeed, defects are present everywhere: in humans, in cars, in bulk Si, and on Si surfaces.

---

[o] UHV methods are too expensive and too complicated for cheap Si technologies.

They cannot be avoided, so they must be controlled. In order to build a realiably working system, the designer must be able to estimate the influence of imperfections on the functionality of the whole. In Si-based systems, the important defects may start on the surface, where they may affect the growth of films and the properties of interfaces. But most importantly, surface defects are the most easily observable atomic-size defects at all.

Some surface defects, such as Si(001) missing dimers and C-defects (Sect. 3.3.1) and surface steps (Sect. 3.3.2), are thermodynamical features of clean surfaces. Other defects are metastable by-products of surface preparation (Sect. 3.4). They may cover the whole surface with a more or less regular pattern, as the $2\times n$ and $c(4\times2)$ reconstructions (Sections 3.4.1 and 3.4.2), or be localized, as translational domain boundaries (Sect. 3.4.3). There are also single atoms or clusters of a few atoms which are either deposited on the surface or emitted from surface steps. They can diffuse on the surface and interact among themselves and with other defects (Sect. 4.2). These interactions control the process of crystal growth from vapor.

Since surface defects are governed by the same physics as bulk defects, a physicist can apply the experience from surface studies to gain insight into less accessible bulk processes. This requires a theory which not only reproduces the experimental data but can also draw valid conclusions from them – a physicallly correct theory with predictive abilities. Such methods exist and are provided by modern *ab initio* techniques (Sect. 1.3.1).

For instance, interstitial atoms in Si bulk have an intriguing tendency to aggregate on {113} planes.[214] This process causes the so-called transient enhanced diffusion of implanted dopants (page 421), a major roadblock in the preparation of shallow junctions for deep submicron CMOS devices.[28] Also Si(113) *surfaces* attract silicon interstitials (Section 1.5.4). In both cases the governing mechanism is the local stress.[215] The comparison between these bulk and surface structures becomes rather straightforward when the surface is passivated with Sb atoms.[215] The direct electronic contribution to the surface-interstital interaction is then switched off, but the surface stress contribution remains active. In consequence, the surface builds up structures topologically identical with the bulk {113} defects.[215]

Another example is that surface phase transition studies might assist the development of process simulation tools for semiconductor technology (page 241). An interesting case is the $5\times5\rightarrow7\times7$ transition on the Si(111) surface (Sect. 1.5.2). It may be a model system to investigate phase transitions which require mass transport, include moving grain boundaries, and depend on external boundary conditions. A practically important process of this type is the C49 to C54 phase transformation of $TiSi_2$ (page 327).

b. *Adsorption and epitaxial growth*

We will now focus on processes in which atoms or molecules arrive at the silicon surface from the ambient and remain attached to the substrate. When the deposited material is Si, this can lead to homoepitaxial growth of the crystal.[p] The probability that a silicon atom colliding at a thermal speed with an atomically clean Si surface will stick to it is practically 1. This is not necessarily so for other adsorbants (Chapter 5). For example, H_2 or N_2 molecules are virtually inert though atomic H and N readily react with the clean surface; H_2O molecules stick to clean Si(001) with the probability of 1 but most of them are reflected from Si(111).

In general, the ability of the surface to accept an adsorbate is controlled by the local electronic structure of the surface, though geometry constraints may play a role as well. In the first approximation, the key factors are: the electronegativity difference between the reacting species and the availability of electrons for the charge transfer caused by this difference. **Electronegativity** measures the tendency of a species to attract electrons. Mulliken electronegativity $\chi = (I + \eta)/2$ is the finite-difference approximation of the chemical potential[216] $\mu = \partial E/\partial N$. Here, I is the ionization potential and η is the electron affinity (Table 6.5, page 387), E is the total electronic energy of the system and N is the number of electrons; N is treated in this formalism as a continuous variable. For example, the highly electronegative oxygen will prefer a Si atom with occupied dangling orbital close to the Fermi energy E_F, so that O can acquire an electron from Si with a maximum energy gain.

The ability of an orbital to change its occupation depends on the amount of charge already transferred. Some orbitals are "soft": the derivative of χ with respect to N is small. Other orbitals are "hard" and do not let charge easily out or in. This leads to the concept of chemical **hardness**.[217,218,219] Hardness ξ is defined as $\xi = -\partial\chi/\partial N$; chemical **softness**[94] is basically the reciprocal hardness, $1/\xi$. Hardness and softness can be defined in terms of the density functional theory (page 20) and have been used succesfully[220] to analyse the chemical activity of Si(111).

---

[p] **Epitaxial films** (from Greek *epi* + *taxis*, upon + arrangement) adopt their crystallography to that of the substrate. **Homo-** indicates that both materials are the same.

The charge transfer is thus driven by electronegativity differences and opposed by hardness. In general, systems with large separation between occupied and empty states are hard; a small energy gap results in soft species. Strictly speaking, hardness is therefore a property of the whole object (molecule or surface), and not of a single orbital. Since softness may be interpreted as a "charge capacity",[221] hard molecules tend to react with hard regions of the surface while soft molecules prefer soft regions.

The character of the reaction between the surface and an adsorbant is likely to change in the course of adsorption. First, there is a trivial dependence of the **adsorption rate** on the exposure time due to the varying coverage. Second, the adsorbed species may stick better when they are agglomerated in islands. Since the islands must first be nucleated, this may lead to such effects as incubation time[222] or active and passive oxidation regimes (Sect. 5.4.4). Third, adsorption of molecular species usually proceeds through several states: the molecule may be free in the gas or physisorbed, molecularly chemisorbed, or dissociatively chemisorbed on the surface. These states can be separated by energy barriers. The transition rates between them determine the **sticking probability**. A proper theoretical analysis of the energetics of the adsorption process requires that **chemical potentials** of the reacting species are taken into account.[223]

A **physisorbed** molecule is weakly bonded with the substrate through Van der Waals or Coulomb forces. This **precursor** makes no chemical bond with the surface and usually can migrate efficiently over the surface. A **chemisorbed** molecule does make a chemical bond (that is, the inter-molecular bonds are likely to be affected), but retains its identity: it does not break into separate pieces (cf. Fig. 5.6g, page 302). Such a chemisorbed molecule may ultimately dissociate into **moieties**, as it is the case for H_2O (Fig. 5.6e on page 302). The energy gained during dissociation stems from the difference between the chemical bond energies in the atomic configurations before and after the reaction. Naturally, the ultimate amount of the adsorbed material X depends also on the ambient pressure p_X and temperature T through the entropy term $(-RT \ln p_X)$ in the molar chemical potential of the unadsorbed phase.

Table 1.4 lists some values of bond energies. These values can be used to estimate the energy gained or lost in a particular reaction. However, it should be kept in mind that the surface responds to the adsorption process with changes in its atomic and electronic structure. This leads to contribu-

Bond	Energy, eV	Length, Å	Bond	Energy [eV]	Length [Å]
Si-Si	2.3[203]	2.35	H-H	4.5	0.74
Si-H	3.1[a]	1.48	H-F	5.8	0.92
Si-O	3.8	1.61	H-Cl	4.5	1.27
Si-F	5.6	1.54[224]	H-N	4.1	1.01
Si-Cl	3.7	2.02[224]	O-H	4.8	0.96
Si-Br	3.0	2.15[224]	O-F	1.9	1.42
Si-I	2.2	2.43[224]	O-Cl	2.1	1.69
F-F	1.6	1.44	O-O	1.4	1.49
Cl-Cl	2.5	2.00	O=O	4.2	1.24
Br-Br	2.0	2.24[224]	O≡O	5.1	1.10
I-I	1.6	2.66[224]	N≡N	9.8	1.09

Table 1.4. Selected bond energies and bond lengths; most data after Reference 225. ([a]) Approximate value for H on Si(001). The bond Si-H strength in SiH_4 is ~3.7 eV.

tions from rehybridization, electrostatic interactions, and elastic stress to the adsorption energy and may affect the energy balance by ~ 1 eV.

Dissociative adsorption is likely to be irreversible. For instance, the volatile desorption products of H_2O adsorption are H_2 and SiO (Sect. 5.5.3). A widely studied reversible reaction is the dissociative adsorption of H_2: the molecules dissociate on surface steps, H atoms occupy surface dangling bonds, and heating leads to desorption of H_2 (Sections 5.2.1 and 5.2.2). Some reactions are partially reversible; for instance, adsorption of Br_2 produces volatile Br_2 $SiBr_2$, and $SiBr_4$ (Sect. 5.3.2).

Researchers studying adsorption often face a dilemma: is the molecule adsorbed as a whole, or does it dissociate? The answer is not easy to find when the interaction with the substrate strongly affects such "fingerprints" as vibrational frequencies of the molecule and its fragments. The problem is aggravated when the dissociation barrier is low. The very act of measurement may then break the molecule or surface defects may promote the decomposition,[q] making the result of the experiment dependent on the sometimes elusive details of sample preparation. Ammonia (Sect. 5.6.2) and water (Sect. 5.5.2) adsorbed on Si are such troublesome molecules.

A special case of adsorption is when the deposited material forms an **epitaxial** layer, that is, a dislocation-free, crystalline film. The best epitaxial layers are obtained when the growth proceeds by incorporating atoms at

[q]See the influence of surface steps on H_2 adsorption, Section 5.2.

kinks of surface steps.[226] Since the probability that an atom from the vapor hits a step is low, atoms are first adsorbed on terraces and then diffuse around until they find a step or another adsorbed atom, or evaporate. Once on a step, they can migrate along its edge to kink sites. Atoms which migrate *across* steps may encounter an additional barrier for diffusion, known as **Schwoebel barrier.**[227] The atoms immobilized or slowed down on terraces by encounters with other atoms form seeds of two-dimensional (2D) islands. The competition between attachment to step edges and nucleation and growth of islands may lead to **denuded zones,**[228] which are stripes of surface free of islands and running along step edges.

We will now illustrate the mechanism by which Schwoebel barriers affect the surface morphology in the **step-flow** growth. In brief, the velocity of a step becomes linked to the size of the upper or the lower terrace associated with this step. This activates a feedback loop which may either damp or amplify step-flow irregularities. Schwoebel barriers on clean Si(001) are observable[229] but apparently low[230] (Sect. 4.2.3); higher barriers may be induced by impurities adsorbed on step edges. In the absence of significant Schwoebel barriers, step flow is stabilized by elastic repulsion between steps (Sect. 4.3.2) which prevents them from accidental bunching.

When the deposition conditions are such that nucleation and growth of islands are not efficient, that is, when the growth proceeds by step flow, a Schwoebel barrier on the upper terrace *stabilizes* the growth. This occurs by stabilizing the terrace size, so that steps do not bump into each other. Since only the lower-terrace atoms can attach to the step, the speed of step movement is proportional to the size of the lower terrace. If a fluctuation advances a step faster than its lower neighbor moved, the area of the lower terrace of the speeding step is reduced; the step slows down. This happened to the leftmost step and to its associated terrace marked with dots:

(The sketches show a side view of a stepped surface. All steps not affected by the fluctuation move at a constant speed from the left to the right. Thin lines mark the previous location of the surface). If a fluctuation slows down a step, its lower terrace will expand and the step will speed up:

If a Schwoebel barrier hinders the diffusion of atoms from lower to upper terraces, the step flow is *destabilized*. The speed of a step is now proportional to the size of the upper terrace, so fluctuations of terrace sizes are amplified. Bigger terraces grow bigger, smaller terraces decay, steps approach each other, form step bunches, and the surface tends to roughen:

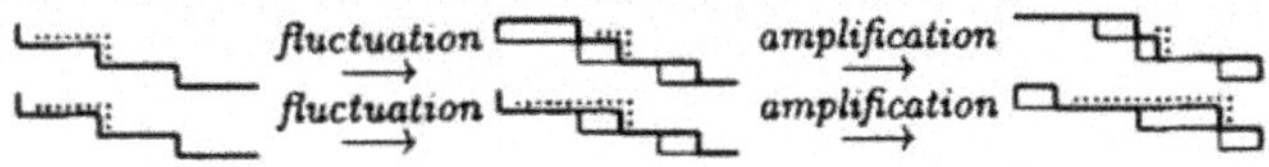

During **sublimation** or removal by sputtering the diffusing elements are dimer vacancies and the situation is reversed: if a Schwoebel barrier prevents upper-step vacancies to annihilate at step edges, so that only the lower-step vacancies can approach steps, the motion of steps is destabilized. Steps retreat with speed proportional to the lower-terrace size, fluctuations in the terrace size are amplified, and surface has a tendency to roughen:

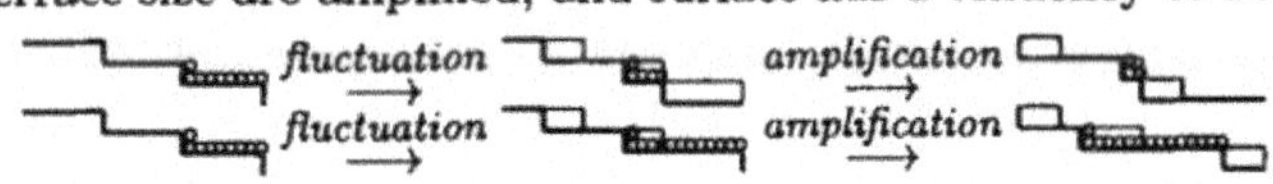

(the steps of interest and the associated terraces are indicated by chains of circles; all steps retreat from the right to the left). On the other hand, a Schwoebel barrier for vacancies approaching from the lower terrace links the speed of step retreat to the upper terrace size, stabilizing the step motion:

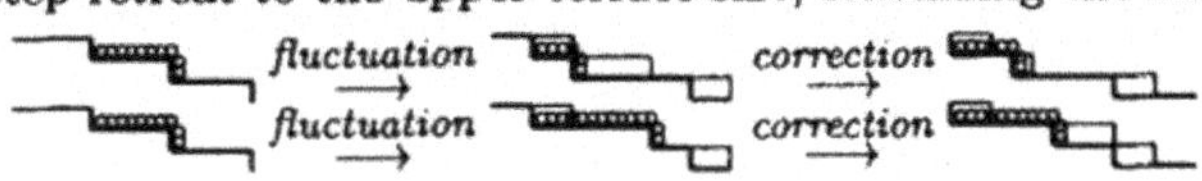

Apart from this direct influence on the step flow, a Schwoebel barrier may affect the growth by forcing the deposited material to stay on the top of an accidentally formed island. If the atoms cannot recombine at the island edge, they will eventually come together to build nucleation centers on the island top. The islands will grow higher, and the surface will roughen.

Such high, three-dimensional (3D) islands are energetically favored in heteroepitaxial growth (for example, SiGe or Ge films grown on Si) when the deposited material has a different lattice constant than that of the substrate. The reason is a more efficient relaxation of the misfit strain in a small 3D island than in a flat, 2D layer. This is the **Stranski-Krastanov** growth mode: though the film wets the substrate and grows epitaxially, thicker films roughen by buildup of islands. Stress-induced islands and facets are generally troublesome, but they may stabilize the epitaxy by disrupting diffusion and driving the system further away from equilibrium.[231]

Epitaxial Si or SiGe films for research purposes are often grown by **Molecular Beam Epitaxy**[232,233] (MBE). Si and Ge evaporized from a source are directed through UHV onto the growing substrate. While MBE allows for atomic-scale control of growth, it is too expensive in production lines. If epitaxy is needed during manufacture of a commercial silicon device, **Chemical Vapor Deposition** (CVD, see Refs 234, 32, and references therein) is used.[r] In CVD, the deposited material is delivered in form of volatile hydrides or chlorides, such as SiH_4 or SiH_2Cl_2. Hydrogen and chlorine passivate the surface and help to keep it clean, obviating the application of UHV. Elevated temperatures take care that the passivating atoms segregate to the surface and that some of them desorb (Sections 5.2.3 and 5.3.2), so that the incoming molecules can adsorb dissociatively at the substrate dangling bonds. New material is thus constantly being deposited.

c. *Desorption, etching, cleaning, cleaving*

Finally, we consider treatments which remove material from the surface. They can etch the substrate, clean it from impurities, or perform both actions simultanously. The etch rate often depends on the dopant type and concentration; this forms the basis of the so-called etch stop techniques used in micromachining technologies.[5] We briefly discuss industrial as well as laboratory methods of etching and cleaning. Since one way to obtain a chemically clean surface for Ultrahigh Vacuum (UHV) studies is cleaving the crystal, we will address this procedure as well.

Many cleaning steps involve aqueous solutions of chemicals. The key concepts are in this case the notions of **acidity** and **alkalinity**. An acid is a substance which has a tendency to lose a proton (H^+) and a base (alkali) is a substance which has a tendency to gain a proton.

Some H_2O molecules in liquid water dissociate, giving a proton away:

$$H_2O \rightarrow OH^- + H^+. \tag{1.7}$$

The hydroxyl ion (OH^-) produced thereby is a typical base. It can accept a proton in the reaction reverse to Eq. 1.7:

$$OH^- + H^+ \rightarrow H_2O. \tag{1.8}$$

The relative strength of acidic or alkaline solutions is measured by the **pH factor**. The pH is defined as $-\log C(H^+)$, where $C(H^+)$ is the concentration (in g/litre) of dissolved H^+. The pH of neutral water is 7

[r] It is different in III-V technologies, which use MBE,e.g., for high-performance microwave and mm wave devices. GaAs MBE is quite versatile while AsH_3 is highly toxic.

(Sect. 5.5.1). Acidic aqueous solutions have pH < 7, that is, $C(H^+) > 10^{-7}$ g/litre. The pH factors of 1% and 10% solutions of HF are, roughly, 2 and 1. Alkaline aqueous solutions have pH > 7, whereby the protons of the dissociated H_2O molecules are accepted by the base ions. Usually, this is the reaction of Eq. 1.8. Since the hydroxyl radicals and the protons are related by the reactions 1.7 and 1.8, it follows from the law of mass action that in the thermodynamic equilibrium the product $C(H^+)C(OH^-)$ is a constant. In other words, pH measures also the concentration of OH^- ions: $pH = 14 + \log C(OH^-)$. The solution contains other related ions, as well. For example, HF solutions contain HF, H^+, OH^-, F^-, and HF_2^-.

Cleaning is a delicate and at the same time critical step in microchip manufacturing.[235] Around 30% of processing steps in contemporary technologies are wafer cleaning steps and 20% of equipment costs are the costs of wafer cleaning equipment.[236] There is also an environmental concern. An important objective is to reduce the use of water and chemicals.[237]

Many things can go wrong if wafers are not clean enough.[238] Nanoparticles and SiC precipitates cause low-voltage breakdown of gate oxides.[239] Excessive levels of carbon or oxygen destabilize silicidation and epitaxy. Metals such as Fe and Cu degrade the gate oxide[240] and reduce the bulk minority carrier lifetimes.[241] By the year 2009, the maximum allowable[22] surface concentration of critical metals (calcium, cobalt, cupper, chromium, iron, potassium, molybdenium, manganese, sodium, nickel, tungsten) must drop to or below 10^9 cm^{-2}, of other metals (such as aluminum, titanium, vanadium, zinc, barium, strontium) to or below 10^{10} cm^{-2}. It is not generally agreed how to comply with these requirements[a] but solutions are being pursued. An important issues is that it is hard to *detect* such small amounts of metals.[242] The critical concentrations of C and O atoms are much higher, above 10^{13} cm^{-2}, but under industrial conditions these elements are more difficult to control than metals.

The methods which are of interest to us include wet cleaning and etching (by solution of chemicals), dry cleaning (as by exposure to Cl_2 or O_2 under illumination with ultraviolet light[243]), dry etching (by reactive gases, Sect. 5.3), sputtering (impact removal of surface atoms by chemically inert ions), and desorption (by high-temperature annealing). Since a bare Si surface contains electronic states easily forming bonds with other species, the cleaning procedure should protect the surface by an easily re-

[a] For comparison, the perfect Si(001) surface has $\sim 7 \cdot 10^{14}$ Si atoms per cm^2.

movable passivation layer. This is achieved either by hydrogen termination (Sections 5.2 and 5.3.1) or by the formation of an oxide film (Sect. 6.3.1). The sample or the wafer can now be transferred through the air. These protective layers either evaporate or are incorporated into the film during the subsequent oxidation, silicidation, or epitaxy. The protective layer is not needed only when the cleaning is conducted in Ultra High Vacuum (UHV) directly before a UHV process.

Wet cleaning procedures are indispensable in the industry[235] because they remove particles. When the surface is wetted and etched by the solution, capillary action lifts the particle 1-2 nm (10–20 Å) above the surface. This breaks the attractive van der Waals force. If the charges brought to the surfaces of the substrate and of the particle by the solution have the same sign, the particle is now removed by Coulomb repulsion.[244] Moreover, dry etching,[245] which is usualy done by fluorocarbons, leaves the surface contaminated by fluorine and carbon species. A wet cleaning step removes the remnant Si–F and Si–C bonds.[246]

Wet cleaning can easily **contaminate the surface with metals**. Of a particular concern is the contamination during the popular SC-1 cleaning (in H_2O:H_2O_2:NH_4OH, Sect 6.3.1). Three major mechanisms are responsible for metal contamination:[247] physisorption through van der Waals forces (bond strengths of ~0.2 eV), chemisorption through forming a chemical bond (bond strengths around an eV), and adsorption through reduction of the metallic cations by the substrate, that is, transfer of electrons from the substrate to the metal. The metal atoms from subsequently cleaned wafers accumulate in the bath. Some of these atoms, notably copper, may become deposited again, so that "cleaning" becomes a source of contamination. The mechanism of Cu deposition from HF solutions has been recently reexamined.[236] Efficient methods to recycle the chemicals used in the cleaning have been suggested.[237]

A procedure which substitutes the alkaline SC-1 by sequences of acidic cleaning steps in H_2O:O_3[248] and dilute HF (IMEC clean[244]) or dilute HF:HCl (Dilute Dynamic Clean, or DDC[249]) aqueous solutions can keep the contamination with metals below 10^9 cm^{-1}. On the other hand, when organic "complexing agents" like $N_3(PO(OH_2))_5$, or, particularly, $N_3(PO(OH_2))_3$ (cTRAMP) are added to SC-1, metals become stabilized in the solution.[250] This method improves the efficiency of cleaning without sacrifying its simplicity. It is noteworthy that cTRAMP is a compound *designed* for complexing of double and triple charged metal ions.[250]

Most of the cleaning solutions involve oxidation of the surface (Sect. 6.3.1). The resulting "native oxide" film (page 359) and the Si surface beneath are relatively rough. This roughness will be a problem for future technologies[251] (page 355). Therefore, there is a tendency to combine the last cleaning step with **etching** of the surface SiO_2 by a procedure which leaves the surface terminated with hydrogen. This is done in baths containing HF (page 279), typically at concentrations around 0.5–1%. Even 0.1% HF can remove the native oxide within a few minutes.

A problem with HF etching, particularly in research, is that it easily leads to contamination with carbon.[252,239] This can produce SiC precipitates.[253,254] A proper pre-treatment, for example, by decomposition of surface carbon compounds by O_3 under UV light and usage of water free from organic carbon[252,255] can minimize this difficulty.

Another emerging problem is that HF etches Si most easily along (111) planes, so that it has a tendency to roughen the Si(001) surface of commercial wafers (Sect. 5.2). Several methods to circumvent this have been proposed, including adding H_2O_2 to very diluted HF[256] or using acidic HF:HCl aquaous solutions.[257,249] HF/ethanol solutions[258,259] can be used as well. It may be interesting to note that, on the basis of *ab initio* calculations, it has been proposed[260] that any alcohol remaining on the surface after the HF/ethanol treatment can be removed from the surface by exposure to diborane (B_2H_6) ambient at temperatures around 300–350°C.

Desorption of atoms and molecules can be used as a means to study the chemical composition of the surface. Different local configurations of the adsorbed atoms can be recognized as far as they are characterized by different bonding energies. Roughly speaking, the bonding energy is reflected in the desorption temperature; this dependence may be obscured by kinetic effects related to microscopic desorption paths, for example, by surface diffusion. Naturally, one can (and does) use thermally activated desorption in UHV as a method to clean the sample from unwanted adsorbates. However, one must keep in mind that not all species can be removed from silicon in this way (for example, Ni is not removed). Finally, desorption of H or Cl from Si is a key event during Chemical Vapor Deposition.

Adsorbed species desorb typically as molecules. Depending on the relative bond energies and kinetic pathways, desorption may have associative character (when molecules containing only the adsorbate are released, such as in H_2 or D_2 desorption) or be accompanied by etching (when the desorbing molecule contains Si atoms, as SiO, SiH_4, or SiF_4). If some of the surface species containing the given adsorbate are volatile and the other are not, the surface may be either etched or not etched by the source gas, depending on the partial pressure of the gas phase, substrate temperature, and atomic details of the reaction paths. For example, high temperatures and low partial pressures of O_2 cause etching by desorption of SiO, while lower temperatures and/or higher partial pressures lead to growth of a SiO_2 film (Fig. 5.3, page 294). Conversely, a silicon surface exposed to atomic H at a sufficiently low temperature will be etched by emission of SiH_4, while at a higher temperature it will be passivated by an atomic layer of hydrogen.

It is convenient to distinguish two types of desorption: Langmuir-Hinshelwood and Eley-Rideal. Reactants in a **Langmuir-Hinshelwood** process are equilibrated with the substrate. When gas phase species participate in the desorption, one speaks of a **Eley-Rideal** reaction.

When the Si sample is locked in a UHV chamber, the simplest way to clean it from the protective H or SiO_2 and from impurities which may have contaminated it during transport is to heat it briefly to a temperature close to that at which the surface melts. Such **flashing** at 1180°C removes SiC from the surface. Though hydrogen and oxygen desorb at lower temperatures (Section 5.2.2 and Fig. 5.4 on page 295), anneals below 1180°C may roughen the surface precisely because SiC remains on the surface: some Si atoms sublimate to the vacuum, surface steps retreat and bunch when they are stopped by a thermally stable SiC precipitate (Section 4.4).

Unfortunately, not all impurities can be removed by flashing. The most troublesome example is nickel (Sect. 5.7.2). To make things worse, dissolved Ni segregates to the surface and becomes stabilized there. This can shorten the lifetime of reusable experimental samples. The only practical way to get rid of nickel is to remove the surface layer, for example, by **sputtering** with Ar^+. Yet lattice defects caused by ion bombardment can never be healed completely, even by prolonged annealing.[261]

If flashing is undesired (for example, when an artificial structure which would be destroyed by high temperatures is studied), SiC can be removed by reaction with O_3. Ozone oxidizes carbon to volatile CO when it is broken by ultraviolet light.[252,262] Note that a surface, even passivated with an oxide layer or, particularily, with hydrogen,[255,253,239] becomes easily contaminated with **hydrocarbons** when exposed to laboratory or cleanroom air. Hydrocarbons may also come from plastic storage boxes.[263] Rinsing with insufficiently pure water (e.g., around a hundred ppb of Total Organic Carbon, or TOC) after an oxidizing etch can also saturate the surface with hydrocarbons. Ultrapure water (17 ppb TOC) keeps the contamination low.[252] Alternatively, in-situ cleaning by a hydrogen plasma processing compatible with UHV equipment has been proposed.[264]

Finally, an atomically clean surface can be prepared by **cleaving** in UHV. Unfortunately, this works well only for the (111) surface.[265] Though other Si surfaces can be also exposed by cleaving, particularly the Si(110) surface,[266] their quality is poor. They are microscopically (and often macroscopically) rough and hardly suited for atomic-scale studies. Even the cleaved Si(111) surfaces show macroscopic marks, often radiating from the point where the cleavage started, and a varying density of surface steps.

1.4.3. *Buried interfaces*

Two silicon interface types have a particular technological relevance: dielectric/Si(001) interface and metal/Si(001) interface.[t] The former borders the active part of a MOSFET and contributes to the electric properties of the MOS capacitor and the transistor channel (Sect. 1.2.2). The interface with a metallic layer connects a silicide contact to a doped electrode and adds to the serial resistance. As device dimensions diminish, films get thinner and interfacial properties become increasingly significant.

The major issues related with the fabrication of **contacts** are:[267,28] consumption of Si, sheet resistance of thin films, contact resistance, and interfacial roughness. These are briefly discussed in Section 5.7. The main challenges in the production of gate **dielectrics**[268,28] are reliability (Sect. 6.6.2a) and homogeneity (page 355 and Sect. 6.3.5) of ultrathin oxides, and control of interactions between the oxides and dopant: dopant pile up and dose loss (Sect. 6.5.5c), and boron penetration (Sect. 6.5.5d).

The Roadmap for Semiconductors[22] compiled by experts from industry and research predicts that the continued miniaturization of electronic devices will soon (within approximately a decade) require MOS structures with such a high capacitance C that dielectric films with thickness t corresponding to approximately one monolayer of SiO_2 will be needed. Naturally, these films will not be made of SiO_2. A still unspecified material with much higher dielectric constant ϵ will be used instead ($C = \epsilon/t$). JVD

---

[t] Moreover, dopant segregation to Si/SiGe interfaces may play a role in SiGe technologies.

silicon nitride (Sect. 5.6.3) is a plausible candidate, but in a more distant future a material with much higher dielectric constant, possibly around $\epsilon = 20$, will have to be employed (SiO_2 has $\epsilon \sim 4$, Si_3N_4 has $\epsilon \sim 6-9$). The new material (Sect. 5.7.4) must at least reproduce the interfacial properties of a good SiO_2 dielectric film. It is expected that an extremely thin, SiO_2-based buffer will have to separate the high-ϵ film and the silicon substrate at least in the advent of the new technologies.[22]

We devote the most attention to SiO_2/Si(001) interfaces (Chapter 6), partially because they are better understood than silicide/Si and Si_3N_4/Si. Atomistic studies of SiO_2/Si may lead not only to an improvement of MOS structures[242] but also to a systematic development of the material which will substitute SiO_2 in its role as a gate dielectric.

Most of the concepts appearing in our discussion of oxidation and of SiO_2/Si(001) interfaces (Chapter 6) are explained either in the preceeding Chapters or on the spot as needed. The reader will be guided to these places by cross-references. Here we introduce some terminology related to the defects in gate oxides (see page 355 for technical details) and make some remarks on tunneling through very thin oxides.

Two types of defects in gate oxides are particularly troublesome: nanoscale inhomogeneities and atomic-sized charge traps. The **inhomogeneities** can be due simply to a local thinning of the oxide, likely to occur at oxide edges. Such places are predestined for electrical punch-through. More sophisticated and more deadly are "pinholes" (Sect. 6.3.5) which, as the name suggests, can short-circuit the oxide already at low voltages. Accumulation of **charge traps** in an old or overbiased transistor can also lead to a dielectric breakdown, but these defects mostly just deteriorate the electric parameters of the device without making it fail altogether.

Charge traps are traditionally divided into interface traps, fixed oxide charge, oxide trapped charge, and mobile oxide charge. The interface traps and fixed oxide traps are created in the process of oxidation (Sections 6.4 and 6.5). Both are related with the atomic details of the reactions at the advancing interface. The main practical difference between them is that the former can exchange carriers with the substrate while the latter can not, possibly because it is located further from the substrate. The oxide trapped charge is due to carriers trapped on (or ionized from) neutral defects in the bulk of the film, chiefly by currents flowing across the oxide. Finally, the mobile charge is caused by impurity atoms, such as Ca^+ or Na^+ ions.

The currents flowing through the oxide can either contain **hot carriers**, that is, carriers which made it to the conduction band of the material, or carriers tunneling under the potential energy barrier of the dielectric. Hot carriers are particularly dangerous because they collect a lot of energy and can cause damage in the interfacial region. **Tunneling** is milder on the oxide, but as the oxide thickness is scaled down, it becomes the major source of gate leakage currents and may adversely affect the lifetime of the dielectric (Sect. 6.6.2a). Atomic-sized defects in the oxide contribute to the tunneling by providing electronic states in the forbidden energy region.

The barrier for tunneling through Si_3N_4 is usually triangular. The current is then $I = A\mathcal{E}^2 \exp(-B/\mathcal{E})$, where $\mathcal{E}$ is the electric field and A and B are independent of $\mathcal{E}$ (**Fowler-Nordheim** emission[269,21]). In SiO_2, this holds for thin films and/or high fields.[270] The barrier in operating MOSFETs with ultrathin oxides ($\lesssim$3 nm) is trapezoidal (**direct** tunneling[271]), because the conduction band offset between Si and SiO_2 exceeds typical gate voltages (Fig. 6.12, page 435). For small biases, I becomes now $I = \alpha V + \beta V^3$, where V is the bias. Defects contribute significantly to the tunneling through SiO_2 (**Trap Assisted Tunneling**).[272]

1.5. Primary Silicon Surfaces and Their Vicinals

Apart from scientific interest, investigations of various surfaces of Si are driven by the hope to make an impact on microelectronics. Si(001) is the dominant substrate for electronic devices, but high-index surfaces are studied as potential substrates for specialized applications[273,274,275] which rely on properties of epitaxial films. Epitaxy on surfaces other than a flat Si(001) often gains stability. The quality of the film may improve. The strain induced in the film by the lattice misfit with the substrate can produce interesting and maybe even useful changes in the electronic structure.

Due to their limited market, exotic technologies are less influenced by the rat race to cost reduction. This makes unorthodox ideas from basic research more attractive to investors. But the interest of the major, most stable industrial players is proportional to the market size. Since our guiding theme is the relation between the interests of science and the interests of big industry, our account of non-(001) Si ends with this Section.

Below we describe the primary surfaces of silicon: (001), (111), (113), and (110). We also mention (11n) surfaces, including (001) and (111) vicinals. Surfaces misoriented towards low-symmetry directions are less frequently studied; for examples, see Refs. 276 and 277. While Si(001) is the main topic of the book and is only summarized here, the other three primary orientations are covered in more detail. Of these, Si(111) is by far the most studied one, being probably the most popular semiconductor surface in academic research. Our further discussion will use them chiefly to illustrate Si(001) features in a broader context.

1.5.1. *Structures of Si(001)*

Si(001) is at the same time the simplest and the only technologically relevant surface of silicon. This fortunate coincidence made us chose Si(001) as the major subject of this monograph. The geometry of the (001)-truncated bulk is described in Section 3.2.1. The main building block which determines the properties of Si(001) is a surface dimer (Fig. 2.3, page 90) which is formed when the atoms of the truncated bulk (Fig. 2.3a) rebond pairwise to reduce the number of surface dangling bonds. The surface made of symmetric dimers (Fig. 2.3c) is metallic (Fig. 3.5, page 144), but the real Si(001) is semiconducting because the dimers are buckled (Fig. 2.3d).

The buckling of dimers lowers the surface energy. This effect, well established experimentally as well as theoretically (Chapter 2 and Section 3.2.2), results from a balance between several quantum-mechanical and classical interactions such as hybridization, exchange and correlation, electrostatics, and surface stress. The ability of a theory to reproduce buckling is thus a credibility benchmark in structural studies. Moreover, buckling of dimers affects step-step interactions (Sect. 4.3.2) through its influence on stress anisortopy (Sect 3.2.2d) and adsorption of molecules (Sections 5.2 and 5.5) through geometry constraints and inter-dimer charge transfer.

Dimerization is a key factor in the dynamics (Chapter 4) and the chemical activity (Chapter 5) of Si(001). For example, atoms and molecules diffusing on the surface use dimer rows as convenient "rails". The existence of the dimer bond is also crucial for a succesful homoepitaxial growth from chemical vapor (CVD). Even some interfaces of Si(001) with SiO_2 (Sect. 6.5.2) or with certain silicides (Sect. 5.7.1b) may be dimerized.

Steps on Si(001) have either monatomic or biatomic height, depending on the misorientation of the surface (Sect. 4.3.2). The interaction of these steps with the deposited material plays a key role in the process of epitaxial growth (Sect. 4.4) and in the adsorption of certain molecules, such as H_2

(Sect. 5.2). In some cases, as during oxidation (Section 5.4.3c), the role of steps seems to depend on the detailed conditions under which the reaction is conducted, while in some other cases, as for adsorption of H_2O (Sect. 5.5), steps play no particular role or are even avoided.

The most frequently encountered surface point defects of Si(001) are missing dimers and the so-called C-defects (Section 3.3.1). They abound in STM images, but can be easily confused with adsorbed impurities.

1.5.2. *Structures of Si(111)*

Being the best cleavage face of silicon, Si(111) is a favorite of experimentalists. Si cleaves easily at (111) planes because each Si–Si bond in the bulk is directed along one of the <111> axes, so that a {111} plane cuts as few (or as many) bonds per area as posssible in the crystal (Table 1.3, page 49). Each atom on the ideal Si(111) 1×1 is backbonded to three subsurface atoms (Fig. 1.6a). Its fourth sp^3 orbital points towards the vacuum and is half-occupied. Atoms with dangling bonds are second-nearest neighbors.

Cleaving produces a metastable[278] 2×1 reconstruction built around π–bonded chains (Fig. 1.6b). This was one of the first semiconductor surface reconstructions predicted on a theoretical basis. It was proposed in 1981 by Pandey[279] as a result of comparison between electronic structures obtained from tight-binding calculations and from experiment. In the next year this assignment was confirmed by *ab initio* theory[280] and by photoemission.[281] The atomic positions were later tabulated by LEED.[282,283]

Upon annealing,[284] the periodicity changes to 7×7 (Fig. 1.6c). Annealing in a certain temperature range can lead to a metastable 5×5.

Many models had been proposed for the 2×1 and 7×7 structures. Most of them involved some form of buckling of individual atoms or other minor rearrangements of the atomic positions from their 1×1 sites. The buckling model of Si(111) 2×1 was devised in 1961 by Haneman[285] who suggested that the energy decreases when the surface atoms change their hybridization state such that they are alternatively raised and lowered. Though the resulting charge transfer from the "down" to the "up" atoms was supposed to open a gap between the dangling orbitals and thus account for the observed semiconducting density of states, this model was not in agreement with the measured electronic structure. In addition, the large charge transfer would induce changes in the binding energies of Si $2p$ (core) electrons; this is incompatible with the small core level shifts deduced from photoemission.[286,287,288] Many other structures, such as the double-bond

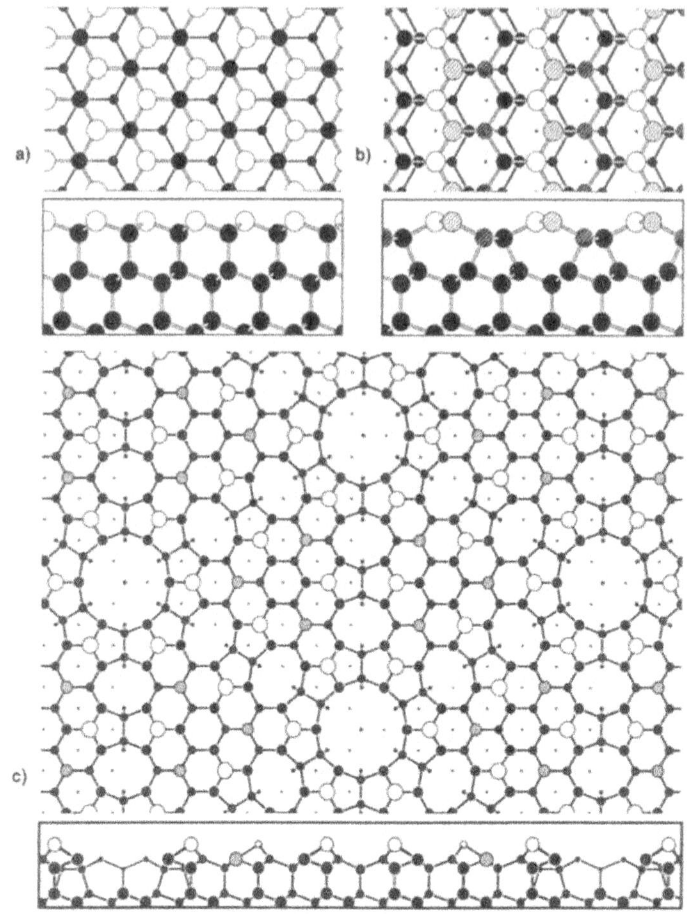

Fig. 1.6. Ball-and stick models of Si(111) reconstructions. Topmost atoms are white, other threefold coordinated atoms are gray, fourfold coordinated atoms are black. Side views are framed. (a) The 1×1 truncated bulk. (b) The 2×1 π-bonded chain. The atoms which have switched bonds are dashed. (c) The 7×7 dimer-adatom-stacking fault. The stacking fault is in the right half of the 7×7 cell limited by the circular "corner holes".

model,[278] the conjugated chain model,[289] or the adatom model[290,291] were suggested, but none could explain all the data.

Yet the Si(111) 2×1 is simple.[279] Each second atom with a dangling orbital and one of its fourfold-coordinated neighbors switch bonds with a subsurface Si (Figure 1.6b). The atom which was previously threefold-coordinated becomes fourfold-coordinated, and vice versa, so that atoms with dangling bonds are now *nearest* neighbors. Their unsaturated orbitals form a π-bonded chain. This opens a band gap and produces a density of states (DOS) compatible with experiment.[281] Also STM images[292] and STS characteristics[293] are in agreement with this structure and its surface-state dispersion. The computed surface energy of Si(111) 2×1 is minimized when the atoms in the chains buckle, that is, when the two chained atoms from the same 2×1 cell have different z coordinates.[294] Such buckling is in agreement with STM topography and spectroscopy.[292]

Since the energy barrier for the 1×1 $\rightarrow$ 2×1 bond switching can be easily overcome (it was computed[280] to be only ~0.03 eV, that is, comparable to the thermal energy at room temperature, kT=0.025 eV) and since this transformation does not require any mass transport across the surface, the 2×1 structure forms spontaneously during the cleavage process. It can be generated also by electron-stimulated desorption of hydrogen from a chemically passivated Si(111) 1×1-H surface.[295]

Annealing transfers the metastable 2×1 irreversibly into the famous, stable Si(111) 7×7. The discovery and some of the studies of the 7×7 were milestones of surface crystallography of semiconductors. Schlier and Farnsworth[284] observed it as early as in 1959 by LEED. Binnig, Rohrer, Gerber, and Weibel[296] reported its real-space STM images in 1983; these were the first STM micrographs of a semiconductor surface ever made.

The structure of Si(111) 7×7 was determined in 1985 by Takayanagi *et al.*[297,298] after a detailed analysis of transmission electron diffraction data. It is very complex with its nine dimer bonds, twelve adatoms of two types, and a subsurface stacking fault in each surface cell (Fig. 1.6c). Additional features of this Dimer-Adatom-Stacking fault (DAS) reconstruction are corner holes and six unsaturated dangling bonds of six exposed "rest atoms" in the layer below the adatoms. Around the corner holes, the atoms are arranged in twelvefold rings, and there are eightfold rings along each boundary between faulted and unfaulted areas. This model combines all the ingredients (except π-bonded chains) which had emerged from previous studies. A complete set of atomic coordinates was later deduced from LEED I–V spectra.[299] The 7×7 surface is metallic, with the corner atoms on the faulted half having the highest DOS near the Fermi energy[300] (Fig. 1.7a).

Due to this structural complexity, *ab initio* calculations for Si(111) 7×7 were initially limited to the individual elements, notably the adatoms.

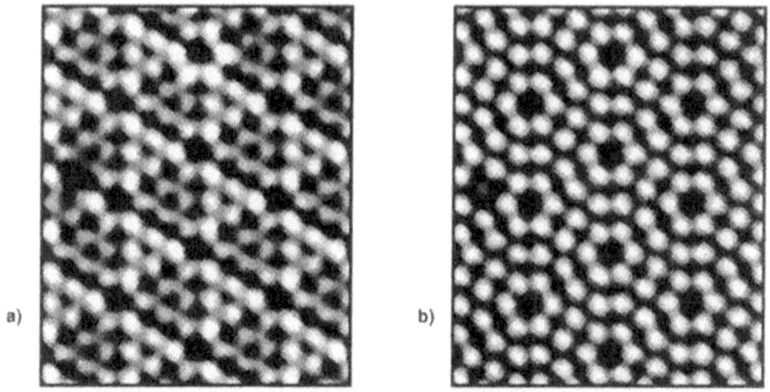

Fig. 1.7. STM images[301] of Si(111) 7×7. The surface unit cell is orientated as in Fig. 1.6c. The bright spots mark the "dangling bonds" of adatoms. **(a)** Occupied states. The adatoms on the faulted half are brighter. Rest atoms can be recognized on both halves as darker features between adatoms. **(b)** Empty states. Only adatoms are visible. The dark adatom which is easy to find on the left side of the empty-state image is possibly due to a boron acceptor atom segregated to the surface.[302,303,304] It can be recognized also in the images of the occupied states.

The first calculation[300] for the complete electronic structure was reported only in 1991. It assumed the structure from LEED.[299] The first optimizations of the atomic positions, published a year later, were done on parallel supercomputers.[305,306] In the meantime, numerical techniques and computers have become so advanced that atomic vibrations of 7×7 can be followed in an *ab initio* simulation run.[307]

It took 26 years for a generally accepted structural model of Si(111) 7×7 to evolve. The DAS structure is now supported by experiment as well as by theory. Today, the Si(111) 7×7 is the standard surface for verification of the operation of STMs in UHV (Fig. 1.7).

Changing the 2×1 cleavage structure into the 7×7 DAS requires mass transport across the surface. This is because each 7×7 unit cell, with its 102 atoms (12 in the adatom layer, 42 in the restatom layer, and 48 in the layer with the stacking fault), has four atoms more than contained in a 7×7 area of the truncated bulk (98 atoms). On the other hand, the 5×5 structure, which has identical components[308] as the 7×7 but a smaller area and somewhat higher energy,[309,305] contains the same number of atoms as the 1×1 and 2×1 surfaces. Thus a local rearangement of atoms can convert 2×1 into 5×5, but cannot convert 2×1 into 7×7. The atoms needed for the

Si(11n)	θ_{111}	θ_{001}	n	Si(11n)	θ_{111}	θ_{001}	n
Si(110)	-35.26°	90.00°	0.0	Si(5 5 12)	24.23°	30.51°	2.4
Si(331)	-22.00°	76.74°	1.5	Si(225)	25.24°	29.50°	2.5
Si(221)	-15.79°	70.53°	1.5	Si(113)	29.50°	25.24°	3.0
Si(111)	0.00°	54.74°	1.0	Si(114)	35.26°	19.47°	4.0
Si(221)	11.42°	43.31°	1.5	Si(115)	38.94°	15.79°	5.0
Si(335)	14.42°	40.32°	1.7	Si(117)	43.31°	11.42°	7.0
Si(112)	19.47°	35.26°	2.0	Si(119)	45.81°	8.93°	9.0
Si(337)	23.52°	31.22°	2.3	Si(001)	54.74°	0.00°	∞

Table 1.5. Angles between surface normals of Si(11n) and (111) and (001) directions.

latter transformation arrive from steps. If the surface is flat (the separation between the steps is well above 200 nm), a metastable 5×5 structure can be formed. A flat 2×1 surface orders into the 5×5 at 330°C.

The metastable 5×5 is irreversibly transformed into the stable 7×7 when the temperature is sufficiently high for an efficient delivery of atoms from the steps. The transformation begins around 600°C and becomes rapid around 800°C. A mixture of 5×5, 7×7, and even 9×9 can be observed under favorable conditions,[310,311,312] but when the surface has too many steps, the 2×1 converts directly into 7×7. Around 830°C, 7×7 changes reversibly into a disordered 1×1 phase.[313,314]

1.5.3. *Si(11n) surfaces*

Si(11n) is the family of surfaces in the (001) to (110) azimuth (Fig. 1.5, page 48). The angle θ_{001} between the (001) axis and the (11n) axis is described by $\cos\theta_{001} = n/\sqrt{n^2+2}$, the angle between (111) and (11n) is $\cos\theta_{111} = (n+2)/\sqrt{3(n^2+2)}$ (Table 1.5). Orientations which do not belong to the (11n) family are studied occasionally.[277] Well-assessed examples are the **Si(17 15 1)** facets found frequently on Si(110) surfaces (Sect. 1.5.5) and surfaces misoriented from (001) towards [100] (Fig. 4.9, page 255).

The surfaces vicinal to Si(001) and Si(111), that is, Si(001) and Si(111) misoriented by a few degrees, have been well characterized. Less is known about surfaces tilted further away from (001) and (111). Clean vicinal surfaces can be described in terms of phase transitions driven by interactions between surface steps (Sect. 4.3.2), with the miscut angle θ being a thermodynamical variable.[315,316,212]

A **vicinal Si(001)** tilted towards (111) consists of (001) terraces and steps. Depending on the miscut θ_{001}, these steps have single and/or biatomic height (Sect. 4.3.2). Around $\theta_{001} \gtrsim 9°$ the surface breaks into Si(117) terraces and stepped Si(001) or Si(115) terraces (Sect. 4.3.2). The tilt of $\theta_{001} = 19.5°$ produces the **Si(114)** surface.[317] The stable Si(114) 2×1 is built of pairs of rebonded and non-rebonded biatomic Si(001) steps (Sect. 3.3.2b), and dimer-wide Si(001) terraces of B type (Fig. 3.17, page 173). Each surface unit cell contains one pair of steps and one dimer. The stability of Si(114) is not affected by adsorption of hydrogen.[318]

Similarly to Si(001), **vicinal Si(111)** with tilts up to 12° towards (001) forms Si(111) terraces and steps of monatomic and triple atomic height. The fraction of triple steps increases with the miscut.[319,320] When the miscut from (111) towards (001) exceeds $\theta_{111} \simeq 14°$, the triple steps are replaced by Si(337) terraces separated by one-unit wide Si(111) 7×7 or 5×5 terraces. A quasi-periodic system of nanofacets is then formed.[213] At $\theta_{111} = 14.4°$, the orientation becomes **Si(335)**; the tilt of $\theta_{111} = 19.5°$ produces the **Si(112)** surface.[274,273,321] The stability of this arrangement is so high that it persists after exposure to atomic hydrogen,[322] at least for Si(112). A flat **Si(337)** corresponds to $\theta_{111} = 23.5°$, or $\theta_{001} = 31.2°$.

When the tilt from (111) towards (001) is further increased, the **Si(5 5 12)** surface is formed[323,324], with $\theta_{111} = 24.2°$ or $\theta_{001} = 30.5°$. The unit cell of the Si(5 5 12) 2×1 reconstruction consists of two Si(337) unit cells and a Si(225) unit cell.[323] The Si(5 5 12) surface resembles therefore a periodic arrangement of nanofacets. But since the angle between (337) and (225) planes is very small, the "facets" are extremely narrow, and the structure is highly periodic, the (5 5 12) orientation of silicon corresponds in fact to a stable, flat surface with a large (7.7×53.5 Å) and complicated unit cell. STM measurements supported by *ab initio* electronic structure calculations revealed the atomic details of the Si(5 5 12) reconstruction;[323] the simulated STM images are in excellent agreement with experiment.

The angle between (001) and (111) is 54.7°. Midway between these two axes, at $\theta_{001} = 25.2°$, there is a primary surface with unusual properties, Si(113). We devote the next Section to this interesting orientation.

1.5.4. *Structures of Si(113)*

The clean Si(113) surface[u] is thermally stable,[325,208] with surface energy close to those of Si(001) and Si(111). The room-temperature periodicity

[u] For a brief review of Si(113) see Ref. 326.

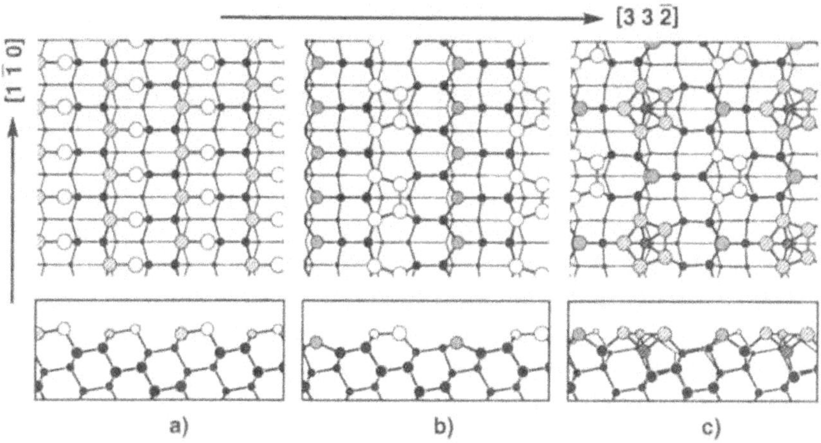

Fig. 1.8. Ball-and-stick models of Si(113) reconstructions. Fourfold coordinated atoms are black, side views are framed. **(a)** {113}-truncated bulk. Surface atoms are white, whereby atoms with one broken bond are dashed and atoms with two broken bonds are plain. **(b)** 2×2 rows of tetramers and adatoms. Each tetramer (plain white) is built from two pairs of white (dashed and plain) atoms of truncated bulk. Each adatom (gray) substitutes for such a single pair of atoms. **(c)** 3×2 array of adatoms (gray), tetramers (plain white), and pentamers (dashed white). Each pentamer is built from a tetramer and a subsurface atom elevated by an interstitial atom (dashed black).

of Si(113) is 3×2.[327] At elevated temperatures it changes reversibly into 3×1.[328] The 3×2 structure is stabilized by subsurface Si atoms located in interstitial sites and bonded to pentamers of Si atoms.[329] The transformation to 3×1 is possibly an order-disorder transition in the sublattice of interstitials.[330] Clean Si(113) has many defects, but their great majority corresponds to antiphase boundaries and excess self-interstitial atoms.[331]

The {113}-truncated bulk consists of alternating rows of atoms in the local geometry of {001}- and {111}-truncated bulk (Fig. 1.8a). This intimate neighborhood of different surface geometries provides a lot of structural freedom. Given that the geometry of Si(113) is relatively simple, facilitating theoretical studies, this makes Si(113) a potential "laboratory" for studies aimed at quantitative understanding and systematic control of dynamic processes which take place on other surfaces, notably on Si(001).

Another observation which links Si(113) to Si(001) is that the latter often develop (113) facets. This can happen, for example, during sublimation of atoms from the surface when moving steps get pinned by defects and form a bunch (Sect. 4.4), or during epitaxy of Ge or SiGe on silicon (for references, see Ref. 326).

But Si(113) is mostly interesting as a **scientific lab.** Let us consider some examples. First, Si(113) might be used as a controllable source of self-interstitial atoms. Its subsurface interstitials are temporarily driven away by adsorbates and return to the surface when the adsorbed atoms desorb.

Second, the specific geometry of Si(113) allows one to follow a dissociative adsorption of molecules. Si(113) is the only surface on which breaking up of O_2 molecules into O atoms was observed directly, whereby each atom could be unequivocally assigned to its molecule of origin.[332]

Third, Si(113) is a sensitive test field for atomistic theories. An interplay between dangling bond density, chemical bond energies, and surface stress may force some atoms into unusual sites. For instance, though Sb atoms are larger than Si atoms, they may become inserted into Si–Si bonds, a behavior not observed in any other Sb–crystalline Si system. The geometry, bonding, and formation mechanism of the resulting new structure are very different from those of the clean surface and are directly related[215] to self-interstitial clusters in implanted or irradiated bulk Si.[333,214]

The direct relation between a surface structure and a bulk defect structure corroborates the idea that the reliability of theoretical methods to study bulk defects can be verified by surface science methods. It is much easier to *measure* atomic-size surface objects than atomic-size objects hidden in the bulk. Studies of surface defects and adsorbed atoms may provide a unbiased source of reference data for validation of theories which may be afterwards applied to approach problems of technological relevance.

We now describe the atomic structures of the clean Si(113). The **truncated bulk** is built of $[1\bar{1}0]$ rows of atomic pairs, each pair containing a (111)-like atom with a broken bond and an (001)-like atom with two broken bonds (Fig. 1.8a). A hypothetical, low-energy 2×2 structure[329] (Fig. 1.8b) contains two of the three elements representative for the clean surface. It can be easily built from the truncated bulk, as described below.

First, the pairs in each second $[1\bar{1}0]$ row dimerize their (001)-like atoms. The resulting **tetramers** are topologically equivalent to the unit of a nonrebonded B-type step edge, (S_B-s, Sect. 3.3.2). A tetramer has a semiconducting density of states (DOS). Its four dangling bonds create two occupied states: one bonding and one weakly bonding (page 160). Since the latter has a node between the dimer atoms, the length of the Si(113) dimer bond is markedly larger than that of an ordinary Si(001) dimer: the computed lengths are 2.32 Å and 2.26 Å.

Second, the remaining pairs are replaced by capping **adatoms**. Each adatom has a dangling bond and three backbonds. An adatom induces a significant tension: one of its backbonds is stretched by more than 5%.[215]

The adatoms and tetramers may be arranged in 3×1 order, such that they alternate in each $[1\bar{1}0]$ row.[334,335] The highest occupied surface band, built from the states of the tetramer and the dangling bond of the adatoms, becomes almost filled, with room for only one more electron per 3×1 cell.

The 3×1 structure has a slightly higher energy than the 2×2. But it becomes stabilized when it captures a Si **interstitial** atom under each second tetramer (Fig. 1.8c). The resulting Adatom-Interstitial-Dimer (AID) reconstruction is in a good agreement with experimental data for the clean surface.[336] Conclusions drawn from this structure are compatible with the behavior of the surface covered with adsorbates.[337,338]

The interstitial atom induces a considerable compressive strain. The tension around the adatoms helps to compensate its lateral component. The vertical component of the strain is relieved as the structure relaxes strongly towards the vacuum, elevating a pentagonal ring of atoms (Fig. 1.8c). This **pentamer**, formed by the original tetramer and the common neighbor of its two (111)-like atoms, is almost flat (within ~0.06 Å) and parallel to the surface. According to *ab initio* data, its shape is close to a regular pentagon, with angles between 106° and 109° (the ideal pentagonal angle is 108°). Recent Transmission Electron Diffraction analysis[339] suggests that the pentamer may be in fact elongated along $[1\bar{1}0]$.

The interstitials affect the **electronic structure** of Si(113), changing the metallic 3×1 into a semiconducting surface. Two electrons move from the occupied surface band to an interstitial-related valence resonance. Two orbitals built from the states of the perturbed tetramer emerge from the surface band to make a narrow subband at higher energies. These two orbitals are empty because the 3×1 surface had two half-unoccupied orbitals per 3×2 cell and two electrons were removed to the resonance. The empty subband appears to the STM as a pentagonal group of spots above the pentamer (Fig. 1.9). Experiment and theory agree[157] that the occupied states in this area are invisible to STM under usual conditions ($V \gtrsim -2$ V).

Since the AID reconstruction has a semiconducting DOS, the dangling bonds of its adatoms must be completely occupied or empty. The adatom

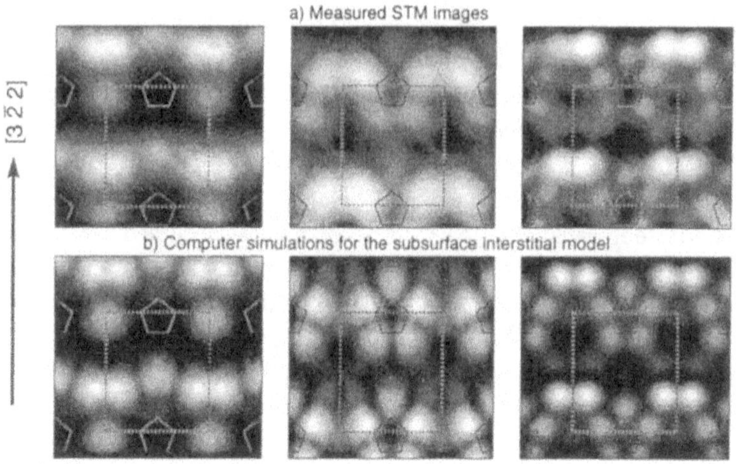

Fig. 1.9. **Measured and simulated STM images of Si(113) 3×2. The rectangular 3×2 unit cell and the position of pentamers are indicated (compare Fig. 1.8c). Simulations from *ab initio* data[329] and Hamann-Tersoff theory.[148] Left**: occupied states, measurement at -2.1 V, 0.7 nA. **Middle**: empty states, measurement at 1.5 V, 1.- nA. **Right**: empty states, measurement at 2.8 V, 2.0 nA. Taken from Ref. 329.

dangling bonds have relatively low energy, so the adatoms are negatively charged and the pentamers are positively charged. The electrostatic interaction between these charges is responsible for the rectangular, $\left(\begin{smallmatrix}3 & 0\\ 0 & 2\end{smallmatrix}\right)$ shape of the AID unit cell, favoring it over the rhomboidal $\left(\begin{smallmatrix}3 & 0\\ 1 & 2\end{smallmatrix}\right)$.

1.5.5. *Structures of Si(110)*

In comparison to other primary surfaces of Si, little is known about Si(110).[v] This is disappointing, if only because Si(110) is exposed when thin Si(001) wafers are broken. If one wants to examine a transistor cross-section (such as one of those sketched in Fig. 1.3, page 9), for example, by STM,[340] the plane of view will be (110). The wafers do not cleave at (111) planes because the angle between (001) and (111) is 54.7°, that is, far from 90°; on the other hand, (001) and (110) planes intersect under the right angle.

Si(110) is an interesting material for heteroepitaxy. Growth of SiGe on such substrates as Si(110) and Si(113) provides new degrees of freedom for bandgap engineering. Also the mode of growth can be interesting. For example, SiGe arranges on Si(110) in self-assembled quantum wires: straight, parallel strips with a regular spacing in nanometer range.[341]

[v]A brief overview of studies on Si(110) up to the year 1993 is given in Ref. 342.

The clean Si(110) is characterized by a $\begin{pmatrix} 17 & 1 \\ 2 & 2 \end{pmatrix}$ periodicity. This reconstruction, usually denoted as "16×2", appears when the sample is cooled below $T_c \simeq 750°C$ after cleaning by a 1200°C flash. Above T_c it transforms reversibly into a disordered 1×1 structure.[343,342]

With some care, Si(001) wafers can be cleaved to obtain Si(110) surfaces of STM quality.[266] Such surfaces are disordered. Annealing at 500–800° C progressivelly introduces order, but on a local scale, within areas smaller than that of the "16×2" unit cell.[266] The "16×2" structure is not generated at such low temperatures, maybe due to kinetic limitations, though the typical Si(17 15 1) facets described later are often formed.

It is now established that "16×2" is built on the basis of a 16×1 array of monatomic steps (Fig. 1.10a). The upper and lower terraces are 25.1 Å (2.51 nm) broad, the steps are 1.9 Å high, and their edges run parallel to $[1\bar{1}2]$ and equivalent directions.[343,344] This builds a multidomain system consisting of areas covered by flat, long ridges and valleys. A ball-and-stick model indicates that atoms on the edges are dimerized (Fig. 1.10a). Passivation of Si(110) with H seems to keep this stepped structure unchanged.[345]

The doubling of periodicity from 16×1 to "16×2" may be due to missing rows (Fig. 1.10b). There is some evidence that these missing rows are decorated with adatoms in a characteristic octet configuration (Fig. 1.10c). These elements seem to be be compatible at least with STM observations (see Refs. 346, 144 and references in Ref. 266) and with D_2 desorption studies.[347] But atomic details of the reconstruction are usually poorly resolved. Most STM images show quite featureless zigzags of rounded protrusions (Fig. 1.11). These protrusions may in principle arise from such octets as in Fig. 1.10c. High-resolution STM micrographs suggest that the reconstruction is the same on the ridges and in the valleys.[144]

Various models of Si(110) "16×2" and the corresponding references are listed in recent Refs. 266 and 347. Several structural elements have been considered in an exploratory tight-binding study.[348] Unfortunately, theoretical results for Si(110) are very scarce.

Typical features of Si(110) are Si(17 15 1) 2×1 facets.[349,350] They contain the same basic elements as the Si(110) "16×2" planes: 25.1 Å wide (110) ridges and valleys with edges along $[1\bar{1}2]$. The reconstruction of these (110) terraces of Si(17 15 1) and Si(110) is probably the same, so the major structural difference is that the steps on Si(110) run up and down in a comb-like fashion, while on Si(17 15 1) they form a staircase (Fig. 1.11). This

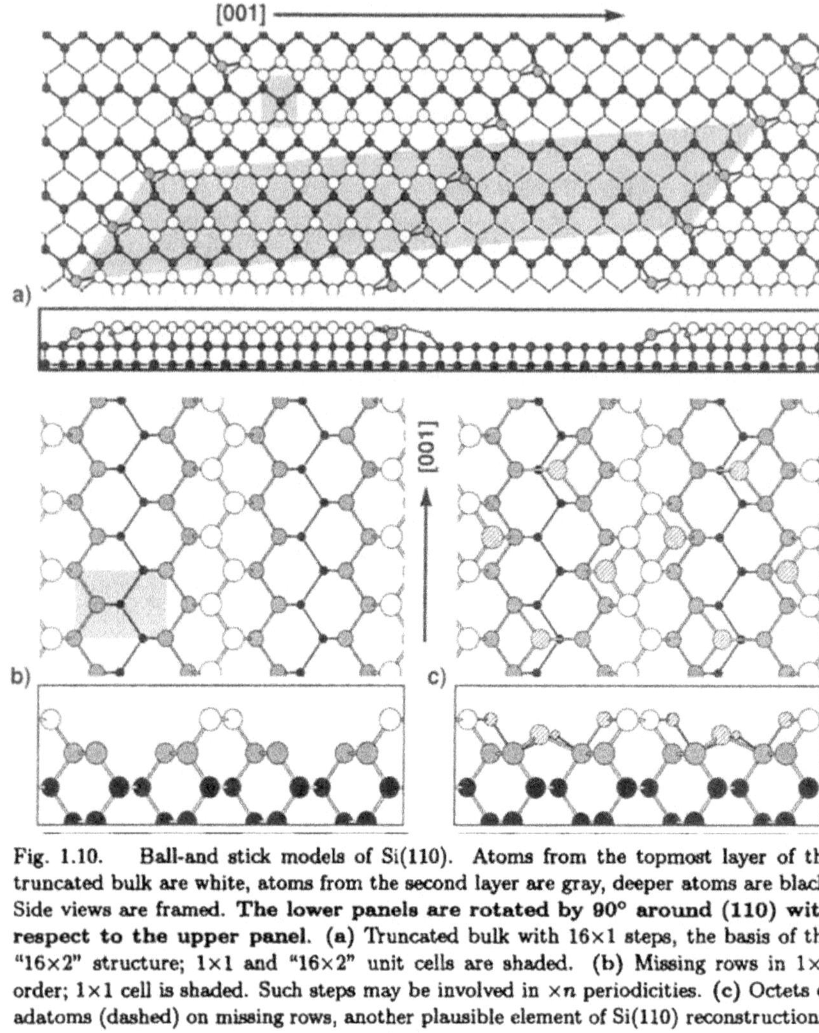

Fig. 1.10. Ball-and stick models of Si(110). Atoms from the topmost layer of the truncated bulk are white, atoms from the second layer are gray, deeper atoms are black. Side views are framed. **The lower panels are rotated by 90° around (110) with respect to the upper panel.** (a) Truncated bulk with 16×1 steps, the basis of the "16×2" structure; 1×1 and "16×2" unit cells are shaded. (b) Missing rows in 1×2 order; 1×1 cell is shaded. Such steps may be involved in $\times n$ periodicities. (c) Octets of adatoms (dashed) on missing rows, another plausible element of Si(110) reconstructions.

indicates that the characteristic width of the "16×2" ridges and valleys is determined by their local structure (maybe by the octets, Fig. 1.10c) and not by step-step interaction or reconstruction of the step edges.

An implication is that theoretical studies need not necessarily consider the entire, large "16×2" unit cell. A lot of information may be obtained already from studies of objects of the octet size. A similar approach

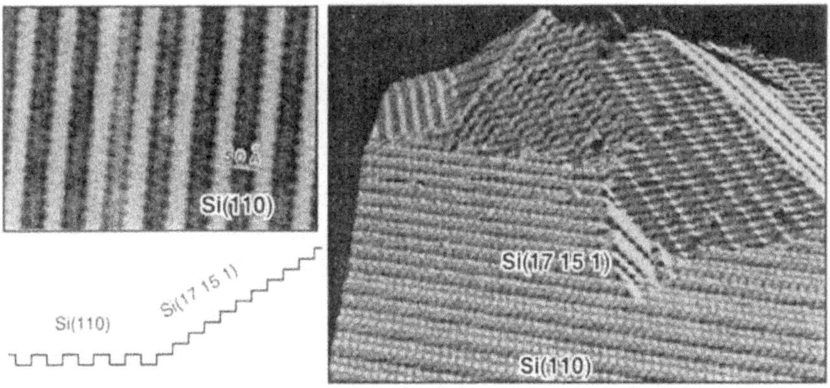

Fig. 1.11. A clean Si(110) surface with a 16×2 terrace and a faceted island. **Right**: a three-dimensional (3D) STM image, from Ref. 351. **Left**: a 2D STM image of a 16×2 terrace (from Ref. 346) and a sketch of the height profile along the leftmost edge of the 3D image. The terrace consists of 16×2 ridges and valleys made by steps running up and down (compare Fig. 1.10a). The rising slope consists of the same steps running only upwards, so that a (17 15 1) facet is formed.

was originally applied to the complex Si(111) 7×7 structure (Sect. 1.5.2, Fig. 1.6c on page 70).

Small amounts of metals have a profound influence[343] on the surface reconstruction (Sect. 5.7.2). This is more dramatic on Si(110) than on the other primary orientations. Upon adsorption of traces of Ni on Si(110) 16×2, LEED reveals structures with 4×5, 2×1, 5×1, 7×1, and 9×9 periodicities.[352] The periodicity depends on the amount of Ni and on the surface temperature. When the Ni coverage varies, reversible phase transitions take place among these structures. The phase transitions can be interpreted as being caused by bulk diffusion and surface segregation of Ni.

A stable 5×1 structure is sometimes observed after annealing at 800°C (for example, see Refs. 353, 354, and 355). It is unclear whether this structure is due to Ni or any other impurity. An STM study[356] revealed that relatively short annealing times (1 min at 1100°C) used for surface cleaning produce a defected 5×2 structure. Since Ni contamination was ruled out in this case, one may argue that the formation of the 5×2 periodicity may be due to a kinetic limitation: a lot of material must be transported far across the surface in order to complete the "16×2" array of steps.

Hands are useless if there are no eyes
to see what is obvious.
– M. V. Lomonosov

CHAPTER 2

The Famous Reconstruction of Si(001)

J. Dąbrowski

How the reconstruction of the clean Si(001) was unraveled and what can one learn from this story.

2.1. Introduction

Besides Si(111), the (001) surface of silicon is probably the most thoroughly studied semiconductor surface. Around a thousand papers relevant to Si(001) appear yearly in major scientific journals. Si(001) is one of the three stable surfaces of Si for which generally accepted structural models exist; the other two are Si(111) and Si(113), described in Chapter 1. Table 2.1 lists some of the publications which shaped our views about Si(001).

This interest has its roots in the success of CMOS (Sect. 1.2.2), the technology which dominated microelectronics. CMOS uses exclusively Si(001) wafers (Chapter 6). But the path from a clean surface to industrial application is long and winding one (Sect.3.1). Instead of trying to pursue it, this Chapter takes a more methodological turn. It illustrates how scientific knowledge is incluenced by the interplay between facts and interpretation, conclusions and assumptions, experiment and theory. For this purpose it employs a historical narrative. A reader who looks for more pragmatic information may want to proceed to Chapter 3.

Before we start, let us make some remarks on the characteristic features of the ideal Si(001) surface obtained by truncation of the bulk by a {001} crystallographic plane (Section 3.2.1). Atoms on such a surface have two broken bonds each, as indicated by the dots in: 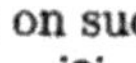. This sketch

Ref	year	Subject relevant to the clean Si(001)
357	1957	Dimerization proposed; LEED
358	1971	**Review** of LEED
359	1979	ARPES showed the presence of surface energy gap
360	1979	Buckling of dimers proposed; tight-binding theory
361	1981	The first *ab initio* optimization of Si(001): the dimers buckle
362	1985	List of **review** articles till 1985
363	1985	The dimers observed directly for the first time; STM
364	1987	**Review** article
365	1987	Empty states of dimers; STM
366	1988	Atomic structures from many calculations listed
367	1988	Measurement of surface stress anisotropy
368	1989	The first correct *ab initio* energy order of dimer structures
369	1990	Dispersion of unoccupied surface states measured
179	1990	Consequences of confusion about the symmetry of dimers
370	1992	Atomic structures from many calculations discussed
371	1992	Surface band folding in $p(2\times2)$ and $c(4\times2)$
372	1992	*Ab initio* favors buckled dimers when converged
373	1992	Low-temperature STM sees many buckled dimers
374	1994	Low-temperature STM falsifies the symmetric dimer model
375	1994	Stress anisotropy explained by buckling; *ab initio* DFT theory
376	1995	Buckling angle measured; X-ray diffraction (XRD)
377	1995	Detailed analysis of *ab initio* convergence
378	1996	Recent measurement of surface stress; TEM
379	1996	Flipping of buckled fimers; *ab initio* molecular dynamics, STM
380	1996	The dimers observed by Atomic Force Microscopy
381	1997	Recent buckling angle measurement, atomic positions; XRD
382	1997	The dimers flip collectively; STM

Table 2.1. History of clean Si(001): selection of references.

represents the ideal Si(001) from the perspective of $Y = [\bar{1}10]$ (see the surface produced by the plane labelled "1" in Fig. 3.1e, page 130). We will use similar graphs for ready reference in the text. The full circles symbolize Si atoms, the dashed line marks the (001) plane which goes through the first subsurface layer, and the solid lines correspond to bonds. Each surface atom has two backbonds connecting it with two subsurface atoms, but only one backbond per atom is visible from the perspective of the sketch.

Since so many bonds are broken, one expects that the ideal Si(001) is structurally unstable. But this is not so clear as it may seem to be. The atoms with the half-occupied broken bonds (·…·) are separated by 3.84 Å (page 131). This is much more than the Si–Si bond length in bulk Si, which equals 2.35 Å. It follows that re-forming the bonds on the surface requires significant atomic displacements from the ideal lattice sites. In other words, any bonds between the surface atoms would weaken the bonds between the surface atoms and the subsurface atoms (backbonds). The loss of energy due to such distortions may exceed the energy gained in the re-formation of the broken bonds. As a consequence, the 1×1 surface may be structurally stable. Although it is now known that the atoms on Si(001) do rearrange in order to reduce the number of broken bonds (they come together in pairs:  → ), it was not immediately obvious that a reconstruction exists which has a lower surface energy than the 1×1.

We will now follow the dramatic efforts which had disclosed the fundamental building block that constitutes Si(001). A dimer is so simple a structure that one may wonder why complicated solutions were proposed. However, a dimer may be simple, but it is not the *simplest* structure imaginable. Even the ideal Si(001) 1×1 may be considered as a plausible candidate for a stable configuration. A seemingly tiny geometric detail, that dimer bonds are not parallel to the surface () but "buckle" (), added so much complexity that even the most careful researchers were confused. Let us not forget that the survey started forty years ago, but only fifteen years ago did we learn how to *see* surface atoms (Sect. 1.3.2a)!

Particular attention will be paid to the interaction between the evolution of surface science methods and the sophistication and accuracy of atomic models. We will observe how experimental and theoretical techniques complement each other to reconstruct the whole picture piece by piece and how successes and mistakes deepen our knowledge and increase the efficiency of surface science methods. We will witness the changes of opinion that have occured during almost half a century of these intensive studies. And we will note how weak and questionable our knowledge can be unless we can really *touch and see* the object we describe. But even then some elementary doubts may remain. Since it is not always possible to prepare a system containing only the ingredients of interest, it is sometimes unclear which of the observed phenomena are intrinsic to the idealized object (here, the perfectly reconstructed surface) and which are caused by common "perturbations" (here, atomic-size defects on the surface).

2.1.1. *Overview: expectations bias our predictions*

Figure 2.1 gives a schematic overview of the efforts to unravel the atomic structure of Si(001). Although it focuses on a single structural aspect, this Figure illustrates the main directions of the research as well as a clearcut relation between knowledge and expectations. To be in a position to analyse

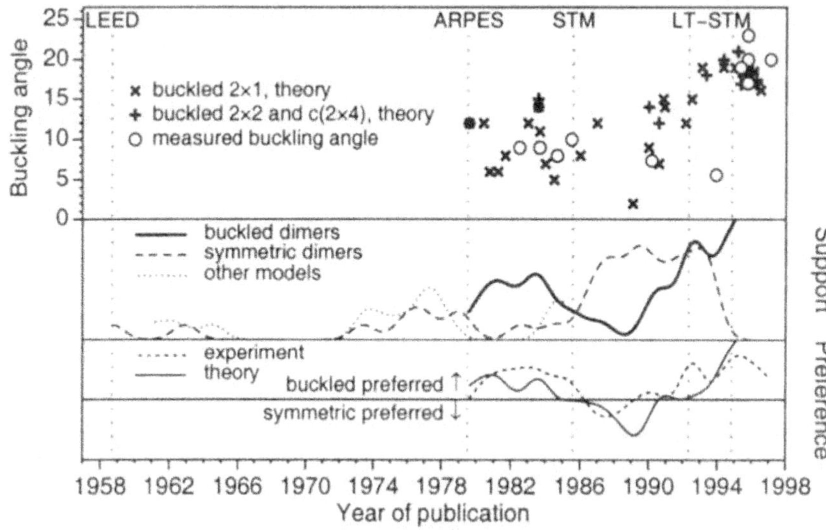

Fig. 2.1. Development of opinions about the structure of Si(001). Experiments which had a particular impact are indicated by dotted vertical lines. The **upper** panel shows a correlation between the values of the buckling angle which were computed and measured between 1979 and 1997. The **middle** panel shows the support for reconstruction models, measured by the number of published papers; each paper is added as a gaussian with half-width of a year. The theoretical and experimental preference (**lower** panel) is measured by the difference between the number of papers supporting either model. The symmetric dimer model dominated in the period between two STM studies: the first observation of mostly symmetric dimers at room temperature[363] (year 1985) and the first observation of mostly buckled dimers at low temperatures[373] (year 1992).

its content, we must begin with a brief summary of the fundamental features of Si(001) which are considered in this diagram.

The atoms on the clean Si(001) surface move pairwise towards each other and form new bonds ( → ; compare Fig. 2.3, page 90). This is the **symmetric dimer** model. In addition, the dimer is buckled: each atom in the pair has a different height above the surface plane (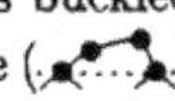). This is the **asymmetric** or **buckled dimer** model.

Symmetric dimerization ( → ) doubles the periodicity to 2×1. Asymmetric dimers () prefer higher periodicities, 2×2 and $c(4\times2)$, which appear because the direction of buckling (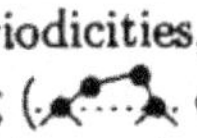 or 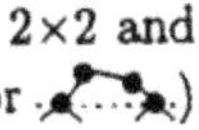) of neighboring dimers is correlated. Unless stabilized by surface defects, this correlation is partially destroyed around $T_{od} = 200$ K. Above T_{od},

LEED[a] sees usually an average 2×1 order. A similar effect makes the most of the dimers appear symmetric in room-temperature STM images: thermal vibrations flip the buckling direction of a dimer (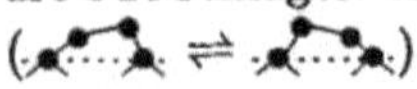) faster than the STM tip can follow. However, even at elevated temperatures the electronic structure of the surface (as seen, for example, by photoemission) is dominated by locally preserved $c(4\times2)$ arrangement of buckled dimers.

Figure 2.1 visualizes the impact of a few key experiments on the interpretation of other experimental *and* theoretical data. The symmetric dimer model (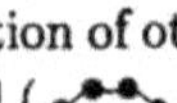) was formulated by Farnsworth and Schlier[383] to explain the 2×1 LEED pattern they observed. Today we know that although this model captured the most characteristic features of the surface reconstruction, it lacked certain details needed to explain the results of experiments performed later. Since the data provided by experimental techniques available at that time were not suited for easy, intuitive interpretation in terms of a surface geometry, various alternative models appeared within the following years. This continued till the late 70's, when a more rigoreous theoretical analysis of the energy distribution of the occupied surface states measured recently by ultraviolet photoemission backed the dimer model.

These measured electronic structures posed new problems. Namely, the computed electronic structure of symmetric dimers (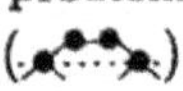) was metallic. However, a surface energy gap was found in an ARPES experiment.[359] Subsequent calculations for a dimer model with buckled dimers (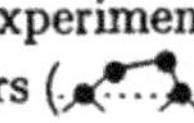) indicated that the buckling opens a gap between occupied and empty surface states and lowers the energy of the dimer.

The idea of buckling was readily accepted and soon supported by data of experimental as well as theoretical origin. Yet, new difficulties with the interpretation of details in the measured LEED and photoemission spectra were reported. This invoked doubts. Papers advocating the symmetric configuration of dimers reappeared. Even new models were designed.

The last of these innovative reconstruction models[384] was conceived shortly before STM studies singled out two structures as the only plausible candidates: the symmetric dimer model and the buckled dimer models.[363] STM is a real-space technique and it is often straightforward to associate the observed patterns with the local arrangement of atoms on the surface (cf. Figs. 3.8a and 3.8b on page 150, and Fig. 3.9 on page 151). The dimers were clearly visible in STM pictures as *symmetric* ovals (Fig. 2.8, page 108).

[a]For experimental methods and their acronyms see Table 1.2 on page 29.

The micrographs were taken at room temperature and it was noted[363] that under such conditions an asymmetric dimer may change the direction of buckling so rapidly that its image looks symmetrical. Nevertheless, seeing is believing and the symmetric dimer model immediately gained support among experimentalists and theorists. Buckled dimers were in retreat.

Still, this did not last. The reliability of experimental and theoretical techniques was constantly improving. It was becoming clear that buckled dimers have a lower energy than symmetric dimers. When the STM images of Si(001), taken[373] for the first time at temperatures in the liquid-nitrogen range, showed a surface with mostly asymmetric dimerization, the popularity of the buckled dimer model once more exceeded the popularity of symmetric dimers. This consensus change happened even though this study could not exclude the stabilizing influence of surface defects on buckling.[373]

Finally, the ultimate argument was brought again by an STM experiment: it was shown[374] that the symmetrical appearance of some dimers in low-temperature micrographs is due to surface defects. The buckled configuration became generally accepted. It is now known that the buckling angle is around 20°. The accuracy of theoretical and experimental determination of this angle is comparable, being somewhat better than 5°, or 25%.

The whole story illustrates how beliefs shape our opinions. Our judgements are biased by our expectations and our expectations are the most efficiently influenced by what we see. This happens even in natural science, even in a field so firmly rooted in the observable reality as solid-state surface science is. It must be so, because every study must stop at a certain point and results must be published. This practical limitation makes scientific knowledge credible rather than absolute (Chapter 7). One might want to recall this while weighting the philosophical impact of some extensively extrapolative scientific conclusions. Important examples are the elegant theories and experiments which shape our opinions about the structure of the Universe, about human mind and consciousness, or about the mechanisms which control the evolution of species.

2.2. Pre-STM Era: Groping through the Dark

The atomic structure of a crystal can be observed almost directly, by X-ray diffraction. But traditional X-ray methods are not useful in surface studies. There are far fewer surface atoms than atoms which contribute to a typical bulk signal. It is not trivial to separate such a weak response from the strong bulk response. Efficient surface-sensitive X-ray techniques exist (Sect. 1.3.2), but are relatively new. When the structural studies of Si(001) began, researchers had to feel their way through indirect data and uncertain models. The first "photographic" method which would clear the view would be based on a different technique, scanning tunneling microscopy. The first hint about the atomic structure came from electron diffraction.

2.2.1. *Early observations and models*

Local structural changes may affect lateral periodicity. When atoms of the upper (001) layers are displaced such that the periodicity remains the same as of the truncated bulk, the diffraction spots represent the reciprocal lattice of the ideal 1×1 lattice. Only integral reflexes are allowed (Fig. 2.2a). Fractional-order spots indicate a higher lateral periodicity. Such simple considerations underlined the first structural analyses of experimental data.

a. *The first idea: 2×1 order of dimers*

It is not obvious that Si(001) 1×1 is unstable (page 83). But when Farnsworth and Schlier[383] from Brown University, Rhode Island, USA reported in 1958 that LEED of clean Si(001) produce *half-integral* diffraction spots arranged in the pattern of a two-domain 2×1 real-space lattice (Fig. 2.2b), it became clear that the clean surface reconstructs, doubling one of its primitive translation vectors. Farnsworth and Schlier wrote:[383]

> Results for clean (001), (111), and (110) germanium faces and the (001) silicon face indicate that the atomic positions in the surface planes are not the same as the corresponding positions in the bulk structure. [...] The observed structures are not due to surface contamination but are characteristic of atomically clean surfaces.

In principle, the doubling might occur either along or across the broken bonds (along [110] or [$1\bar{1}0$], compare Fig. 3.1 on page 130). Schlier and Farnsworth proposed that atoms form dimers (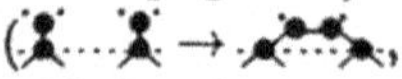, Fig. 2.3b). After pairing, the periodicity along [110] is doubled and a 2×1 reconstruction domain is built. View from the top, the surface resembles: , where dashes symbolize dimers (Fig. 2.5b, page 93). The dimerization reduces broken bonds to one per surface atom. According to Table 1.3 (page 49), this could bring the surface energy down to γ_{001}=0.075 eV/Å^2, below the lowest value estimated for ideal surfaces (γ^0_{111}=0.080 eV/Å^2).

Schlier and Farnsworth's experiment stemmed from efforts to develop

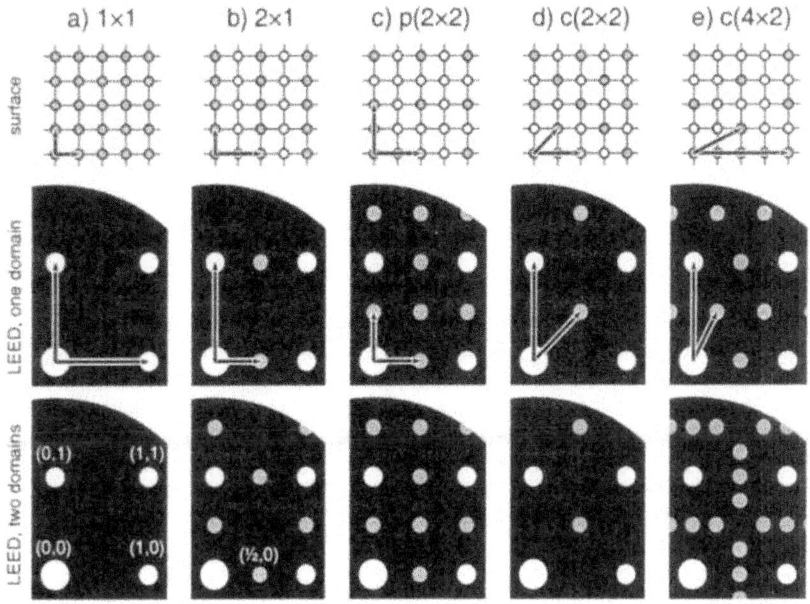

Fig. 2.2. A schematic representation of (a) the 1×1 surface cell of Si(001) and (b-e) of the periodicities observed on the clean surface: (b) 2×1, (c) $c(2\times2)$, (d) $p(2\times2)$, (e) $c(4\times2)$. The corresponding one- and two-domain LEED patterns are also shown.

efficient cleaning,[385] prompted in turn by severe problems of contemporary semiconductor industry. Dirty surfaces caused a wide spread of device parameters. Ready products had to be examined individually and sorted into classes according to their properties. The finding that only clean surfaces produce 2×1 spots was hoped to assist the development of clean processing.

The first analysis of dimerization was given a couple of years later, in 1962, also by scientists from Brown University. Green and Seiwatz[386] pointed out that dimerization does not have to change the backbond *lengths* (Fig. 2.3c). Since bond-bending force constants in Si are weaker than bond-stretching force constants, the bending of bonds is relatively easy and the energy gain due to rebonding may exceed the energy loss due to the strain.

Green and Seiwatz used experimental bond energies and bulk elastic constants to estimate that the dimerization gains $\Delta E_0 \simeq 0.9$ eV/dimer. They concluded that the dimerized surface is favored over the truncated-bulk surface. But ΔE_0 was only about a half of the Si–Si bond energy, assumed then to be $E = 1.95$ eV. A significant part of the energy released

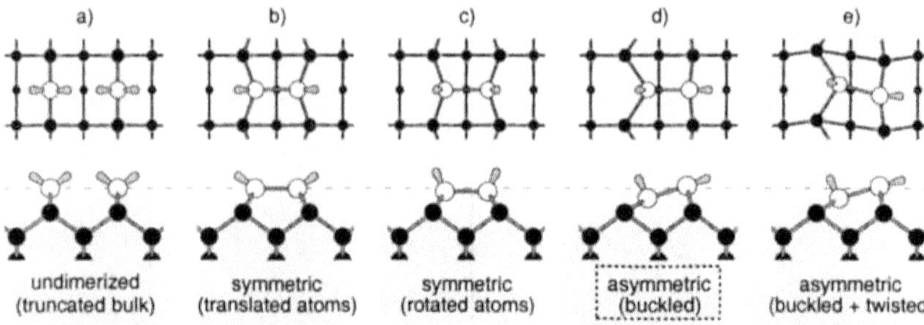

Fig. 2.3. Dimerization of Si(001). Top view (upper pannels) and side view (bottom panels). Surface atoms are white, subsurface and bulk atoms are black. "Broken bonds" of the surface atoms are symbolically indicated as gray clubs. (a) Truncated-bulk surface. (b) Schlier and Farnsworth (1957): dimerization by lateral displacement of surface atoms; backbonds of the dimer atoms are stretched. (c) Green and Seiwatz (1962), and Levine (1973): bending of backbonds is sufficient for dimerization. (d) Chadi (1979): the dimer may be buckled (the correct structure). (e) Jona, Shih, Jepsen, and Marcus (1979): the dimer may be buckled and twisted.

during bond reformation was consumed for the bending of the backbonds. It is interesting to note that the dimerization energy calculated nowadays by sophisticated *ab initio* methods (page 135) is ~2 eV/dimer, or ~80% of the bond energy (in fact, $E \approx 2.3$ eV: Table 1.4, page 57). However, the estimate was qualitatively correct: Si(001) can gain energy by dimerization.

b. *The first alternative and the first puzzle*

The estimate of the dimerization energy presented by Green and Seiwatz might have been plausible, but it was only an estimate. As we argued on page 83, the distortions needed to achieve re-formation of bonds on the surface are significant. In fact, only a modern self-consistent calculation can *prove* that dimerization is energetically favorable.

A year before Green and Seiwatz published their work, Haneman[285] noted that the 2×1 periodicity may reflect an interplay between bond strengths, atomic displacements, and hybridization of atomic orbitals. He suggested that atoms in each second row in the [110] direction may be *raised* with respect to their neighbors (Fig. 2.4b). This was a generalization of his model designed for (111) surfaces of the diamond lattice.

To follow the reasoning of Haneman, consider an isolated silicon atom. The configuration of its valence electrons is $3s^2 3p^2$. The energy of a Si(3s) electron is by ~5 eV lower than the energy of a Si(3p) electron.[387] However,

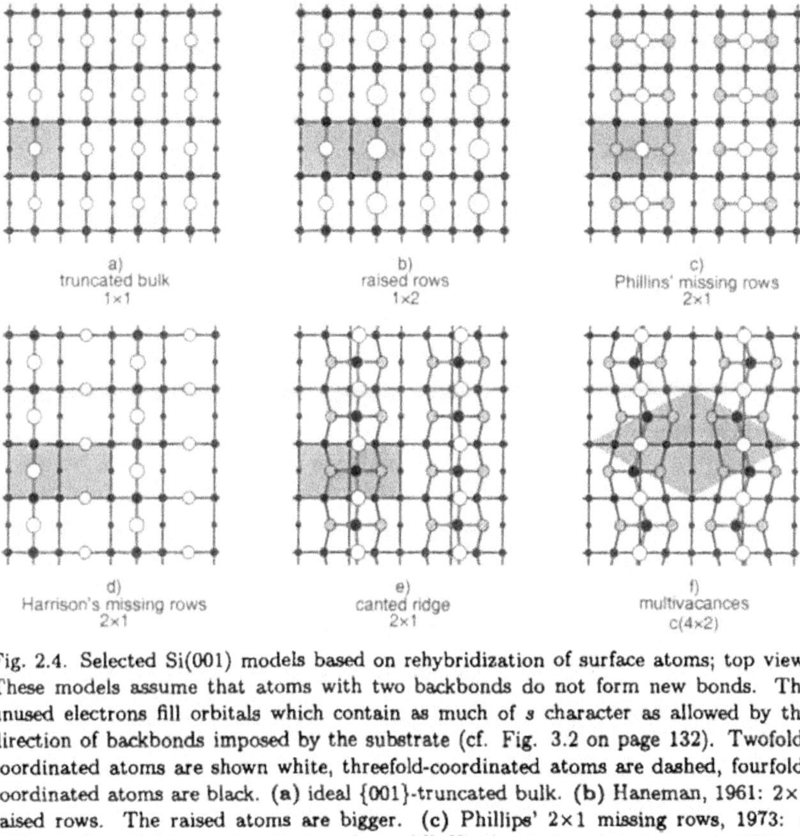

Fig. 2.4. Selected Si(001) models based on rehybridization of surface atoms; top view. These models assume that atoms with two backbonds do not form new bonds. The unused electrons fill orbitals which contain as much of s character as allowed by the direction of backbonds imposed by the substrate (cf. Fig. 3.2 on page 132). Twofold-coordinated atoms are shown white, threefold-coordinated atoms are dashed, fourfold-coordinated atoms are black. **(a)** ideal {001}-truncated bulk. **(b)** Haneman, 1961: 2×1 raised rows. The raised atoms are bigger. **(c)** Phillips' 2×1 missing rows, 1973: a half-monolayer *added* to the ideal surface. **(d)** Harrison's 2×1 missing rows, 1976: a half-monolayer *removed* from the ideal surface. **(e)** Harrison, 1976: 2×1 "canted ridge". **(f)** Poppendieck, 1978: $c(4\times2)$ multivacancies, a derivative of Harrison's "canted ridge".

this configuration is not suited for bonding in a covalent crystal. In order to saturate all its valences by making four strong bonds with Si neighbors, each Si atom mixes its atomic "spherical" s^2p^2 orbitals into "tetrahedral" sp^3 hybrides. This corresponds to the promotion of one of the low-energy s electrons to the high-energy p shell. The energy gained from the formation of crystalline bonds overcomes the energy cost of this $s \rightarrow p$ promotion.

The situation is somewhat different on the Si(001) surface: there are only two neighbors available for bonding. If a surface atom sits in the ideal bulk lattice site, it has two electrons in sp^3 backbonds. The remaining two electrons occupy an sp lone pair orbital (cf. the orbital D in Fig. 3.2b, page 132), while one of the high-energy p orbitals remains unoccupied (the orbital B in Fig. 3.2b). The effective valence electron configuration of the surface atom becomes $3s^{1.5}3p^{2.5}$. Only $\frac{1}{2}s$ electron is now promoted to p.

But note that the s^2p^2 configuration of the free atom (that is, without the costly promotion) allows for two bonds, exactly as many as needed on the surface. The p electrons could be shared in the backbonds. This requires, however, that the angle between the backbonds changes from the sp^3 angle of 109.5° towards 90°. This would push the surface atoms upwards and the backbonds would stretch. So it makes sense that only every second atomic row rises, allowing for more efficient strain relaxation in the second layer. Clearly, this mechanism differs from that of dimerization. The structural transformation is now driven by a tendency to balance between hybridization and strain, not by the tendency to saturate broken bonds.

The two reconstructions differ quite substantially, so one might expect that experimental studies would soon decide between them. Instead, the following years brought new ideas and models. Not without good reason: in 1962, Lander and Morrison reported[388] that $\frac{1}{4}$-order streaks appeared in LEED patterns of carefully prepared Si(001). The streaks, apparently produced by short-range fourfold periodicity, coalesced into $\frac{1}{4}$-order spots when the sample was patiently annealed. This indicated that the stable surface reconstruction has a fourfold periodicity. Yet the largest surface cell which can be built of Schlier and Farnsworth dimers, the centered $c(2\times2)$ cell (Fig. 2.5c) has only a *twofold* periodicity.

The position of the $\frac{1}{4}$-order spots revealed that the fully ordered surface contains $c(4\times2)$ and $c(2\times4)$ domains (Fig. 2.2e, page 89). Lander and Morrison constructed the $c(4\times2)$ periodicity placing an array of dimers on top of the dimerized surface (Fig. 2.6c). This halves the number of dangling bonds with respect to that in 2×1 dimers. They simulated the diffraction of electrons on this structure;[389] the agreement with experiment was good.

In 1964, Ruth Seiwatz[289] noted that π bonds may be built between the dangling bonds of dimers if the dimers twist to arrange in "conjugated chains" (Fig. 2.6d). The chains arranged out of phase, like a sine and a cosine, make $c(4\times2)$ periodiciy. When the phase changes at random, the diffraction pattern is 2×1, reflecting the spacing between the chains.

The conjugated chain model was originally formulated to account for

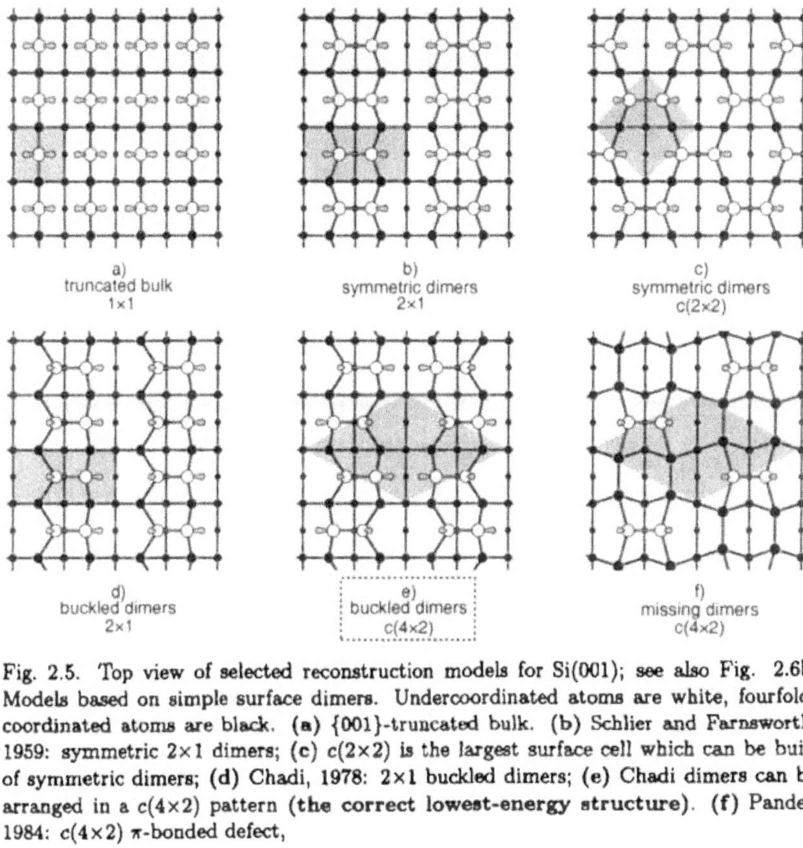

Fig. 2.5. Top view of selected reconstruction models for Si(001); see also Fig. 2.6b. Models based on simple surface dimers. Undercoordinated atoms are white, fourfold-coordinated atoms are black. **(a)** {001}-truncated bulk. **(b)** Schlier and Farnsworth, 1959: symmetric 2×1 dimers; **(c)** c(2×2) is the largest surface cell which can be built of symmetric dimers; **(d)** Chadi, 1978: 2×1 buckled dimers; **(e)** Chadi dimers can be arranged in a c(4×2) pattern **(the correct lowest-energy structure)**. **(f)** Pandey 1984: c(4×2) π-bonded defect,

reconstructions of Si(111). Seiwatz described only some qualitative ideas how the model could work for the Si(001) surface and announced intention to publish more details. However, her untimely death made this impossible.

2.2.2. *Dimers, chains, vacancies, or maybe something else?*

After this first wave of interest a period of relatively little activity followed. The year 1973 brought a new wave of results and ideas.

a. *Soft phonons, double bonds, and rediscoveries*

A completely novel approach[390] was presented by Trullinger and Cunningham. They noted that the interaction between surface-induced dipole mo-

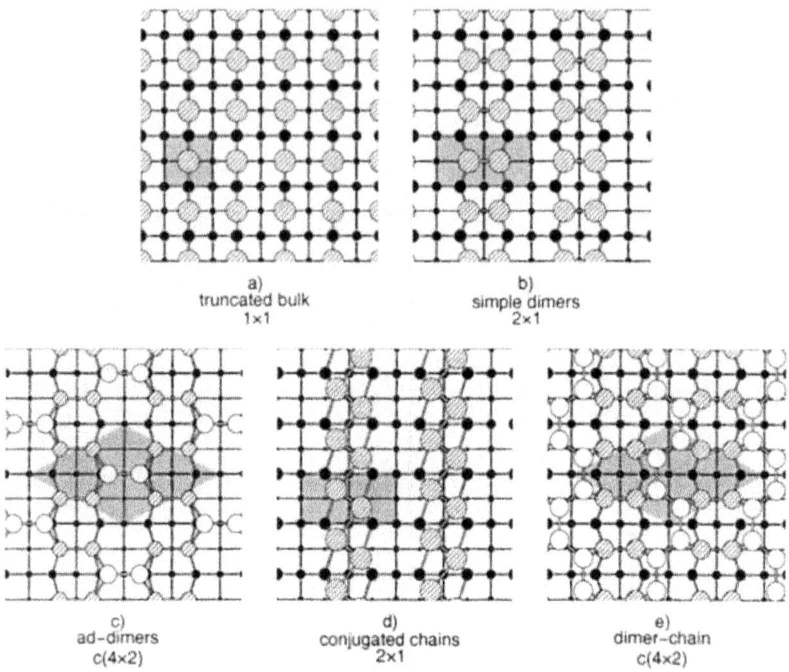

Fig. 2.6. Top view of Si(001) reconstruction models based on ad-dimers and/or dimer chains. Ad-dimer atoms are white, truncated-surface and dimer atoms are dashed, bulk atoms are black. **(a)** {001}-truncated bulk. **(b)** Dimerized surface, compare Fig. 2.5. **(c)** Lander and Morrison, 1962: $c(4\times2)$ ad-dimers. **(d)** Seiwatz 1964: Conjugated chains, 2×1 configuration. **(e)** Northrup 1985: Dimer-chain, $c(4\times2)$ configuration.

ments may soften frequencies of phonons localized near the surface. It can happen that the phonon frequencies $\omega(\vec{k})$ at certain points $\vec{k}$ of the surface Brillouin zone become imaginary ($\omega(\vec{k})^2 < 0$). This means that the atomic structure is unstable. The atoms must move to resume stability. The periodicity of the resulting reconstructed surface depends on the $\vec{k}$-point which drives this process. Trullinger and Cunningham suggested that their theory gave a general reconstruction mechanism of solid surfaces. This was the fourth mechanism proposed as an explanation of the 2×1 structure.

However, Ibach and Rowe[391] soon demonstrated that the 2×1 periodicity is not due to a simple surface mode softening. They observed that hydrogen adsorption did not destroy the half-order LEED spots.[b] But the

[b]This was fortunate, because the periodicity of H/Si(001) depends on the conditions of

saturation of dangling bonds should significantly reduce the dipole moment, hence the adsorption of H would restore the 1×1 pattern of the ideal surface. So the concept of soft mode reconstruction was not helpful in this context. However, it was *not* forgotten and remains a useful criterion by which the local stability of a particular surface structure can be verified.[392,393,394]

Phillips[395] proposed another new mechanism of surface energy lowering: microfaceting stabilized by double bonds formed between the neighboring atoms. He added rows of atoms to the bulk-terminated surface. In that manner, 2×1 trenches or "missing rows" appeared (Fig. 2.4c), but the number of broken bonds remained unchanged with respect to 1×1. The side walls of the trenches are (113)-oriented microfacets (Fig. 1.5, page 48), where atoms with unsaturated orbitals are the nearest neighbors. When these orbitals mix and become bonds, the surface energy should be lowered.

Another idea that could be assigned to the fruitful year 1973 was only rediscovered. On basis of his chemisorption studies,[396] Levine pointed out, as Green and Seiwatz had done (page 89) that dimerization only requires bending of the backbonds. He also argued that the remaining unsaturated orbitals of dimer atoms (dots in ) may form a π bond. Green and Seiwatz had written that "a preliminary examination would seem to indicate that full double-bond formation [...] is energetically unfavorable". But their paper was unknown to Levine. In the meantime it had been forgotten, and future authors usually attributed the introduction of the bond rotation into the dimer model to Levine.

Levine's argument helped us to realize that the dimer model may be energetically competitive to the missing-row model of Phillips which also invoked double bonding but did not introduce any significant strain. But Ibach and Rowe[391] used it to argue against dimerization. They reasoned that adsorbed H would use one orbital per each dimer atom and thus change the double bond between the dimer atoms into a single bond. If the dimer model had been right, the resulting change in the bond length would have been reflected in LEED intensities, contrary to the experiment.

b. *New data: electronic structure*

Apart from numerical fits to LEED, the argumentation had thus far been speculative. But the same year 1973 opened a new possibility to quantify the discussion. EELS results by Rowe and Ibach[397] demonstrated electronic

exposure (Sect. 5.2.1). White and Woodruff[398] were puzzled when in 1976 they repeated the experiment of Ibach and Rowe and found a 1×1 structure!

states of the clean Si(001) for the first time. In the following year their UPS study delivered the energy distribution of the occupied states.[399]

This was really helpful. Electronic structure is much easier to grasp than complex LEED spectra. The density of states, $\rho(E)$, can be interpreted by simple tight-binding models. Knowing $\rho(E)$ calculated for postulated atomic positions, one can guess how it is affected by changes in the atomic structure. This helps to judge whether a particular class of models is promising or not. Generally speaking, the effect of structural changes on LEED I–V curves is too obscure for such intuitive evaluation.

Small wonder that this motivated numerous theoretical studies. The objective was to understand the character of the surface states and to compare the computed and measured $\rho(E)$ But the required numerical techniques were only emerging. Even the results obtained so far for the band structure of such a fundamental and well-defined object as the ideal 1×1 surface were in a mutual conflict (see references in Ref. 400).

Between 1972 and 1974, Appelbaum and Hamann developed a self-consistent, semiempirical pseudopotential approach allowing one to compute electronic structures much more reliably. Together with Baraff they performed a series of calculations for Si(001). They began in 1975 with the 1×1 surface () and found two partially overlapping bands covering the whole energy range of the bulk forbidden gap.[400,401] The lower band has a "dangling bond" *sp* character (orbital D in Fig. 3.2b, page 132). The upper band has a "bridge bond" p character (orbital B in Fig. 3.2b).

Phillips' model (Fig. 2.4c) was examined next.[402] But the calculated and measured $\rho(E)$ did not correlate. The structure was found to be potentially unstable: the second bond built by the half-empty orbitals of the neighboring atoms in the {113} facets seemed weak.[c] Much stronger bonds would be made if atoms *dimerized* along the ridges. When the model also failed to reproduce the LEED I–V spectra,[403] it was eventually abandoned.

The dimer model performed better. The computed[402,404] $\rho(E)$ compared well with the UPS data. It was confirmed that the dangling orbitals of the dimer form an intra-dimer π bond, albeit weak.

c. *New models and new arguments against dimers*

Nevertheless, surface energies of the discussed geometries had not been computed and intuition was the leading factor in the interpretation of the

[c] Indeed, these orbitals point in different directions.

calculated band structures. For example, Harrison[122] also concluded from his tight-binding results that atoms on the truncated-bulk Si(001) dehybridize. But he viewed this as an advocation for vacancy models (Fig. 2.4d and 2.4e, page 91). In his excellent book[387] he argued that surface atoms on the truncated-bulk Si(001) have only two neighbors, so their most efficient electronic configuration would be s^2p^2, with two p electrons in the backbonds and the other two localized on an s "dangling bond".

Since the angle between p orbitals is 90°, that is, by nearly 20° less than the angle between sp^3 hybrides, complete dehybridization would invoke elastic stress. The stress can be more easily relaxed when the surface is corrugated by removing every second row of atoms (Fig. 2.4d, page 91). This structure, named "simple ridge", differs from Phillips' model (Fig. 2.4c) by the direction of the missing rows. They run parallel to the "rabbit ears" (Fig. 3.2, page 132) of the topmost surface atoms in Phillips model, and perpendicular to the "rabbit ears" in the simple ridge.

Another structure proposed by Harrison was the "canted ridge" shown in Fig. 2.4e. It was constructed by removing another row of atoms from the bottom of trenches between the ridges or by adding atoms on top of the ridge in Fig. 2.4c. Each 2×1 cell has now two threefold-coordinated atoms and one twofold-coordinated atom. Electrons are transferred from the threefold-coordinated atoms on one side of the ridge to the threefold-coordinated atoms on the other side. The atoms which accept electrons tend towards s^2p^3 (dehybridization), while the atoms which donate electrons tend to rehybridize towards sp^2. The changes in the angles between the hybride orbitals distort the ridge, leading to a canted geometry.

Harrison's opinion that dimerization is implausible was in accord with the interpretation of some experimental data. A LEED study by Jona *et al.*[403] had found little correlation between theory and experiment when the dimer model was assumed. Dimers had never been succesful in LEED simulations, but now the problem was serious: the analysis went beyond a simple kinematic theory and used a new, dynamic theory,[405] that is, it accounted for multiple scattering of electrons. To explain their results, Jona *et al.* adopted the idea of Seiwatz (Fig. 2.6d, page 94) that surface atoms with single dangling bonds may form π-bonded chains. LEED simulations for this conjugated chain model yielded I–V curves in acceptable agreement with experiment. But when in 1978 Kerker, Louie, and Cohen[406] calculated the corresponding density of states, it appeared that the electronic structure was dominated by dangling-bond states and did not resemble the measured $\rho(E)$. In addition, the presence of many dangling bonds and a relatively large strain suggested that such chains are unfavorable energetically.

In such a climate more and more models were proposed. Mitchell[407] suggested that Si(001) constitutes a closely packed overlayer. White and Woodruff found no satisfactory correlation between their calculated and measured LEED spectra.[408] White and Woodruff suspected a breakdown of their numerical analysis but were puzzled that it had worked[409] for H adsorption. The H adsorption study had revealed a 1×1 structure with only a small relaxation, so they speculated that the problems might be due to changes in atomic structure which extend many layers under the surface.

Also Poppendieck *et al.*[410] reported that LEED spectra indicate a distortion of many layers. They considered more than a thousand structures designed in such a way that they might yield some of the experimental I–V features. The data was best described by a "multilayer vacancy" model (Fig. 2.4f, page 91), a $c(4\times2)$ version of Harrison's canted ridge (Fig. 2.4e). But they concluded by saying that

> The preferred model is also the most complicated one tried, allows the most variational freedom, and so would be expected to yield the best qualitative fits. It should be pointed out that the most important freedom in such a fitting effort is when to quit.

Opinions had thus diverged. In the middle of 1978, after two decades of research, conflicting data existed on the two most promising structures, dimers and chains. The dimers, favorized by band structure measurements and calculations, were dismissed by electron diffraction measurements and calculations. The chains, favored by diffraction, were incompatible with band structure results. Some LEED studies found no compelling structure.

The "simple" Si(001) began to be more confusing than instructive. This was intriguing and irritating also because it had been known that (001) surfaces of two group-IV semiconductors, Si and Ge, exhibit the same 2×1 pattern.[383] Such a close analogy between different materials suggested a general reconstruction mechanism. A complex structure reaching deep under the surface seemed incompatible with this generality.

d. *LEED* and *subsurface strain*

The puzzle was solved by the end of 1978 by Appelbaum and Hamann,[411] who minimized the elastic energy using a two-parameter bond force model, an adaptation of Keating's model for bulk semiconductors. They observed that the dimerization introduced a *strain field* extending deep below the surface. The positions of atoms differed from the ideal bulk sites even in the fifth layer. These displacements changed the simulated I–V curves drastically. Agreement with the experiment was good even though no fit was made! Some experimental features appeared in the simulation only when the displacements were included. New LEED analyses performed in 1978 by Tong and Maldono[412] and in 1979 by Jona *et al.*[413] confirmed that such distortions produce a significantly better description of I–V spectra.

The dimer model triumphed. It could explain UPS and LEED. This was made possible by a "global" approach combining different techniques: LEED measurement and analysis, UPS measurement and band structure calculations, and numerical optimization of atomic positions by elastic energy minimization. Key by-products included a deeper understanding of factors crucial in LEED analysis of semiconductor surfaces, a deeper understanding of relations between electronic and structural properties of surfaces, and advent of numerical techniques for band structure calculations.

2.2.3. *The mysterious fourfold periodicity strikes back*

Was the structure of the Si(001) determined? Not necessarily. Previous LEED studies had found a two-domain $c(4\times2)$ pattern (page 92). This was confirmed by He scattering done in 1978 by Cardillo and Becker.[414] The scattered atoms formed $\frac{1}{4}$-order diffraction maxima,[d] even on surfaces not sufficiently ordered to produce visible $\frac{1}{4}$ LEED spots. But dimerization alone cannot account for fourfold periodicity. Surprisingly then, numerical simulations of He scattering on 2×1 dimers, reported in 1979 by Grimmelmann and Tully,[415] reproduced the measured intensities in the directions in which there are no $c(4\times2)$ spots (cf. Figures 2.2b and 2.2e, page 89).

The electronic structure of dimers also posed a problem. The dimerized Si(001) should be metallic, with a high density of states $\rho(E_F)$ at the Fermi energy E_F. But when Himpsel and Eastman[359] determined with ARPES their $\vec{k}$ dispersion, it became clear that something was wrong.

The surface turned out to be semiconducting! The top of the occupied surface band was found about 0.3 eV above the bulk valence band. No photoemission from states near E_F could be detected. Himpsel and Eastman concluded[359] that "in their present form, none of the [current] models

[d]Cardillo and Becker were the first to observe a meaningful structure in diffraction from a *semiconductor* surface. The first experimental observation of a diffraction structure in light gas scattering was reported already in 1930 by Estermann and Stern.[416]

describes the surface states for Si(001) 2×1 even qualitatively" and hinted that an atomic distortion might cause a separation of the surface bands.

2.2.4. *Buckling of dimers*

After Himpsel and Eastman's work had been presented at a conference in 1978, the authors appended to the proceedings a note on an improved model discussed at the meeting by Jim Chadi. In the next year, Chadi published the details.[360] Independently, LEED studies have led Jona, Shih, Jepsen, and Marcus[417] to a similar idea. The solution was based on the old rehybridization arguments of Haneman[285] and Harrison,[122] though viewed from a different perspective.

a. *Chadi's buckled dimers*

The new model formulated by Chadi[360] was constructed along the same line of thought as Himpsel and Eastman had presented[359] – that a distortion might open a gap in the surface density of states. So far, both atoms of the dimer had been considered equivalent (): the dimer was symmetric. The highest occupied (lowest empty) state of a symmetric dimer is the bonding π (antibonding π^*) combination of the dimer dangling bonds (the dots in ). The interaction between the neighboring dimers broadens these states into partially overlapping energy bands (Fig. 3.5, page 144).

A key argument for the dimerization was that a new bond can be formed between two surface atoms without stretching the backbonds. It suffices to rotate one of these atoms by $\alpha \simeq 35°$ around its substrate neighbors, the other one by $-\alpha$, and the two atoms will make a bond of usual length ( → , Fig. 2.3c, page 90). However, the same can be achieved through rotating each atom by a different angle (, Fig. 2.3). The dimer buckles: its atoms sit at different heights above the surface.

Chadi[360] used his tight-binding scheme to optimize the total energy of the dimerized surface without imposing any symmetry constraints. The result was that the simulated dimers buckled spontaneously. The symmetric configuration turned out to be unstable. While the dimers buckled, the occupied surface states decreased their energies, the empty surface states moved to higher energies, and an energy gap opened between them.

b. *Buckling supported by experiments*

Buckling promotes charge transfer from the "down" to the "up" atom. To a certain extent, these two atoms behave as two different species. With the buckling angle increasing, the states of the "up" atom become more s-like (that is, the "up" atom becomes more electronegative). The "down" atom hybridizes in a more sp^2-like manner, becoming less electronegative.

Though core charges are unaffected, the valence charge transfer influences the *binding* of core electrons, so that the $2p$ photothreshold depends on the buckling angle. A symmetric dimer has a single photothreshold, because both dimer atoms are identical. A buckled dimer has two photothresholds, and the energy difference between them measures the charge transfer. Therefore, one may expect that experimental determination of Surface Core Level Shifts (SCLS) would verify the buckled dimer model. Himpsel *et al.*[286] reported in 1980 that cores of ~0.5 monolayer of surface atoms are shifted toward smaller binding. Studies of channelling and blocking in ion beam scattering, combined with Monte Carlo simulations for various structures, also indicated asymmetric dimers.[418,419]

c. *Buckling supported by theory*

Motivated by the success of the asymmetric dimer model, Ihm, Cohen and Chadi[420] performed in 1980 self-consistent semi-empirical pseudopotential calculations for the electronic structure of the 2×1 buckled dimers. The pseudopotentials were "semi-empirical" because they were fitted to reproduce experimental spectra of isolated ions and atoms.

Total energy minimization by calculation of forces on atoms was still unavailable. The calculations of Ihm *et al.* assumed the atomic positions determined by Chadi. The main aim of the calculations was a careful comparison between electronic properties of symmetric and asymmetric dimers.

Though self-consistent distribution of electrons reduced the energy gain from Chadi's value of 0.46 eV/dimer to 0.12 eV/dimer, the asymmetric dimers remained energetically competitive to the symmetric dimers. Also confirmed was Chadi's prediction that the bonding and antibonding π dimer states become separated in energy when the dimer buckles. But at the same time, the calculations yielded a pretty dispersive antibonding state, so that the surface band gap was reduced to 0.1 eV. Another problem was too wide (though narrower than for symmetric dimers) an occupied surface band.

Mazur and Pollmann[421] found a plausible a solution as soon as in 1982. They calculated the electronic structure adopting the atomic positions given by Chadi and using a new tight-binding scattering method for semi-infinite solids.[422] The computed surface band gap and the dispersion of the bonding π state was compatible with the experiment. Previous discrepancies were interpreted as a technical problem caused by the choice of pseudopotentials. Too large a bandwidth in Chadi's tight-binding calculations turned out[423] to be a shortcoming of the first-nearest-neighbor approximation, which overestimates the interaction between dangling bonds. Also Bowen, Dow, and Allen[424] found that a more realistic tight-binding potential can reproduce the dispersion of the occupied surface states.

LEED[425] and ion backscattering[426] measurements published in 1982 indicated that the dimer bond length is longer than the energy optimization had suggested. Mazur and Pollmann[421] found that this was in fact consistent with the observation that the calculated energy position of the occupied surface band respectively to the valence bulk band was not exactly the same as in the experiments. When the dimer bond length was increased by 10% to 2.57 Å, that is, close to its measured [425] value of 2.54 Å, the surface band moved to the right position.[421] A similar bond length of 2.46 Å had been recently obtained in semiempirical, self-consistent Hartree-Fock calculations by Verwoerd.[427,428] Also the first *relaxation* of atoms in a self-consistent way by Yin and Cohen[361] favored buckling. This first parameter-free optimization of the Si(001) structure was done by a recently developed, *ab initio* pseudopotential code.

It seemed that any remaining inconsistencies were secondary and would disappear as numerical and experimental techniques would improve.

2.2.5. *Simple models lose some credit*

The superiority of the buckled dimer was settled. In February 1982, Haneman[429] wrote: "The magnitudes of the atom shifts [calculated by Yin and Cohen] are probably not far from the true structure".

a. *Quantum chemistry reports problems*

But in the same year a suprise came. All-electron calculations by Redondo and Goddard[430] questioned the asymmetric dimer model.

This study accounted for spin effects; the surface was approximated by a small, hydrogen-terminated cluster. Correlation of electronic motion stabilized symmetric dimers. For example, buckling by 0.5 Å (~12°) costed ~1 eV. Redondo and Goddard had already argued[431] that the closed-shell approximation used in other total-energy calculations is problematic.

Consider two electrons, a and b, distributed in two single-electron orbitals $\phi_1(r)$ and $\phi_2(r)$, such as the two dangling-bond orbitals of a dimer. Let the system be in the spin singlet state, meaning that the two-electron wave function $\Phi(r_a, r_b)$ is symmetric in respect to exchange of the electron coordinates r_a and r_b. A closed-shell bond is formed when the electrons share the same bonding orbital $\phi_1 + \phi_2$:

$$\begin{aligned}
\Phi(r_a, r_b) &= [\phi_1(r_a) + \phi_2(r_a)][\phi_1(r_b) + \phi_2(r_b)] \\
&= \Phi_i(r_a, r_b) + \Phi_c(r_a, r_b), \quad \text{where} \\
\Phi_i(r_a, r_b) &= \phi_1(r_a)\phi_1(r_b) + \phi_2(r_a)\phi_2(r_b) \\
&\quad \text{(ionic contribution)}, \\
\Phi_c(r_a, r_b) &= \phi_1(r_a)\phi_2(r_b) + \phi_1(r_b)\phi_2(r_a) \\
&\quad \text{(covalent contribution)}.
\end{aligned}$$

In other words, electron motion is not correlated. In the open-shell form of the wavefunction $\Phi(r_a, r_b)$, the electrons occupy different orbitals:

$$\Phi(r_a, r_b) = \phi_1(r_a)\phi_2(r_b) + \phi_1(r_b)\phi_2(r_a) = \Phi_c(r_a, r_b) \quad \text{(covalent contribution)}$$

which means that the motion of electrons is fully correlated.

The closed shell description is reliable when the orbitals strongly overlap. But when the overlap is small, a closed-shell wavefunction retains the repulsive "ionic" contribution Φ_i with two electrons on the same atom. Nature correlates the motion of electrons in order to eliminate this effect.

The correlation affects the buckling via a competition between the π bonding and the buckling angle. The dimer atoms may form a π bond. In a buckled dimer this bond is weakened, because the overlap between the two dangling-bond orbitals is smaller. As a result of this, if π bonds are strong, the symmetric dimer has lower energy than the asymmetric dimer. If the π bonding is switched off, the dimer may be buckled by Jahn-Teller distortion (page 136). This is because the two dangling bonds form a half-occupied, degenerate state. This degeneracy is lifted when the dimer buckles.

Thus, if the π bond is weak, the asymmetric geometry may be favored. A closed-shell wavefunction with its spurious ionic term underestimates the strength of the π bond between distant orbitals, giving preference to buckling. The generalized valence bond scheme used by Redondo and Goddard was designed for a smooth interpolation between a closed-shell description reliable for small separation between orbitals and a valence bond wavefunction which has the correct behavior for large separations. This approach was expected to treat both types of dimers on the equal footing.

Redondo and Goddard found also that the singlet and triplet spin states of the symmetric dimer were close in energy but had different bond lengths. They suggested that ordering like c(4×2) may be driven by strain relaxation and realized as a mixture of spin-triplet and spin-singlet dimers.

But Verwoerd re-examined his calculations and in early 1983 he reported[432] that the method which he had used, MINDO/3, did account for the correlation to an extent comparable with the method applied by Redondo and Goddard. Though MINDO/3 was based on a simplified Hartree-Fock scheme and as such it did use the closed-shell approximation, it was nevertheless a semiempirical approach. The electron correlation was accounted for when matrix elements were fitted to experimental data.

Verwoerd's new calculations[432]did yield the symmetric dimer in the spin triplet state as the lowest energy configuration. But the singlet spin state of the asymmetric dimer was only 0.08 eV higher in energy. Verwoerd argued that the correlation energy of in the triplet state of the symmetric dimer was overestimated with respect to that of the singlet state by some tenths of eV. The true ground state of the symmetric dimer was in fact the spin singlet, with energy at least 0.07 eV higher than of the asymmetric dimer. He concluded that the asymmetric dimer, with the bond length of 2.46 Å and a buckling of 0.48 Å, is the real lowest energy configuration.

Verwoerd recalled also that recent work function measurements[433] supported the buckled model. Work function depends on the surface dipole moment, that is, on the buckling angle. The charge redistribution calculated[434] for the asymmetric dimer suggested a work function change of −0.37 eV, close to the measured value of −0.34 eV. Although symmetric dimers do possess a dipole moment of the correct sign, the calculated dipole strength was too small in this case to account for the experimental data.

So what went wrong with the generalized valence bond calculations? This method apparently overestimates Coulomb repulsion, which artificially increased the energy of the partially ionic asymmetric dimer. Once more investigations of Si(001) have deepened the understanding of the existing methods, this time of quantum chemistry calculations.

Yang *et al.*[425,435] pointed out another issue. Though they concluded that the dimer atoms buckle by as much as 0.40 Å, their LEED simulation yielded the best results when the dimers were *twisted* (Fig. 2.3e, page 90). A twist requires that some of the backbonds are stretched and some are compressed. This uses a lot of energy and bonds would avoid such distortions unless there is no other option. Twisted dimers had been once suggested by Seiwatz, as components of π-bonded chains (page 92). But according to Chadi's tight-binding calculations,[419] twisted dimers should be unstable. The only symmetry breaking that lowered the energy was buckling. Was this an indication that the dimer model is intrinsically inconsistent? After all that effort, this would be a disappointing conclusion.

The ICPS meeting in San Francisco in 1984 brought a re-examination of the LEED data. The discussion was hot. Duke, Paton, and Holland[436] showed a very good fit with 0.36 Å of buckling, no twist, and claimed to have established "a structure which is manifestly compatible with the ion scattering data, the LEED scattering data, the qualitative features of the various model predictions for the top two layers, and the quantitative predictions of Chadi and of Yin and Cohen for the third and deeper layers". But Shu, Yang, Jona, and Marcus[437] strongly advocated for the twisted dimer model. It yielded a better fit than achieved by Duke, Paton, and Holland. The situation was ripe for some twists of thinking...

At about the same time, Farrell *et al.*[438] applied high resolution EELS to obtain $\vec{k}$ dispersion of unoccupied states. The minimum in the empty surface state was found at the Γ point. According to the calculated band structures, Si(001) dimers should have empty states which have energies below the energy at Γ. This might mean that neither the symmetric nor the asymmetric dimers with a 2×1 unit cell describe the reconstruction correctly. Still, since some diffraction studies had suggested a larger cell, $c(4\times2)$, band folding might have been invoked to explain the minumum at Γ. But the recent doubts were reflected in the last sentence of this paper:

> [it cannot be excluded] that some radically different atomic arrangement exists on the clean, reconstructed Si(001) surface.

b. *Pandey's missing dimers*

During the same San Francisco conference the asymmetric dimer model was challenged by Pandey,[439] who had previously criticised the idea of buckling as only being applicable to ionic materials.[440] He now reported a series of *ab initio* calculations of the same type as the calculations of Yin and Cohen[361] but with a more complete basis set, in short, with more precission.

The atoms were allowed to move according to the computed forces until a minimum of total energy was reached. The dimer acquired either a nearly symmetrical configuration or got stacked in a metastable, buckled geometry with higher energy. Pandey concluded that a large buckling

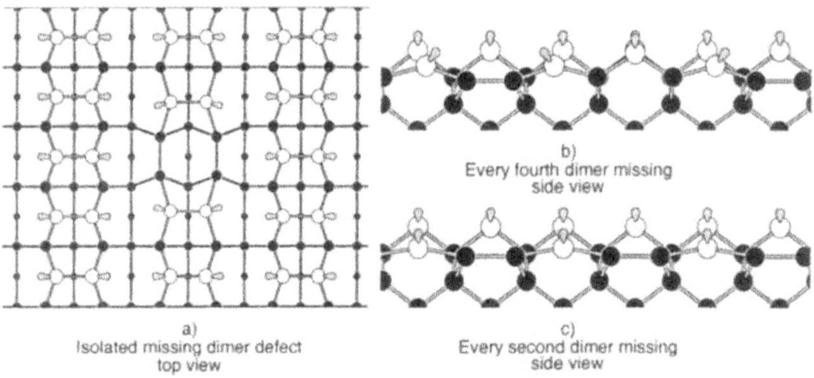

Fig. 2.7. Pandey's "missing dimer" or "π-bonded defect" model. Top view: (a) an isolated missing dimer. The bottom of the dimer vacancy is rebonded, forming two dimers perpendicular to the surface dimers. Side view: (b) removal of every fourth dimer afects mainly bond angles; (c) removal of every second dimer affects bond lengths.

(above 10°) is energetically unfavorable. Energy changes associated with a small buckling (within 5°) were below the numerical accuracy, estimated as 10 meV per dimer. This almost negligible energetical effect of small buckling[e] and a clear loss of energy for any significant buckling angles indicated that buckling plays no important role as a reconstruction mechanism.

Pandey proposed an alternative to the buckling (Fig. 2.7). When a single dimer is removed from an otherwise perfectly dimerized Si(001), the two dangling bonds of the dimer are removed with it and four new dangling bonds are created (four backbonds are broken). The new dangling bonds point pairwise towards each other and may be eliminated as on the ideal surface: by dimerization. Each dimer vacancy thus *reduces* the number of dangling bonds by two (Fig. 2.5f, page 93). The resulting elastic strain (Fig. 2.7c) becomes mainly angular when the distances between the missing dimers in a row increase (Fig. 2.7b), so moderately abundant missing dimer *defects* could be energetically favorable (Sect. 3.3.1a). Pandey speculated that the energy gained by rebonding the vacancy bottoms might overcome the lost elastic energy even in the dense, $c(4\times2)$ packing of the vacancies.

c. *Northrup's chain-dimer model*

A radically different arrangement of atoms was proposed in 1985. John Northrup reported [384] that his *ab initio* DFT calculations indicated insta-

[e] **Room-temperature thermal energy is around $kT = 25$ meV.**

bility of the dimerized surface with respect to adsorption of Si dimers on its top. The ad-dimers were energetically competitive to the ideally dimerized surface when they formed $c(4\times2)$-ordered, π-bonded chains running parallel to the dimer rows (Fig. 2.6e, page 94). The new model was thus a hybrid of a model with dimers and ad-dimers (Fig. 2.6c) and the conjugated chain model (Fig. 2.6d). For this reason it was dubbed "chain-dimer". The usually observed 2×1 pattern could be due to disorder in the arrangement of the ad-dimers in the surface chains. Diffraction patterns would then reflect the 2×1 periodicity of the dimerization in the first subsurface layer.

The difference between the energy computed for the new model and the dimer model was close to the estimated numerical accuracy. But the electronic structure of the new reconstruction was quite reasonable. The density of states was semiconducting and the unoccupied surface band had minimum in Γ. The model accounted for the observed periodicities of the clean surface without any need to invoke the controversial buckling. Taking this all together, it represented a serious challenge to the dimer model.

2.3. To Buckle or Not to Buckle?

As Haneman put it in his contemporary review,[364] "experiments have lead to practically every arrangement of surface atoms conceivable having been proposed". But the paper published by Notrhrup was the last one in a nearly thirty-year long series of attempts to *design* the model of Si(001). New experimental techniques soon supplied crucial data which, though still not free from ambiguity, have limited the allowed surface structures to two: symmetric dimers () and asymmetric dimers ().

2.3.1. *The first group photo of dimers*

Surface science had just obtained a powerfull tool, the Scanning Tunnelling Microscope (STM) designed by Binnig, Rohrer, Geber, and Weibel.[146] This invention won Binning and Rohrer the 1986 Nobel Prize for Physics. When STM was used to study Si(001), it seemed that the moment of truth came.

a. *What did STM reveal*

In July 1985, two years after STM produced the first atomically resolved images of a semiconductor surface,[f] Tromp, Hamers, and Demuth[363] published

[f] This surface was the (111) surface of silicon.[296]

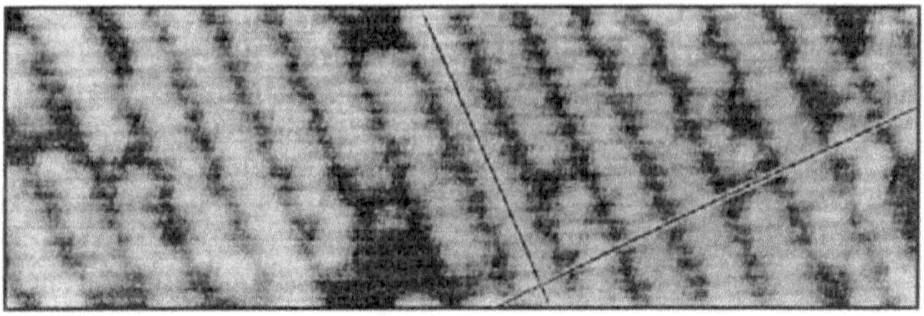

Fig. 2.8. One of the first STM micrographs of Si(001), after Tromp, Hamers, and Demuth (from Ref. 363). Occupied states. The basic structural element is imaged as a *symmetrical* bright oval: O. Each oval corresponds to a dimer: 8, where circles represent atoms. These ovals are arranged in rows: OOOO≡ 8888. Though many defects are present, 2×1 periodicity can be still recognized. Some of the rows appear as zig-zags; they are composed of asymmetric dimers buckled in alternating directions: 8888 (black circles symbolize "down" atoms, invisible in the image). Compare Fig. 3.8, page 150.

lished the first STM topographs of Si(001). They represented the occupied surface states (Fig. 2.8). The authors concluded:

> We have established the basic nature of the clean Si(001) surface: Both buckled and nonbuckled dimers are present, in roughly equal proportions, suggesting that their energies are nearly degenerate. Defects occur in rather large numbers. Some of these defect are similar to the π-bonded defect proposed by Pandey. [...] There is little doubt that [they] are present on any Si(001) 2×1 surface.

Any remaining doubts about the absolute supremacy of the dimer model were disposed in the following year by a systematic STM survey[441] which included estimates of STM images for various reconstruction models. Though the simulated pictures simply displayed surfaces of constant electron density obtained by superposition of atomic charges, which is a very crude approximation, the result of the analysis was quite impressive. Only the dimer model could consistently account for the characteristic rows of oval protrusions, sometimes smoothly evolving into zigzags of round protrusions (Fig. 2.8). Each oval could be very well described by the overlapping electron clouds of a symmetric dimer. The smooth transformations of the ovals into the zigzag lines was easy to understand if this pattern represented upper atoms of buckled dimers, with the direction of buckling alternating along the dimer row, as in Chadi's $c(4\times2)$ model.

The first etape of the extensive work to unravel the atomic structure of the "simple" Si(001) surface was completed. The original idea of Schlier and Farnsworth, that is, reduction of unsaturated bonds by dimerization, turned out to be correct. Results of a blind though sophisticated search spanning three decades were verified by one real-space image.

b. *What STM could not reveal*

But the problem of buckling remained unresolved. In the following decade the symmetry of dimers would be a major issue. Though some dimers appeared symmetrical in STM and some buckled, no good interpretation was found. The problem was twofold. First, scanning tunneling microscopes worked only at room temperature. One could argue that the dimers seen as symmetric are buckled, but the direction of buckling rapidly changes due to thermal vibrations. Second, the imaged surfaces contained many defects. One could argue that buckling is induced by the defects.

Nor could STM distinguish between twisted and straight dimers. Yet, total energy cluster calculations recently reported by Brink and Verwoerd[442] repeated the conclusion[419] that twisting is unlikely.

The relative position of atoms in a Si(001) surface dimer is a tiny but essential detail. First of all, the stability against buckling – or the lack of it – is a result of a balance between a number of factors. These are hybridization, electrostatic interaction, strength of π bonding, correlation of electron motion, and surface stress and opportunities for its relaxation. Thus, the ability to give the proper answer to the question of buckling was (and is) a benchmark of theoretical methods. On the experimental side the ability to detect and determine the local structure of dimers was (and is) also a tough test. Second, knowledge about the equilibrium surface structure is important for correct theoretical interpretation of surface processes. Although at non-zero temperatures even intrinsically symmetric dimers vibrate and may be significantly buckled most of time, the resulting phonon bands, surface strain fields, and time spent by atoms in specific configurations are not the same as when the dimers are intrinsically asymmetric.

2.3.2. *Electronic structure attracts more attention*

The year 1986 brought the ultimate vindication of the dimer model. At the same time, there was an impressive progress in the understanding of the electronic structure of Si(001). This was in part enabled by LEED studies due to Olshanetsky and Shklyaev[443] and to Kaplan[444] conducted by the end of the 70's. These authors had investigated silicon surfaces with orientations differring from the exact (001). They had noticed that when (001) surfaces are somewhat misoriented towards <110>, electron diffraction yields a single-domain 2×1 pattern instead of the two-domain 2×1 pattern typical to nominally flat surfaces (see Section 4.3.2 for phase transitions between one- and two-domain surfaces).

A band structure measured for a single-domain surface would be easier to interprete than that of a two-domain surface. But a single-domain pattern is indicative of biatomic heights (page 131). Any steps, biatomic or monatomic, may influence the results of measurement. To minimize the problem, band structure investigations prior to 1986, that is, prior to the work of Bringans *et al.*[445] had used two-domain surfaces as having more exact (001) orientations. Their electronic structure was expected to be less affected by surface steps. Yet Bringans' ARPES study of the single-domain Si(001) 2×1 was not really troubled by biatomic steps. Instead, it clarified how the two-domain structure should be unfolded into two single-domain structures (Fig. 3.6, page 146).

Another essential piece was delivered by Mårtensson, Cricenti, and Hansson.[446] They measured $\vec{k}$ dispersion of the occupied surface band and, simultaneously, minima of another band which pinned the Fermi level of n-doped samples. The minima, observed by ARUPS in Γ and J' points of the Brillouin zone ~0.7 eV above the Γ maximum of the upper filled surface state, could come from the bottom of the lowest empty surface band.

STM also continued to contribute important data. Hamers, Tromp, and Demuth[365] imaged the *unoccupied* states of the dimers and found that the energy gap is indeed well formed, at least 0.5 eV. A more detalied analysis of STM spectroscopy (STS) published around the same time by Hamers, Avouris, and Bozso[447] confirmed that the energy position of peaks in electron density of states as obtained from STM analysis were in agreement with photoemission (Himpsel and Eastman[359]) and inverse photoemission (Himpsel and Fauster[448]) spectra.

The experimental situation was clear: there is a well-formed surface energy gap. Now the theory should catch up.

a. *Problems with the band gap*

So far, theory had been focused on the highest occupied surface band. A very reasonable agreement with experimental data was reached (page 101). But the lowest unoccupied surface band posed at least one unsolved problem: as theoretical descriptions gained sophistication, the computed distance between the occupied and the empty states was systematically downgraded! The original tight-binding result was that the buckling opens a gap between two surface bands so that one band is occupied and the other band is empty (page 100). In 1986, Pollmann *et al.*[423] concluded that though the buckling does shift the two bands in opposite directions, the surface remains metallic when *ab initio* calculations are converged.[g]

The old dilemma came back. The buckling had been invoked because the electronic structure of symmetric dimers had been calculated as metallic. Now it appeared that buckled dimers were metallic as well.

But maybe something went wrong with theoretical descriptions of empty states? They were based on the Density Functional Theory[58] (DFT, page 20), that is, on an approach designed to yield the correct spatial distribution of electrons in their ground state in order to compute the total energy of the system. DFT single-particle electron states, particularily the empty states, are only a technical by-product. Since the DFT potential resembles the real potential felt by electrons, DFT electronic structures often resemble real electronic structures, but this has its natural limits.

One might attribute the problem with the energy gap of buckled dimers to an oversimplified treatment of electron exchange and correlation in standard DFT calculations. Electron exchange and correlation was (and often still is) routinely described in the so-called local density approximation[43] (LDA). This approach yields quite accurate total energies of atomic structures but it underestimates the energy gap of bulk semiconductors. Pollmann *et al.*[449] argued that the empty surface band is mainly derived from conduction-band like states. The empty states are therefore likely to follow the bulk conduction band to higher energies when a proper description of exchange and correlation effects is used.

The remedy might be the "quasiparticle correction" (page 22) used already by Hybertsen and Louie to correct the LDA band structure of bulk germanium[450] and of arsenic on the Ge(111) surface.[54] Still, it was not obvious that it would put the the empty surface band at the proper energy.

For this reason the origin of the gap of the dimerized surface remained unclear. In particular, it was not shown that buckling really is the right way of solving the energy gap problem. Since a significant correction of the computed energy gap was necessary, it could well be the case that the same type of correction would open a sufficiently wide gap for symmetric dimers.

---

[g]That is, when a sufficiently complete basis set is used to describe the wave functions.

2.3.3. *Seeing is believing*

The view that buckled and symmetric dimers coexist was gaining popularity. The evidence for prevailing asymmetry was being undone.

The strongest theoretical argument for buckling, that the buckling distortion opens the gap between the bonding and the antibonding surface states, had lost its magic. After it was calculated that there are two possible lengths of a dimer bond, one might have argued that symmetric dimers can make a $c(4\times2)$ pattern. Most dimers in STM images appeared to be symmetrical, unless located close to defects. Further work would bring a reinterpretation of core level shifts (page 112) in favor of symmetric dimers.

In this situation, a recent LEED observation by Tabata *et al.*[451] of an order-disorder phase transition between 2×1 and $c(4\times2)$ around 200 K was not a strong argument for the buckled dimer model. One could try to explain the transition by a rearrangement of dimers with short and long bonds. Or one could argue that though the phase transition is between symmetric and buckled dimers, it is induced by the presence of defects.

a. *SCLS revisited: dimers are symmetric*

Since 1980, one of the key arguments in favor of buckling had been the SCLS results (Sect. 2.2.4b): $2p$ core states of a half of surface atoms are shifted towards smaller binding energy by 0.52 eV and produce a surface photoemission peak S. This peak was assigned to negatively charged "up" atoms; positively charged "down" atoms would then produce a S' peak at larger binding energies (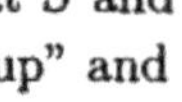
).

The S' component was indeed observed[452] at about −0.4 eV. Even so, it was partially masked by the strong bulk emission (B). It was hard to tell how many atoms contributed to it. The uncertainty in the determination of the fraction x of atoms contributing to S was also fairly large. This made it difficult to argue that S and S' have the same intensity (as expected, if they originated from "up" and "down" atoms of buckled dimers).

It is not easy to determine the number of atoms responsible for a particular SCLS structure. Firstly, the surface emission overlaps with much stronger bulk emission. Secondly, some estimate of the escape depth must be done in order to translate the relative intensities of the bulk component B and the surface components into areal concentrations.

In 1987 and 1988, a group or researchers from the University of Illinois, Urbana[453,454] recalibrated the escape depth on the basis of results for

Si(111) and concluded that a *whole monolayer* of atoms ($x = 0.92 \pm 0.07$) contributes to S. The message was that

> All Si dimer atoms, whether buckled, non-buckled, or dynamically buckled, contribute to the emission for the surface-shifted core-level component, ruling out large asymmetrical charge distributions about the buckled dimer atoms.

The S' peak should come from subsurface atoms: 

Together with recent STM confirmation[365] that a buckled dimer experiences a charge transfer,[h] the message was clear: the dimers are symmetric!

b. *Surface stress: symmetric dimers are OK*

So far, experimental information on the structure of Si(001) mainly came from studies of the atomic geometry by electron and ion diffraction and from investigations of electronic states (surface bands, core levels, electron tunneling). In 1988, Men, Packard, and Webb[367] used LEED to extract a new type of data, surface stress anisotropy. In the same year, Alerhand *et al.*[56] developed the theory needed to quantify the results.

Stress σ measures the total energy change caused by distortions of the lattice (Sect. 3.2.2d). Since each surface lattice vector can be changed in two independent lateral directions, stress is a tensor, σ_{ij}. A distortion parallel to the dimer bonds is not equivalent to a distortion perpendicular to them, so the stress of Si(001) is anisotropic: $\sigma_{\parallel} = \sigma_{xx} \neq \sigma_{yy} = \sigma_{\perp}$.

The stress anisotropy was accessed by bending slightly misoriented, two-domain samples. A two-domain surface has 2×1 and 1×2 domains of reconstruction, separated by steps of monatomic height (cf. Fig. 3.1e, page 130). The dimer bonds in 2×1 domains are formed along the [110] crystallographic direction (X), while in the 1×2 domains they run along the [$\bar{1}$10] direction (Y), that is, perpendicular to [110]. This means that deformations which stretch the dimer bonds in 2×1 domains compress such bonds in 1×2 domains. The relative energy of dimers in 2×1 and 1×2 domains is changed thereby, and this change is a measure of the stress anisotropy. Indeed, an isotropic surface (with 1×2 and 2×1 domains responding to the external strain in the same way) would experience no change in the relative energy of the domains. The temperature was high enough to allow for exchange of atoms between domain boundaries in response to the applied strain; the changes in domain population were monitored by LEED. The

---

[h] The maximum density of the occupied states is above the "up" atom, but the maximum of the unoccupied states is above the "down" atom.

result was that the domains compressed along dimer bonds grew larger and the domains stretched along dimer bonds grew smaller.

Alerhand *et al.*[56] made a tight-binding estimate which yielded the surface stress anisotropy of the correct sign and with a reasonable value of $\Delta\sigma = \sigma_{\parallel} - \sigma_{\perp} = 70$ meV/Å^2. The anisotropy computed *ab initio* by Payne *et al.*[455] and by Roberts and Needs confirmed a this qualitative agreement. Payne *et al.* used a new, efficient method of solving DFT equations, recently designed by Car and Parrinello[100]. At the same time, these new *ab initio* calculations validated also the previous assessment of Pandey that buckling is not stable energetically.

c. *Energy minimization: you get what you want*

By the late 80's most theoretical data witnessed against buckling. New self-consistent empirical tight-binding calculations[366] resulted in a conclusion similar to that of Pandey: the buckling is small, the difference between the buckled and symmetric configurations is also small, around 50 meV per dimer, hence both structures may coexist at room temperature. In 1989, all-electron *ab initio* cluster calculations which included spin effects[456] also confirmed this trend. The electrons were found to be correlated, occupying a singlet state. A tight-binding estimate employing a Hubbard-like hamiltonian led to the conclusion that inter-dimer interactions between the individual spins opens an energy gap between the surface-state bands. At the same time, Craig and Smith[457] studied the dimers by a newly formulated, slab version of the semi-empirical MINDO method[i] They as well found the symmetric configuration more favorable.

According to Craig and Smith, symmetric dimers constitute the absolute minimum in the total energy surface, but defects could induce buckling of dimers in the $c(4{\times}2)$ order. This is would be compatible with the STM images: most dimers are symmetric in STM micrographs and buckling is seen chiefly near to defects (cf. Fig. 2.8, page 108). The computed dependence of total energy on the buckling angle flattened out around 3.5°. The energy of such dimers was only 10 meV/dimer higher than symmetric dimers. One could imagine that an external perturbation stabilizes them.

But already in the same year 1989, Zhu, Shima, and Tsukada[368] re-

[i] The MINDO method, discussed on page 104, had been originally used for cluster calculations. The new formulation was implemented for slabs: lateral periodic boundary conditions were applied so that the cluster expanded into a slab with two parallel surfaces. This was more suited to treat surfaces than the old cluster version had been.

ported a DFT study yielding the $c(4\times2)$ buckled dimers as the *lowest* energy structure, 0.32 eV below the symmetric dimers. The density of states of the $c(4\times2)$ geometry was semiconducting and the minimum of the lowest unoccupied band was at Γ, though the energy gap was narrow (0.01 eV).

Did this indicate a comeback of buckling? Had the suggestive STM images biased interpretation of experiments and encouraged improvement of calculations until the dimers became symmetrical, but *not* any longer?

Some DFT results published in 1990 did not confirm this view.[458,459] The symmetric dimers were found to be stable. But other studies conducted at about the same time lead to the opposite conclusion. Using a similar DFT method, Roberts and Needs[460] computed that 0.057 meV/dimer is gained by 2×1 buckling, and extra 0.037 meV is released when it changes to $p(2\times1)$. Their previous results,[455] favoring symmetric dimers, were found to be not converged. Independently, Kobayashi, Terakura, and Blügel performed DFT calculations and observed buckling.[461,462] This debate had motivated Badziag, Verwoerd, and Van Hove to reconsider the problem. Their new calculations[463] employed an unrestricted semi-empirical Hartree-Fock approach allowing for open and closed shell wavefunctions (page 102). Clusters of two different sizes (Si_9H_{12} and $Si_{15}H_{16}$) were used to control possible artifacts due to the cluster approximation. The conclusion was the same as in the early work of Verwoerd: the dimers do buckle.

The scale tipped again. Buckled dimers were back in play.

However, opinions were still divided. For instance, discussing the origin of symmetric dimers in STM images, Badziag *et al.*[463] suggested that the symmetric configuration may be stabilized by the charge acquired from the STM tip. To the contrary, an *ab initio* Hartree-Fock study[464] which used clusters of the same size found *neutral* dimers stable as symmetric, whereas *charged* dimers tended to buckle; the message was that since defects are often electrically charged, they could initiate buckling. A simillar interpretation was given by Kochanski and Griffith.[465] On the basis of STM measurements and a simple cast of interaction between buckled dimers on an Ising-type hamiltonian (Sect. 4.3.1a), they proposed that the ground state of dimers is the symmetric configuration, the buckling is caused by surface defects, and the observed amount of buckling is affected by the tip.

2.3.4. *The sobering variety of opinions*

The overall impression an independent observer studying the literature published around 1990 would get was that neither theory nor experiment could decide what the dimers really look like. Interpretations of the numerical results appeared to be strongly biased. Even "*ab initio*" calculations are not completely free from approximations and "parameters". *Ab initio* is termed "parameter-free" because, in contrast to empirical or semi-empirical methods, its "parameters" are not fitted to experiment. They describe the numerical accuracy of the method. These "parameters", like the size of the basis set or the theory used to calculate the exchange-correlation energy, are chosen such that the code fits in the computer and terminates within acceptable time. The researcher must decide whether the results are reliable enough or detailed "convergency tests" are needed (Sect. 1.3.1b). Limited resources or pressing time can make "reliable" mean "reasonable". Which reminds us of the remark once made by Poppendieck, Ngoc, and Webb that the most important variational freedom is the freedom to quit.

This illustrates the rule which sets the standards of any investigation and any reasoning: "Do not stop when you arrive at the result you personally believe is right". It is essential to obey this rule particularly in basic, non-technological disciplines, as in fundamental physics and in philosophy. The more applied research is, the more easily misconceptions are detected. This is simply because a failed commercial application leads to painful financial losses. Money has always been efficient as a feedback! The other extreme is marked by philosophy. By their virtue, truly philosophical statements have subjective backgrounds, hence mistakes are hard to spot.

If science eventually became commercialized, this dichotomy would lead to a real danger (Sect. 7.3.1). Science has, and will have, an overhelming influence on our worldview. One of the main "philosophical" experiences from commercial scientific investigations would then be: all valuable statements can be verified, they can be proven by demonstrating a working application. This would then be readily taken into the sphere of philosophical thought, leading to a new wave of naive positivism with a "strong", "scientific" support. It is due to the multitude of extrapolative theories about phenomena which are not easily verifiable experimentally[466] that an orthodox positivistic program can be still recognized as an oversimplification. If the curiosity of science gave way to practical commercialism, the limits of human perception and reasoning would become hidden behind the financial and linguistic barrier of the word "useless". Few would be then aware of these intrinsic limits and comprehend the intellectual freedom they offer (Sect. 7.3.4). The rest would be bound by their "omniscience".

It is precisely because of this danger that a strong *academic* science is needed. A science which needs knowledge for the sake of knowledge. A science with a broad interest and weak coupling to money. A science prone to serious mistakes it can be proud of because they reflect the creativity and the unconventional thinking of scientists. A science which can not only teach us about the Universe, but can provide the experiences we need to understand and appreciate our selves. A full development of a human personality requires more than a full stomach and an access to Internet.

2.4. A Clear Picture Finally Emerges

The asymmetry of dimers was finally accepted because they were *seen* in STM images to buckle independently of defects (page 120). This time the interpretation of the key experiment was supported by other techniques, which have sharpened their accuracies, reconsidered their conclusions, and reached a well-earned consensus.

2.4.1. *Reconciliations*

Apart from a rather discouraging discussion between theoreticians, the early 90's brought a heated debate about the interpretation of surface core level shifts. It was also unclear whether the dimers appear symmetric in STM images because they vibrate or because symmetric dimers have the lowest energy. It took a couple of years to reconcile these different views.

a. *SCLS revisited: polemics*

The recent SCLS study by the Urbana group (Sect. 2.3.3a) had concluded that the surface peak S, originating from Si($2p$) core electrons which are less bonded (by 0.52 eV) than Si($2p$) cores in the bulk, is due to a complete monolayer of surface atoms, that is, to all dimer atoms.[453,454] In other words, any significant asymmetry of the dimers had been ruled out. The S' peak, originating from Si($2p$) cores more bonded than in the bulk, had been attributed to subsurface atoms. But this interpretation 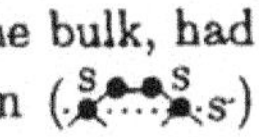 was not based on a *direct* observation but rather on an *indirect* calibration technique. As such, it was not accepted by all researchers.

Precisely this relation between assumed atomic structures and calibration of SCLS made a proper assignment of the S and S' peaks very important. The intensity ratio between bulk and surface photoemission peaks is a function of electron escape depth.[453] Along with the assumption that the S component arises from a full monolayer or a half of a monolayer, different escape depths are adopted. The resulting ambiguity (a factor of two) in the calibration of SCLS caused serious difficulties in the interpretation of chemisorption and intitial stages of growth.[179,467]

The old assignment of the S and S' components (
) was then reinforced in 1991 by Wertheim *et al.*,[468] who performed SCLS at the liquid nitrogen temperature. The quality of the data was good enough to allow for a lineshape analysis of the S' component. Since bulk electrons are excited to bulk states while surface electrons are excited to surface states which differ from those of the bulk, different lineshapes are expected for bulk and surface emission.[469] Wertheim *et al.* concluded that the S and S' peaks must come from *surface* atoms. The S peak, assigned to "up" atoms by Wertheim *et al.*, appeared split into two parts separated by 190 meV. This double peak was interpreted as a crystal-field splitting of the $2p$ core state of the "up" atoms. (This was the first report in the literature which claimed to resolve a crystal-field splitting in a SCLS spectrum. It was, however, not confirmed by further studies.[470])

But a few days before this paper appeared in Physical Review Letters, the editors received a manuscript from the Urbana group with a report on a study ingeniously designed to distinguish between 
 and 
. Lin, Miller, and Chiang[471] evaporated Ge on clean Si(001) and followed the evolution of Si($2p$) and Ge($3d$) core level shifts. If the S and S' peaks originated from the dimer atoms, as expected for a buckled dimer (
), they should be quenched by an increasing amount of Ge. But if the S peak came from surface atoms and the S' peak from subsurface atoms (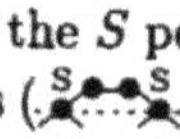
), the intensity of the S component should decrease with Ge coverage from the very onset of deposition, because S sources are converted into S' sources: 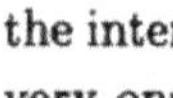
 → 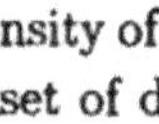
. The S' component would remain approximately constant until the surface is completely covered by Ge atoms, that is, as long as all subsurface atoms are still silicon. When the second Ge layer begins to grow (all S sources are gone), S' should gradually disappear.

The SCLS signal behaved precisely as expected for symmetric dimers. The calibration of the Urbana group was thus defended. The dimers were shown to be virtually symmetric.

However, this seemingly straightforward experiment did not terminate the discussion. Rowe and Wertheim wrote[472] that the different behavior of

S and S' may be related to different electronegativities of Si and Ge atoms. Since Ge is more electronegative than Si, the first Ge atoms which arrive at the surface may tend to substitute the negatively charged "up" atoms in the dimers (). If this is the case[j] then the "down" atoms are being substituted only when all the "up" atoms become Ge atoms. This would lead to the same behavior as observed in the expermient of Lin *et al.*; in other words, the observation of Lin *et al.* is compatible with buckled dimers. The old calibration of Himpsel *et al.*[286] could be correct.

Himpsel[467] proposed another interpretation: S may come from the "up" atoms, S' may be due to the subsurface atoms, and the signal from the "down" atoms may be hidden under the strong bulk peak B (). He pointed out that the calibration which leads to the assignment of S to the full monolayer of atoms causes inconsistencies in the interpretation of adsorption on Si(001). In addition, the analysis by Yang *et al.*[473] suggested that S comes from only a half of a monolayer. The S peak was found to be only half as intense as the allegedly full-monolayer S' peak.

Still, the counter-arguments were not decisive and Lin *et al.* could rightfully maintain their view[474]. The problem remained unsolved.

b. *Theory, STM, and SCLS* agree

The theory suffered from the same indecision as SCLS. Several groups applied similar techniques, but the conclusions diverged, reflecting too low a numerical accuracy and differed expectations. This called for a careful analysis of the influence of numerical conditions. Such analysis[372] was presented in 1992. It was concerned with the most popular approach, DFT-LDA, and demonstrated that while fully convergent calculations yield buckling, lack of convergence can lead to either symmetric or buckled dimers.

This conclusion was in accord with novel STM measurements reported at the same time by Robert Wolkow.[373] Previous STM studies had been done at room temperature, due to technical difficulties in the design of a microscope with temperature control. Wolkow presented the first images of Si(001) obtained at 120 K. At low temperatures virtually all dimers buckled and were arranged in $c(4{\times}2)$ and $p(2{\times}2)$ domains.

It was now possible to maintain a view fully compatible with all STM studies and with *ab initio* DFT calculations. Buckled dimers are energetically stable. At higher temperatures the direction of buckling may ocas-

[j] Later *ab initio* calculations, published in 1998, confirmed this tendency.[475]

sionally flip there and back between  and . Calculations[372] confirmed the suggestion of Tromp *et al.*[363] that at room temperature this flipping is so rapid that an STM image represents an average configuration which is, naturally, symmetric. A defect may impair flipping of the neighboring dimers, acting as a "pinning center" for buckling.

How did core level shifts fit into the picture? *Ab initio* calculations by Pehlke *et al.*[476,180] and SCLS experiments by Landemark *et al.*[470] made this clear. Two new peaks were measured: SS at -0.061 eV and C at 0.200 eV (Fig. 3.10, page 152). The SS component behaved in the same way as the S peak at different temperatures and photon energies. It was therefore natural to link both features to surface atoms: the S peak to "up" atoms and the SS peak to "down" atoms. The S' peak was identified with the second-layer atoms (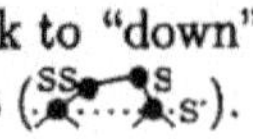). The intensity ratios were consistent with these assignments. The peak C was attributed to atoms in the third layer.

The calculations showed that the energies of core states of subsurface atoms differ from those in the bulk.[476] Core electrons are more strongly bonded in the second layer than in the bulk. The energies of electrons emitted from dimer atoms depend on the dynamics of screening (Sect. 1.3.2e). The peaks assigned in experiments to "up" (S) and "down" (SS) atoms are located within the limits computed by the initial state theory[476] (the screening charge responds slowly to the removal of an electron) and the final state theory[180] (fast response of the screening charge).

2.4.2. *Completing the picture*

New low-temperature STM studies (Tochihara,[374] Band,[477] Munz,[478] Kipp,[479] Uchikawa,[480] Shigekawa[382]) confirmed and expanded the work of Wolkow.[373] In 1994, Tochihara, Amakusa, and Iwatsuki[374] presented STM micrographs which explained why patches of symmetric dimmers appear in low-temperature STM images. Namely, the combined influence of the so-called C-defects (which pin the direction of buckling) and the $c(4\times2)$ matrix of buckled dimers frustrates the buckling of some dimers (page 150) and makes them flip even at low temperatures. Tochihara *et al.* concluded:

> [Defects] affect buckling of dimers only locally and [do not cause] the extensive formation of the $c(4\times2)$ arrangement. The present paper suggests that a defect-free surface will exhibit a complete $c(4\times2)$ arrangement of buckled dimers [at low temperatures].

The local structure of clean Si(001) was finally established. Theory and experiment could focus on converging their results.

Already in 1993, John Northrup demonstrated[481] that the electronic structure of the $c(4\times2)$ buckled dimers agrees almost perfectly with the measured bands.[369] He applied a quasiparticle correction (page 22) to the *ab initio* DFT-LDA electronic structure. Two groups[482,72] verified independently that quasiparticle corrections put the surface bands of buckled dimers in the right place, whereby Rohlfing, Krüger, and Pollman showed[483] that the corrected electronic structure of symmetric dimers remains metallic. This solved the problem (page 111) with the partial overlap of the π and π^* energy bands of buckled dimers computed with DFT-LDA.

Ab initio studies (Wang,[484] Northrup,[481] Bogusławski,[485] Garcia,[486] Dąbrowski,[487] Shkrebtii,[488] Krüger,[489,490] Ramstad,[377] Fritsch,[491] Cho,[492] Milman,[493] and others) obtained buckling routinely now. Problems with the quantitative explanation of the measured stress anisotropy[494,455,495] disappeared: it sufficed that the buckling was reproduced.[486,487]

At the same time, the influence of a tungsten tip on the STM image was exploited, also *ab initio*.[496] The tip could flip the dimer, though an energy barrier had to be surmounted. Cho and Joannopoulos speculated later[492] that this may have fascinating implications for massive memory devices. Since a dimer can store one bit of data (=0, =1), the density of 420 Gbyte per mm^2 might be reached. The state of each dimer would be read and written by STM. This sounds like science fiction, particularly because an Ultrahigh Vacuum STM would make a very clumsy and slow a "disk drive". But in the not so distant past of vacuum tube electronics the idea that a computer thousands times more powerful than the famous ENIAC could fit into one's pocket would have been seen as science fiction as well. ENIAC weighted a few tons and it was said that when it was turned on, the lights of the city of Philadelphia dimmed...

The first *direct* observation of buckling was made in 1992 by Fontes, Patel, and Comin.[497] Their tool was the X-ray standing wave method. In order to make the separation of the signal from the dimer atoms and from the subsurface atoms straightforward, Ge dimers on Si(001) were imaged. Since *ab initio* data suggest the same behavior of Si and Ge dimers, Fontes *et al.* argued that their result is relevant for Si dimers as well. Two years later an international, Swiss-Italian-Japanese-Swedish team reported[498] angle-dependent surface core-level photodiffraction measurements and simulations showing that the S peak is compatible only with one assignment: 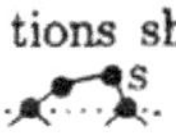

. The buckling angle Θ was 19°. X-ray diffraction yielded[376,381]

$\Theta = 20 \pm 3°$. *Ab initio* c(4×2) buckling angles fit in very well; for example, Northrup[481] computed 17.5° while Ramstad *et al.*[377] obtained 18.8°.

Perhaps the most direct snapshot of dimers was published in December 1995 by Hong, Shyu, Chou, and Wei.[499] This is a *holographic* image, with the resolution in all directions better than 1 Å. A visual inspection of Fourier transformed data (the data are collected in $\vec{k}$-space) is sufficient to appreciate the reconstruction. Unfortunately, the buckling is averaged out: the applied method, Kikuchi electron holography, is a kind of lensless Fraunhofer holography and produces a spacially averaged 3D image.

Jona and his coworkers who have so fervently advanced the idea that the dimers not only buckle but also twist (page 104) may have also had a point. *Ab initio* molecular dynamics simulations[500] indicated that a dimer can lower the barrier for flipping between  and  by making a twisting motion. A twisting mode was identified in the phonon spectrum, also on the basis of *ab initio* DFT calculations.[491]

2.5. Conclusions and Summary

In this historical essay on the investigation of atomic structures of Si(001) we illustrated the complementary role of different experimental and theoretical techniques in composing a detailed and unique atomic-scale picture of even a simple structure. No single technique, not even STM, can do it alone. A number of times a quasi-consensus was reached, but the investigations went on, driven by curiosity of those who were not convinced.

Variety of opinions is the main factor which motivates researchers to make elaborate tests of their assumptions. It is one of most important forces which advances studies on the fundamental level. Non-commercial research moves to another subject when the community reaches a happy consensus. A scientifically-minded person must thus take care that his or her audience is broad enough to provide this precious variety. Otherwise, the consensus is likely to get stuck at a local minimum. This is true in natural science just as in philosophy, religion, or sociology.

2.5.1. *What did we learn*

Chapter 3 collects the structural results for clean Si(001) and extends them by technical details. Here we will focus on some general issues.

Certainly, the most direct thing we learned is that the clean Si(001)

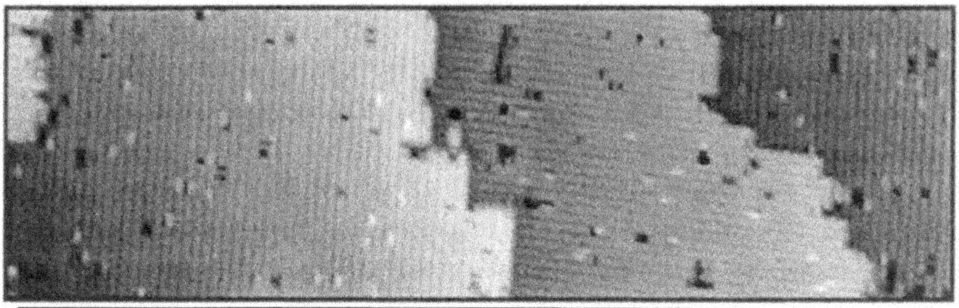

Fig. 2.9. Miniaturization drives technology into the atomic regime. In this STM image (from Ref. 28) of clean Si(001), atomic details like dimer rows, buckled dimers, defects, and monatomic steps can be recognized. The length of the scanned area is 0.1 μm. Within a decade, this will be the minimum feature size of state-of-the-art CMOS devices. See also Fig. 1.2 on page 8 and Fig. 1.4 on page 11.

surface dimerizes in order to increase the number of interatomic bonds of the surface atoms (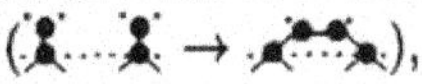), that these dimers are buckled (), and that the buckled dimers have the lowest energy when arranged in a $c(4\times2)$ array: 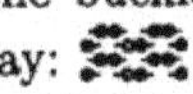 (top view; the dashes are dimers, the circles mark their "up" atoms"). But this would not be too much, given that it took four decades of intense studies to find out, and that one can hardly use this data to build a device or to enchant a broad publicity. The most practical result was the one from which the whole affair started: that only clean surfaces produce a 2×1 LEED pattern. A commercialized research program would *stop* at this point, viewing the continuation as a waste of resources!

But many practical crafts would have been thrown away with these "wasteful" studies. This becomes increasingly evident as the miniaturization drives the size of microelectronic devices into the atomic-scale regime (Fig. 2.9). One of the major achievements of materials science is that it has established experimental and theoretical methods to investigate complicated solid state systems on the atomic scale. This was only made possible by investigations such as those described in this Chapter.

We believe the most valuable results of these 40+ years are those which contributed to the development of surface science as a whole. On the theoretical side, this includes refinement and verification of *ab initio* methods for calculations of electronic and atomic structures, benchmarking of other quantum chemical approaches and classical potentials, and establishing ways of reliable interpretation of experiments like LEED, SCLS, and STM. On the experimental side, one should name advances in scanning probe techniques, like STM and AFM. STM can see through ultrathin oxide layers and can be used to study SiO_2/Si(001) films and interfaces. Its combination with AFM is particularily promising for this purpose.[501] Other examples are core level spectroscopy and ion backscattering.

Knowledge about the clean surface is the basis for studies of composed systems. It is remarkable that dimerization, the characteristic feature of the clean surface, is present also at certain SiO_2/Si(001) interfaces and, possibly, at certain silicide/Si(001) interfaces. The knowledge on reactions between various elements with surface dimers can be a prerequisite for the understanding of much less observable reactions at buried interfaces.

Even the dimer buckling, a feature limited exclusively to the *atomically clean* surface, has its practical value. Methods sensitive to such details may provide key data which (in conjunction with numerical simulations) can be used to prove or disprove entire classes of structural models. This stimulates benchmarking and refinement of the existing and developement of new experimental and theoretical techniques. These techniques, when matured, will assist simulation and design, examination and mastering of fabrication of new classes of structures, devices, and materials. Viewed from this perspective, the paramount success of *ab initio* calculations in the description of the structure of the surface up to the finest details is definitely not a mere Guiness-book record, but a validation of the most reliable theoretical method of contemporary computational materials science.

Naturally, we do not claim that all this progress was due solely to the reasearch on the clean Si(001) surface. Yet, given the volume of the work done, the basic character of the encountered and solved problems, and the simplicity of this system, the role of Si(001) has been significant.

2.5.2. *What is still missing*

It seems that the investigations of the atomic and electronic structure of the clean Si(001) surface are largely completed. New data will be needed mostly as a reference database for calibration or verification. For example, results for clean Si(001) are now often quoted in theoretical papers which deal with adsorption on Si(001). Observation of Si(001) dimers and defects is a test of the STM tip and of the surface preparation.

It is surface preparation which provided an intriguing question about Si(001) dimerization. In 1996, Xu *et al.*[502] reported that laser-induced desorption of Si leaves some of the surface atoms **undimerized**! The authors suggest that the atoms were frozen in the 1×1 sites due to the rapid cooling. But no calculation known to us reported a barrier for dimerization.[k] It is imaginable that subsurface damage is responsible for this effect.

Another open issue is the **dimer bond length**. Some experiments favor longer bonds than theory and other experiments (page 137).

A somewhat puzzling observation related to buckling of Si(001) dimers was made for the (001) surface of **germanium**. The two IV-group elements, Si and Ge, are similar in many aspects. In particular, Ge(001) and Si(001) surfaces reconstruct in analogous structures, whereby the energy difference between buckled and symmetric dimers appears to be larger on Ge(001). But it was found that if the temperature is sufficiently high, Ge(001) dimers spend sufficiently long time in the symmetric configuration to induce metallic (that is, partially occupied) surface states.[503,504] At least 10% of the surface states appear at room temperature as metallic.[505] Partially occupied electronic states of dimers can be observed also on Si(001), but only when the sample is rather heavily doped[369] (n^+ type, ~0.005 Ωcm) The samples used in the Ge(001) experiment[505] were nearly intrinsic (n type, ~50 Ωcm), that is, the observed metallicity is not likely to be an effect of doping. No metallic states of comparable concentration have been observed on nearly intrinsic Si(001) samples[359] (n type, ~10 Ωcm).

The existence of partially occupied dimer bands on Ge(001) may be related to another effect which seems to be specific to this surface, namely, to spontaneous formation of an ordered domain pattern consisting of buckled $c(4\times2)$ and symmetric 2×1 strips oriented along the dimer rows.[506] Observed at room temperature by STM, the $c(4\times2)$ and 2×1 domains cover approximately the same area of the surface. The strips have a width of several dimer row spacings. The symmetrical appearance of dimers in the 2×1 strips has most probably the same origin as on Si(001): all dimers are buckled, but each buckled dimer of the symmetric 2×1 domain flips rapidly between its two configurations,  and . It is possible that this domain pattern is formed as a result of the difference in the surface stress anisotropy of the $c(4\times2)$ and 2×1 phases (see also Sect. 4.3.2, particularly page 253) and to an exceptionally low density of defects on the Ge(001) surfaces.[506] On the other hand, it may also indicate that the dynamics of Ge(001) dimer flipping is different from the Si(001) dynamics, so that the average time a dimer spends in a metallic (that is, nearly symmetric) configuration is longer on Ge(001) than on Si(001). Last but not least, the

[k]The only exception is the case of a *single* dimer vacancy. See Section 3.3.1a.

band gap of bulk Ge is smaller that that of bulk Si. This may have its influence on the geometry and dynamics of surface dimers.

Finally, we note that one can still ocasionally read theoretical papers which claim that the true ground state of the Si(001) surface corresponds to symmetric dimers and that **buckling** must be caused by the presence of defects. Personally, we do not suppose that such claims are correct. The failure to compute buckling dimers as the ground state seems to indicate a numerical problem caused, for example, by too small cluster sizes[116] (Sect. 1.3.1b) or insufficient convergence of the exchange and correlation contribution to the total energy (Sect. 1.3.1a). There is an overhelming theoretical and experimental evidence for buckling.

However, let us recall Figure 2.1 (page 85): correlation between theory and experiment does not necessarily mean that the correct solution was found! Therefore, one should not depreciate the value of a theoretical or experimental work on the grounds of "strange" conclusions.

The truth is not necessarily there where the majority sees it.

The impact of an acute angle of a triangle on a circle produces an effect no less powerful than the finger of God touching the finger of Adam in Michangelo...

- W. Kandinsky

CHAPTER 3

Geometries of Clean Si(001)

J. Dąbrowski

Atomic and electronic structures of perfect and defected Si(001).

3.1. Introduction

Clean Si(001) is the simplest of Si surfaces. At the same time, Si(001) wafers are used for industrial fabrication of integrated circuits. What one learns from studies of this surface may be thus simple enough conceptually and close enough to the technological reality that it might assist in solving practical problems. The most fundamental investigations are those which address the structural properties of the surface: the atomic geometries and their electronic states. We devoted this Chapter to them.

The structures of clean Si(001) are understood to a significant degree. There is little doubt left about the atomic geometry of perfect Si(001). Down to the finest details, it can be predicted by *ab initio* calculations (that is, by a theory that uses no experimental input, Sect. 1.3.1) with accuracy comparable to experimental. Surface defects are well assessed. Plausible statements can be made about local atomic structures which cannot be determined by a direct measurement. Atomic-scale resolution of STM combined with the understanding of the underlying arrangement of atoms allows us to identify certain features (such as the step "kissing sites", see Fig. 3.24, page 185) as relics from the time when the surface was formed.

But given that four decades were needed to clarify the geometry and energetics of Si(001) dimers, it is not particularily strange that the knowledge about objects more complex and more difficult to study, such as defects and metastable structures, is less developed. Even though contemporary research tools have progressed dramatically, some rather fundamental questions remain open. And some of them are even about the *perfect* surface!

Is there a barrier for dimerization (Sect. 2.5.2)? Where are the backbond states of $c(4\times2)$ dimers, what is their energy dispersion (Sect. 3.2.3d)? What are the exact formation energies, diffusion barriers, and interaction potentials of dimer vacancies (Sect. 3.3.1)? Why are C-defects metallic (Sect. 3.3.1b)? What is the electronic structure of steps (page 176)? Why are translational domain boundaries so rare (Sect. 3.4.3)? These are sample questions awaiting answers. We hope that this Chapter will help the reader to pinpoint the interesting issues and place them in a broader context.

There is little doubt that one of the major motivations to investigate Si(001) is that contemporary Si microelectronic technologies, which control about 95% of the total microelectronics market, rely on (001)-oriented Si. The first commercial Si devices were produced on Si(111); the industry switched to Si(001) because SiO_2/Si(001) films and interfaces have a smaller number of hazardous defects. Nowadays, the technology is so advanced and so much experience has been gathered that Si(001) will nearly certainly remain the industrial choice. Even though other directions, such as Si(113), might ultimately turn out to be somewhat better substrates.

What technological relevance can be ascribed to studies of *clean* Si(001)? Sometimes it is pointed out that knowledge about the topology of surface reconstruction and about the rules which govern its formation is needed for understanding, predictive theoretical modeling, and efficient control of epitaxial growth. While this is certainly correct, epitaxy is viewed by technologists as a costly, complex, and unstable process. It is avoided whenever possible (which is usually the case). And when applied, it is *not* made by deposition on the atomically clean Si(001) surface, which exists only under ultra-high vacuum (UHV) conditions. UHV equipment is too expensive and too complicated to be installed in production lines. So the *direct* impact of such studies on the industrial practice is low, if any.

But the knowledge of the geometries of the clean surface, when augmented by the konwledge on surface diffusion and interactions between surface Si atoms and such species as hydrogen (Sect. 5.2), certain halogens (Sect. 5.3), or oxygen (Sect. 5.4), helps us to understand, simulate, and design such diverse processes as homoepitaxial growth of silicon films by Chemical Vapor Deposition (CVD), plasma etching, and formation of ultrathin oxide layers. This is possible because the characteristic Si(001) dimers are formed not only in UHV but also, for example, when silicon is exposed to SiH_4 vapor during CVD homoepitaxy, or when the surface cleaned from the native oxide by HF etch is heated in a furnace. The dynamical processes typical for the clean surface contribute then to the dynamics of the surface reactions. If one wants to describe physics and chemistry of these reactions, one should therefore begin from dynamics of the clean surface (Chapter 4). Knowledge on atomic arrangements and electronic structures of the perfect surface and its typical defects is prerequisite for such studies. More so when direct imaging methods derived from STM[a] are used.

Moreover, Si(001) surfaces are easily accessible to measurements and directly related to systems which, like SiO_2/Si(001) and silicide/Si(001) interfaces, are crucial to MOS technologies (page 7). Combined experimental and theoretical studies of interactions between Si(001) and atoms of various species may affect the evolution of methods useful in the simulation of technological processes. For example, comparison between experimental observations and atomic geometries computed by *ab initio* techniques and by higher-level (semiempirical and empirical) approaches helps to visualize, rationalize, and quantify the cases when these simpler methods fail (page 227). Reliable higher-level techniques can help in sorting out the physical mechanisms of implantation or segregation of dopants to interfaces, for instance. This knowledge can be converted into equations needed to model a particular technological process accurately and efficiently.[28]

Investigations of clean Si(001) are the first, the least applied link in this long chain. They are focused on understanding the elementary physical interactions between silicon atoms confined to the specific environment of the crystal boundary. Studies that purposedly explore a realistic path to technology are uncommon at this stage. We believe that they will grow in number to the satisfaction of researchers and technologists alike.

---

[a] See Sect. 1.3 for experimental and theoretical methods, Table 1.2 (p. 29) for acronyms.

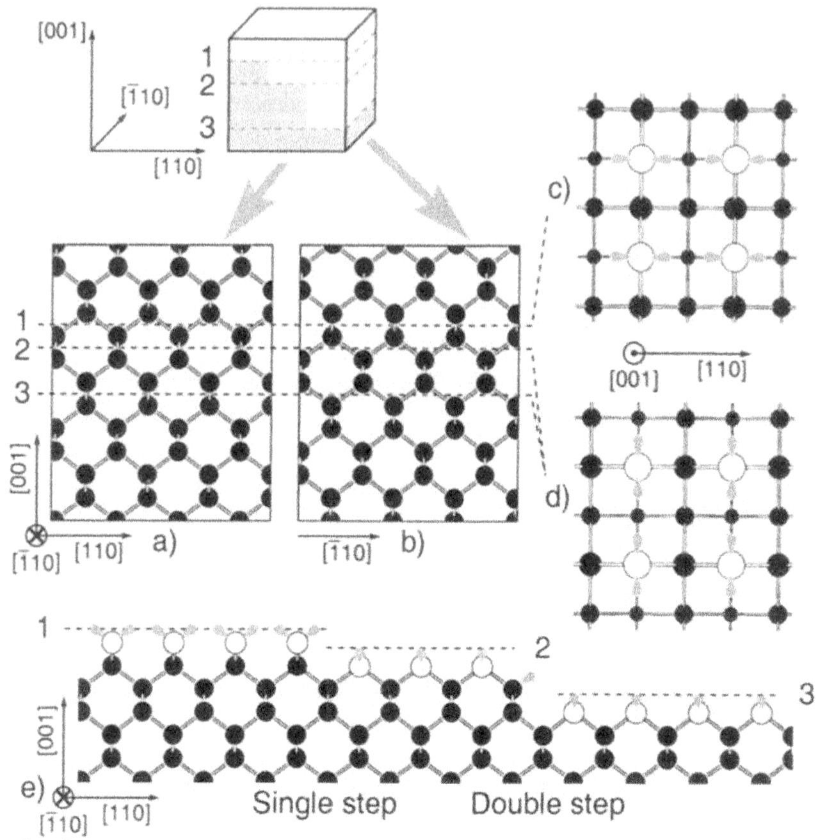

Fig. 3.1. {001}-truncated silicon. Surface atoms are white, bulk atoms are black, gray clubs are the broken bonds. **(a)** Bulk viewed along $Y = [\bar{1}10]$ and **(b)** along $X = [110]$. **(c)** Top view (more distant atoms are smaller) of the bulk truncated at the plane "1" and **(d)** at the plane "2" or "3". **(e)** Side view of a truncated-bulk surface with terraces built of planes "1", "2", and "3". The terraces are separated by steps. The step from "1" to "2" has monatomic height, the step from "2" to "3" has biatomic height.

3.2. Perfect Si(001)

We begin with the {001}-truncated bulk and the ideally reconstructed surface. Attention is paid to the buckling of dimers and its influence on properties such as surface stress anisotropy and the electronic structure.

We will often use sketches such as . This sketch depicts the ideal Si(001) viewed from the perspective of $Y=[\bar{1}10]$ (cf. the surface produced by the plane labelled "1" in Fig. 3.1e). Si atoms are symbolized by full circles; the dots mark the electrons of the broken bonds (the gray clubs in Fig. 3.1e). The dashed line, not shown in Fig. 3.1e, is drawn through the second layer. The solid lines correspond to bonds. A surface atom is backbonded to two second-layer atoms; only one backbond per atom is visible from this perspective.

3.2.1. *The {001}-truncated bulk*

Consider the ideal Si(001) 1×1 surface obtained by truncation of the bulk by a {001} plane (Fig. 3.1). Atoms with two broken bonds are arranged in a square array (Fig. 3.1c). The lattice constant is $a_s = a_b/\sqrt{2} = 3.84$ Å and corresponds to the length of the side of the square marked by four neighboring surface atoms; $a_b = 5.43$ Å stands for the bulk lattice constant (Fig. 1.1, page 5). The primitive translation vectors connect two nearest neighbors on the surface. If the normal to the surface has the direction of the [001] axis, these vectors run along [110] and $[\bar{1}10]$ directions of the bulk silicon lattice. Each surface atom is connected to the substrate with two backbonds; the bond length is $b = a_b\sqrt{3}/4 = 2.35$ Å. The distance d between subsequent atomic planes parallel to the surface is $d = a_b/4{=}1.36$ Å.

Though the 1×1 cell is a square, the point group symmetry is that of rectangle, C_{2v}. This is because the broken bonds are aligned along a single <110> axis, [110] or $[\bar{1}10]$, depending on the truncation plane (Fig. 3.1).

Macroscopic areas of any real surface must contain **surface steps**, caused either by a global misorientation from a low-index plane or by a local roughness. So imagine a surface consisting of (001) terraces separated by steps of monatomic height (Figure 3.1e). The plane "1" in Fig. 3.1 is chosen such that the broken bonds are aligned along X=[110]. Truncation at the next atomic layer (plane "2") produces an identical surface but with broken bonds oriented along $Y=[\bar{1}10]$ (compare Figs. 3.1c and 3.1d). It follows that a realistic Si(001) exhibits an intrinsic orientational **domain structure**. The domains are separated by steps of monatomic height and can be mapped onto each other by 90° rotation and translation. If a single-domain structure is observed, it means that all steps have biatomic height.[b]

The electronic orbitals of the ideal Si(001) 1×1 are often described as sp^3 "rabbit ears" (Fig. 3.2a). This is a convenient simplification which should be used with care as it describes a *basis set*, not the *eigenfunctions*.

---

[b]On Si(001), steps higher than two layers must be stabilized by defects or impurities.

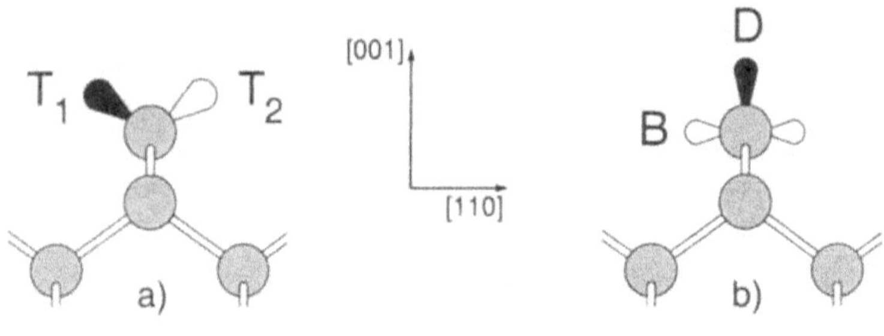

Fig. 3.2. Atomic orbitals of the ideal, {001}-truncated surface of silicon. (a) Two broken-bond sp^3 orbitals, T_1 (black) and T_2 (white). (b) Symmetrized orbitals, D (sp "dangling bond", black) and B (p_x "bridge bond", white).

Consider sp^3 hybrids projecting from each silicon atom:

$$\begin{aligned} T_1 &= \frac{1}{2}(s+p_z)+\frac{\sqrt{2}}{2}p_x \\ T_2 &= \frac{1}{2}(s+p_z)-\frac{\sqrt{2}}{2}p_x \\ T_3 &= \frac{1}{2}(s-p_z)+\frac{\sqrt{2}}{2}p_y \\ T_4 &= \frac{1}{2}(s-p_z)-\frac{\sqrt{2}}{2}p_y \end{aligned} \tag{3.9}$$

where p_x, p_y, and p_z are p orbitals pointing in the direction of $X = [110]$, $Y = [\bar{1}10]$, and $Z = [001]$. On the surface, one can assign T_1 and T_2 to the orbitals indicated by the dots in  (Fig. 3.2a). They are equivalent by symmetry and carry one electron each. The orbitals T_3 and T_4 are used by the surface atoms in the backbonds.

Bonding combinations of hybrids from Eq. 3.9 form the valence band of bulk Si. The conduction band is built of their antibonding combinations. The energy of T_1 and T_2 falls therefore within the energy gap. But T_1 and T_2 are artificial hybrides designed to describe the *bulk*. They even do not have the right symmetry to represent the surface: the symmetry of Si(001) 1×1 is too low to support degenerate states. The natural basis functions *on the surface* are the hybrides D and B built as symmetric and antisymmetric combinations of the "broken bonds" T_1 and T_2:

$$D = \frac{\sqrt{2}}{2}(T_1+T_2) = \frac{\sqrt{2}}{2}(s+p_z) \tag{3.10}$$

$$B = \frac{\sqrt{2}}{2}(T_1 - T_2) = p_x.$$

The "dangling bond" sp orbital D protrudes vertically from the surface while the "bridge bond" orbital B lies in the surface plane (Fig. 3.2b).

Both unused electrons of a surface atom tend to occupy the state D. This is because D has 50% of s character, while B is p-type (Eq. 3.10). The promotion of an electron from Si(3s) to Si(3p) costs[387] 5 eV. In the bulk, this is overcompensated by the formation of four directional bonds (page 49). The interatomic distances on the ideal Si(001) 1×1 surface are much larger than the equilibrium Si–Si bond length, too large to make the promotion profitable without structural changes.

The interaction between the orbitals of neighboring surface atoms, though not too strong, broadens them into bands. Calculations[400] show that these bands overlap. The ideal Si(001) 1×1 surface is therefore metallic, with bands spanning over the whole range of the bulk energy gap.

The metallic character of Si(001) 1×1 calls for some care if the computed energy of this surface is used as a reference value in *ab initio* slab calculations. A $\vec{k}$-point set good for a reconstructed semiconducting surface may be too small for a metallic surface (page 26). It is easy to overestimate reconstruction energies by underestmating the reference energy.

3.2.2. *Atomic structures*

We begin with a description of Si(001) dimerization and its consequences for physical and chemical properties of the surface. Next, we focus on the surface stress anisotropy $\Delta\sigma$. Structural data quoted below are obtained from DFT-LDA *ab initio* calculations (Sect. 1.3.1), unless stated otherwise. Table 3.1 lists some papers relevant to the reconstruction of clean Si(001).

a. *Dimerization*

The 2×1 reconstruction of Si(001) is caused by the fundamental geometric property of {001} planes in the diamond structure: the presence of two broken bonds per surface atom on {001}-truncated bulk (Fig. 3.1, page 130). In order to reduce the dangling bond density, the surface atoms dimerize ( → 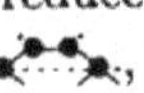, Fig. 2.3 on page 90). The dimers form a 2×1 array, being arranged in rows (Fig. 2.5b, page 93). Viewed from top, the dimerized Si(001) surface appears as: 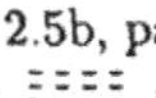, where the dashes symbolize the dimers.

The dimer rows are clearly visible in STM micrographs (Fig. 3.3a) and can be resolved[380] even by AFM (Fig. 3.3b). Holographic images of sur-

Ref year	Subject relevant to dimerization of clean Si(001)
357 1957	Dimerization proposed; LEED
411 1978	Dimerization induces deep subsurface strain; theory (valence force)
412 1978	LEED favors the dimer model
360 1979	Buckling of dimers proposed; tight-binding theory
361 1981	First *ab initio* optimization of Si(001)
286 1980	Buckling consistent with surface core level shifts
425 1981	First measurement of the buckling angle; LEED
363 1985	Dimers observed directly for the first time; STM
365 1987	Empty states of dimers; STM
368 1989	First correct *ab initio* energy order of dimer structures
507 1990	Buckling angle measured; X-ray diffraction
372 1992	Convergency analysis of *ab initio*; dimers buckle and flip fast
373 1992	Low-temperature STM sees many buckled dimers
508 1993	Buckling angle measured; transmission electron diffraction
180 1993	*Ab initio* final-state theory of core level shifts
496 1993	Interaction between dimers and W tip; *ab initio* DFT theory
374 1994	Low-temperature STM shows that buckled dimers are stable
375 1994	Stress anisotropy explained by buckling; *ab initio* DFT theory
509 1994	**Review** of stress effects
510 1995	Dimerization and epitaxial growth; Monte Carlo simulations
377 1995	**Review** and convergency analysis of *ab initio* results; DFT theory
511 1995	High-resolution core level spectra
498 1995	Buckling angle measured; core-level photoelectron diffraction
376 1995	Buckling angle measured; X-ray diffraction
499 1995	3D image of a dimer; Kikuchi electron holography
378 1996	Recent measurement of surface stress; TEM
379 1996	Flipping of buckled fimers; *ab initio* molecular dynamics, STM
380 1996	Dimers observed by Atomic Force Microscopy
381 1997	Recent buckling angle measurement, atomic positions; XRD

Table 3.1. Dimerization of clean Si(001), structural aspects: selection of references.

face atoms can be retrieved from diffraction patterns of Kikuchi electrons (Fig. 3.3c): scattering by atoms of a single class (such as dimer atoms) produces the reference beam, so that the holograms carry collective information about relative positions of surface atoms as seen from the emiter atoms. This interesting approach, though easy to implement, is rather exotic; usually, atomic coordinates are obtained from X-ray data (Sect. 1.3.2).

Dimerization (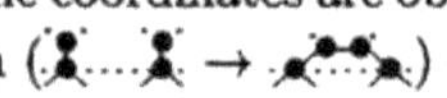) releases energy because a σ bond is created within each pair. The remaining two broken bonds of the dimer (the dots in ) are brought closer to each other and build a π bond, albeit not too strong. The effective energy of this π bond (including the lattice relaxation effects discussed later) is 0.3–0.5 eV (page 271). Dimerization distorts the backbonds, but this costs relatively little energy because only

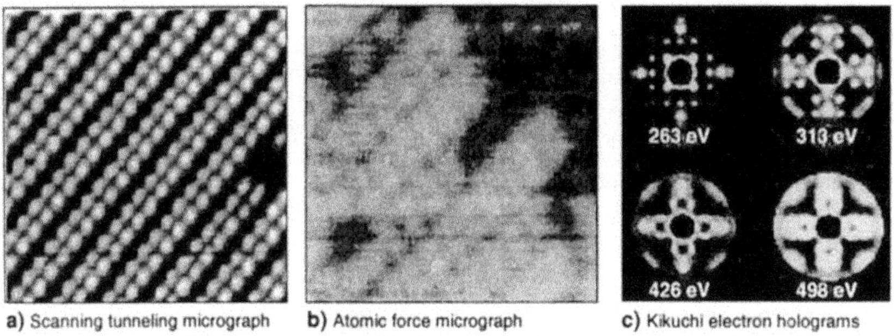

Fig. 3.3. Si(001) dimers can be detected by various methods. **(a)** A typical STM image of dimer rows; empty states.[301] **(b)** The rows resolved by AFM. A step of monatomic height is in the middle of the image. From Ref. 380. **(c)** Momentum-space diffraction patterns of Kikuchi (inelastically scattered) electrons carry holographic information. From Ref. 499.

the angles between the bonds must be altered. The backbond lengths can remain unchanged (Fig. 2.3, page 90). The forces which oppose bond bending in Si are weaker than bond-stretching forces;[c] so angular distortions are preferable energetically over similar changes in bond lengths.

According to *ab initio* data, 2×1 dimerization reduces the energy by ~2 eV per dimer. This is 80 times more than the room-temperature kT. The dimers persist up to at least 1200°C. At higher temperatures, the 2×1 order disappears, but the dimers may prevail (Section 4.3.1).

The strain field caused by dimerization extends several atomic layers below the surface. This is well established by theory and experiment. For instance, a realistic interpretation of the basic features in LEED energy spectra of Si(001) 2×1 requires that displacements of atoms in four or five subsurface layers are considered (page 98). A strain field reaching down to at least the sixth layer has been deduced from an XRD study.[376]

The dimerization is crucial for the chemical activity of Si(001), particularly in the early stages of adsorption and film growth[510] (Chapter 5). The arriving species can easily attach themselves to dimer "dangling" bonds, that is, to the orbitals involved in the weak π bonds: . The interaction with the dangling bonds may be strong enough to break the impinging molecules (Sections 5.5.2 and 5.6.2). In general, Si–Si bonds are only attacked when dimers are saturated (Sections 5.2.1 and 5.3), though notable exceptions are fluorine (Sect. 5.3.1) and oxygen (Sect. 5.4.3). This

[c] The corresponding force constants differ by roughly a factor of three.[512]

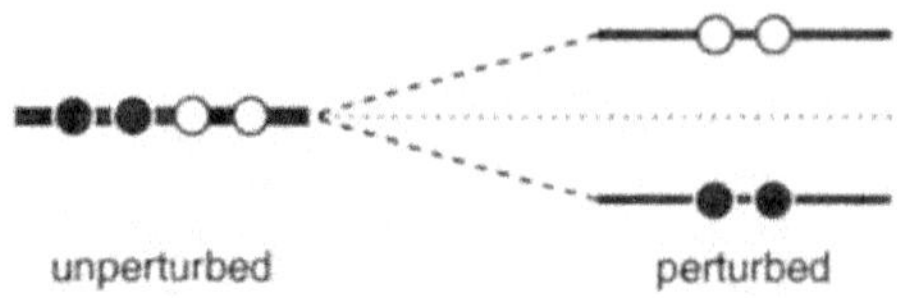

Fig. 3.4. **The mechanism of Jahn-Teller distortion. Left: a partially occupied, degenerate state. Right: Symmetry lowering lifts the degeneration, leaving the center of mass unchanged. The occupied state decreases its energy, which may lead to a decrease of the total energy and, in this way, to a spontaneous symmetry breaking.**

dependence on the adsorption mechanism on the adsorbate coverage may be reflected in such parameters as sticking coefficients or activation energies, contributing to deferred growth (incubation) or self-limited growth.

b. *Buckling of dimers*

On top of the dimerization there are some subtle but remarkable structural features. They are caused by electronic effects which break the symmetry of the dimers and by elastic effects which introduce order in the asymmetric distortions of the neighboring dimers. Though on the energetic scale these effects are approximately ten times weaker than the dimerization, they are important if only because they affect the *morphology* of the whole surface. Small as they are for a dimer, they add up and influence the long-range interaction between mesoscopic objects, surface steps.

The symmetry is broken by asymmetric buckling (, Fig. 2.3 on page 90). The dimer bond makes an angle of $\sim 20°$ with the surface. This distortion is driven by rehybridization: after buckling, the orbitals occupied by the electrons unused in bonds become more s-like. The "down" atom moves closer to the plane of its three nearest neighbors, so that its orbitals become more sp^2-like. Its "dangling" orbital acquires more p character; its energy increases, that is, its electronegativity decreases. The opposite happens to the "up" atom, which moves away from the plane of its neighbors. Its "dangling" orbital becomes more s-like, that is, more electronegative. As a consequence, the two electrons in the two "dangling" orbitals buckle the dimer in an attempt to stay longer on the "up" atom. This is opposed by the weak π bond between the "dangling" orbitals and by electrostatic forces. Buckling does not require changes of bond lengths.

While symmetric dimers are metallic, buckling opens a gap between occupied and empty surface states.[360,483] This behavior resembles Jahn-

Teller distortion.[513] Consider a point defect in the bulk, or a small atomic cluster. When its Highest Occupied Molecular Orbital (HOMO) is partially occupied and happens to be degenerate due to symmetry of the atomic coordinates, the structure may be unstable against symmetry-breaking distortions. A small symmetry-breaking perturbation would lift the degeneracy of the HOMO orbital, leaving – in the first approximation – the center of weight of the split orbitals at the unchanged energy (Fig. 3.4). The original HOMO was partially occupied, thus the sum of energies of its electrons drops. This may decrease the total energy of the whole object.

When the dimers on a dimerized terrace buckle in the same direction (Fig. 2.5d, page 93), the surface periodicity remains 2×1: , where the "up" atoms are indicated by open circles. *Ab initio* calculations estimate that in this case the surface energy drops by ~0.1 eV per dimer.

The 2×1 buckling is artificial. The direction of buckling on real surfaces alternates within each dimer row (→). This reduces the energy by about 50 meV per dimer[377] due to relaxation of local stress and, possibly, to improved electrostatics. Depending on the registry of buckling in the neighboring dimer rows, this alternation leads to 2×2 (, Fig. 4.3c, page 239) or $c(4\times2)$ periodicities (, Fig. 4.3d). The $c(4\times2)$ structure has slightly lower energy than the 2×2 structure,[514] probably by a few meV/dimer.[d] This energy difference is sensitive to the strain field below the surface. Calculations must relax at least five (001) layers of atoms in order to reproduce the surface energy of $c(4\times2)$ as lower than that of 2×2.

c. *Bond angles and dimer bond length*

Symmetric dimerization rotates each of the surface atoms by about 35° (from to). The **angles** between the bonds of surface atoms remain close to the ideal bulk value of 109.5°, whereby the angle between the two backbonds increases to ~112°, and the angle between a backbond and the dimer bond slightly decreases, to ~108°. The angles between bonds of subsurface atoms are more affected. The angle between a backbond and a bond of the neighbor in the subsurface layer decreases to ~92°.

When the dimers buckle (→) within the 2×1 periodicity (→), the angles between backbonds remain nearly as in a symmetric dimer. The angle between a backbond and the dimer bond changes to ~90° for the "up" atom and ~120° for the "down" atom. Only

[d]The thermal energy, kT, is about 25 meV at room temperature.

when the dimers relax to 2×2 or $c(4\times2)$ arrangement (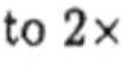 or 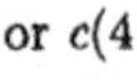), the subsurface atoms are not constrained by the symmetry and yield to the forces exerted by the bent backbonds (Section 3.2.2d). The angles between the backbonds on Si(001) 2×2 change to ~99° ("up" atom) and to ~120° ("down" atom). The angles on Si(001) $c(4\times2)$ are nearly the same.

It is well established by theory and experiment that the buckling angle (the angle between the dimer bond and the surface plane) is close to 20° (see Section 2.4.2, which also lists some recent studies). The situation appears to be more complicated when the dimer **bond length** is concerned.

X-ray spectroscopy studies reported in Ref. 498 yielded $\alpha = 19°$ for the buckling angle and $d = 2.25$ Å for the dimer bond length. Similar results were obtained in Ref. 376: $\alpha = 20 \pm 3°$ and $d = 2.37 \pm 0.06$Å. These values coincide with converged *ab initio* calculations for the $c(4\times2)$ structure. For example, Ref. 481 finds $\alpha = 17.5°$ and $d = 2.29$ Å, while Ref. 377 finds $\alpha = 18.8°$ and $d = 2.29$Å. The latter work, published in 1995, contains an elaborate discussion of numerical convergency and lists converged atomic positions computed for various arrangements of dimers.

Atomic positions in the six topmost layers have recently been determined by X-ray diffraction.[381] The study yielded the buckling angle $\alpha = 20°$ and the bond length $d = 2.64$ Å. This value of d exceeds the lengths determined by the other methods by about 10%; the reason for this is unclear.

d. *Surface stress*

Elastic interactions have an important share in the instability of the 2×1 order of buckled dimers. Dimerization strains the surface anisotropically. The anisotropy of the stress tensor σ_{ij} is strongly reduced by subsurface relaxation in 2×2 and $c(4\times2)$ structures.

Stress is a concept similar to pressure. It measures how the total energy of the surface, E, changes with the strain ϵ, that is, with the relative deviation of a lattice dimension from its equilibrium value. Each of the two lattice dimensions can be distorted in two independent lateral directions, so the surface strain is a tensor: $\epsilon = \epsilon_{ij}$. The surface stress becomes:

$$\sigma = \sigma_{ij} = \frac{1}{A}\frac{dE}{d\epsilon_{ij}}, \tag{3.11}$$

where A is the area of the surface, $E = A\gamma$ is the total surface energy, and γ is the surface energy density (Sect. 1.4.1). The positive value of σ_{ij} corresponds to tensile stress, that is, to the situation when the surface would like to shrink (it feels tensed).

The stress anisotropy can be measured[367] by monitoring the intensity of LEED 2×1 and 1×2 spots when the sample is bent such that the strain is directed along [110] or [1$\bar{1}$0]. The bending distorts the terraces in 2×1 and 1×2 domains in mutually perpendicular directions, across and along dimer bonds. When the temperature is high enough to allow the exchange of atoms between the domain boundaries (that is, between monatomic steps, Sect. 3.3.2) and when the surface stress is anisotropic ($\sigma_{\parallel} = \sigma_{xx} \neq \sigma_{yy} = \sigma_{\perp}$), the domain favored energetically by the strain grows at the cost of the other domain. The theory describing the population changes was developed in Ref. 56 on the basis of Ref. 515.

First, we consider the surface with **symmetric** dimers:  (side view),  (top view, dashes mark dimers). It feels stretched along the dimer bonds: , that is, $\sigma_{\parallel}$ is tensile ($\sigma_{\parallel} > 0$). This reflects the elastic reaction of backbonds opposing the dimerization: . The arrows indicate the force exerted by the bent backbonds. The surface would like to shrink along the dimer bonds in order to minimize the bending.

In the direction perpendicular to the dimer bonds, the backbond bending produces stress of the opposite sign. The surface feels compressed across dimer rows: , $\sigma_{\perp}$ is compressive ($\sigma_{\perp} < 0$). This is because the surface atoms, having three neighbors, prefer sp^2 hybridization to the sp^3-like hybridization imposed by the dimer geometry. The angles between sp^2 hybrides (120°) are larger than the angles between the backbonds (~112°). The surface would like to expand across the dimer bonds in order to accommodate the tendency of the backbond "scissors" to open.

The anisotropy $\Delta\sigma = \sigma_{\parallel} - \sigma_{\perp}$ of symmetric dimers computed[487] within DFT-LDA reaches 150 meV/Å^2. When electrons are confined to the π band[375] (Fig. 3.5, page 144) in order to simulate a semiconducting surface in a DFT-LDA calculation,[e] it increases to 195 meV/Å^2.

The anisotropy is sensitive to the **buckling** angle. The simplest structure built of buckled dimers (, side view) is 2×1, with all dimers buckled in the same direction: . In this sketch of the top view of the surface, dashes are dimers and full circles are their "up" atoms. 2×1 buckling makes the $\sigma_{\parallel}$ component even more tensile, by 30 meV/Å^2. This happens because the dimer must pull harder on the backbonds to maintain a shortened projection of the dimer bond on the surface plane. The

[e] See page 20, 120, and Sect. 3.2.3c for DFT-LDA problems with electronic structures.

$\sigma_\perp$ component becomes less compressive, by 75 meV/Å^2. This reduction occurs because the "up" atoms tend to dehybridize towards s^2p^3, and the angle between p orbitals is 90°, less than the angle between backbonds of the "up" atoms.[f] The net effect is that the stress anisotropy $\Delta\sigma = \sigma_\parallel - \sigma_\perp$ decreases[487,486] to 105 meV/A^2.

When the surface relaxes to 2×2 () or $c(4\times2)$ (), the buckling angle is nearly unaffected, hence the tensile stress $\sigma_\parallel$ is barely changed. But the compressive (negative) $\sigma_\perp$ is reduced[487] in 2×2 by +50 meV/Å^2. This is due to relaxation of atoms in the first subsurface layer. The angles between the backbonds change to ~99° for the "up" atom and to ~120° for the "down" atom. These values are closer to the ideal values of 90° (s^2p^3) and 120° (sp^2) than the angles of 107° and 111° in the buckled 2×1. Since the relaxation is chiefly confined to dimer rows, the stress anisotropy for the $c(4\times2)$ surface[486] () is virtually the same as for the 2×2.

The alternation of the buckling direction along the dimer rows has thus a profound influence on the stress anisotropy. In the 2×2 and $c(4\times2)$ superstructures, the anisotropy falls down to 55 meV/Å^2, which is only about a third of that in the 2×1 array of symmetric dimers.

The calculations quoted so far considered perfectly ordered surfaces at 0 K. When the temperature is increased, the dimers vibrate and may flip between  and . Around 200 K (−73°C), the ground-state c(4×2) structure transforms[451,516] into a disordered 2×1 surface (the transition takes place in the temperature range of $\Delta T \simeq 100$ K; see page 238 for details). Since the dimers tend to flip collectively,[382] the disorder appears as an increased number of antiphase boundaries between different $c(4\times2)$ domains rather than as dimers randomly buckled. Some order prevails even at temperatures at which Si atoms are exchanged between steps. *Ab initio*-based estimates indicate that even at such high temperatures (~500°C[367]) the stress anisotropy is two times smaller than it would be for symmetric dimers. The calculated[375] $\Delta\sigma$ = 70 meV/Å^2 at 500°C coincides with $\Delta\sigma = 70 \pm 10$ meV/Å^2 measured in Ref. 494.

Interestingly, the same value of $\Delta\sigma$ = 70 meV/Å^2 was obtained in tight-binding calculations which provided the first theoretical interpretation of the experimental data. This was a rather fortuitous agreement. Subsequent numerical estimates arrived at larger anisotropies $\Delta\sigma$, chiefly

[f] However, the tendency of the "down" atom to become sp^2 is now stronger than in a symmetric dimer, so the sign of the $\sigma_\perp$ change is not obvious without a calculation.

because they used a wrong assumption about the dimer geometry — symmetric dimers. The agreement between theory and experiment was restored when the calculations became able to predict the correct dimer geometry.

And **how exact** is the measured anisotropy? The answer is not straightforward, and we will devote the rest of this Section to this problem.

The measured[494] $\Delta\sigma$ was quoted above as "70 ± 10 meV/Å^2". But this "experimental" value depends on certain theoretical assumptions. The related uncertainty is *not* included in the ±10 meV/Å^2 error bar.

Ref. 494 did not measure $\Delta\sigma$, but rather $\sigma_0/\Delta\sigma = 49 \pm 7$ Å^{-1} which determines the initial slope of the dependence of the domain population on the strain.[494] From elasticity theory, $\sigma_0 = 2\pi l\mu/(1-\nu)$, where μ and ν are the bulk modulus and the Poisson ratio and l is the spacing between two domains of the same orientation.[56] The catch is that it is unclear to what extent bulk parameters are valid close to the surface. Calculations with classical potentials found[517] an effective $\mu/(1-\nu)$ by factor 0.64 smaller than in the bulk. This would scale $\Delta\sigma$ down to 45 meV/Å^2.

Assuming $\Delta\sigma/\lambda_\sigma = \sigma_0/\Delta\sigma = 49 \pm 7$ Å^{-1}, one can use the constant $\lambda_\sigma = (\Delta\sigma)^2/\sigma_0$ to verify the value of $\Delta\sigma$. This constant measures interactions between monatomic steps. It can be deduced from STM images (Sect. 4.3.2). The results for λ_σ reported in Ref. 518 correspond to $\Delta\sigma$ ranging from 26 to 35 meV/Å^2. But with λ_σ obtained in Ref. 519 the anisotropy $\Delta\sigma$ becomes 60 to 175 meV/Å^2.

It appears that the range of $\Delta\sigma$ compatible with the available experimental data (from 26 to 175 meV/Å^2) is so broad at the moment that it covers and extends beyond the anisotropies computed for 2×2, c(4×2), and 2×1 buckled dimers, and even for symmetric dimers.

The spread of experimental data notwithstanding, the calculated value of $\Delta\sigma$ is only an estimate from a static, 0 K case. The real situation at 500°C may be more complex. For example, *ab initio* molecular dynamics calculations[500] show that the dimers which are about to flip acquire a configuration with small buckling but with significant lateral distortions (shift along the dimer row and twist with respect to the dimer row axis, reminescent of Fig. 2.3e, page 90). Contribution from such configurations to the stress anisotropy is unknown. Moreover, the surface contains numerous defects (page 155) which may influence the stress tensor.[495] The defects may also screen the elastic interaction between steps, adding more complication to the experimental determination of surface stress properties.

In brief, such a tiny structural detail as the buckling of dimers has a profound effect on surface stress anisotropy $\Delta\sigma$. Since this anisotropy is responsible for long-range interactions between monatomic steps, and since the effect of buckling on $\Delta\sigma$ prevails at temperatures at which the steps can move, it becomes clear that the buckling – possibly along with surface defects and dynamical effects of dimer flipping – influences processes which depend on step-step interaction. By reducing the stress anisotropy, it reduces the instability of large flat terraces against roughening of the surface by wavy monatomic steps (Sect. 4.3.2). It is also responsible for unequal sizes of 2×1 and 1×2 domains on surfaces under elastic stress (Sect 4.3.2). This manifests itself on surfaces strained by external load,[494] on surfaces strained by thermal treatment,[520] and around macroscopic surface defects which pin step motion during step flow.[520]

3.2.3. *Electronic structures*

As expected for an effect driven by rehybridization of surface electrons, the buckling of dimers strongly affects the electronic structure of the surface. Here we focus on the electronic features energetically close to the Fermi level E_F. Some of them affect the measured core level energies. Spatial distribution of electronic states close to E_F is reflected in STM images. Table 3.2 lists selected papers related to the elecronic structure of the clean Si(001), including STM and surface core level spectroscopy (SCLS).

The electronic structure of the clean Si(001) surface has been extensively studied. There are good reasons for this interest. First, electronic and atomic structures are closely related. Investigations of electronic properties have led to the rejection of a number of Si(001) models and initiated and later supported the idea of dimer buckling (Chapter 2). Second, the frontier surface orbitals, particularly those in the vicinity of the Fermi energy, are responsible for the chemical activity of the surface. In order to trigger a reaction, electrons must be exchanged between the incoming species and the substrate. Reactions on the surface are therefore influenced by the location of the occupied or empty orbitals of the surface atoms and by their extension in space and in energy[521,522] (page 55).

The electronic band structure is routinely measured by photoemission (occupied states) and inverse photoemission (empty states). Scanning tunneling microscopy and spectroscopy provide spatially-resolved data. Core level spectroscopy yields (indirect) information on the atomic geometry through the influence of the local environment on the energy of core elec-

Ref year	Subject relevant to Si(001) electronic structure
397 1973	Surface states identified; EELS
399 1974	Density of filled states measured; photoemission spectroscopy
402 1976	Dimers consistent with measured density of occupied states; theory
359 1979	The surface is semiconducting; Angle-resolved photoemission
360 1979	Buckling opens surface band gap; tight-binding theory
286 1980	The first core-level investigations of Si(001)2×1; SCLS
420 1980	The first *ab initio* results for buckled dimers
523 1981	Two-domain 2×1, occupied states, ΓJ, ΓJ′; JK; ARUPS
369 1990	Single-domain 2×1, occupied states, ΓJ, ΓJ′; ARPES
524 1990	Single-domain c(4×2), occupied states, ΓJYY′Γ, ΓJ′; ARUPS
525 1991	**Review** of angle-resolved photoemission results
372 1992	Dependence of the indirect band gap on buckling angle; *ab initio*
371 1992	Single-domain 2×1, empty states, ΓJ, ΓJ′; inverse photoemission
470 1992	Identification of the four main core level peaks; SCLS
476 1992	Initial-state theory of Si(001) SCLS; *ab initio* LDA
180 1993	Final-state theory of Si(001) SCLS; *ab initio* LDA
481 1993	Occupied and empty states along $\Gamma J_2'$; *ab initio* GW
522 1993	*Ab initio* hardness and softness; study of Si clusters
131 1994	**Review** of angle-resolved photoemission results
482 1994	GW correction to LDA is basically a rigid shift; ΓJ and ΓJ′
72 1995	Buckled 2×1, *ab initio* LDA and GW bands along $\Gamma J K J' \Gamma J_2'$
483 1995	Surface band gap is due to buckling; *ab initio* GW theory
491 1995	Occupied and empty states along ΓJYY′Γ; *ab initio* LDA
511 1995	Occupied states measured along $\Gamma J_2'$ consistent with theory
479 1996	Reflectance Difference Spectroscopy is sensitive to buckling order

Table 3.2. Electronic structure of Si(001), including core levels: selection of references.

trons. Reflectance Difference Spectroscopy (RDS) is sensitive to local ordering of buckling dimers.[479] RDS spectra were computed *ab initio* with a quasiparticle correction[481] to LDA for 2×1 () and 2×2 () buckling and compared to measured room-temperature data. A mixture of 2×1 and 2×2 was needed to reproduce the experimental results.[479]

In our discussion we will focus on the electronic structure near the Fermi level. First, we will analyze the band structure of the 2×1 reconstruction. The c(4×2) reconstruction, considered next, introduces only small (though important) changes to this band structure. We will not address states which, like the backbond states, lay deeper in the valence band. These features are covered in the review article by Uhrberg and Hansson.[525] Here we will only mention some unresolved problems.

a. *Electronic states of a dimer and the 2×1 bands*

Let us return for a while to Figure 2.3 on page 90. When two neighbor-

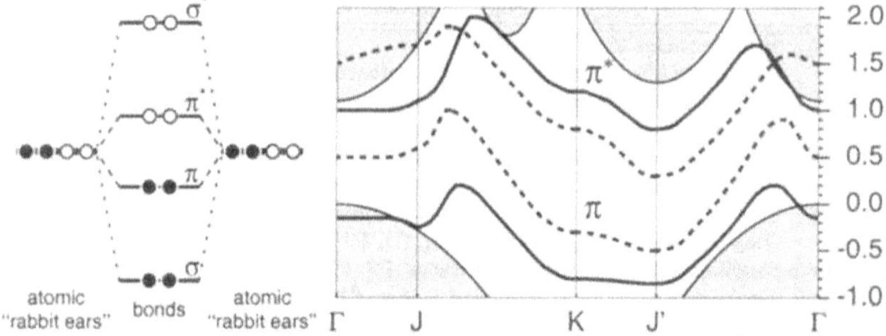

Fig. 3.5. **Electronic structure of Si(001) dimers. Left: Interaction between dangling orbitals of an isolated dimer produces bonding σ and π, and antibonding σ^* and π^* states. Right: Interaction between dimers broadens these states into bands. A sketch of the computed electronic structure of Si(001) 2×1 (see, for example, Refs. 72, 491, 483). Broken lines: symmetric dimers. Solid lines: buckled dimers. Thin solid lines are the bulk band edges. Names of Brillouin zone points as in Fig. 3.6.**

ing atoms of the {001}-truncated surface move towards each other, two of their sp^3 hybride orbitals overlap. They mix and build two combinations: bonding σ and antibonding σ^* (Fig. 3.5, left). The overlap is large and the mixing potential is strong, so the splitting ΔE_σ between the bonding and the antibonding states is considerable. The bonding σ state is occupied with two electrons and becomes a resonance in the valence band. The empty antibonding σ^* state is a resonance in the conduction band.

The other two "dangling bonds" of the atoms (the dots in ) are roughly parallel to each other (Fig. 2.3c) and mix into a π-type bond. Since the overlap between these orbitals is small, the energy splitting ΔE_π between the bonding π and the antibonding π^*, $\Delta E_\pi \approx 0.5$ eV, is much less than ΔE_σ. The electronic structure of an isolated dimer is thus such as in the left panel of Fig. 3.5. The four broken bonds bring four electrons, so the bonding states are occupied and the antibonding states are empty.

When the dimers are arranged periodically, these sharp states broaden into bands. The small energy splitting ΔE_π is less than the dispersion of the π and π^* bands in the direction along the dimer rows. The result (Fig. 3.5, right panel, broken line) is that the bonding and the antibonding π bands partially overlap. The surface becomes metallic, with E_F crossing both energy bands. This is in conflict with the measured band structure of Si(001) which shows no states at the Fermi energy E_F. The clean, perfectly dimerized Si(001) surface is semiconducting, that is, the π band is completely occupied and the π^* band is completely empty. These two bands are separated by a band gap of ~0.5 eV. State-of-the-art *ab initio* GW calculations confirm that this gap is caused by buckling of the dimers (); the π and π^* bands of symmetric dimers () partially overlap.[483]

The electronic structure of clean Si(001) close to the Fermi level is thus dominated by the "dangling" orbitals of the dimer atoms. While the states due to the surface bonds (the dimer bond and the backbonds) are almost completely hidden as broad resonances in the valence band,[369,525] the two "dangling" orbitals form two bands which are partially localized in the energy range forbidden for the bulk electrons (Fig. 3.5, right). Both bands are dispersive along the dimer rows (the longer edges of the 2×1 Brillouin zone) and flat along the dimer bonds (the shorter edges of the zone). This shows that the electrons in different rows are virtually decoupled.

b. *The c(4×2) surface bands*

The ordering of dimers which has the *lowest* energy is c(4×2), not 2×1. That is, the dimers are arranged as 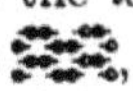, not as 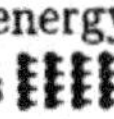 (see also Figures 2.5e and 2.5d on page 93). The direction of buckling in a c(4×2) domain alternates from dimer to dimer along and across the dimer rows.

The c(4×2) LEED pattern changes into 2×1 at ~200 K (Sect. 4.3.1), but the c(4×2) order is locally preserved.[382] The 2×1 phase is in fact a disordered c(4×2). The higher-order reflexes disappear from the LEED pattern due to a high density of antiphase domain boundaries. Since the local periodicity of the surface remains c(4×2), the room-temperature band structure of Si(001) should be interpreted in terms of the c(4×2) Brillouin zone (BZ). When represented in the 2×1 BZ, it shows not only the two bands (π and π^*) expected for 2×1 dimers, but also additional two bands. These are "echos" caused by a higher local periodicity, as explained below.

Rearranging the dimers from 2×1 to c(4×2) changes the electronic structure. The most dramatic of these changes have their origin in the reorganization of the reciprocal lattice. In particular, there are reciprocal vectors that are distinct in the 2×1 Brillouin zone but equivalent in the c(4×2) BZ. As an example, consider the J and J′ points (Fig. 3.6). The electronic states $\psi_\pi(\mathrm{J})$ and $\psi_\pi(\mathrm{J}')$ from the π band of the 2×1 surface have different momenta $\vec{k}(\mathrm{J})$ and $\vec{k}(\mathrm{J}')$ and different energies. When the periodicity becomes c(4×2), we get $\vec{k}(\mathrm{J}) = \vec{k}(\mathrm{J}')$. The $\psi_\pi(\mathrm{J})$ and $\psi_\pi(\mathrm{J}')$ states are now folded into the same $\vec{k}$-point, that is, a photoemission experiment sees them in different energies but at the same momentum.

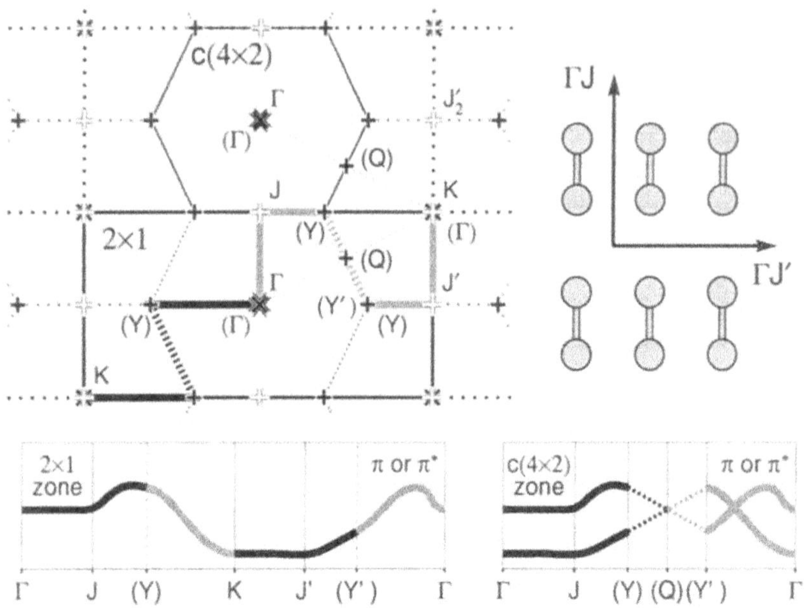

Fig. 3.6. **Upper panels**: Relation between Brillouin zones of 2×1, thick lines, and c(4×2), thin lines. Names of c(4×2) special points are given in brackets. In the c(4×2) periodicity, (Γ)≡K, (Y)≡(Y′), and J≡J′. The heavy lines joining Γ and K follow a path equivalent to one that circles around the irreducible wedge ΓJKJ′Γ of the first 2×1 zone. They go also around two equivalent (Γ)J(Y)(Q)(Y′)(Γ) wedges of neighboring c(4×2) zones. **Lower panels**: A sketch of a 2×1 π or π^* band on the edge of a 2×1 irreducible zone wedge (left) and folded to the edge of a c(4×2) wedge (cf. the heavy lines in the upper panel). The section (Y)(Q)(Y′) is walked twice, the pairs ΓY, KY′ and YK (Y′)Γ are equivalent in c(4×2). The c(4×2) potential repels the bands at the crossing points.

Figure 3.6 illustrates the relation between the Brillouin zones of 2×1 and c(4×2). The centers of c(4×2) zones coincide with the centers Γ and corners K of 2×1 zones. In the c(4×2) zone, the time reversal symmetry maps the point Y onto Y' and the point J onto J'. Hence, when the 2×1 zone is folded into the c(4×2) area, we get $K \equiv \Gamma$, $Y \equiv Y'$, and $J \equiv J'$.

Band structures are often shown on the edges of Brillouin zone irreducible wedges. Fig. 3.6 explains how the edge of the c(4×2) irreducible wedge unfolds into a 2×1 wedge edge and two equivalent Y–Y' sections.

Apart from the BZ folding effects mentioned above, three prominent changes occur in the band structure when the dimers reorder from 2×1 to c(4×2). Firstly, the band gap opens slightly due to a marginally greater buckling angle. Secondly, the folded band branches (subbands) repel at Γ. This repulsion is substantial in the π^* band, for which the Γ and K energies are very close in 2×1 geometry (Fig. 3.5, right panel). This effect is far less pronounced in the π band, for which this distance is already big in 2×1. Thirdly, the points where the subbands cross become anticrossing points where the subbands repel. All these details can be recognized in the experimental and theoretical spectra, as discussed in the following Section.

c. *Interpretation of photoemission data*

A selection of experimental and theoretical results is displayed in Fig. 3.7. Below we explain how this figure was compiled from different sources of data. We will then analyze the origin of the energy bands.

Figure 3.7 compares results of three different experiments. The location of the surface bands with respect to the bulk valence band maximum (VBM) is known only for one of these studies,[369] done by ARPES.[g] The positions of the bands measured by ARUPS[524] are defined with respect to sample-dependent Fermi level. We thus shift the ARUPS bands such that the A states obtained by ARUPS and ARPES coincide at Γ. Similarily, the empty surface states measured with IPE[371] are shifted in such a way that the IPE peak of the S_1 subband at Γ coincides with the bottom of the empty surface band at Γ (F state) which was detected in n-type samples in the ARPES study. This results in a band gap of 0.54 eV between the bulk VBM and the π^* band, in agreement with various spectroscopic methods which place the lowest empty state 0.3-0.5 eV above the VBM.[526]

Fig. 3.7 compares these data to the results of *ab initio* DFT-LDA calculations.[491] A comparison of LDA eigenvalues with the measured electronic states may be difficult because the position of the bands is sensitive to some details of the exchange and correlation potential which are not properly represented by the LDA. This is largely offset by a quasiparticle correction, for example, in the GW approximation (pages 22 and 120). *Ab initio* GW calculations[481] of the electronic structure of the clean Si(001) c(4×2) along the $\Gamma J_2'$ line indicate that the necessary correction to the LDA eigenvalues is mainly a rigid shift; the width of the surface bands is affected by less than 0.1 eV. This is corroborated by comparison of LDA and quasiparticle-corrected band structures for the 2×1 surface.[482,72] We shifted the π and π^* pairs of the LDA bands so that the upper occupied

[g] See Section 1.3 for experimental methods and Table 1.2 (page 29) for acronyms.

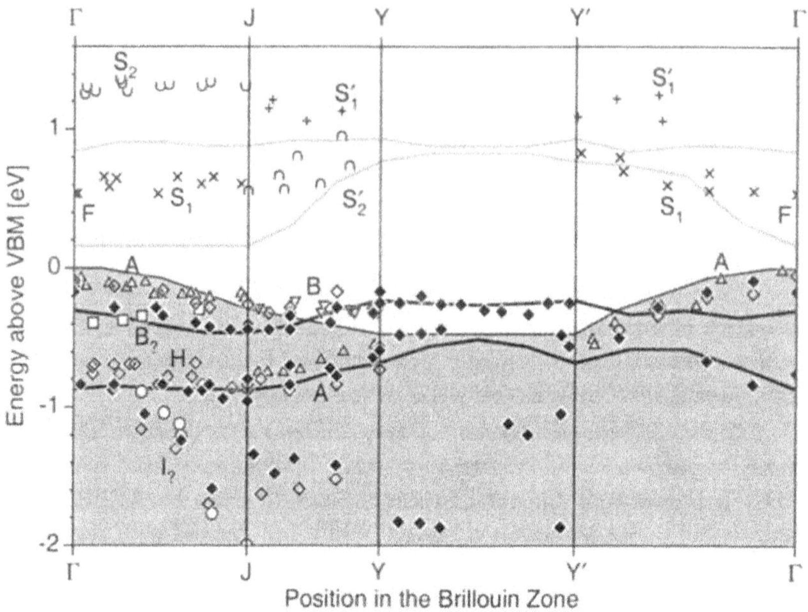

Fig. 3.7. Measured (symbols, photoemission data) and calculated (lines, *ab initio* DFT-LDA) surface band structure of the clean Si(001) surface in vicinity of the Fermi level. The states are shown on the edge of the irreducible part of the $c(4\times2)$ Brillouin zone (Fig. 3.6). For meaningful comparison, the measured as well as the calculated bands are aligned as described in the text (page 147). The shadowed area marks projected valence bulk states. Heavy diamonds[524] (occupied surface states) represent data acquired between 200 and 120 K on a sample with a single-domain $c(4\times2)$ LEED pattern. The remaining data was collected at room temperature on samples showing a single-domain 2×1 LEED pattern. Closed symbols (triangles, squares, and circles after Ref. 524, diamonds after Ref. 369) refer to occupied states, "open" symbols[371] (crosses and hoofs) refer to empty states. Sections of the measured bands are labelled following the traditional 2×1 convention; the subscript "?" indicates bands of unclear origin (page 149). The DFT-LDA dangling-bond bands were computed[491] for the $c(4\times2)$ reconstruction.

and the lowest unoccupied states at Γ coincide with the GW results.

It is evident from Fig. 3.7 that the occupied π states are properly described by the $c(4\times2)$ dimers. The average energy and the dispersion of the measured and calculated bands agree within the accuracies of the experiment and the theory. These accuracies seem to be comparable.

The quantitative agreement is worse for the empty states. Although theory and experiment yield bands of comparable shapes, the computed states are about 0.2 eV too low in energy in the ΓY section. Yet a difference of 0.2 eV is not critical: the accuracy of the GW data which we used to locate the LDA states at the Γ point of Fig. 3.7 was estimated[481] as 0.1-0.2 eV. Moreover, quasiparticle corrections depend usually on $\vec{k}$. Shifting the LDA bands by a $\vec{k}$-independent value is therefore a rough approximation. It introduces an additional uncertainty of approximately 0.1 eV.

The difference between the electronic structures measured at room temperature on nominally 2×1 surfaces and measured at low temperature on a c(4×2) sample is minimal. This is consistent with the opinion that the c(4×2) persists locally even when electron diffraction shows a 2×1 pattern.

d. *Photoemission data: open questions*

Is the electronic structure of Si(001) understood? Largely, but not completely. At least two details remain unclear in Fig. 3.7 (page 148). Firstly, it is not obvious what is the origin of the feature labelled "$B_?$" (four white squares, ΓJ line). The $B_?$ branch, found in the 2×1 sample examined in the ARPES study, was not detected in the 2×1 sample examined by ARUPS. On the other hand, $B_?$ coincides with the position of the A band of the c(4×2) sample. It is thus possible that the ARPES samples for some reason contained larger c(4×2) domains than the ARUPS samples. This is plausible, because the dynamics of buckling is known to be affected by surface defects, that is, the domain structure is dependend on the sample preparation method (see Fig. 4.4 on page 244).

The second problem is the origin of the branch labelled "$I_?$" in Fig. 3.7. It splits off the H band on the ΓJ line and goes steeply down, probably levelling out on the JY line. It may have its source in the backbond B_1 band calculated for 2×1 buckled dimers[423] but, though the B_1 band appears at approximately the same energy as the $I_?$ band, its dispersion is qualitatively different: it bends *upwards*, not downwards. Can this discrepancy be removed in the c(4×2) structure? Alternating direction of buckling does affect the backbonds strongly and may change the dispersion of the B_1 band. Yet, no calculations for the dispersion of this band have been reported for the 2×2 or for the c(4×2) reconstruction.

e. *Scanning tunneling microscopy*

At room temperature, a dimer flips rapidly[363,372] between  and . This occurs as migration of soliton waves representing boundaries between c(4×2) phase domains.[382,481] Room-temperature STM shows therefore symmetric dimers.[363,372,379] Only near to certain surface defects

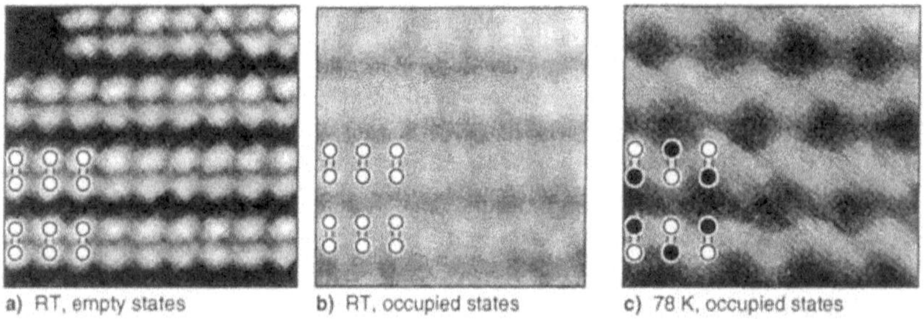

Fig. 3.8. STM of Si(001); some dimers are indicated as dumbbells. (a) Empty states,[301] π^* dangling bonds. Room temperature (RT). The dimers look symmetric. (b) Occupied states,[301] π bonding state. Room temperature. The dimers look less symmetric. Enhanced symmetry of empty states may be induced by the STM tip. (c) Occupied states at 78 K, from Ref. 480. The dimers buckle, "down" atoms are sketched as black circles.

(C-defects, page 165; steps, Sect. 3.3.2) is this flipping suppressed and buckling can be imaged. It is difficult to distinguish between an average between two images computed for dimers buckled in both direction, as "$\frac{1}{2}$(+)", and an image computed for a symmetric dimer ( ). In both cases the empty states resemble pairs of double, circular spots (Fig. 3.8a) and the occupied states produce ovals (Fig. 3.8b). The circular spots mark the "dangling" orbitals (dots in); more precisely, they are the images of their empty, antibonding combination π^*. The ovals correspond to the π bonds formed by the "dangling" orbitals.

STM can observe buckled dimers at low temperatures.[373] The images then display areas of $c(4\times2)$ and $p(2\times1)$ ordering (Fig. 3.8c, page 150), whereby the $c(4\times2)$ dominates and the $p(2\times1)$ mainly appears as antiphase boundaries between $c(4\times2)$ domains. Bright spots in the images of occupied states correspond to "up" atoms of the dimer ( ). Bright spots in the images of empty states (not shown) correspond to "down" atoms ( ).

Low-temperature images prove[374] that buckling of dimers decreases their total energy. The key observation is the influence of C-defects on buckling. A C-defect (Sect. 3.3.1b) contains a pair of dimers buckled in the same direction. Consequently, it imposes the buckling phase in the chains of alternatively buckled dimers which extend from both sides of the defect. It acts as a stiff antiphase boundary between two halves of a dimer row, each buckled with the opposite phase. The buckling phase imposed on one side of a C-defect in a $c(4\times2)$ surface must thus mismatch the phase

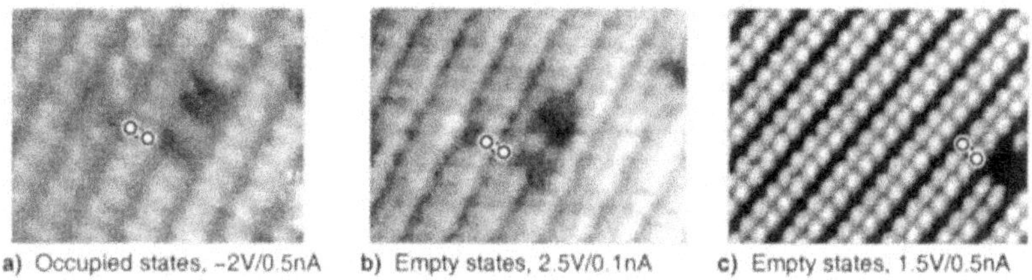

Fig. 3.9. Voltage-dependent STM micrographs[301] of Si(001) dimers at room temperature. The location of dimers can be deduced from the location of the image of a dimer vacancy complex. (a) Occupied states, −2 V/0.5 nA: dimer bonds are centered below the middle of the ovals. (b) Empty states, 2.5 V/0.1 nA: since the "dangling" bonds of dimer atoms (Fig. 2.3c, page 90) are imaged far away from the surface, the images of bonds belonging to dimers which are neighbors across the rows nearly merge into bands. Dimer bonds are centered between the bands. (c) Empty states, 1.5 V/0.5 nA: the tip is now closer to the surface, the images of the "dangling" bonds of a single dimer are less split, and dimer bonds are centered in the middle of each pair of spots.

of the matrix. The dimers in the mismatched row feel frustrated and are more prone to flipping than other dimers. The result is that the frustrated dimers are observed as symmetric even at low temperatures. This effect decays with distance, being finally overcome by the influence of the matrix.

The relation between the position of the bright rows observed in STM micrographs and the position of dimers depends on the imaging conditions. This is illustrated in Figure 3.9, in which a surface defect (dimer vacancy, Section 3.3.1a) can be used as a reference point. When the occupied states are imaged, the STM rows coincide with the dimer rows (Figure 3.9a). The same is true for the empty states, provided that the tip is kept close to the surface (Figure 3.9c).

But when the tip scans the empty states from a large distance, the bright rows are centered *between* the dimer rows (Fig. 3.9b). This is because the empty state is the antibonding π^* surface band (Sect. 3.2.3a). It has a node above the center of the dimer and its shape resembles the "dangling" bonds (the clubs in Fig. 2.3c, page 90). The distance between the "clubs" of neighboring rows decreases with increasing height, so that in the image the spots belonging to each dimer move apart.

Scanning tunneling spectroscopy (STS), that is, I–V characteristics of the sample-tip tunneling diode, is a source of information about the energy distribution of the surface density of states, $\rho(E)$. Though it is difficult to extract from the I–V curves the exact $\rho(E)$, one can easily measure

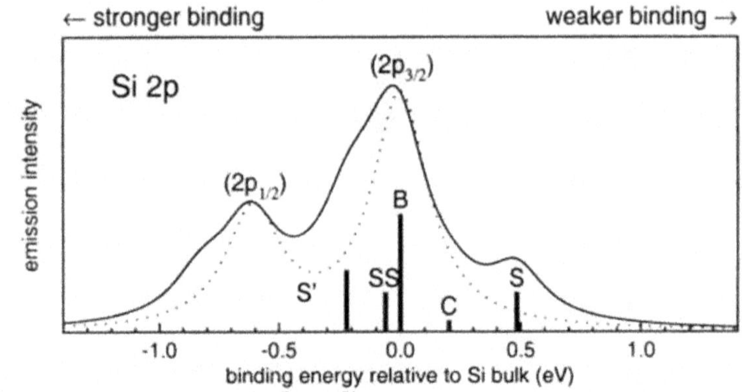

Fig. 3.10. A schematic representation of the surface core leve shift spectrum (solid line) of the clean Si(001). A positive shift means that the core state is less bonded than in the bulk. Solid line: $c(4\times2)$ spectrum. The bars indicate positions and relative heights individual lines recorded for the Si $2p$ level with a photon energy of 130 eV at 0° emission angle.[470] The dotted line and the labels in brackets show the bulk $2p$ component. The spin-orbit splitting between the $2p_{1/2}$ and $2p_{3/2}$ partners is 0.608 eV.

positions of the energy band edges.[527] The surface band gap obtained in this way for the clean Si(001) is 0.5 eV.

f. *Core states and surface core level shifts*

Surface Core Level Spectroscopy (SCLS) studies have made an important contribution to the unraveling of the atomic structure of the Si(001) surface (see Chapter 2). Here we summarize the main results of these studies.

Four SCLS peaks[470] (Fig. 3.10) are typical for clean Si(001). The peak at $+0.483 \pm 0.08$ eV, labelled S, comes from the "up" atom. The positive sign means that the state of the surface atom is less bonded than the corresponding state of a bulk atom. The peak at -0.061 ± 0.001 eV, labelled SS, corresponds to the "down" atom. The S' peak at -0.220 ± 0.005 eV is due to atoms in the second layer. If a buckled dimer is represented by a simple sketch, the first three SCLS peaks are thus: 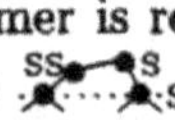

. The fourth peak, C, located at $+0.200\pm0.005$ eV, has been attributed to deeper atoms, but a contribution from surface defects cannot be excluded.

This assignment is corroborated by a number of observations:

1. The intensity ratios of the SCLS peaks are consistent with this model.[470] If the subsurface peak S' is assumed to come from a full monolayer of Si atoms, each of the dimer-atom peaks (S and SS) comes from half a

monolayer of surface atoms. The intensity of the C peak corresponds to half a monolayer of third-layer atoms.

2. The intensity of S varies with the azimuthal angle as predicted for . A fully converged multiple scattering simulation supports only this interpretation.[498] There was absolutely no agreement with simulations for symmetric dimers (): the observed and computed intensity minima and maxima are out of phase. Likewise, the assumption that S comes from the "down" atom was ruled out. An excellent agreement was obtained when the dimer was buckled by 19°.
3. The "down"-atom peak, SS, behaves in the same way as the "up"-atom peak, S, (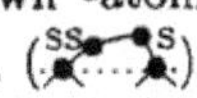) when the photon energy (that is, the escape depth) and the temperature are varied.[470] In particular, the intensities of SS and S retain the same ratio at 300 K and at 120 K, that is, when the (2×1) reconstruction is observed and when $c(4\times2)$ is observed.
4. The intensity of the C peak diminishes with decreasing escape depth. This is consistent with the interpretation which locates the sources of this peak deeper under the surface.[470]
5. The origins of S and S' () are illustrated by their dependence on the coverage Θ_{Ge} of the surface with Ge atoms[471] (Sect. 2.4.1a, page 117). The intensity of the subsurface S' is proportional to the number of Si atoms in the second layer. It does not change with Θ_{Ge} until $\Theta_{Ge} = 1$ monolayer. Subsequently, it is linearly quenched with Θ_{Ge} to disappear when $\Theta_{Ge} = 2$. The "up"-atom peak S is proportional to the number of Si surface atoms and vanishes when $\Theta_{Ge} \geq 1$.

The S peak (+0.483 eV, 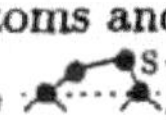) is clearly separated from the bulk emission. According to *ab initio* results,[180] the energy of this peak is well described by the final-state theory, which locates it at +0.48 eV. The final-state correction to the initial-state energy[476] is +0.23 eV. One work[468] reported a splitting of its maximum (by about 0.19 eV) and assigned it to an influence of the crystal field, but further studies did not confirm this observation.[470] The decomposition of the observed signal into individual components might have been affected by the C peak.

On the other hand, the SS peak (−0.061 eV, ) is hidden under the bulk emission and is therefore difficult to detect. The initial-state theory overestimates the binding energy of $2p$ electrons of the "down" atom by about 0.4 eV, placing the SS maximum at −0.41 eV. The final-state theory still slightly overestimates this energy, placing the SS maximum at −0.09 eV. This is within the numerical error of about 0.03 eV for energy differences between surface features and below 0.15 eV for the energy differences between surface and bulk features.[180]

No significant screening effects are expected for the subsurface S' peak (-0.220 ± 0.005 eV, 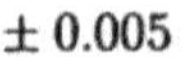). The dielectric constant of Si is large ($\varepsilon = 11.9$) and the core hole is nearly completely screened, making the final-state effects small: the screening charge in the bulk is $(\varepsilon - 1)(\varepsilon + 1)e = 0.85e$. Indeed, the initial-state and the final-state theories yield the same value for the binding energy in the second layer. The calculated value is -0.13 eV, which agrees with the experiment within the numerical accuracy.

Calculations indicate that different atoms in the second and third layers have different core level energies, scattered around the bulk core energy within ~0.3 eV range.[476] It is plausible that experiment resolves these states only partially, as the S' and C components.

It is interesting to note that the final-state correction for the "down" atom is significant, about 0.5 eV, and that it is about two times larger than for the "up" atom. An analysis of this effect revealed that the site dependence of screening is caused by pecularities of the surface electronic structure. A core hole produces a perturbation potential attractive to electrons. This potential lowers locally the energies of electronic states. When a hole is created in the "down" atom, the unocuppied orbital of this atom draws electrons from the Fermi level. The newly acquired electrons screen the hole efficiently and are responsible for the enhancement of screening.

This discussion suggests the following physical picture of the core level shifts. There are two main contributions to the measured binding energy of core electrons. First, the surface atoms are immersed in a background electrostatic potential which originates from the spatial distribution of ions and valence electrons. This potential changes with the distance to the surface and becomes an important contributor to the shifts of core binding energies in the subsurface region. Second, final-state effects are dominated by the local electronic structure of the surface and have a major influence on the core level shifts of dimer atoms. The contribution of final-state effects to level shifts is negligible for electrons from cores of subsurface atoms.

3.3. Surface Defects

Defects observed on a clean surface may have their origin on the surface or may be indicators of extended bulk defects. The latter are exit points of screw or edge dislocations. Dislocations split into partials create characteristic domains of dimerization. These domains are displaced relatively to the 2×1 and 1×2 domains of the perfect surface. Instructive ball-and-stick models of dislocations at surfaces, diagrams of interactions between steps/terraces and dislocations, and LEEM images can be found in Ref. 528.

Below we analyze the defects intrinsic to Si(001). The surface "point" defects: missing dimers, their small clusters, and subsurface vacancies (C-type defects), have a size of a few lattice constants. Thermodynamics at high temperatures and limited diffusion at low temperatures ensure that they exist on all atomically clean Si(001) surfaces. Also the "extended" defects, surface steps, are ubiquitous. This is not only because all real surfaces are misoriented. Surface stress anisotropy destabilizes flat surfaces against formation of wavy steps of monatomic height (Sect. 4.3.2).

3.3.1. *Point defects*

Clean Si(001) surfaces usually contain many microscopic flaws. The smallest of these "native" surface point defects are the so-called A-defects. They correspond to missing dimers, and are also known as dimer vacancies (DV). If we visualize the 2×1 dimer rows as vertical "files" in the sketch:  (Fig. 2.5b, page 93), an isolated missing dimer is represented by: .

A few per cent of surface dimers are missing on surfaces prepared in a standard way. These dimer vacancies often come together in complexes ranging from vacancy pairs to real vacancy "islands". The simplest and usually numerous complexes are dimer vacancy pairs (2DV), with two neighboring dimers missing in the same row (). When discovered, they were labelled B-defects. A complex 1DV-2DV, denoted usually as (1+2)DV, is also abundant. As its name suggests, it is a pair of a single vacancy, 1DV, and a double vacancy, 2DV. These two constituents are located in the same dimer row and separated by a single ("split-off") dimer. The (1+2)DV looks therefore like: . Other combinations of missing dimers, as 3DV, (1+2+1)DV, and complexes of more irregular shape or containing more vacancies are also easy to find in many STM images.

High concentration of DV and their complexes may indicate contamination, particularly with nickel; Cu and Ag contamination have a similar effect.[529] In certain circumstances, some of the "dimer vacancies" may be in fact adsorbed H_2O molecules (page 302). DV supersaturation on clean Si(001) may be achieved by bombardment with rare gas ions or by etching.

The dimer vacancies and their small complexes appear as dark patches in STM scans. Their atomic structure is simple: the topmost surface atoms are missing and the exposed atoms of the subsurface layer tend to re-create their broken bonds. Interestingly, this rebonding can only be deduced from calculations; STM tips cannot penetrate into such tiny pits to reach the electronic states at their bottoms.

The missing dimers are important not only because they are widespread but also because they can move, so they contribute to mass transport across the surface. Being mobile, they can produce interesting structures: surfaces which contain only highly energetic D_A steps (page 189) and surfaces which contain parallel trench lines (page 194). Missing dimers may influence the surface stress anisotropy, affecting the long-range elastic interaction between monatomic steps (Sect. 4.3.2).

A completely different but also abundant point defect, known as the C-defect, occupies an area of two neighboring dimers from the same dimer row. It appears bright in STM pictures, where it looks like a pair of extremely buckled dimers:  (or , if we zoom in). Only the two "up atoms" are visible, building a bright pair of spots oriented along the dimer row. The electronic structure of C-defects is metallic (amphoteric, in the bulk defect terminology): they introduce partially occupied states in the surface energy gap, pinning the Fermi energy. Consequently, C-defect sites are very active chemically. It is at C-defects where oxidation begins[530] or where Li atoms are initially adsorbed.[531] Many objects which appear to be intrinsic C-defects may be, however, associated with impurities.

Defects on the initial surface have a negative impact on the performance of microelectronic devices.[532] Several techniques have been proposed to improve the surface quality. Methods like low-energy ion sputtering[533] or laser-induced desorption[502,534] remove the defected layer, exposing atomically flat, defect free subsurface layer. However, efficiency of such methods and their usefulness in production lines is disputable. For example, laser desorption may lead to homogenity problems: whole wafers, which are large (200 mm in diameter now, 300 mm soon, 400 mm within a decade[22]), must be treated simultaneously. Moreover, UHV conditions used to study defects on clean surfaces are incompatible with production lines. Inexpensive

Ref year	Subject relevant to the structure of missing dimers
439 1984	Existence of dimer vacancies (DV) postulated; discussion of *ab initio* data
363 1985	Missing dimers observed for the first time; STM
535 1989	Total energy calculations of missing dimer reconstructions; *ab initio*
460 1990	Total energy calculations of missing dimer reconstructions; *ab initio*
458 1990	Interstitial dimer model for the A-defect; *ab initio* theory
495 1992	Missing dimers and surface strain in Si(001) and Ge/Si(001)
536 1993	Missing dimer formation on Si(001) 2×1; classical potentials
484 1993	Formation energies of DV complexes; *ab initio* theory
537 1993	Some "missing dimers" are related to water; STM
538 1993	Some "missing dimers" are related to water; STM
539 1995	Formation energies of 2DV; classical potentials
540 1995	How to obtain a low density of DV; STM
541 1996	Split-off dimer (1+2)DVs correlate with Ni contamination; STM, AES
542 1996	Missing dimer formation energy in a 4×7 cell; *ab initio*

Table 3.3. Structure of missing dimers: selection of references.

chemical approaches which leave the surface oxidized or hydrogenized seem to be more promising.

Below we will discuss the atomic and electronic structure of the fundamental point defects, their interactions among themselves and with the surrounding surface, as well as mechanisms of their formation.

a. *A- and B-type defects: missing dimers*

This Section describes atomic and electronic properties of surface dimer vacancies. Table 3.3 gives a selection of references relevant to this subject.

The missing dimers,[363] often described as "dimer vacancies" or "surface vacancies", are also known as "π-bonded defects": the dimer bonds at the defect edge may be shortened, indicating a stronger π bond. Defects appearing in STM images like single and double missing dimers are sometimes denoted as A- and B-defects. We will represent a dimer vacancy by the abbreviation "DV", as described in the introduction to this Section.

The idea that some dimers may be missing was conceived at the time when dimers as such were still a hypothetical model, though a widely accepted one. It is due to Pandey (Section 2.2.5b) who noted[439] that the number of broken bonds can be locally reduced by creating a dimer vacancy and rebonding its bottom (Fig. 3.12b, page 159 and Fig. 2.7, page 106).

Together with C-defects, missing dimers and their complexes are the **most frequent** defects on clean Si(001). It is easy to prepare a surface with as much as 10% of dimers missing. Special care is needed to keep the

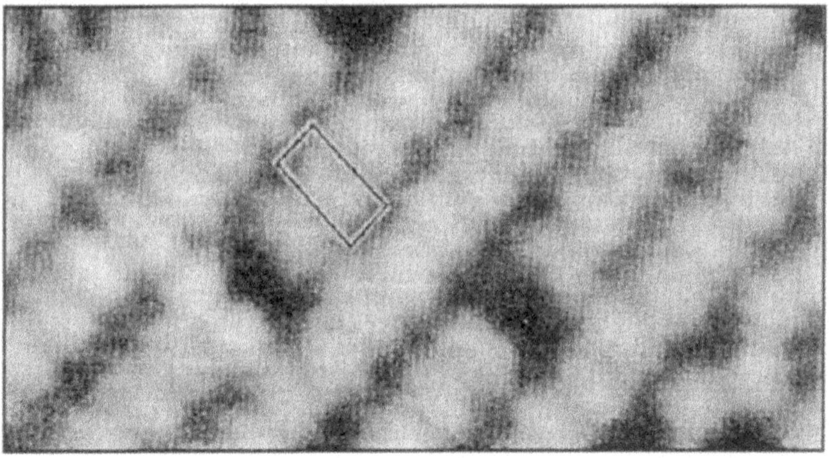

Fig. 3.11. STM image of the occupied states of missing dimers. From Ref. 441.

concentration of these defects low, such as 1% (for example, see Refs. 540 or 543). Controlled amount of missing dimers can be created by sputtering with Ar or Xe ions or by etching (for more information, see references 1 to 11 in Ref. 544). Already in 1986 it was pointed out[441] that a few percents of defected sites may affect LEED I–V profiles, He diffraction patterns, or measured electronic band structures. However, no systematic study of this influence has been done. The notable exception is the case of the order-disorder phase transition (page 238). So defects are easily blamed when the agreement between the outcome of an experiment and the favored theoretical model is not as good as desired.[h]

The **assignment** of A- and B-defects to missing dimers is confirmed by the observation of similar defects induced by sputtering or sublimation. But the STM tip, sharp as it is, cannot penetrate into the small vacancy to study its bottom. So can one tell what happens to the atoms there? Can STM really show that these defects *are* dimer vacancies at all? (For interesting examples, see pages 235 and 296.)

The problem is even more tricky than it may seem. Namely, STM does *not* observe atoms; it only gives hints about the spatial distribution of electron density of states (Sect. 1.3.2a). That no dimer is seen does not immediately mean that no dimer is there. STM is visually very suggestive,

[h] We have done it too! For example, when discussing the electronic structure, page 149.

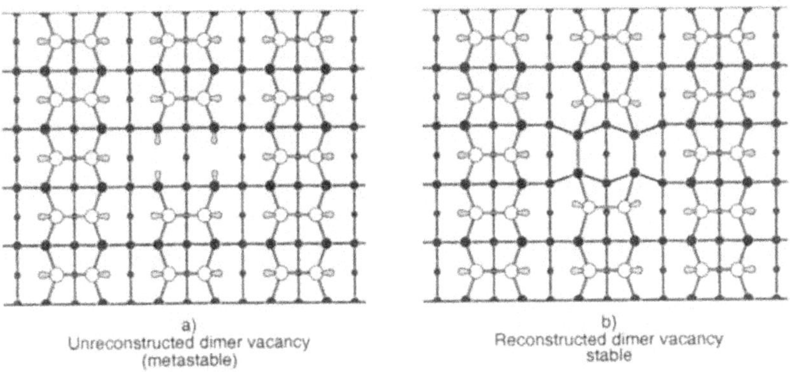

Fig. 3.12. Schematic representation of atomic configurations of a single dimer vacancy. (a) Unreconstructed dimer vacancy, DV*; metastable (compare Fig. 3.13b). (b) Reconstructed dimer vacancy, DV; stable (for side view see Fig. 3.13a, page 164).

but it can be interpreted conclusively only in conjunction with calculations of atomic and electronic structures.[153,157] Theoretical *ab initio* studies did challenge the missing dimer model. Depressed dimers[458] were suggested instead. But further calculations did not confirm this idea.[484]

The least biased method to extract structural details from STM images is by *ab initio* computation of total energies and electronic structures. And here comes a suprise. Such calculations[484] show that the process of rebonding in a *single* vacancy is hindered by an energy barrier. The rebonded configuration, DV (Fig. 3.12b), with its formation energy of ~0.3 eV, is indeed the **ground state**, but the vacancy has a **metastable**, non-rebonded configuration, DV* (Fig. 3.12a) with weaker π bonds and the formation energy higher by 0.4 eV (room temperature kT is 0.025 eV).

Can one test this prediction?

For easier reference in the text, we will now represent a dimer row as a series of thick dumbbells: , the non-rebonded substrate atoms by dots, and pairs of rebonded substrate atoms by thin horizontal lines. The metastable vacancy configuration, DV*, is thus given by: , and the stable, rebonded configuration by:  (compare Fig. 3.12).

Unfortunately, no simulated STM images of a metastable, non-rebonded vacancy have been published so far. In order to estimate its appearance, we discuss the electronic structure of this defect. Next, we check our expectations against the available data.

A characteristic feature routinely computed for DV* is that the π bonds of the dimers at its edge are *weaker* than on the unbroken surface. The physical reason is as follows. On a perfectly dimerized surface, the dangling bonds d_1 and d_2 of atoms in each dimer form occupied bonding ($\pi \sim d_1 + d_2$) and empty antibonding ($\pi^* \sim d_1 - d_2$) combinations. When the dimer is on the DV* edge, its dangling bonds mix with the dangling bonds e_1 and e_2 of the two edge atoms and a new object, a **tetramer** of atoms, is created.[i] Instead of just one interatomic bond in the dimer ($d_1 + d_2$), the "dangling" bonds of the tetramer atoms may produce three bonding segments: $e_1 + d_1$, $d_1 + d_2$, and $d_2 + e_2$. The edge atoms are quite far apart, so we ignore the fourth bonding segment, $e_1 + e_2$. When the dimers are symmetric, the interaction between the four orbitals produces two symmetric (+) and two antisymmetric (−) tetramer orbitals:

$\pi_+ \sim e_1 \widehat{+} d_1 \widehat{+} d_2 \widehat{+} e_2$, three bonding segments indicated by hats ($\widehat{\ }$),
$\pi_- \sim e_1 \widehat{+} d_1 - d_2 \widehat{-} e_2$, two bonding segments,
$\pi^*_+ \sim e_1 - d_1 \widehat{-} d_2 + e_2$, one bonding segment, and
$\pi^*_- \sim e_1 - d_1 + d_2 - e_2$, no bonding segments.

The energy of each combination is a monotoneous function of the number of bonding segments. The π_+ and π_- states are thus fully occupied by the four electrons from the four dangling bonds, while π^*_+ and π^*_- are empty.

The conclusion is that the formation of a tetramer makes the dimer bond weaker:  (we marked the weaker bonds by thin lines). In STM images of the occupied states, the dimers on DV* edges should look like something between the occupied ($d_1 + d_2$) and the empty ($d_1 - d_2$) states of the perfect surface. These edge dimers may look symmetrical in STM images because STM records a time average over many flips of buckled dimers, or because the edge dimers are intrinsically symmetric. In either case, the weakened bonds should resemble a bean (π) with enhanced ends (π^*). Such objects can indeed be observed. For example, a similar state is visible on one side of one of the vacancies in Fig. 3.11.

But at least two observations appear to be inconsistent with the identification of this defect as a metastable vacancy of Fig. 3.12a. First, both tetramers of the metastable vacancy should look the same, while in the observed vacancy they look different. Second, the computed[484] energy of DV* is by 0.4 eV higher than the ground state. This is an order of magnitude

[i] Such tetramers are typical not only for the metastable surface vacancies. They can be found on S_B-s, S_B-dv, and D_B steps on Si(001) and they are one of the main building blocks of the Si(113) surface. See Sections 3.3.2 and 1.5.4.

more than the thermal energy at room temperature ($kT = 0.025$ eV), so unless a huge barrier for rebonding exists, the metastable atomic configuration should be extremely rare in room-temperature STM images.

A possible explanation is that the metastable configuration is stabilized by **adsorbates**. For example, a H_2O molecule may have opened the strained bonds at the bottom of a stable vacancy:

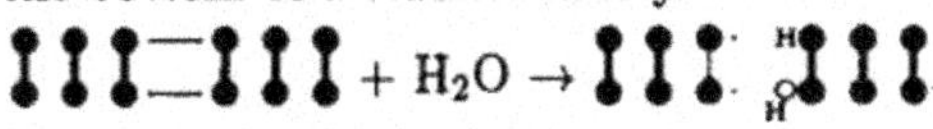

Since the tetramer on the right-hand side has its edge atoms passivated, the STM image of its dimer resembles a regular dimer on the perfect surface. The other tetramer is clean, so its dimer appears as two spots in Fig. 3.11.

Adsorption of H_2O at DV defects may be favored over dissociative adsorption on isolated dimers ( → 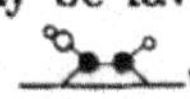, Sect. 5.5.2), because in the latter case the intra-dimer π bond is broken, which costs 0.3–0.5 eV (page 134). This energy cost may be even higher for adsorption on a dimer at the vacancy edge, because π bonds appear to be stronger there. But in the case of adsorption at the vacancy bottom, the energy loss due to bond breaking may be smaller. According to *ab initio* data,[484] the energy needed to excite DV to DV* is 0.44 eV, with uncertainty ~0.1 eV (see the discussion below). Therefore, it cannot be excluded that the dissociated H_2O is bonded at DV* bottoms stronger than on terraces. The fact that adsorbed H_2O is less frequently found near to steps than on terraces (Sect. 5.5.2) may in fact reflect a similar π-bond effect: step edges may possibly deplete the terraces from water by preferentially bonding it at ledge atoms.

At any rate, the **message** from *ab initio* theory is that the A-defects are likely to be dimer vacancies. The bottom of a stable vacancy is rebonded: 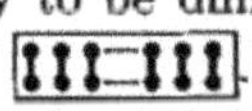. This rebonding leads to a local tensile stress in the direction along the dimer rows. Formation of a DV defect costs energy, but this cost is small,[460,484,542] probably less than 0.5 eV. If we estimate the equilibrium concentration $\Theta(T)$ of dimer vacancies from the Boltzmann factor, we obtain that the formation energy of 0.5 eV corresponds to $\Theta(500°C)=0.04\%$ of a monolayer and to $\Theta(1000°C)=2\%$ ML. The formation energy of 0.22 eV (Fig. 3.13a) corresponds to $\Theta(500°C)=3\%$ ML and to $\Theta(1000°C)=18\%$ ML.

As next we will consider **agglomerates** of surface vacancies. Clustering of vacancies into complexes is driven by the additional strain field generated by an isolated missing dimer. However, a word of caution is needed here. Our discussion of the energetics of vacancy clustering is mostly based on *ab initio* calculations[484] in which the dimer vacancies and their complexes were simulated in supercells with surface periodicity of maximum 2×9. This ensures that the vacancies in adjacent supercells are separated along dimer rows by more than 10 Å. Across the rows, however, they form trenches: . More effort was made to separate the vacancies along the rows, because the observed clusters which occupy more than one dimer row appear to be built of a few well-defined single-row patterns resembling those occuring as separated, single-row defects. While the configuration used in the calculations may capture the fundamental differences between these single-row patterns, the cross-row interaction between the defects is definitely not negligible. Indeed, when the surface is oversaturated with dimer vacancy defects, they tend to arrange in lines oriented perpendicular to the dimer rows (page 194). This leads to an approximate $2\times n$ periodicity, with $n \sim 10$. The interaction is, however, not strong enough to make $2\times n$ dimer vacancy rows a stable feature of the surface. Too high concentrations of randomly created vacancies (such as 1/3 of a monolayer[545]) or annealing temperatures outside the approximate range of 500–950°C (page 198) prevent $2\times n$ structures from being formed.

Three sets of *ab initio* calculations for DV with similar accuracies have been reported.[460,484,542] Comparison of the computed formation energies (we assume the buckled 2×1 surface as the reference system) indicates that the interaction across dimer rows is smaller than along dimer rows and not easy to resolve, but it possibly does not vanish:

2.43 eV	0.51 eV	0.35 eV	0.39 eV	0.45 eV
2×2	2×4	2×5	2×7	4×7
Ref. 460	Ref. 460	Ref. 484	Ref. 542	Ref. 542

The open circles in the graph on the left are the formation energies of 2×4, 2×5, and 2×7 vacancies. The full circle represents the 4×7 arrangement. The dotted line is only to guide the eye; it is a function of type $(\text{period})^3$+const, fitted to cross the four $2\times n$ data points. The graph suggests that the numerical uncertainty is around ± 0.1 eV/DV. Although this is comparable to the energy difference $\Delta E = 0.06$ computed between the 2×7 and 4×7 structures,[542] the accuracy of ΔE may be actually higher because it was obtained in one set of calculations. Namely, our comparison of numerical results from different sources most probably reveal errors of systematic origin, such as those due to uncomplete relaxation of atoms or too small basis sets (Sect. 1.3.1b). Errors of this type are likely to cancel out in results reported by a single research group.

Keeping in mind the potential influence of the interactions between dimer vacancies on the computed formation energies, we proceed to dimer vacancy complexes. The simplest stable complex, 2DV, contains two vacancies in the neighboring sites in the same row, like: 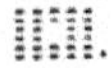. The formation of a 2DV complex from two isolated DV defects releases 0.11 eV. The atoms exposed at the bottom of the pair rebond on one side:  (Fig. 3.13c). In the direction along dimer rows, the rebonded half of the defect is under tensile stress while the non-rebonded half is under compression.

When a third isolated vacancy approaches the divacancy from the non-rebonded side, the stable 3DV configuration is formed: 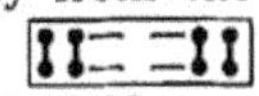 (Fig. 3.13e), with an energy gain of 0.1 eV. This fundamental 3DV configuration introduces local tensile stress, as in the case of a rebonded DV. It is not favorable for the approaching vacancy to attach itself to the rebonded side (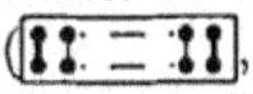, Fig. 3.13f). In this case, a metastable configuration 3DV* would be created and the energy would be lost: an isolated DV and an isolated 2DV have energy by 0.33 eV lower than 3DV*.

It is interesting that a complex of three vacancies which has a *lower* energy than 3DV exists. When the approaching DV stops short off the non-rebonded side of the divacancy (, Fig. 3.13d), so that the two defects are separated by a "split-off" dimer, the energy gain is by 0.03 eV larger than when the regular trivacancy is formed. This (1+2)DV complex is indeed very often observed in STM images.

However, an experiment designed to study the distribution of vacancy complexes has revealed that the most abundant defects are not (1+2)DV but single vacancies.[540] Assuming that the theory[484] has got the relative order of the formation energies of the different vacancy complexes right, one can explain this excess of 1DV over the energetically more stable (1+2)DV complexes as an entropic effect. The pressure of the 1DV lattice gas maintained in the experiment (defect density 1.7%, annealing temperature 1250°C) may be too low for the gas to condensate into (1+2)DV droplets. Indeed, a previous publication[441] reported STM images of samples with about five times higher dimer vacancy concentrations and prepared by annealing at 1100°C and subsequent slow cooling. The vacaancies did con-

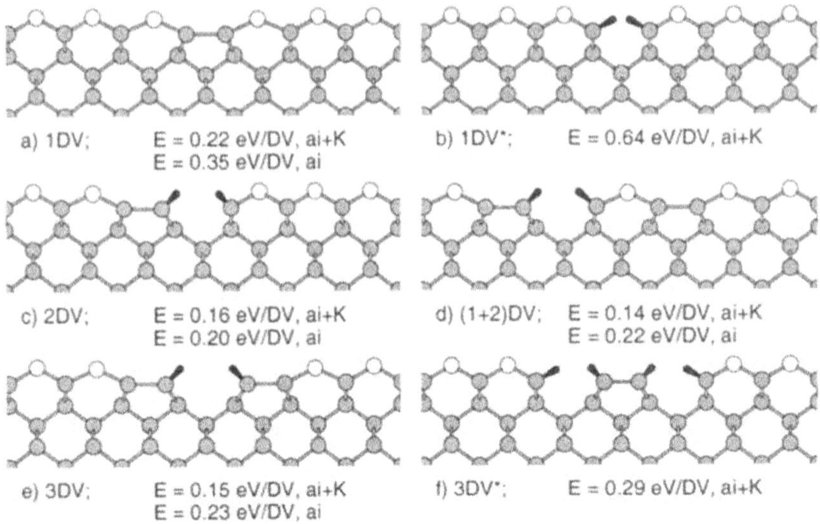

Fig. 3.13. Dimer vacancies; side view. Surface dimer atoms are white, the dimer bonds are perpendicular to the picture and not shown. Dangling bonds created at dimer vacancies are indicated by black clubs. Formation energies[484] are in eV/missing dimer. The values labelled "ai+K" correspond to *ab initio* energies corrected for deep subsurface relaxation within the Keating model (page. 17). The values labelled "ai" are raw *ab initio* results with only three subsurface layers relaxed. **(a)** A stable dimer vacancy, DV. **(b)** A metastable dimer vacancy, DV*. **(c)** A stable dimer divacancy, 2DV. **(d)** A stable triple dimer vacancy with a split-off dimer, (1+2)DV. **(e)** A stable dimer trivacancy, 3DV. **(f)** A metastable dimer trivacancy, 3DV*.

dense in this case, mainly into the low-energy (1+2)DV complexes.[j]

Addition of further vacancies results in more and more complicated agglomerates. An interested reader is referred to Ref. 484. Here we only note that nucleation of vacancies may lead to formation of "vacancy islands". The anisotropic shape of these broad depressions is determined by the interplay between anisotropic diffusion of vacancies, their anisortopic adsorption and detachment rates at the island edges (that is, at surface steps), and surface-step formation energies (a vacancy island must be sur-

---

[j] On the other hand, it is conceivable that the higher density of the split-off dimer states may reflect a possible nickel contamination of the latter samples. A recent experiment has shown that the concentration of the (1+2)DV complexes is proportional to the AES nickel signal from the surface.[541] The reason for this dependence remains elusive. It may be that Ni stabilizes the (1+2)DV complex, but it is also possible that Ni simply increases the density of the vacancies. See Sections 3.4.1 and 5.7.2 for more discussion.

Ref year	Subject relevant to C-defects
546 1989	Local electronic structure of C-defects: they are metallic; STM
134 1990	**Review**: Surface defects and Fermi level pinning
537 1993	Some C-defects are related to water; STM
538 1993	Some C-defects are related to water; STM
374 1994	C-defects frustrate the buckling in $c(4\times2)$ domains; low-temperature STM
547 1994	Two H atoms are (nearly) stable at low coverages as *cis*; *ab initio*
548 1994	Extraction of atoms at and around C-defects; STM
549 1995	A single H atom (1/2 *cis*-H) resembles 1/2 of a C-defect; STM
540 1995	C-defects are about as abundant as dimer vacancies
550 1996	Model with missing "down" atoms is unstable; *ab initio*
551 1996	Rebonded subsurface vacancy model, coordinates listed; *ab initio*
542 1996	C-defect as a subsurface vacancy saturated by H; *ab initio*
552 1996	Some C-defects can be reconfigured by STM tip
553 1997	Some C-defect like objects have semiconducting LDOS; STM

Table 3.4. C-defects: selection of references.

rounded by steps of S_A and S_B types; cf. Sect. 3.3.2).

No *ab initio* calculations of the energy barrier for generation of DV on the perfect surface have been reported so far. The available data[536] comes from classical potentials and predicts that the formation process is so slow that the vacancies observed in STM images are most probably not created thermodynamically. Instead, they may be formed during cutting and polishing and survive the annealing process. Classical potentials were used to investigate also other aspects of vacancy energetics. For example, 2DV formation energies assuming steps as sinks for the ejected dimers was computed.[539] Unfortunately, the accuracy of classical potentials in the description of Si surfaces is difficult to control (page 227).

b. *C-type defects: Subsurface vacancies?*

Along with missing dimer defects, C-type defects are the most frequently encountered atomic-size flaws on clean Si(001) surfaces.[540] Some of them may be related to adsorbed impurities, notably water (Sect. 5.5.2), but some of them may be intrinsic to the clean surface. These "clean" C-defects are the main subject of this Section. Table 3.4 lists some papers which discuss C-defects or contain information helpful for their identification.

The Fermi energy E_F on clean Si(001) is pinned in the energy band gap. This implies the presence of electronic states induced by some defect sites. C-type defects are the only abundant native point defects which introduce local metallic electronic states.[546] They therefore become a natural candidate for a pinning center.[134] Their metallic or nearly-metallic character is corroborated by the high chemical activity of the defect sites during the initial state of Si(001) oxidation[530] and adsorption of Li.[531]

In STM images, C-defects look like two highly buckled dimers occupying neighboring sites in the same row and buckled in the same direction, as in: 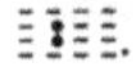. But the **atomic structure** of these defects is not yet clarified. Many extrinsic defects may appear in STM images as objects similar to C-defects. It suffices that impurities saturate one side of two neighboring dimers. Good candidates are here molecular water (page 303) or OH radicals (page 304). In both cases, the force which stabilizes the impurities on one side of the dimer is a hydrogen bond (Sect. 5.5.1). But there are also well-supported models which do not invoke any impurities. The difference between a "real" C-defect and a "fake" C-defect may be in their electronic structures: while a C-defect should be metallic, impurity-related objects may have no electronic states at the Fermi level.

Let us assume that C-defects are intrinsic to clean Si(001). The stability of C-defects and the fact that their local electronic structure (metallic) differs from that of the usual buckled dimers (semiconducting) may suggest that the "down" atoms are indeed missing (Fig. 3.14a). But experiments in which Si atoms were deposited show that the C-defects are not filled preferentially.[134] Though the model of Fig. 3.14a is sometimes invoked, *ab initio* calculations confirm that it is not correct.[551] The formation energy computed[550] for the missing atoms model, 1.79 eV, is markedly higher than the formation energy of 1.4 eV calculated by the same group for another model[542] of the C-defect (Fig. 3.14b). In addition, since some of the broken bonds created by removal of the "down" atoms become rebonded in a very asymmetric way, STM images simulated for the resulting structure cannot account for the experimental data.[551]

The model of Fig. 3.14b assumes that an atom is removed from the subsurface layer, so that a single *subsurface* vacancy is created.[551] The four broken bonds around the vacancy site become reorganized, as indicated in Fig. 3.14b and in the insert in Fig. 3.15 labelled C^0. Two of the four atoms surrounding the vacant site are on the surface. These are the "down" atoms of the C-defect, invisible in STM images. They move towards one another and dimerize. The stress induced by their pairing is relatively easily accommodated by relaxation of other surface atoms.

The third and fourth atoms of the vacancy behave in a complex way.

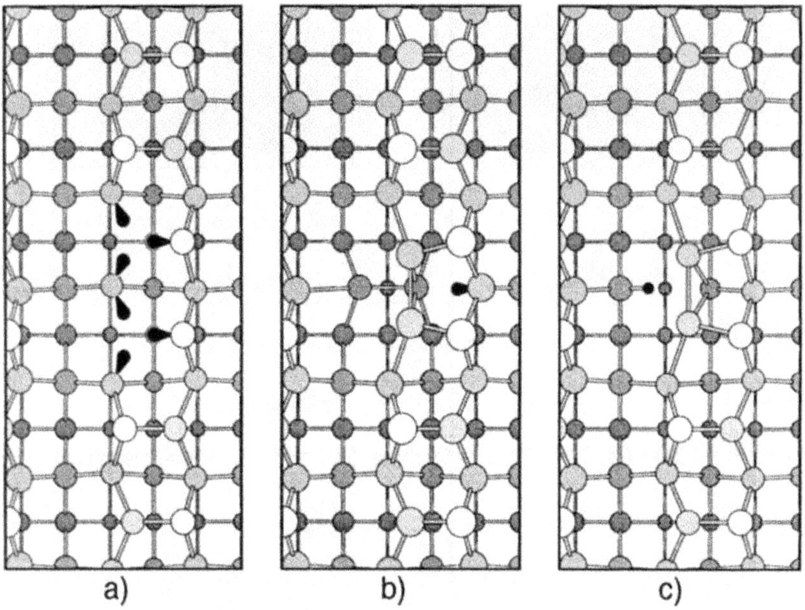

Fig. 3.14. Ball-and-stick models of the C-defect. Atoms further from the surface are shown smaller and darker. (a) Position of surface atoms as suggested by STM: two consecutive "down" atoms appear to be missing.[546] The broken bonds created by actually removing these atoms are indicated by black clubs. This structure is unstable.[550,551] (b) Rebonded subsurface vacancy (Uda-Terakura model[551]). A subsurface bond remains broken (black club) after this rebonding. See also Fig. 3.15, middle panel, charge state C^0. (c) H-saturated subsurface vacancy.[542] The H atom is black.

The third one moves relatively deep into the vacancy and becomes *fivefold* coordinated. It makes a σ bond with the fourth atom. Most remarkably, it becomes involved in a complex bonding with the orbitals of the dimer formed by the "down" atoms. But the bond with one of the original neighbors breaks (the black club in Fig. 3.14b and in the "C^0" insert of Fig. 3.15). Effectively, this bond has switched by nearly 180°, to the atom on the other side of the vacancy. The atom which gave the bond away retards into a planar geometry, donating its now unused electron to the cluster of atoms with the fivefold-coordinated silicon.

The most interesting part of this rather unusual group involves a triplet consisting of the fivefold-coordinated atom and two dimerized "down" atoms. The bonding between the atoms in the triplet can be rationalized by analogy to bonding in polyatomic rings, as in benzene.

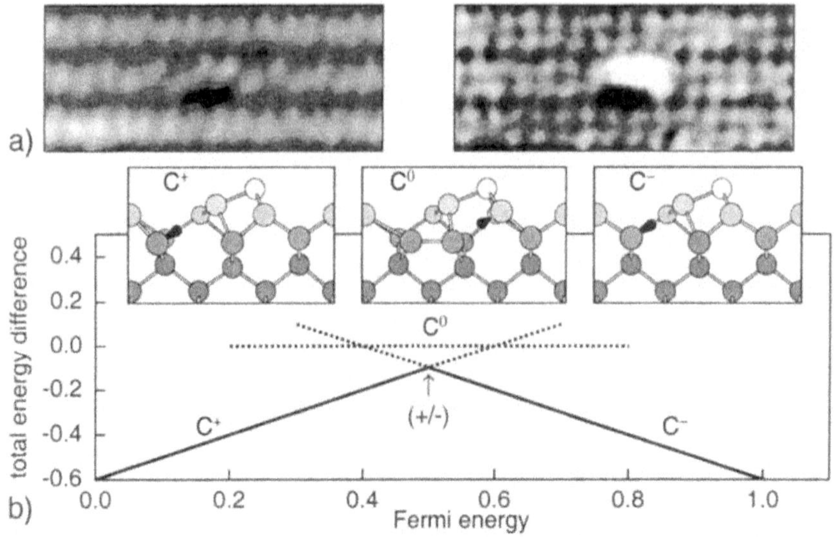

Fig. 3.15. **(a)**: Occupied (left panel, −1.8 V/0.5 mA) and empty (right panel, 2.0 V/0.5 nA) states of a C-type defect.[301] **(b)**: A possible charge-dependent relaxation mechanism of Fermi level pinning by a C-defect.[301] Cross-sections of atomic structures of positively charged, neutral (Fig. 3.14b), and negatively charged defects and alleged dependence of their total energies on the Fermi energy. A similar charge-dependent relaxation of the subsurface vacancy as the one responsible for the negative-U property of bulk vacancies[554,555] is assumed. This relaxation may render the neutral C-defect energetically unstable. See text for details.

The three atoms contribute three orbitals to the bonds. Three interacting orbitals can produce one strongly bonding, one weakly bonding, and one antibonding state. This system can localize four bonding electrons. Exactly four are available: three from the orbitals delivered by the atoms of the triplet, and one transferred from the planarized, abandoned neighbor.

It follows that the electronic structure of this model is semiconducting. Indeed, such is the outcome of *ab initio* calculations.[551] The splitting between the occupied and empty states was computed to be 0.3 eV. However, a C-defect should be *metallic*. Can this be reconciled?

The presence of the gap is the only apparent problem with the electronic structure of the model; otherwise, the STM image of this structure can reproduce the experimental STM micrographs. It was suggested[551] that vibrations of the dimerized "up" atoms broaden the states of the defect such that the bonding and antibonding states partially overlap, producing a metallic density of states. If this model was right, C-defects should be semiconducting at low temperatures. To our best knowledge, no measurements to check this hypothesis have been reported.

Another possible explanation of the metallic character of C-defects comes from the similarity to a bulk vacancy. Reconstruction of a bulk vacancy depends on its charge state.[554,555] The atoms at a *neutral* vacancy rebond, forming two pairs. A positively or negatively charged vacancy opens one or two of its "surface dimers" in order to accomodate the charge. This is associated with energy gain due to relief of the strain invoked by the pairing of atoms. The gain is substantial and destabilizes the vacancy charged with a single hole. The summed up energy of a neutral and a double-positively charged defect becomes lower than the energy of two single-positively charged centers: electron-phonon interaction changes the electrostatic repulsion between two holes into attraction mediated by the lattice relaxation. This is a case of the **negative Anderson** U effect.

How does it apply to the subsurface vacancy? The calculations for the model of Fig. 3.14b were done only in the *neutral* charge state. Since the rebonding of the subsurface atoms is weak[551] and induces a lot of stress, it is imaginable that the influence of electron-phonon coupling in this system is similarly strong as in the bulk vacancy. This can place the electron (+/0) and (0/-) transition levels[k] close on the energy scale. If the relaxation effects are strong enough, the neutral defect may even become destabilized (Fig. 3.15). The defect state observed in the gap becomes the (+/-) transition state. And if the charged configurations do not switch the bond in the vacancy, as indicated in Fig. 3.15, one can imagine that the defect changes between the relaxed positive and negative charge states without a barrier. Recharging can thus readily occur, for example, when the sample bias changes its sign in a scanning tunneling spectroscopy experiment.

Though speculative, the idea of a charge-mediated relaxation of a subsurface vacancy[301] is substantiated by *ab initio* calculations reported for the Uda-Terakura model with an adsorbed H atom[542] (Fig. 3.14c). When the bond switching is suppressed by saturation of one of the broken bonds of the vacancy by hydrogen, the electronic structure of the defect changes little and can still account for the topography measured in STM.

H adsorption is in fact a third possibility to explain the metallic char-

[k] An (n/m) transition level corresponds to the Fermi energy at which the total energies of the defect in n and in m charge states are the same.[556]

acter of C-defects. One can see this by counting electrons. The bond with the H atom takes away one of the electrons needed to fill the bonding states, and the structure must turn metallic.

Interestingly, not all C-defects appear as metallic in scanning tunneling spectroscopy. Some C-defects are seen as **semiconducting,**[553] that is, there is an energy gap separating their occupied and empty states. This indicates that not all C-defects are the same. It may be that only hydrogenated subsurface vacancies are metallic. Impurities as water may be also responsible for semiconducting appearance of some C-defects (Sect. 5.5.2, see in particular Fig. 5.6h on page 302). But note that we assumed that the semiconducting character of the defects is not an artifact caused by a blunt STM tip. Such a tip may collect current also from neighboring dimers. While a topology measurement may view it as a roughly constant, background current, a spectroscopic mesurement will be dominated by it.

Compelling evidence that not everything that resembles a C-defect is a subsurface vacancy is given by the observation that at least some of objects which appear to be C-defects could be rearranged[552] by STM tip, from  through  to . Such behavior is incompatible with any of the subsurface vacancy models and with the molecular water model (Sect. 5.5.2). But it can be easily accounted for by the model describing this C-type defect as a *cis* pair of, for example, OH radicals (Sect. 5.5.2).

Water-related defects might have been observed[548] in an earlier study which used STM tip to affect the atomic structure at and around C-type objects. The original surface contained numerous defects (around 5%). It was found that scanning at high voltage and current[l] changed the appearance of some of the C-type sites. Most of the **transformation** events (38 out of 46) changed the defect into a pair of missing dimers, (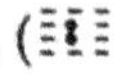 → ), in one case a triple dimer vacancy was created, and in seven cases two dimers disappeared in the nearest neighborhood (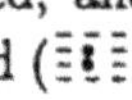 → 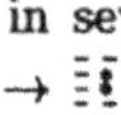). One may interpret the transformation of a C-type site into a pair of missing dimers as a tip-induced dissociation of the two water molecules constituting the defect (Fig. 5.6h, page 302) and subsequent passivation of the two underlying surface dimers by the H and OH reaction products. The passivated dimer pair becomes practically invisible on the unpassivated surface. Alternatively, one may view the transformation of the C-type objects into double missing dimers as extraction of a single dimer, followed by desintegration

[l] The surface was scanned at −2.85 V sample voltage and 0.45 nA tunneling current.

of a "true", *native* C-defect into a real dimer vacancy through emission of a mobile silicon monomer (Sect. 4.2.1a): 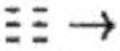 →  → 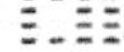.

The double missing dimers which appeared at some of the C-type objects (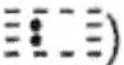) might have been "true" 2DV defects in the neighborhood of "true" C-defects. The atomic structure of the subsurface vacancy model of the native C-defect (Fig. 3.14b) is compatible with the experimental observation that the double vacancies are created on the "down" side of the defect rather than on its "up" side (, not ). The "down" side is strained by the anisotropic reconstruction of the defect site. The presence of a nearby double vacancy may help to relax this strain.

In summary, the origin and the atomic structure of C-type defects have not been clarified completely. Some C-type objects are due to surface contamination, probably with water-related species. Other C-type objects may be native to the surface.

3.3.2. *Surface steps and vicinal Si(001)*

So far we have been neglecting deviations from the ideal (001) orientation. Yet it is clear that perfectly oriented surfaces do not exist. Even the smallest misorientation leads to atomic-size steps (Fig. 3.1e, page 130). Steps are the typical features of surfaces vicinal to Si(001), that is, surfaces with slightly misaligned orientation in relation to the [001] axis (Sect. 4.3.2).

Steps interact via elastic forces. Two elastic media carry these interactions: the two-dimensional surface and the three-dimensional substrate. If the temperature is sufficiently high, steps can respond to the elastic forces by exchanging atoms across terraces. In this manner, steps can move on the surface. Steps also interact through entropic repulsion. Both interactions have the same dependence on the spacing between steps and their strengths are comparable for typical growth temperatures.[557] The thermodynamics of step-step interactions leads to a suprisingly rich phase diagram of vicinal Si(001), as discussed in Section 4.3.2. For the time being, however, we focus on the structural properties.

Let us estimate a typical distance between steps. A surface with misorientation (or "miscut") angle $\theta \simeq 0.5^{\circ}$may be accepted as "well oriented". Assuming that the steps are monatomic, that is, that their height is equal to the distance d_1 between subsequent (001) layers, $d_1 = a_b/4 = 1.35$ Å, we obtain that at $\theta = 0.5^{\circ}$ the average separation between two neighboring steps is $l = \frac{1}{2}a_s\sqrt{2}/\tan\theta \approx 155$ Å. This corresponds approximately to 40 surface lattice constants a_s ($a_s = a_b/\sqrt{2} = 3.84$ Å). A typical terrace

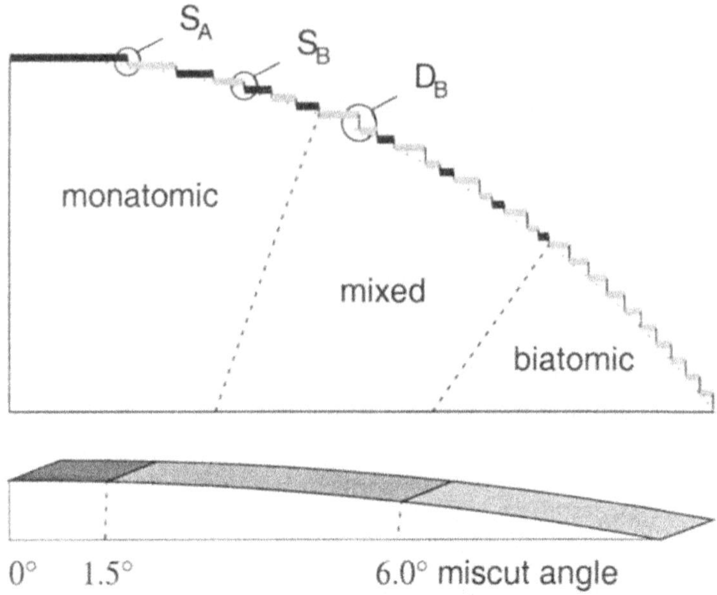

Fig. 3.16. Schematic representation of stepped Si(001) surface morphologies. View along $[1\bar{1}0]$. In the upper diagram, heavy black and gray lines indicate 1×2 and 2×1 surface domains on the rounded surface. The miscut is sketched with exagerrated angles. Different types of steps are indicated; Fig. 3.17 explains the nomenclature. The lower diagram displays the miscut angle ranges within which the pure monatomic, mixed, and pure biatomic phases are stable at $T \approx 500°$C. The fraction of the minority 1×2 orientation in the mixed phase, shown constant in the diagram, is in reality a contiuous function of the misorientation angle. See Section 4.3.2.

between such steps can thus hold a chain of 40 dimers oriented parallel to the step edge, or a chain of 20 dimers oriented perpendicular to the edge. About 2.5% of surface atoms belong then to step edges.

In reality, the height of steps depends on the temperature and misorientation (Fig. 3.16). For θ below 1–2°, steps have monatomic height of $d_1 = a_b/4 = 1.35$ Å. At $\theta \simeq 1$–$2°$ their separation is $l \simeq 15a_s$ (~50 Å). For larger miscuts, monatomic steps give way gradually to steps of biatomic height,[558] $d_2 = a_b/2 = 2.7$ Å and a mixed phase[559,560] appears. When θ exceeds 6–8°, steps become almost exclusively biatomic, at least below ~800°C.[561,558] When $\theta = 6°$, the distance between steps is thus $\sim 7a_s$.

An interesting effect occurs for very *small* miscut angles. It is caused by stress anisotropy (Sections 3.2.2d and 4.3.2). All steps at $\theta \approx 1.5°$ are

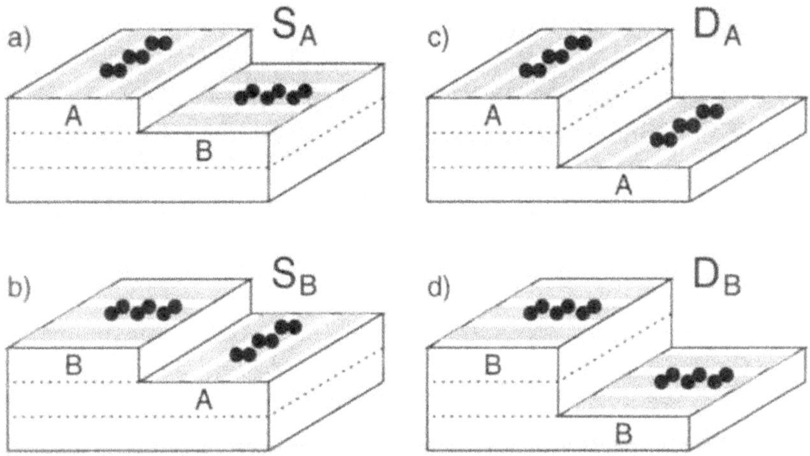

Fig. 3.17. A schematic illustration of the popular Chadi's classification of Si(001) steps. See Fig. 3.18 for the corresponding atomic structures. **(a)** S_A, single layer height (monatomic) step of type A; **(b)** S_B, single layer height (monatomic) step of type B; **(c)** D_A, double layer height (biatomic) step of type A, all terraces are of "type A"; **(d)** D_B, double layer height (biatomic) step of type B, all terraces are of "type B".

monatomic, separated by ~50 Å. When the surface is made flatter, the steps move apart. But for "flat" surfaces ($\theta < 0.1°$) the average distance between steps saturates in the range of thousands Å, remaining roughly independent of the miscut. *Steps on precisely oriented Si(001) do not disappear.* They bend into waves and closed loops instead.[562,563] A nominally flat surface is full of hills and valleys. Since steps increase the surface entropy, the tendency to build such hills and valleys is enhanced at elevated temperatures. This entropy effect leads eventually to a roughening transition.

We will now discuss the general structural and electronic properties of the steps, assuming that all dimers are symmetric (). Buckling () is addressed in the technical Sections that follow. Note that to neglect the buckling *is* a significant simplification. For example, *ab initio* calculations show that the registry of dimer buckling, 2×1 versus $c(4\times2)$, may affect the energy difference between two atomic configurations of a monatomic step by a value comparable to the step formation energy.[485]

Si(001) steps are **classified** according to the nomenclature introduced by Jim Chadi.[564] Monatomic steps ($d_1 = a_b/4 = 1.35$ Å) are labelled "S" (single) and biatomic steps ($d_2 = a_b/2 = 2.7$Å) are labelled "D" (double). Both types are divided into subclasses A and B following a simple observation. Since Si(001) is built of dimers, it is natural that steps are atomically straight when their direction is either parallel or perpendicular to the dimer bonds. If the dimers on the *upper* terrace are *perpendicular* to the edge of a straight step, the step and the terrace are said to be of type "A". If the dimers on the *upper* terrace are *parallel* to the step edge, the step and the terrace are of type "B". Thus, step "S_A" is a step of single layer (monatomic) height with dimers on the upper terrace oriented perpendicular to the step edge: , where dimers on the upper terrace are marked by thicker dashes. Figure 3.17 illustrates this terminology.

How much energy does a step cost? The answer is not straightforward. Due to interactions between steps (page 249), the step formation energy depends on step-step separation (see, for example, Fig. 3.26, page 190).

The first impression about the energetics is given by the examination of the atomic structures in Fig. 3.18. When the step is **monatomic** and the dimers of the upper terrace are perpendicular to the edge (S_A step), no new broken or strained bonds are created. Steps of any other type invoke bond breaking or straining. It is thus plausible that S_A steps are the most favorable energetically. That they cost *any* energy at all is because a strain field is generated at the step edge and the surface bands are disrupted.

A *miscut* surface cannot contain exclusively S_A steps. Monatomic steps separate 1×2 and 2×1 domains of dimerization, therefore the direction of upper-terrace dimers in a staircase of monatomic steps must vary from terrace to terrace, being alternatively perpendicular (S_A) and parallel (S_B) to the edge (Fig. 3.16). Similarly, a rough monatomic step or a monatomic-height edge of a surface island must contain both S_A and S_B segments.

S_B steps are more complicated than S_A steps. They exist in three variations. A simple step, which we denote S_B-s, introduces additional dangling bonds at the edge (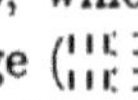, Fig. 3.18). Another version of S_B, a rebonded step (S_B-r), has these bonds saturated by threefold-coordinated atoms (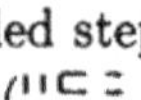, Fig. 3.18). A common modification of S_B-s is S_B-dv, a step with rebonded dimer vacancies,[m] 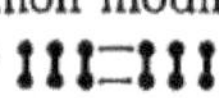, located on the upper terrace next to the edge dimer: 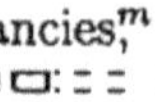. This structure, reminescent of the split-off dimer configuration of surface vacancy complexes of the (1+2)DV type (Fig. 3.13d, page 164), restores the dangling bond density to the value of the perfectly dimerized surface. The S_B-dv variation of S_B-s may be viewed as S_B-r with a dimer vacancy trapped next to the edge. Calculations show

[m] See Section 3.3.1a and Fig. 3.20, page 179.

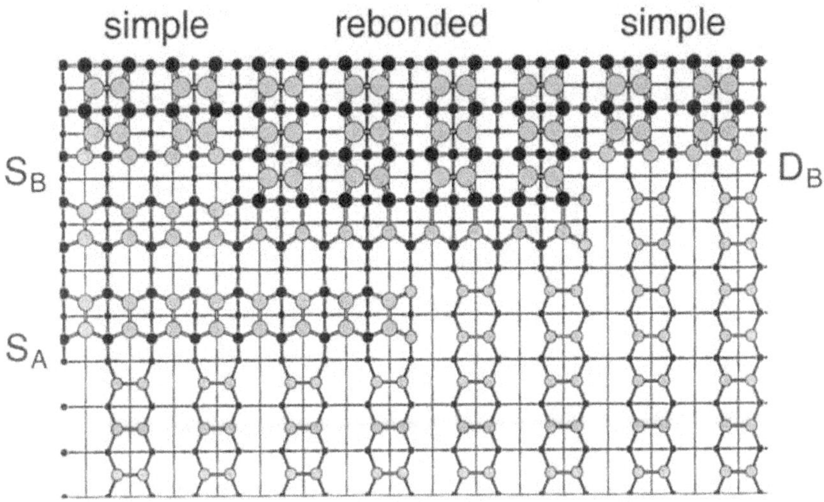

Fig. 3.18. Ball-and-stick models illustrating structural relations between the fundamental configurations of of Si(001) steps. For clarity, the dimers are shown as symmetric. Left side: monantomic steps S_B-s, S_B-r, and S_A. Right side: biatomic steps D_B-s and D_B-r. The monatomic and biatomic steps of B type meet in the middle of the picture as rebonded steps, S_B-r and D_B-r, and continue as simple S_B-s and D_B-s steps towards the sides of the figure. The monatomic S_A step is depictured below the S_B and S_B-r steps. Fourfold coordinated atoms are black. Threefold coordinated atoms are gray; these are the dimer, step edge, and step rebonding atoms. The dimer and edge atoms on simple steps of B type form characteristic tetramers with bonding described on page 160.

that S_B-s, S_B-dv, and S_B-r have different formation energies. All three types are observed in STM images, apparently due to kinetic limitations.

A straightforward observation confirms that formation energy of S_B is higher than that of S_A. Monatomic steps are never perfectly smooth, even on surfaces tilted exactly toward [110] or [1$\bar{1}$0]. They are thermally roughened (Fig. 3.19). S_B steps are mocu more rough than S_A steps. Since the roughness is caused by segments of the other step type (“**kinks**”), the step with a lower formation energy should be smoother. Hence, S_A steps are energetically more favorable than S_B steps.

The density of kinks is a function of temperature ($e^{-E/kT}$) and of the surface orientation. When Si(001) is tilted toward [110] or [1$\bar{1}$0], the kinks are purely entropic; steps would be straight at the temperature of

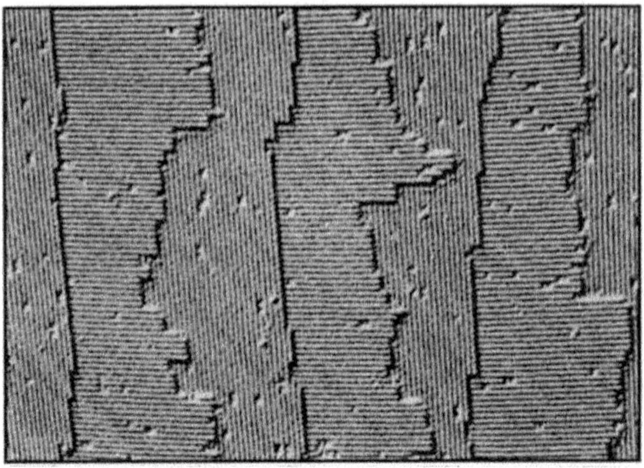

Fig. 3.19. A typical STM image (occupied states) of a vicinal Si(001) with monatomic steps. The dimer rows are rendered as straight, grey lines. It is evident that S_B steps are much rougher than S_A steps. (From Ref. 565).

absolute zero.[n] That kinks are observed at any temperature even for very exact [110] and [1$\bar{1}$0] tilts is due to the lack of thermodynamic equillibrium. The exchange of atoms between step segments becomes imposible at *low* temperatures (Fig. 4.1, page 230) and further cooling does not change the kink distribution. When the surface is tilted along [100] or [010], the kinks are predominantly geometric. Since the steps on such surfaces must run at an angle of 45° to the dimer bonds, each "straight" step has the same amount of S_A and S_B segments.

Biatomic steps of type B may be also either simple or rebonded. The rebonded structure, D_B-r (Fig. 3.18), may be viewed as a S_B-r–S_A pair: (⁞⁞⊏⁞⁞) ≡ (⁞⁞⊏ ≡) + (≡ ≡ ⁞⁞⁞). As was the case for monatomic steps, the rebonded D_B-r configuration has lower energy than the simple D_B-s (⁞⁞⁞⁞⁞). However, in contrast to the monatomic steps, the D_A structure has a high formation energy and is difficult to observe (page 189). Even kinks on D_B steps rarely have the D_A form. Instead, "D_A" segments are split into S_A and S_B separated by a small, dimerized terrace. D_B steps are much stiffer than S_B steps: on surfaces tilted along one of the dimerization directions, [110] or [1$\bar{1}$0], D_B run very straight, with few thermal kinks.

The **electronic structure** of Si(001) steps is not completely under-

[n] Unless the surface is very flat (Section 4.3.2)

stood. Steps are usually not believed to pin Fermi level on the clean surface. Yet the detailed changes which they induce in the electron density of states are elusive. Scanning tunneling spectra taken at S_A steps[566] show no extra states in the surface band gap. Similar measurements performed at S_B-r steps indicate[530] that the band gap is narrower at the step, which means that some new states are created in the gap above the occupied π and below the empty π^* dimer bands. On the other hand, recent *ab initio* LDA studies of monatomic steps[485] did not find any edge-induced states in the gap. It cannot be excluded that the additional states observed in experiment are associated with the very high density of kinks on S_B steps.

But in this case the theoretical results may be not conclusive. The computed surface band gap is sensitive to details which are difficult to obtain accurrately – sensitive to the buckling angle, for example. In addition, LDA calculations suffer from the so-called band-gap problem: the separation between occupied and empty states comes out too small. The necessary corrections to the LDA energy positions, (quasiparticle corrections, Sect. 1.3.1a), depend on the wave functions of the orbitals involved. Computation of these corrections for surface steps would still be expensive.

a. *Monatomic steps*

Here we will discuss STM images, atomic and electronic structures, energetics, morphology, and step-induced buckling of dimers of S_A and S_B. Table 3.5 lists selected work. Since step formation energies have been best determined from measured step shapes, some remarks on their dynamics are made. Step interactions are treated in Sect. 4.3.2. The role of steps in diffusion, epitaxy, and sublimation is addressed in Sections 4.2.3 and 4.4.

First, we consider **atomic geometries** with no regard to dimer buckling. When the dimers are **symmetric** (), the S_A structure is trivial (Fig. 3.18), but S_B steps remain rich enough to warrant a closer analysis.

S_B-s and S_B-r steps differ in the position of the step edge with respect to the dimers on the lower terrace (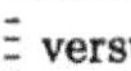 versus 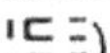). Take a needle-like, two-dimers wide island on top of a 2×1 surface (Fig. 3.20a). Since the substrate periodicity is 2×1, the needle has two different S_B edges. The dimers of one edge sit above the place where the dimer rows were. The dimers of the other edge sit above the trough between the rows.

With the edge dimers above the dimer row site, the step has the **simple** structure, S_B-s. In this case, the dimers of the substrate are either destroyed or unperturbed. Of the other dimers, those on the needle edge feel

Ref year	Subject relevant to monatomic steps on Si(001)
441 1986	High-resolution STM images of S_A S_B-r, and S_B-s
564 1987	Clasification, fundamental analysis; TB slab calculations
567 1990	Buckling of dimers at S_A steps, STM
568 1990	Energetics of S_A, S_B, and D_B steps; classical potentials
569 1990	Steps on Si(001) miscut in the [010] zone; STM
134 1990	**Review**
565 1990	S_A and S_B step formation energies; STM
570 1991	Kissing sites; STM
566 1991	Electronic structure of S_A; STM
571 1991	Electronic and atomic structures; cluster calculations
572 1992	S_A and S_B step formation energies; STM
530 1992	Electronic structures of dimers, C-defects, and S_B-r compared; STM
573 1992	Electronic structures; STM, surface photovoltage
574 1993	Only rebonded S_B survive annealing and slow cooling; STM
485 1994	Electronic and atomic structures; *ab initio* slab calculations
575 1994	Step stiffness, step and kink energies; LEEM, capilary wave analysis
576 1995	Structures, energies, interactions; *ab initio* slab calculation
577 1995	Buckling at S_A, S_B-s, and S_B-r; *ab initio* slab calculation
578 1995	S_A step and kink energies; high-temperature STM
579 1996	S_A, S_B-s, S_B-dv, and S_B-r structures, buckling; STM
580 1996	Step-induced 2×2 ordering; Monte Carlo simulation
581 1997	Step and kink energies, island shapes; Monte Carlo simulation
123 1998	Step and kink energies, large-size TB calculation
582 1999	Atomic structure of S_A and S_B steps; *ab initio* DFT slab

Table 3.5. Si(001) monatomic steps: selection of references.

the presence of the step most strongly. Their orbitals mix with the orbitals of the open dimers, so that characteristic edge tetramers are formed (like the element 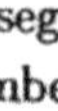 in 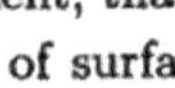). We have described such tetramers in detail on page 160, in the context of the vacancy defects. If we count as "dangling bonds" the orbitals involved in the π bonds within the S_B-s tetramers and the orbitals involved in the π bonds of unperturbed dimers, then it is apparent from Fig. 3.20a that each S_B-s step produces two additional dangling bonds per its segment, that is, per tetramer. Since dimer vacancy defects reduce the number of surface dangling bonds, a dimer vacancy located near the step edge as in Fig. 3.20b restores the dangling bond density to the value for the flat surface. At the same time, the stress arond the vacancy can be relaxed at the step. S_B-s edges are usually observed in this S_B-dv form.[573]

With the edge dimers above the trough site, the step has the **rebonded** structure, S_B-r (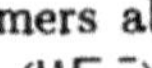). In this case, the substrate dimers adjacent to the edge do not open. The dimer bond survives, and the dimer atoms on the step side are incorporated into the step wall. The other dimer atoms

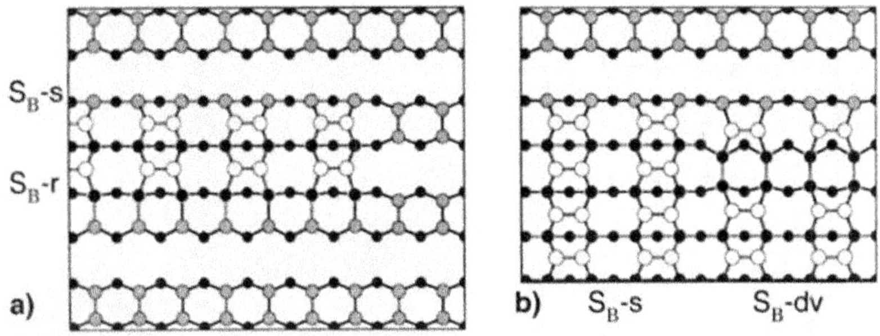

Fig. 3.20. S_B steps on Si(001). Threefold-coordinated atoms are white and gray, fourfold-coordinated atoms are black. Deeper laying atoms are smaller. **(a)** Illustration of the difference between S_B-s and S_B-r steps. **(b)** S_B-dv modification of the simple S_B-s step. The step acquires a dimer vacancy (DV) and the effective dangling bond density at the step edge is thus reduced.

become "rebonding" atoms. The bond in such a dimer may be viewed as the bond in a dimer in which one of the atoms (the wall atom) is frozen in a bulk-like site. It is clear that this bond is stretched more than the bond in a free dimer. The orbitals of the edge dimer on the needle and the orbitals of the rebonding atoms are now quite far apart and the edge dimer may be viewed as unperturbed, at least in the first approximation.

We now recall that the dimers are **buckled** ( ). On a perfectly dimerized surface the direction of buckling can be easily changed. At room temperature, it flips between and so rapidly that the dimers appear symmetric when viewed by STM. Only below ~200 K can STM observe the buckling. Certain defects stabilize the buckling, so that some dimers can be seen as buckled even in room-temperature STM images.

S_A steps ( ) exemplify such pinning centers. The "up" atoms ( ) are always in registry with dimer rows on the lower terrace (Fig. 3.21a). This is in accord with the strain fields induced by the dimers along the edge. Consider the force exerted by the dimer bonds of the lower terrace on their subsurface neighbors at the edge line. If the buckling is as in Fig. 3.21a, this force has the same sign as the force exerted on these atoms by the backbonds of the "up"and "down" atoms of the upper edge.

The same strain compatibility stabilizes wide terraces against formation of translational domains of dimerization (Section 3.4.3). The important diffference is that the energy gain per S_A edge dimer is small, within uncertainty of contemporary *ab initio* methods.[485] We speculate that this may be

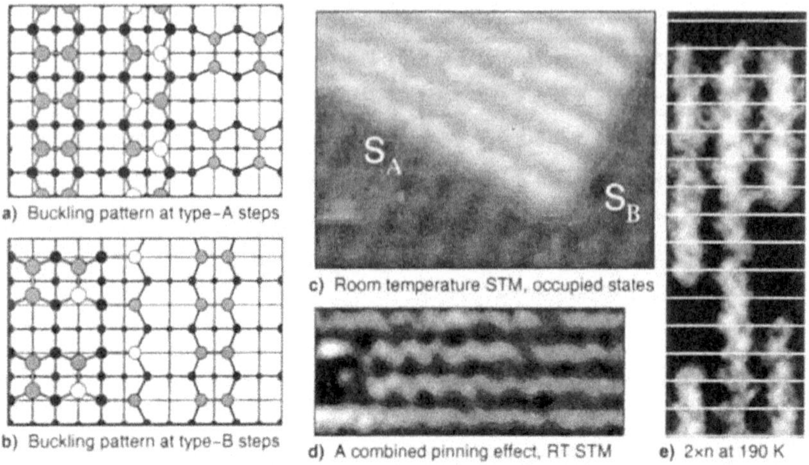

Fig. 3.21. Buckling at monatomic steps. **Left**: Ball-and-stick models. "Up" atoms are white, other surface atoms are gray, fourfold-coordinated atoms are black. **(a)** S_A edge dimers are buckled even at room temperature. **(b)** S_B edge dimers are buckled if stabilized by defects (cf. STM images in Fig. 3.22a and 3.22b). **Middle**: Room-temperature STM.[301] **(c)** Enhanced buckling along S_A and nearly symmertic dimers along S_B. **(d)** A C-type defect (Sect. 3.3.1b) and S_B edge of a dimer vacancy complex combine their pinning strength.[301] **Right**: **(e)** Adapted from Ref. 583: STM image of $2{\times}n$ (Sect. 3.4.1) taken at 190 K, that is, below the temperature at which buckling freezes on wide terraces. Dimers between two closely spaced S_B steps continue flipping. Dimers in a long "bridge" are buckled. The bridge possibly covers a boundary between translational domains of dimerization (Sect. 3.4.3). The horizontal lines indicate the $\times 1$ surface lattice spacing.

because the coupling on the edge is through second neighbors and because dimer buckling is itself a second order effect when compared to dimerization (Section 3.2.2). The correlation between the buckling at S_A and the registry of dimer rows on the lower terrace may be reinforced by the strain field of the whole $c(4{\times}2)$ domain of buckling. The S_A edge stabilizes such a domain against flipping (Fig. 3.21). On the other hand, the same S_A buckling pattern, apparently invoked by the same strain-relief mechanism, is strong enough to support formation of *single-dimer wide* dimer strings on top of translational domain boundaries (Section 3.4.3). It is thus plausible that it is the *vertical* strain field that correlates S_A buckling with the lower terrace rows. Note that it is the relaxation of atoms in at least five subsurface layers that makes the $c(4{\times}2)$ arrangement of buckling dimers more favorable than the $2{\times}2$ pattern (Section 3.2.2b).

Some STM images reveal another interesting detail of the S_A edge:

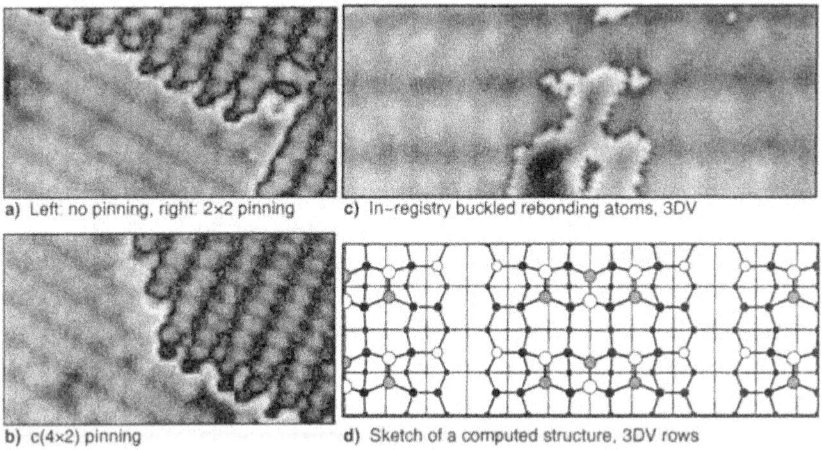

Fig. 3.22. Buckling at S_B steps. STM images[301] show occupied states, the gray scale is split to highlight the structure at step edges. **Left**: STM images show that when S_B-r edge dimers buckle, then their "up" atoms are *out of registry* with the buckling of the rebonding atoms. **(a)** 2×2 buckling, possibly stabilized by the kink or by a defect in the lower terrace. Edge dimers on the leftmost part of the step are symmetric. **(b)** c(4×2) buckling, possibly reinforced by a nearby defect. **Right**: Neighboring S_B-r steps favor a different buckling pattern. **(c)** Both S_B-r edge dimers of a triple dimer vacancy (3DV, compare Fig. 3.13e on page 164) are buckled. The rebonding atoms buckle *in registry* with the edge dimers. **(d)** Sketch of *ab initio* step structures reported in Ref. 577. "Up" atoms are white, other surface atoms are gray, fourfold-coordinated atoms are black.

the lower-terrace dimers adjacent to the step appear to move slightly towards the step edge.[579] This may reflect strain relaxation in the direction *perpendicular* to the edge. In this direction the upper terrace is under tensile stress and tends to shrink, while the lower terrace is under compressive stress and tends to expand. This is consistent with the increase of the tensile stress component with the buckling angle (page 138): stress compensation at the edge caused by the expansion of the lower terrace allows for enhanced and more stable buckling on the upper terrace. However, no corresponding simulated STM images or detailed atomic positions are available in the literature. It may be that the apparent shift of the lower-terrace dimer is a purely electronic effect, or even a distortion of the tunneling current caused by the presence of a nearby protrusion – the step edge.

S_B-r steps (¦¦⊑ ⋮) appear symmetric to STM (Fig. 3.21) even at low temperatures (Fig. 3.21e). Yet according to *ab initio* data,[577] edge dimers and the rebonding atoms should buckle as well. *Ab initio* calculations indicate that buckling is enhanced also on S_B-s steps.[485] No theoretical work is available for the buckling at S_B-dv.

The enhanced flipping of dimers at S_B may be due to the buckling phase conflict between the step edge an the terrace. The terrace prefers the c(2×4) periodicity, in which the buckling direction alternates across dimer rows: $\rightarrow \leftarrow \rightarrow \leftarrow$. But the dimers on S_B edges seem to prefer the opposite pattern: $\rightarrow \rightarrow \rightarrow \rightarrow$; this is the conclusion from *ab initio* calculations.[485] This incompatiblity can frustrate the buckling direction of edge dimers.

If S_B steps have their preferred buckling pattern, they should act as pinning centers (albeit weak) for the direction of buckling. The STM micrograph in Fig. 3.21d indicates that such a pinning exists. It shows the buckling induced by a C-defect in the vicinity of an S_B edge of a dimer vacancy complex. The C-defect pins the buckling efficiently in the neighboring row terminated by the S_B edge, but has less effect on the buckling in the neighboring "perfect" row. The enhnanced flipping at S_B edges may thus be seen as an analogue to the enhanced low-temperature flipping on one side of an isolated C-defect embedded in a *c*(4×2) matrix (page 150).

Buckling at S_B edges can be observed in a neighborhood of step or terrace defects, even at room temperature[584,301] (Fig. 3.22). Interestingly though, its pattern is the *opposite* to that predicted *ab initio*:[485,577] the "up" atoms of the edge dimers are out of registry with the "up" rebonding atoms (Fig. 3.22a and 3.22b), while the calculations produced them in registry (Fig. 3.22d). We suppose that this discrepancy is induced by the elastic interaction between S_B-r step edges which was present in the calculation (Fig. 3.22d) but is absent at a misorientation step. Indeed, if dimers on the edges of a dimer trivacancy (Fig. 3.13e, page 164) become buckled, the rebonding atoms buckle in phase with them (Fig. 3.22c). A dimer trivacancy is a nearest-neighbor pair of two S_B-r sections in the same dimer row. The calculations did in reality describe 2×6 rows of dimer trivacancies.

Our next subject is the step **formation energy**. It is not *a priori* obvious which of the B steps, S_B-s, S_B-dv, or S_B-r, has a lower formation energy. Calculations suggest that S_B-r steps cost less than S_B-s. That S_B-r steps have significantly lower formation energy than S_B-s steps is corroborated by recent observation[585] that the motion of kinks on S_B-r steps proceeds in units of two dimers in a row. Steps try to avoid the formation of S_B-s segments even at the price of enhanced barrier for kink migration. Roughly equal abundancy of S_B-r (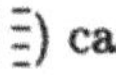) and S_B-s (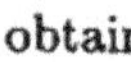) can be obtained during homoepitaxial growth at low temperatures.[573]

Theory and experiment agree that the formation energy of S_A steps (≡≡¦¦¦) is lower than the formation energy of S_B steps. But the strong interaction between monatomic steps (Sect. 4.3.2) through surface and subsurface strain fields makes computer simulations difficult. Particularly troublesome is the long-range interaction mediated by the surface strain field. Its origin is in the partial cancellation of the anisotropic surface stress: on a two-domain surface, the tension along the dimer bonds and the compression across the dimer bonds may compensate. As a consequence, the formation energy depends strongly not only on the distance between the steps but also on the details of the surface reconstruction (dimer buckling!). It is not easy to perform accurate calculations with unit cells or clusters so big that the step-step interaction contribution to the step formation energy is exactly included and reliably isolated. Such size restrictions do not apply to measurements. With the rapid development of the performance of computers, this drawback of simulations may disappear in near future.

Formation energies of steps are hidden in the equilibrium shape of steps as well as in the equilibrium shape of monolayer-high islands. Thermodynamical equilibrium requires that different segments of the same step exchange atoms (it is not necessary that atoms or dimers *detach* from the steps). Room-temperature STM images show the shape of steps frozen at 350-450°C (Sect. 4.2.3b); this is the temperature at which the diffusion along the steps stops. The effective freeze-in temperature T_f depends on such conditions as the cooling rate or the density of surface defects.

Energies of S_A and S_B step formation, λ_{SA} and λ_{SB}, can be determined from room-temperature STM images.[565] To minimize interaction between steps, the samples are misoriented by 0.5° or less. Since the distribution of kink lengths follows Boltzmann statistics, a statistical analysis of thermally excited kinks on S_B steps provides information on the S_A step energy, and vice versa. The experiment measures $E(l)/kT_f$, the formation energies of kinks with l segments, where T_f is the temperature at which the (presumably) equilibrium distribution of kinks was frozen in:

$$n_{+l}n_{-l}/n_0^2 = \exp\left(-E(l)/kT_f\right),$$

where n_{+l}, n_{-l}, and n_0 are the probabilities to find kinks with l segments running in the two possible directions and the probability of finding no kink (see Ref. 586 for practical details). The energy of a kink is proportional to the kink length plus a constant, interpreted as the **corner energy** C.

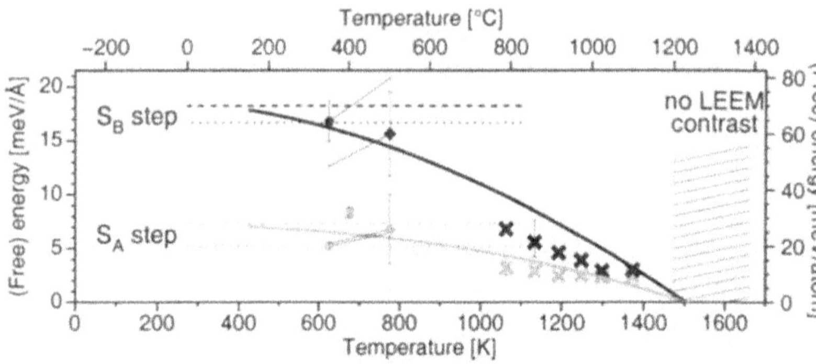

Fig. 3.23. Energetics of S_A and S_B steps. **Crosses and solid lines**: Temperature dependence of *free energy* of step formation; the lines represent analytic model with the partition function employing the step *energies* indicated by **dashed lines** and 80 meV as the corner energy.[575] The **dotted lines** mark the step energies suggested by an analysis with an Ising interaction hamiltonian.[581] **Full symbols**: Step *energies*, plotted at the freeze-in temperature T_f which was assumed in data interpretation: Refs. 565 and 518 (circles), Ref. 586 (diamonds). The slanted lines indicate rescaling of the measurement of one group to the T_f assumed by the other group. The square shows the energy obtained assuming independent kinks; data collected around 400°C.[578]

Step free energies $\beta_{SA} = \lambda_{SA} - S_{SA}T$ and $\beta_{SB} = \lambda_{SB} - S_{SB}T$, where S_x is the configurational entropy of the step x, can be found from LEEM by an analysis of Fourier components of step shape and from the equilibrium shape of Si islands of monatomic height.[575] An analysis of the LEEM data in terms of an Ising hamiltonian is given in Ref. 581.

These results are summarized in Figure 3.23. The temperature dependence of free energies is compared to the step energies deduced from STM. A consistent set of energies is obtained when the freeze-in temperature around 400°C is assumed. Approximately, $\lambda_{SA} = 25 \pm 10$ meV/atom, $\lambda_{SB} = 65 \pm 10$ meV/atom, and $C = 70 \pm 10$ meV. These values agree with results of recent Tight Binding (TB) calculations[123] performed for surfaces with relatively big step-step separations. Since the TB hamiltonian was fitted (Sect. 1.3.1c) to reproduce the geometries and electronic properties of bulk Si and the dimerized Si(001) surface, this agreement confirms that the energetics of S_A and S_B is not determined by unusual electronic effects.

Step free energies tend to vanish around 1230°C. A slight increase of measured free energies, seen in Fig. 3.23 near to this temperature, is an artifact caused by the neglect of overhangs, vacancies, and islands in the interpretation of the experimental data.[581] An interesting observation is

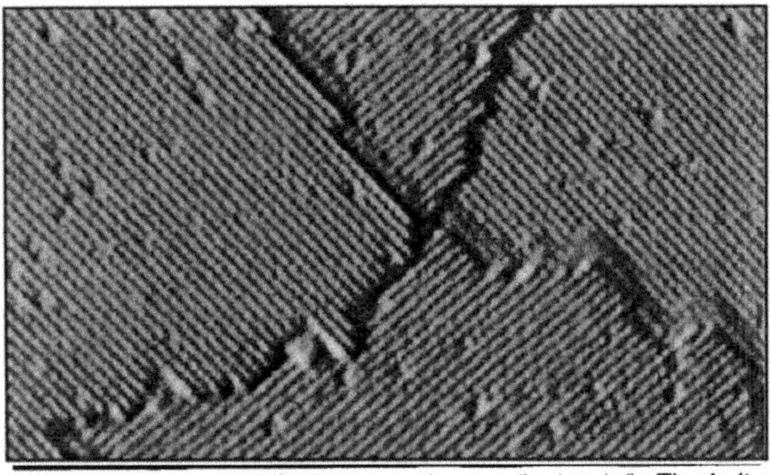

Fig. 3.24. Two monatomic Si(001) steps meeting at a "kissing site". The shading is keyed to slope. After Ref. 570; adapted from Ref. 134.

that the corner energy is of the same order of magnitude as kink energies. This increases the energy of short kinks, stabilizing steps against thermal disorder. In other words, the corner energy increases the temperature T_R at which the entropy term, ST, brings down the free energy of steps to zero (Fig. 3.23). Formally, T_R is the temperature at which a thermodynamic roughnening transition occurs. At about the same temperature $T_R \simeq 1230°C$, LEEM contrast is lost.[563]

Kinked steps sometimes meander so much that two neighboring monatomic steps touch at a single point, known as a **"kissing site"** (Fig. 3.24). Such sites are often observed on surfaces tilted toward [100] or [010] and can be also found when the tilt is toward [110] or [1$\bar{1}$0]. In the latter case it has been shown[570] that they are a nonthermodynamic feature, caused by boundaries between transitional domains of dimerization (Sect. 3.4.3). These domain boundaries are created during surface cleaning at high temperatures. During the subsequent cooling, the boundaries become preferential sites for the condensation of the two-dimensional gas of silicon detached from steps at high temperatures. A step advances along a domain boundary line utill it reaches the end of the terrace and touches the next step at a kissing site. Since the boundaries are metastable, the density of kissing sites is higher when the surface is cooled down faster.

Having explored the geometric and energetic details, we turn to elec-

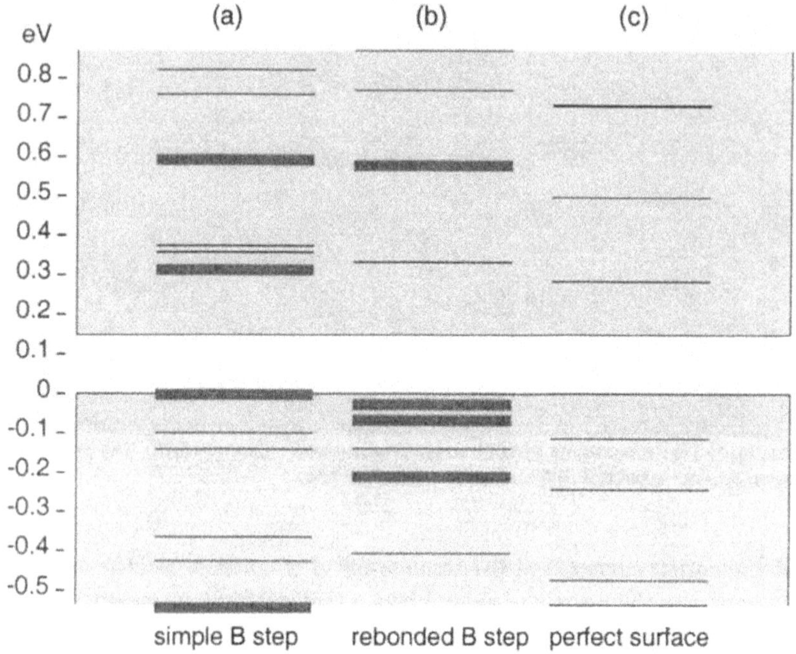

Fig. 3.25. Electronic states of Si(001) monatomic steps, as calculated[571] by the DV-Xα method. The surface was modeled by a small Si cluster with "bulk" bonds simulated by hydrogen passivation of the appropriate bonds on the cluster surface. Step states are indicated by heavy lines. **(a)** S_B-s steps, **(b)** S_B-r steps, **(c)** S_A steps (since the cluster can describe the dimer states only approximately, we interpret the S_A states as describing the perfect surface. The boxes indicate the dimer surface bands.)

tronic structures. Given that S_A steps hardly perturb the dimers, it is natural to suppose that they do not produce states in the surface band gap. As for simple S_B-s steps (⁝⁝⁝ ⁝ ⁝), the tetramers terminating them have fully occupied bonding and empty antibonding states (page 160). Studies of Si(113), on which such a tetramer appears in each surface cell, show[301] that these states are separated by a gap wider than that of Si(001). That these steps have no states in the gap is also indicated by the fact that 2DV, defects which contain a tetramer of S_B-s kind (compare ⁝⁝⁝ ⁝ ⁝ and ⁝⁝⁝ ⁝⁝⁝), have not been reported to induce such states.

Yet the existing theoretical evidence does not exclude that S_B-s steps have their highest occupied states (Highest Occupied Molecular Orbitals, or

HOMO) above the dimer π band. As an example, consider Fig. 3.25a, which shows the calculated[571] states of S_B steps. The HOMO of S_B-s tetramers are close to the top of the occupied π band of the surface dimers. These calculations modeled the surface as a small cluster; this alone could affect the eigenvalues by more than 0.1 eV. An energy shift of this size is sufficient to move states from the surface band gap into one of the surface bands, or vice versa. Also note that *ab initio* LDA bands[485] should be quasiparticle-corrected (page 111); this correction depends on the wave function.

A study of surface charge dynamics by STM measurement of surface photovoltage gives insight into this electronic structure.[573] The lifetime of the charge transferred from the STM tip to S_A and S_B-dv is comparable to (shorter than) the lifetime of the charge transferred to terrace dimers. This indicates that the electronic states of S_A and S_B-dv steps are degenerate with states which can easily dissipate the charge, that is, with the terrace energy bands. The charge transferred to C-defects which have states in the surface gap remains localized two times as long as on the perfect surface. S_B-s sections are rare, so the results for S_B-s cannot be interpreted easily.

What about rebonded (S_B-r, 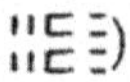) steps? One might expect that the dangling bonds of the rebonding atoms, being half-occupied and relatively isolated from other surface orbitals, could make band-gap states, possibly of metallic character. Scanning tunneling spectroscopy[530] does indicate that electrically active states exist at S_B-r. This is supported by surface photovoltage studies:[573] the lifetime of excess electrons on S_B-r is comparable to that on C-defects. On the other hand, the theoretical results displayed in Fig. 3.25b suggest that S_B-r steps do not produce states in the surface band gap. Similarly, recent *ab initio* studies of monatomic steps[485] did not find any edge-induced states in the gap, neither for S_B-s nor for S_B-r. It cannot be excluded that the measured states come from kink corners, or that the atomic structure of S_B-r steps is different than it is currently believed. But it is also possible that contemporary computational approaches fail to predict the position of the step electronic states with sufficient accuracy.

b. *Biatomic steps*

Next, we address atomic structures, morphology (kinks, splitting), step-induced buckling of dimers, and STM images of the biatomic steps on Si(001). Step interactions are treated in Section 4.3.2 devoted to phase transitions and the role of steps in surface diffusion, homoepitaxy and sublimation are discussed in Section 4.2.2. Table 3.6 lists some references.

Ref year	Subject relevant to biatomic steps on Si(001)
587 1986	Different models of D_B steps; TB slab calculations
564 1987	D_B-r steps proposed, D_A kinks discussed; TB slab calculations
588 1987	D_B-r reconstruction confirmed; STM
589 1988	D_B steps on Si(001) and Ge(001), edge buckling visible; STM
134 1990	**Review**
566 1991	Split D_A kinks; STM
590 1991	A metastable D_A surface can be prepared; STM
591 1994	D_B formation energy; equilibrium crystal shape
576 1995	Buckling, formation energies, interactions; *ab initio* slab calculation
592 1996	D_A and D_B formation energies; STM
582 1999	Atomic structure of D_B steps; *ab initio* DFT slab

Table 3.6. Si(001) biatomic steps: selection of references.

Much of the interest in the vicinal Si(001) stems from the tendency of these surfaces to form double steps. Si(001) surfaces with exclusively biatomic steps are *primitive*: they consist entirely of atoms from one of the two inequivalent fcc sublattices I and II of the diamond structure (Fig. 3.1, page 130). In consequence, the direction of dimerization is the same on all terraces (Fig. 3.16, page 172). This means that the surface can be regarded as a single-domain 2×1, in contrast to the usual two-domain, 2×1 + 1×2 surfaces with monatomic steps. The possibility to prepare single-domain surfaces played an important role in the unraveling of the reconstruction of Si(001). In addition, double steps may be formed during the step-flow mode of epitaxial growth. Due to the different speed of S_A and S_B steps, the S_B steps catch up with the S_A steps and form biatomic edges (Sect. 4.4).

Last but not least, the primitive character of double-stepped Si(001) has its consequences for the epitaxial growth of compound overlayers with diamond structure, like GaAs/Si(001). In such compounds, cations C occupy one of the two sublattices and anions A occupy the other one. Both sublattices are equivalent in silicon, so growth on a non-primitive Si(001) results in antiphase disorder: the film contains both possible phases, C(I)-A(II) and C(II)-A(I), reflecting the local termination of the Si(001) with the sublattice I or II. A primitive, double-stepped Si(001) was thus seen as a natural candidate for the growth of ordered compound films.[593] Though practical integration of silicon and III-V technologies in microelectronics is still not accomplished, GaAs(001) grown on highly misoriented Ge(001) substrates is now a commercial product for photovoltaics used to power communication satellites.

Like S_B steps, D_B steps (Fig. 3.18, page 175) appear in a simple (D_B-s) and in a rebonded (D_B-r) form, the rebonded one being typical. The difference between them is conceptually the same as the difference between S_B-s and S_B-r. Fig. 3.18 illustrates this similarity and visualizes that the D_B-r structure may be viewed as an S_B-r step located so close to an S_A step that halves of the edge S_A dimers are incorporated into the S_B step wall as the rebonding atoms.

The energy of D_A steps is so high that they are rare. D_B steps tend to be straight, but even if an A-type kink is forced by entropy or by misorientation, the A edge splits into two monatomic steps, S_A and S_B. This means that *stable* double-stepped Si(001) surfaces have nearly exclusively terraces of B type (dimers arranged as in: or cf. Fig. 3.17d, page 173). A-type domains appear only at split kinks. A *metastable*, single A-domain surface (as in:) can be prepared by sputtering of a two-domain surface.[590] Sputtering creates dimer vacancies; these attach preferentially to B-type steps, eating the B terraces away, so that B terraces eventually become A-type and steps become D_A steps. The A-domain surface reverts to the two-domain surface when annealed at 750°C.

Predominantly single-domain, A-type Si(001) can be produced also by stressing the sample mechanically or thermally (Section 4.3.2) or by passing DC current through the sample to heat it (Section 4.2.3c). A-type domains can then be forced to grow at the cost of B-type domains, or reversely. But repulsion between S_A and S_B keeps in this case the monatomic steps from coalescing into biatomic, D_A edges.

The linear energy density λ_{DB} of D_B-r steps was found[592] in a similar way as the energies of S_A and S_B. In a certain range of miscut angles, D_B steps mix with S_A and S_B (Sect. 4.3.2) and are sometimes split into S_A+S_B. Knowing the energies of S_A and S_B, one obtains λ_{DB} from Boltzmann occupancy ratio.[o] This yields the energy in units of thermal energy kT_f at the temperature T_f at which the thermal distribution was frozen in (page 183). With $T_f = 500$°C, one obtains[592] $\lambda_{DB} = 13$ meV/Å, or $\lambda_{DB} = 50$ meV/a_s, where $a_s = 3.84$ Å is the surface lattice constant.

STM revealed[592] a small fraction of unsplit kinks, each with a single D_A segment. Their energy was estimated as $\lambda_{DA} = 150$ meV/a_s (39 meV/Å), assuming $T_f = 500$°C. This estimate may be inaccurate,

[o] Strictly speaking, the interaction between S_A and S_B edges should be taken into account. It can be estimated[592] that the contribution from this interaction to the energetics of the steps is in this case practically negligible (4 meV per B-type edge dimer).

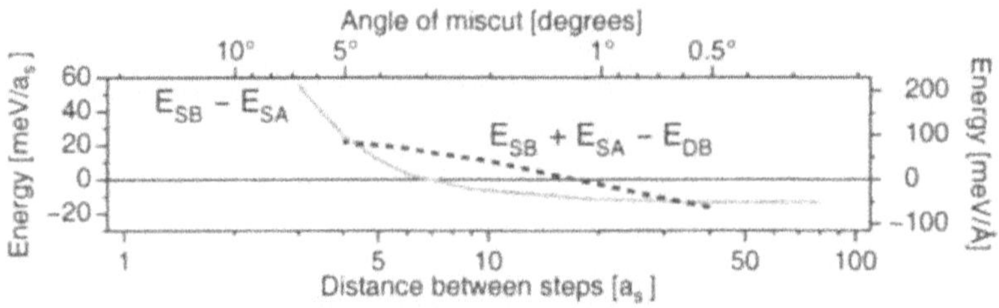

Fig. 3.26. Dependence of step energy differences ΔE on step-step separation l obtained with classical potentials.[568] Gray solid line: $\lambda_{SB}-\lambda_{SA}$. Broken line: $\lambda_{SB}+\lambda_{SA}-\lambda_{DB}$. The units are meV/$a_s$ and a_s, where $a_s = 3.84$ Å is the surface lattice constant. Details such the absolute values of ΔE, in particular, the favoring of S_B over S_A for small miscuts (compare Fig. 3.23, page 184), seem to be an artifact of the classical approach (pages 210 and 227), but the character of $\Delta E(l)$ should be properly rendered.

first because the statistics were poor (33 D_A kinks), and second because the kink energy may be dominated by the (unknown) corner energy C_D.

How do these energies compare with other data? A similar value of $\lambda_{DB} = 42\pm20$ meV/a_s was extracted[591] from the equilibrium shape of voids in Si. For computed energies, experimentalists like to refer to pioneering tight-binding results.[564] The value of λ_{DB} obtained for miscut of $\alpha = 5.4°$ (not much different from the $\alpha = 4.5°$ of surfaces used in the STM study[592]) is $\lambda_{DB} = 50$ meV/a_s, in a remarkably good agreement with experiment. Yet the computed $\Delta E_{sd} = (\lambda_{SB} + \lambda_{SA}) - \lambda_{DB}$ was $\Delta E_{sd} = 110$ meV/a_s, which is much larger than the experimental $\Delta E_{sd} \simeq 40$ meV/a_s. Recent *ab initio* calculations[576] yielded for the same $\alpha = 5.4°$ a realistic $\Delta E_{sd} = 38$ meV/a_s, but the absolute value of $\lambda_{DB} = 170$ meV/a_s seems too large. As for the D_A steps, they are more energetic than D_B by 490 meV/a_s (tight-binding) or 260 meV/a_s (*ab initio*). Both numbers exceed the experimental estimate, $\lambda_{DA} \simeq 100$ meV/a_s.

Assuming that the *ab initio* theory reproduced the physics, these discrepancies may be due to the complex elastic interactions on a stepped surface. A proper account for these interactions is difficult in calculations as well as in the interpretation of experimental data. Results obtained with classical potentials show[568] that inclusion of deep subsurface relaxation is crucial for a realistic computation: when the number of relaxed layers was increased from 12 to 40, step formation energies dropped by a *factor of two* (!). In addition, the difference between λ_{SB} and λ_{SA} depends strongly on step-step separation, particularly for miscuts above $\alpha \simeq 5°$ (Fig. 3.26);

it saturates at small miscuts. The difference $\Delta E_{sd} = \lambda_{SB} + \lambda_{SA} - \lambda_{DB}$ is less sensitive to α; it tends to saturate at large miscuts.

$\Delta E_{sd} > 0$ means that a D_B-r step (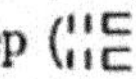) is preferred over a S_B+S_A pair (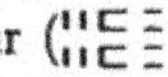 + 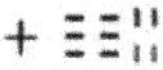). Fig. 3.26 illustrates how this is reconciled with the instability of biatomic steps for small miscut angles (Fig. 3.16, page 172). Strong surface stress anisotropy gives two-domain surfaces (surfaces with monatomic steps) adventage for small miscuts by facilitating surface stress relaxation (Sect. 4.3.2). An additional effect, not included in Fig. 3.26, is that D_B steps are smooth while S_B steps are rough. Entropy favors S_B steps, particularly on wider terraces, where they can meander freely.

Kinks on D_B steps are an interesting subject. They are relatively unfrequent on surfaces tilted exactly in the [110] azimuth, but can be easily found when the tilt deviates from the [110] azimuth so that the line of steps rotates from the ideal [110] direction toward [100]. The kinks are then forced by geometric constraints. The immediate observation, apparent in STM images and easily understandable from the energetics of D_A, S_A, and S_B, is that virtually all kinks are split into monatomic pairs: $D_A \rightarrow S_A + S_B$.

This splitting causes an intriguing correlation between the dimerization phases on terraces of bi-stepped samples. Dimers on the lower terrace may be either shifted to the left (R_+) or to the right (R_-) with respect to the dimers on the upper terrace (Fig. 3.27a and 3.27b). When the steps are straight (the surface is tilted in the [110] azimuth), both configurations are equivalent. However, when the tilt deviates from the [110] azimuth, a correlation in the dimerization phase is seen. Indeed, misorientation kinks discriminate between R_+ and R_- (Fig. 3.27, four lower panels).

It has been observed[592] that when the misorientation kinks are oriented such as in Fig. 3.27, the shift is to the right (R_-). Note that an unsplit D_A kink of this orientation favors R_+: the number of dangling bonds at the kink edge is then minimized (compare Figs. 3.27c and 3.27d). When the kink splits, the same mechanism makes the favored shift depend on the number of dimers in a dimer row of the split-off terrace. The terrace ends at a low-energy S_B-r step when this number is even ("even kink") and the shift is R_+ (Fig. 3.27e) or when it is odd and the shift is R_- (Fig. 3.27f). Otherwise, the terminating step would be the high-energy S_B-s.

A split kink could easily adapt its structure to any shift, by switching between even and odd split-terrace lengths. The existence of a preferred shift may indicate that dimerization of terraces begins at split kinks and

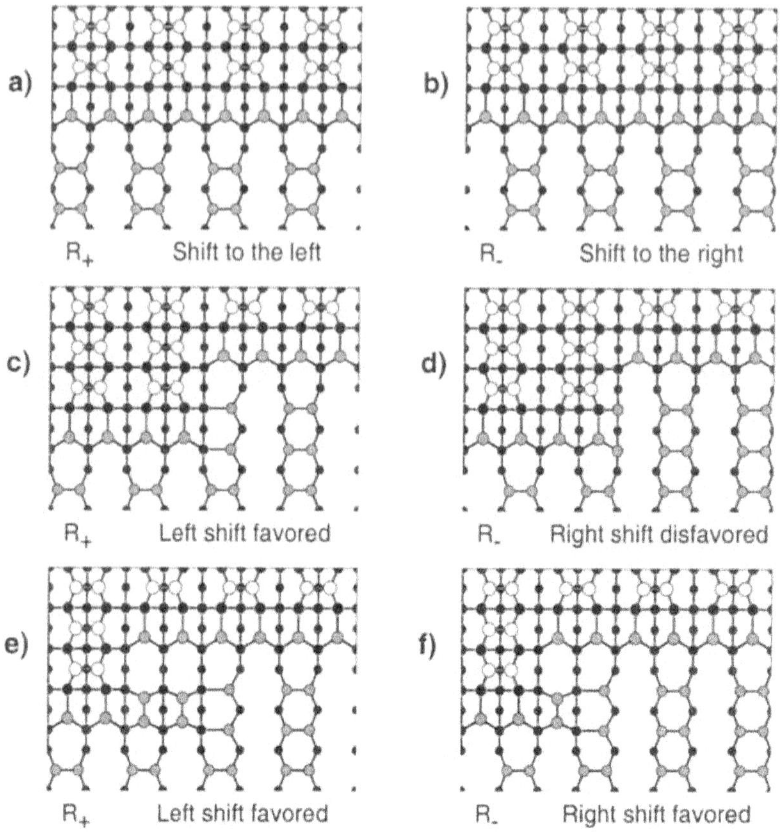

Fig. 3.27. Correlation between dimerization phases when D_B steps are inclined from [110]. Threefold-coordinated atoms are white and gray. No correlation is expected for straight steps: (**a**) shift to the left, R_+, is equivalent to (**b**) shift to the right, R_-. A non-split D_A kink (**c**) favors R_+ and (**d**) disfavors R_- due to the number of broken bonds. A split kink (**e**) with an *even* number of dimers favors R_+; (**f**) an *odd* number of dimers favors R_-, which is the observed configuration for this direction of the inclination.

proceeds on already existing *undimerized* terraces. But alternatively, one can imagine that the sites seen later as kinks act as nucleation centers from which *dimerized* terraces grow by sublimation or adsorption of Si atoms.

The nucleation of dimerization at odd split kinks can be a kinetic effect. In principle, one cannot exclude that odd kinks are more stable than even kinks during the nucleation process.[p] Strain relaxation at step

[p] The observed split kinks are odd.[566,592] But this is an *a posteriori* effect. Indeed,

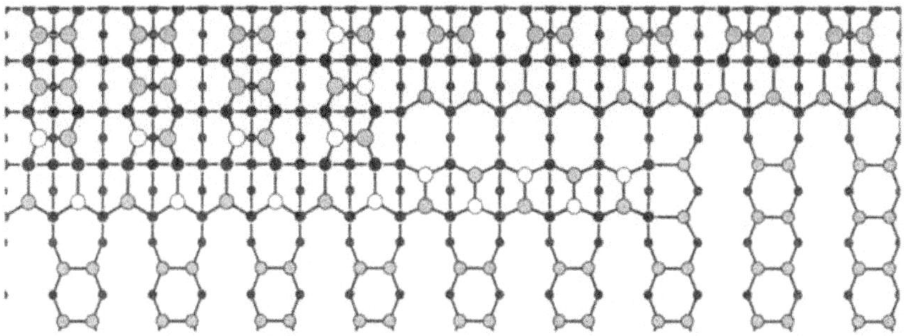

Fig. 3.28. Buckling of dimers on D_B-r step edges and on a split D_A kink. Surface "up" atoms of interest are white, other threefold-coordinated atoms are gray, fourfold coordinated atoms are black. The D_A kink is split, extending into a one-dimer wide and five-dimers long terrace. Buckling of atoms on the S_A edge of the terrace is the same as on a usual S_A edge. The long terrace is terminated by a S_B-r step. The buckling on D_B-r follows the rules for S_A (Fig. 3.21a, page 180) and S_B (Fig. 3.22, page 181).

and kink edges may play a role in this stabilization. Can this relaxation be related to dimer buckling? Maybe, but it is not clear that an effect due to buckling, even a collective effect, can remain influencial at temperatures as high as those at which the dimerization phase is determined (~1200°C).

This brings us to the **buckling** of dimers. A D_B-r step is a sum of S_B-r and S_A. If the rebonding atoms are viewed as parts of S_A edge dimers, the rebonding "up" atoms should be in registry with the dimer rows on the lower terrace (Fig. 3.21, page 180). On the other hand, the neighborhood of D_B-r edge atoms is nearly identical with the neighborhood of S_B-r atoms (Fig. 3.18, page 175). One thus expects the same correlation between the edge dimers and the rebonding atoms of D_B-r as it is known for S_B-r (Fig. 3.22, page 181): "up" rebonding atoms out of registry with the "up" dimer atoms. The buckling pattern should be thus such that the dimer rows on the lower terrace are in registry with "up" atoms on the D_B-r edge (Fig. 3.28). STM images[588,589,566] agree with this prediction.

In summary, our knowledge about atomic structures of Si(001) steps seems to be quite detailed and self-consistent. The related electronic structures are less certain, and the step formation energies are known only approximately. Certain structural features related to steps, such as kissing sites of monatomic steps (page 185) and the above described correlation of

---

if even kinks nucleated the dimerization, the dimerization phase would be the opposite and even kinks would terminate with S_B-r. That is, even kinks would then dominate.

the dimerization phase on upper and lower terraces of biatomic steps, may carry information about the high-temperature dynamic processes which take place on the surface near to the melting point of silicon.

3.4. Metastable Structures

There are two classes of metastable reconstructions of the nominally clean Si(001): 2×n, due to missing dimer defects arranged in rows running *perpendicular* to the dimer rows, as in: :::::, and reconstructions like $c(4\times4)$, related to ad-dimers. Such structures are generated by specific annealing/cooling procedures, by homoepitaxy, or by ion bombardment.

The 2×n structure may indicate contamination,[594,595] particularly with Ni, but also with Cu or Ag.[529] The dependence of surface morphology on the surface Ni concentration[q] has been recently calibrated.[541] But this reconstruction may be a feature of the clean surface.[596,597,539,543] In our opinion, Ni atoms are not a basic ingredient of the 2×n structure. Their influence is indirect, through amplification of properties intrinsic to Si(001).

3.4.1. *Si(001)* 2 × n

This Section describes the atomic and electronic structure of the characteristic "stripes" on Si(001) 2×n and the mechanisms of their formation, including contamination aspects. Table 3.7 gives a selection of references.

There are several ways to obtain the 2×n structure ($6 \leq n \leq 12$) on a nominally clean Si(001) surface. They include:

- Single[598] or repetitive[543] anneal and quenching;
- Sputtering with rare gas ion beams;[572,596,599]
- Hydrogen etching[599] (Section 5.2);
- Oxygen etching[597] (Section 5.4.4);
- Low-dose halogen etching[600,601] (Section 5.3);
- Electron-simulated desorption.[128]

The common factor shared by these methods is that they produce high densities of dimer vacancies (DV). When the temperature is high enough for DV to migrate on the surface, they line up in regularly spaced trenches *perpendicular* to the dimer rows, as in Fig. 3.29. The vacancy lines may sometimes extend for thousands Å without interruption. The density of

[q] Interactions of Ni with silicon surfaces are discussed in Section 5.7.2.

Ref year	Subject relevant to Si(001) 2×n
439 1984	Dimer vacancy (DV) model proposed; suggestion
594 1984	Si(001) 2×n reported for the first time; LEED
602 1985	Trace amount of Ni leads to 2×8 reconstruction; LEED
598 1986	No Ni found, initial conditions essential; LEED and AES
512 1986	Ni contamination is not the sole reason for 2×n; LEED
595 1988	2×n is a sign of Ni contamination; STM
590 1992	Onset of DV rows formation
596 1992	Clean surface can show 2×n; Ar bombardment
597 1993	Clean surface can show 2×n; SiO_2 desorption
603 1993	Anisotropic diffusion of DV; empirical theory
484 1993	Anisotropic diffusion of DV; *ab initio* theory
604 1995	Ordering and disordering od single DV; empirical theory
539 1995	Basic building blocks of vacancy lines; STM, classical potentials
605 1995	Anneal/quench cycles may build up surface contamination
540 1995	The role of is Ni to increase the density of DV; STM
606 1995	Narrow DV lines have mostly S_B edges, wide lines prefer S_B; H etch
599 1996	2×n by Ar^+ bombardment, O_2 etching, H etching; STM
541 1996	How to avoid Ni contamination during anneal/quench cycles
607 1997	**Reviev**
237 1997	How to avoid Ni contamination by HF purification
128 1999	DV lines created by electron beams; STM

Table 3.7. Si(001) 2×n: selection of references.

the vacancies must be high enough to allow for their condensation before they annihilate at step edges. However, this density should not be *too* high: when as many as ~ 30% of the dimers are removed at random, the vacancies agglomerate into two-dimensional "islands".[545]

It may seem that the simplest recipe to make Si(001) 2×n is to repeat cycles of annealing (~1200°C) and quenching.[543] But the vacancies are then stabilized by nickel atoms which segregate to the surface during this process (Sect. 5.7.2). When special care is taken to keep the surface free from Ni, the 2×n does not appear.[607] The other recipes produce dimer vacancies in a cleaner way. A simple test can verify if a 2×n structure is due to Ni contamination: annealing around 950°C destroys the "clean" 2×n, while the nickel 2×n survives even at 1200°C. Moreover, the Ni-induced lines are more regular than the 2×n vacancy trenches of the clean surface, as evident particularly by high surface concentrations of nickel.

The trenches contain two basic blocks. The simpler one is a double dimer vacancy, 2DV (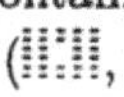, Fig. 3.13c, page 164). The other one, (1+2)DV, is composed of a double and a single vacancy separated by a single dimer (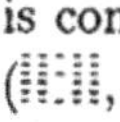, Fig. 3.13d). These blocks often alternate. Some line fragments are

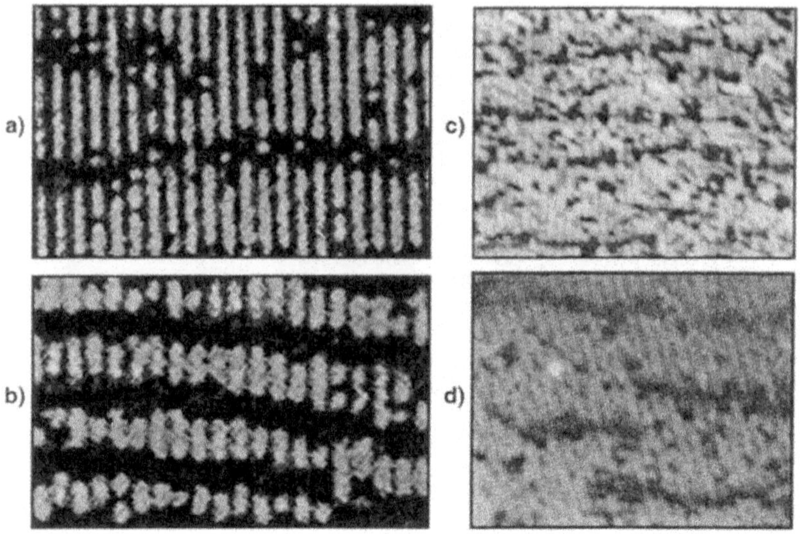

Fig. 3.29. STM images of Si(001) 2×n reconstructed surfaces. Dimer vacancy rows run perpendicular to dimer rows. The structures on the **left** (from Ref. 543) were promoted by trace Ni contamination. They were obtained after repetitive anneal/quench cycles, are highly ordered, and thermally stable. **(a)** Lower density of vacancies, many split-off dimers. **(b)** Higher density of vacancies. **Right**: Dimer vacancy rows on a clean surface are less ordered and thermally unstable (see text). **(c)** 2×n after Br_2 etch at 830°C. From Ref. 600. **(d)** 2×n after Ar^+ bombardment and 750°C annealing. From Ref. 596.

built of such complexes as 3DV, (1+2+1)DV, (1+3)DV, or (2+2)DV.

The abundancy of these basic blocks is in accord with the *ab initio* result[484] that the lowest formation energy (per missing dimer) have the lines of (1+2)DV defects (0.14 eV/DV) and that the formation energies of lines of complexes as 2DV, 3DV, or (1+3+1)DV are only 0.1-0.2 eV higher. The structures at trench bottoms are difficult to measure, but the calculation indicates that the dangling bonds of the 1DV, 2DV, and 3DV elements rebond such as in Figures 3.13d and 3.13e.

Let us now consider the mechanism by which the 2×n trenches are formed. The tendency of isolated missing dimers to arrange in tiny complexes like 2DV or (1+2)DV was discussed in Section 3.3.1a. A dimer vacancy introduces stress which is relieved more efficiently when two or three such defects come together. Elastic interactions seem to drive the lining up of defects and keep the individual lines appart so that the quasi-regular 2×n structure appears. A short-range attraction between the vacancies in neighboring rows, taken together with their preferential diffusion along the dimer rows (Sect. 4.2.1c), can easily lead to $2\times n$ lines crossing the dimer rows at right angles – provided that the density of the defects is sufficiently high, that the defects are sufficiently mobile, and that the self-assembling attraction and repulsion between the defects is strong enough to overcome thermal disordering. According to results of classical potential calculations,[539] two isolated dimer divacancies gain 0.08 eV when they form a short "line", becoming nearest neighbors in adjacent dimer rows: 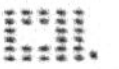. The energy gain for two (1+2)DV split-off dimer defects is 0.28 eV. These numbers should be compared with *ab initio* data for vacancy lines (Fig. 3.13, page 164): 0.12 eV and 0.24 eV. The *ab initio* and classical potential results are remarkably similar in this case.

Is there another mechanism which keeps the trenches straight and which makes them run across and not along the dimer rows? The edges of a trench are a close pair of monatomic steps (Sect. 3.3.2a), but the direction and their straight strike are *not* determined by the step formation energy. First, the trench edges are of S_B type (Fig. 3.17, page 173), though S_A steps have a lower formation energy. Second, free-standing S_B steps are very rough (Fig. 3.19, page 3.19), while the trench walls run relatively straight. If the similarity with steps is invoked, then the tendency to build trenches across the dimer rows compares to the intrinsic instability of large, flat areas of Si(001) against the formation of an array of steps (Sect. 3.3.2). This instability is driven by the tendency to relieve the stress anisotropy of 2×1 domains. Since the clean 2×1 surface is compressed along the dimer rows and tensed across the dimer rows, (page 138), any analogy with monatomic steps suggest that the formation of the $2\times n$ is caused by the opportunity to compensate the *compressive* component of surface stress, possibly by the strong tension due to the reconstruction at the bottom of the trenches.

Further clues are provided by a comparison with surfaces of ultrathin Ge films on Si(001). Ge bulk has the same crystallographic structure as Si, but its lattice constant is larger by about 4%. The lateral lattice constant of very thin (a few atomic layers) Ge films is confined by the Si substrate, and the film is subject to compressive stress. The surface of the film responds to the stress by a spontaneous buildup of *stable* 2×8 surface vacancy rows, a $n = 8$ single-vacancy analogue of the Si(001) $2\times n$. The germanium between the trenches can now easier expand in the direction perpendicular to the trench lines.[495] Similarly, interstitial Ni atoms diluted in the Si matrix

(Section 5.7.2) tend to expand the lattice, creating additional compressive stress which may be partially relieved in the $2\times n$ geometry.

Finally, let us consider the interactions between missing dimers from the thermodynamical point of view. Increased temperature leads to a higher concentration of surface point defects: atoms sublimate from the surface, leaving missing dimers behind them. The missing dimers, mobile already at 200-300°C (Section 4.2.1c), behave like a two-dimensional gas. The pressure and the density of this gas increase with temperature. When they reach beyond a critical limit, the gas begins to condensate and a new phase appears. The condensation competes with evaporation from the existing "droplets" and annihilation of the mobile defects at surface steps.

The tendency of randomly created DV's to arrange in lines perpendicular to the dimer rows is pronounced only after annealing above 500°C.[596] The vacancies on surfaces annealed at 650°C and 850°C are lined up. These lines are destroyed at 950°C.[572] The orientation of the vacancy rows created around or below 500°C is determined either by the vacancy formation mechanism or by the interaction between the vacancies and step edges.[601,608] Moreover, when the concentration of vacancies exceeds ~0.3 ML, the trenches grow wider instead of becoming more densely spaced.[r] Their bottoms are now similar to regular terraces, the stress changes, trenches perpendicular to the upper dimer rows are destabilized[607] and change their direction by 90°. This tendency can clearly be seen in STM images obtained after hydrogen etching (Sect. 5.2): narrow vacancy lines run perpendicular to dimer rows, wider vacancy islands are elongated parallel to the dimer rows.[609]

To summarize, the $2\times n$ structure is a metastable ordering of 2DV defects in lines across the dimer rows. It is most probably intrinsic to clean Si(001) and reflects the tendency to relieve the compressive stress of the dimerized surface. It is formed around 800°C and requires a relatively high concentration of dimer vacancies. While usually quite disordered, it is promoted by Ni contamination, which makes it astonishingly regular. The appearance of this reconstruction during a repetitive cycles of annealing and quenching is a clear sign of Ni enrichment of the surface region.

3.4.2. *Si(001) c(4×4)*

The $c(4\times4)$ is a group of structures. It was observed as early as in 1967 but

[r] Possibly due to repulsion between step edges, to excessive number of S_B steps which would have to be created, or to both effects.

Ref year	Subject relevant to Si(001) $c(4\times4)$
610 1967	$c(4\times4)$ reported for the first time; LEED
439 1984	Missing dimers proposed to account for $c(4\times4)$
611 1987	$c(4\times4)$ is stable in a narrow range of temperatures; LEED
612 1988	Hydrogen exposure before annealing helps to produce $c(4\times4)$
613 1991	Growth from disilane can also lead to $c(4\times4)$
614 1992	A missing-dimer like $c(4\times4)$ during annealing in H_2; STM
615 1992	Mixed dimer model proposed; STM, *ab initio*
597 1993	A missing-dimer like $c(4\times4)$ after SiO_2 desorption; STM
128 1999	$c(4\times4)$ after electron-simulated desorption of Si; STM

Table 3.8. Si(001) $c(4\times4)$: selection of references.

has been investigated in detail only recently. See Table 3.8 for references.

When Si(001) is annealed at 600–700°C for a few minutes and then cooled down to room temperature, LEED shows $c(4\times4)$ spots.[611] Some authors[612,614] observed that $c(4\times4)$ is more easily obtained when the sample is exposed to hydrogen before annealing, so that the initial state is a 1×1-H surface.[s] The $c(4\times4)$ LEED pattern appears as well when Si is grown on Si(001) at substrate temperatures in the same range or somewhat higher. The growth can proceed by deposition of either Si atoms[610,616] or Si_2H_6 molecules.[613,617] A $c(4\times4)$ structure has been observed[597] after desorption of ultrathin oxides around 730°C. Electron-simulated desorption may also lead to the formation of $c(4\times4)$-reconstructed areas.[128]

A regular array of missing dimers (Fig. 3.30a) was proposed[614,597] as a plausible candidate for the $c(4\times4)$ reconstruction. According to *ab initio* data,[615] such an array has a relatively low energy, at least the lowest among the structures shown in Fig. 3.30. But the computed electronic structure cannot account for certain STM images.[615] On the other hand, a combination of two other models of Fig. 3.30 (Fig. 3.30b "parallel dimers") and Fig. 3.30c "mixed dimers") can explain these features. Such a mixture can possibly lead to a more efficient relaxation of the surface stress and lower the energy of the combined structure below that of the missing dimers.[615]

Since the observed structures appear only in a narrow range of temperatures, $c(4\times4)$ may be a consequence of a kinetic bottleneck in surface diffusion. Its atomic structure may depend on the initial conditions, such as the density of atoms in the topmost layer.[t] Note that in some cases $c(4\times4)$ areas have been observed together with $2\times n$ rows of missing dimers[597] but,

[s] Such a surface is disordered and atomically rough due to H etching (Section 5.2).
[t] See the 5×5 reconstruction of Si(111), Section 1.5.2.

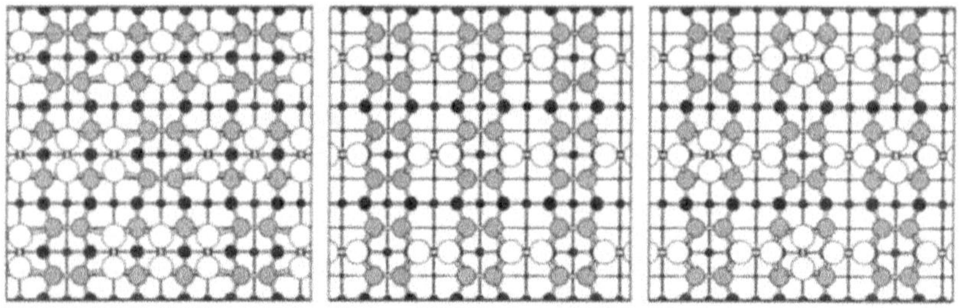

a) Missing dimers; 40 meV/1×1 b) Parallel ad–dimers; 90 meV/1×1 c) Mixed ad–dimers; 110 meV/1×1

Fig. 3.30. Schematic models of metastable c(4×4) reconstructions of clean Si(001). **(a)** missing dimers, compare Fig. 3.12b, page 159, **(b)** parallel ad-dimers (buckling not shown), and **(c)** mixed ad-dimers. Surface atoms are white, subsurface atoms are gray. *Ab initio* formation energies[615] are with respect to the dimerized, buckled 2×1 surface.

for example, Ref. 615 reports no such rows. Different surface preparations may lead to metastable c(4×4) structures with differently arranged atoms. Since the c(4×4) structures are metastable and kinetic factors may dominate their formations, a simple comparison of their total energies cannot provide conclusive arguments about their relative abundances. But it is important that all structures of Fig. 3.30 have energies *higher* than the perfectly dimerized surface, that they represent local minima of the total energy, and that STM images can be explained with their help.

The mixed dimer model[615] is build on the basis of the parallel ad-dimer atomic configuration sketched in Figure 3.30b. The parallel dimers are 2×2 ad-dimers on top of a symmetric, 2×1 dimerized surface. Their bonds are parallel to the substrate dimer bonds. The c(4×4) symmetry is then obtained by buckling the ad-dimers (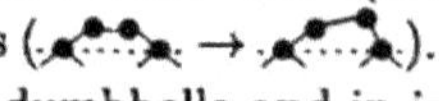 →). The ad-dimers appear in images of occupied states as dumbbells and in images of empty states as pairs of circular spots. Aproximately 50% of the c(4×4) surface area observed in the STM study of Ref. 615 can be described by this model.

The remaining area, though also showing a local c(4×4) periodicity, is more complicated. In addition to the features typical for the parallel ad-dimer model its STM images have new elements. The model proposed in Ref. 615 accounts for these elements by pairs of Si atoms which break the strained dimer bonds of the substrate and form ad-dimers with bonds perpendicular to the dimer bonds of the substrate (Fig. 3.30c). In our opinion this is a plausible combination, particularily given the observed behavior of Si dimers during growth of Si on Si(001): the deposited silicon

Ref year	Subject relevant to Si(001) translational domains
567 1990	Structure of undecorated antiphase boundaries; homoepitaxy, STM
570 1991	Nucleation on antiphase domains leads to "kissing" of steps; STM
563 1992	Screw dislocations initiate "kissing" sites; LEEM
618 1993	Structure of decorated boundaries; Xe sputtering, STM, *ab initio*
619 1993	Antiphase boundaries dominate nucleation during LT CVD; STM
620 1994	Antiphase boundaries dominate nucleation; Monte Carlo simulation
621 1998	Nucleation at antiphase boundaries, atomic structures; tight binding

Table 3.9. Translational domain boundaries: selected references.

atoms form parallel ad-dimers which, when the density of atoms in the top layer increases beyond a certain limit, begin to rotate to acquire the usual, perpendicular orientation (Section 4.2.1b).

3.4.3. *Translational domain boundaries*

In the following, we overview atomic structures of the antiphase boundaries of dimerization. Table 3.9 lists selected references.

Translational domain boundaries, or antiphase domains of dimerization, are metastable structures which form when silicon is deposited on (or removed from) the surface. They act as nucleation centers during growth, by MBE[567] as well as by CVD[619]. Thus, if layer-by-layer growth is desired, formation of translational domain boundaries should be avoided.

It is somewhat surprising that the dimerization of Si(001) surfaces prepared by usual procedures (flashing to 1200°C and cooling) is so perfect. The only imperfections in an otherwise ideal 2×1 array of dimers are occasional missing dimers or C-defects and, naturally, the surface steps which terminate each flat (001) terrace. All surface atoms in a terrace dimerize in the same phase, forming dimer rows like: , where the horizontal lines represent dimer bonds, and dimer rows are vertical "files" of dimers. The only type of antiphase boundaries observed on a usual Si(001) terrace are boundaries between domains of different buckling phases (page 238). Translational domain boundaries are absent: no undimerized atoms are left between dimer rows, as: , and dimer rows run straight across the whole terrace, never ending in the middle of the way as: .

This perfectly ordered dimerization may be surprising if one assumes that it begins independently in many places on a terrace as the surface is cooled down from the temperature at which dimers are unstable. In fact, it is not obvious that such a view of the dimerization process is correct. The perfectness of dimerization may simply reflect the way the surface is formed: the dimers may be "zipped" together by step edges speeding over the surface. The correlated dimerization between terraces on single-domain surfaces (Sect. 3.3.2b) corroborates this view. Since well-defined steps and terraces appear at the same temperature at which the 2×1 LEED pattern indicative of dimerization sets off (Sect. 4.3.2), the observed uniformity of dimerization on the whole terrace may be imposed when the step/terrace system is formed from a thermally roughened surface.

Translational domains are best observed in the initial stage of homoepitaxy, when they are freshly formed. The temperature and growth rate should be such that the growth proceeds via nucleation and coalescence of 2D islands. When 2D islands meet, there is 50% chance that their dimerization phases do not match. The coalesced island will then contain two translational domains separated by an antiphase boundary.[567]

One type of antiphase boundary, "vertical" (AP1), corresponds to dimer rows separated by three surface lattice constants instead of two. This would leave a row of undimerized atoms between two dimer rows, as: . In reality the undimerized atoms are absent, and the actual boundary is a missing row: 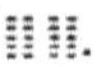. The atoms at the bottom of the trench dimerize,[u] so that the dimerization pattern becomes: , with the "vertical" dimers belonging to the lower layer. Thus, AP1 is in fact a case of two extremely close S_A steps (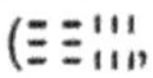, Section 3.3.2a): one step down and one step up.

The structure of undecorated AP2 boundaries is more complicated. They appear when the dimer rows of two growing islands are out of phase and do not match each other, so that one island's dimer rows are in register with the other island's "trenches" between the dimer rows, as: 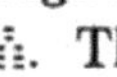. The mismatched rows do not approach each other; this would lead to excessive stress in the surface layer. Instead, they are separated by a dimer vacancy line similar to the lines forming the 2×n structure (Section 3.4.1). As usual also for the 2×n trenches, the AP2 boundary is formed as a line of (1+2)DV "split-off" dimers, like:  (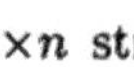 is the perfect dimerization). Thus, the AP2 boundary is in fact built by two extremely close S_B steps (Section 3.3.2a): one step down and one step up. One of these steps is rebonded (S_B-r: 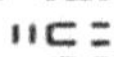, compare Fig. 3.20a, page 179), the other one is split by a dimer vacancy trapped next to the edge (S_B-dv: , compare Fig. 3.20b, page 179). The row of dimer divacancies separating the edge atoms

[u] This dimerization cannot be resolved in STM images, but it is clear that the position of the bottom atoms is nearly identical as of lower-terrace atoms adjacent to an S_A step (compare Fig. 3.18, page 175).

of both steps helps to relieve the stress produced by the mismatch in the registry of dimerization in the translational domains.

The atoms in the split-off line () seem to be weakly dimerized. Some authors interpret this as a stress-induced weakening of the dimer bond: the vacancies in the S_B-dv edge have a dimerized bottom, which introduces additional stress. In our opinion, this weakening is an electronic effect caused by hybridization of the surface orbitals at the edge (page 160).

Translational domain boundaries can also result from layer-by-layer *removal* of surface silicon atoms, as by Xe sputtering at temperatures above 450°C. AP2 boundaries produced in this way are decorated with silicon atoms. The antiphase domains of dimerization are formed in a way similar to the formation of the domains during deposition: dimer vacancies nucleate into dimer vacancy "islands" which then expand and meet. Since the phase of dimerization on bottoms of the islands is random, antiphase boundaries are readily formed. Atoms on top of the boundary line appear to be more stable than atoms on a usual step edge. This enhanced stablility is consistent with the role of the antiphase boundaries as nucleation centers during epitaxial growth. The result is that strings of silicon dimers are left on top of the boundary.[618] Such strings are stable up to 710°C.

What mechanism stabilizes the strings? An AP2 boundary decorated by the string can be sketched as , with the vertical lines representing the dimers forming a mini-terrace directly above the antiphase boundary. An analogous dimer string located on top of a perfectly dimerized surface is represented by: . In both cases, all atoms in the string are placed on top of a complete surface layer and are bounded to the surface atoms in the same way. The difference is in the relaxation of the dimerization stress.

The mechanism of stress relief in the dimer string on top of the antiphase boundary closely resembles that of the stress relief on S_A step edges (page 179). A dimerized surface feels compressed along the dimer rows and stretched across the dimer rows (page 138). The dimer-string atoms in registry with the rows get a push upwards, while the atoms in registry with the troughs between the rows are pulled down. If the string sits on a boundary, this interaction promotes a 2×2-like buckling pattern within the string:  (enlarged, it looks like: 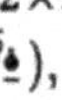), where the circles indicate the "up" atoms of the buckled dimers (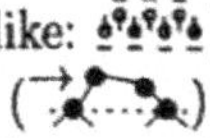). As one can infer from the behavior of dimers on the flat surface (page 140), this buckling pattern is of advantage for the stress relief also inside the string. On the other hand, the same interaction between the string and the stress field of the perfectly dimerized terrace suppresses this buckling. *Ab initio* results[618] confirm this intuitive picture.

We mentioned that no antiphase boundaries of dimerization can be observed on terraces of a surface prepared in a usual manner. Yet, it is still possible to find *traces* of their former presence. These traces exist at "kissing sites" of steps (Fig. 3.24 on page 185, discussion on page 185), that is, at points where two monatomic steps touch. Monatomic steps tend to repell each other, but a simple mechanism involving translational domain boundaries may keep them in proximity: since such a boundary must end on a step, increased stability of ad-dimers on top of the boundary helps to overcome the repulsion. The high-energy domains become covered, at the price of formation of a double atomic step. Close examination of dimer row registry at "kissing sites" shows that steps can be pinned together by AP1 as well as by AP2 boundaries.[570] This indicates that both types of boundaries act as nucleation centers. Domain boundaries may also be formed as screw dislocations emerge from the bulk at high temperatures.[563]

All is change.
– Heracleitus

CHAPTER 4

Evolution of Clean Si(001)

J. Dąbrowski

Dynamic properties. Diffusion, phase transitions, morphology evolution.

4.1. Introduction

The common feature of the objects treated in this Chapter is their dynamic behavior. We will discuss material transport on the surface, thermodynamic transitions between buckling phases and between phases of surface steps, and evolution of surface morphology during deposition and sublimation of silicon. Research in this field, theoretical as well as experimental, often applies novel techniques. For example, a diffusing ad-dimer can be followed by STM tip, which circles around it in tiny loops a few Å wide.[149]

A great deal of knowledge on the dynamics of Si(001) has been already collected. Most of this knowledge is internally consistent, which means that the basic understanding is there. We know that single Si atoms adsorbed on the surface ("monomers") diffuse efficiently even at room temperature, we know how to spot them on the surface, and we can describe them theoretically. We know that adsorbed Si pairs (ad-dimers) and missing dimers can migrate as well, and we have a good idea about their mutual interactions and their interactions with surface steps. We can use *ab initio* input to simulate transitions between buckling phases. We know how to relate the shape of steps on a well-oriented surface to their structure on a misoriented surface. We know how to interpret such "strange" effects as the presence of wavy steps on extremely flat surface or the formation of a nearly single-domain surface on a thermally stressed substrate which, according to its low miscut angle, should in principle have a double-domain surface.

But as in the previous Chapter, we will note that there are numerous unsolved problems or controversial issues of fundamental importance. Why are ad-dimers trapped at sites which are consistently predicted by *ab initio* calculations to have particularily *high* energy (page 213)? What are the exact formation energies and diffusion barriers of mobile surface defects (page 207) and how to describe the interaction of these defects with themselves (page 216) and with surface steps (page 222)? Are non-rebonded biatomic steps good or poor sinks for Si monomers (page 228)? How do step kinks flow on the atomic scale, and how does this affect homoepitaxial growth (page 228)? How to link *ab initio* methods to model calculations, so that the latter could properly account for dynamical effects during the order-disorder transition between buckling phases and simulate it with experimental accuracy (page 241)? Playing with such problems may be an inspiring scientific exercise on one hand; on the other hand, it may lead to refinement or development of tools which can be used in a different context, for example, to assist simulations for semiconductor technology.

4.2. Surface Diffusion

Silicon surface diffusion is responsible for a number of phenomena. Both epitaxy and sublimation proceed through an intermediate state in which mobile elements migrate on the surface after being adsorbed from the vapor or released from such weak spots as steps or dislocation exit points. It is the interplay between surface diffusion, sticking coefficients, and vapor pressure that determines the growth mode. When there is neither deposition nor sublimation, surface diffusion also plays a role because it is through exchange of material over terraces that the mesoscopic elements (steps, two-dimensional islands, facets, three-dimensional moulds) communicate and respond to external stimuli (temperature, strain, electric field).

Surface diffusion helps to arrange atoms in an artificial way. Well-oriented surfaces, which normally have 2×1 and 1×2 dimerization domains separated by steps of monatomic height, can be transformed into nearly single-domain surfaces by mechanical stress[a] acting at temperatures at

[a] This stress can come either from bending in a special sample holder, or simply from

which steps exchange atoms. Heating a sample by passing DC current through it favors diffusion of material in one direction ("electromigration") and invokes an effect similar to that of mechanical stress. Diffusion of excess dimer vacancies created, say, by bombardment with Xe ions, allows them to arrange into lines spaced at regular intervals (page 194). When the vacancies have a chance to annihilate massively at steps, they push monatomic S_B steps () onto monatomic S_A steps (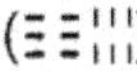), eventually producing an energetically unfavorable D_A configuration (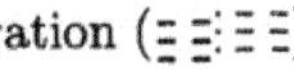) of biatomic steps.[590]

4.2.1. *Isolated mobile objects on clean Si(001)*

We now collect information on the structure and diffusion properties of the simplest mobile elements: isolated Si monomers (page 209), isolated Si addimers (page 212), and isolated dimer vacancies (page 214). Diffusion of all these elements is highly anisotropic, particularly at low temperatures.

The **monomers** (single silicon atoms adsorbed on top of the surface) are the most mobile. They diffuse on dimer row crests. The energy barrier for this diffusion is about 0.5–0.7 eV, low enough to make monomers migrate extremely fast even at room temperature. The barrier for diffusion across dimer rows is higher by 0.2–0.3 eV. This difference is sufficient to disable jumps across the rows at room temperature. The DFT-LDA (page 20) formation energy of a monomer[577] is $\sim$1.8$\pm$0.3 eV (error estimated from the difference between DFT-LDA and measured cohesive energies of silicon). Formally, it corresponds to the energy (per atom) needed to remove atoms from kinks on surface steps into the most stable monomer sites on terraces, whereby the energy of the step edge remains unchanged (kink flow).

The **ad-dimers** (dimerized pairs of Si atoms adsorbed on top of the surface) are less mobile. They begin to move at about 65°C. Like monomers, they prefer to diffuse on top of dimer rows. The barrier on this path was measured to be 1.0–1.2 eV. *Ab initio* calculations indicate that the barrier across dimer rows is by 0.3–0.4 eV higher than the barrier along the rows. The formation energy of an ad-dimer (with respect to kinks on surface steps) can be obtained from *ab initio* data[577,622] as $\sim$2.6$\pm$0.6 eV per dimer; we estimated the error as above. The binding energy of an ad-dimer with respect to monomers was computed as 0.76 eV per two atoms.[622]

The **vacancies** are the least mobile. They also move along dimer rows, but the corresponding barrier is estimated to be 1.7 $\pm$ 0.4 eV. Tem-

deformations generated by the heating-and-cooling procedure.

peratures of 200-300°C are necessary to activate their motion. The barrier for diffusion across the rows is higher by at least 0.2 eV.

How are these barriers measured? Thermal diffusion of ad-dimers and dimer vacancies can be observed in real time by high-temperature STM. It therefore seems natural to obtain diffusion coefficients and energy barriers from Arrhenius plots of the observed jump rates. In a simple kinetic model, the jump rate r is expressed as $r = f \exp(-E/kT)$: the system attempts to surmount the barrier E with frequency f, the probability of success is proportional to the Boltzman factor $\exp(-E/kT)$. However, the temperature windows can be quite narrow.[b] A more practical approach is to *assume* the attempt frequency f and to calculate the barrier from the observed rates.

The diffusion rate f and the macroscopic diffusion constant D differ by a multiplicative factor: $D = rl^2 = D_\circ \exp(-E/kT)$, where $2l$ is the distance covered in a single successfull jump.[623] Since the magnitude of l is similar to the surface lattice constant $a_s = 3.84$ Å, we have $D_\circ \approx 10^{-15} f$, where $D_\circ$ is measured in cm^2s^{-1} and f is measured in s^{-1}.

Usually, f is taken to be of the order of the bulk optical phonon frequency, $f = 10^{13}$. It can be also estimated from computed energy surfaces.[624] Measurements of f for monomer diffusion yield similar values. For example, Mo *et al.*[625] report $D_\circ = 10^{-3}\ cm^2s^{-1}$, that is, $f = 10^{-12}\ s^{-1}$. If we assume that the choice of $f = 10^{13}$ matches f with accuracy of approximately one or two decades, then the uncertainty ΔE is 0.1–0.2 eV at room temperature. This uncertainty scales linearly with the temperature.

When an atom or a molecule is adsorbed, a lot of heat can be released. For instance, adsorption of a Si atom can produce nearly 3 eV, or an equivalent of around 30 thousand Kelvin. Until this energy is dissipated, the adsorbate may move without regard for the temperature.

Determination of **adsorption sites** and silicon diffusion paths is a typical problem which calls for *ab initio* treatment. First, this information is not accessible directly from experiment, hence calculations are needed. Second, since various physical effects (as rehybridization and charge transfer) determine the shape of the actual energy surface, calculations which contain arbitrary parametrization lead to arbitrary results. Model calculations may yield an absolute energy minumum at a barrier on the diffusion path! Some related references can be found in Ref. 626.

---

[b] When the barrier is 2 eV, the change of temperature by 100 K changes the jump rate more than 20 times. This makes real-time observations problematic.

Ref year	Subject relevant to Si monomers
627 1990	Monomer diffusion parameters computed; *ab initio* calculations
625 1991	Monomer diffusion parameters; room-temperature STM
628 1991	Monomer diffusion parameters estimated; growth simulation
629 1991	Monomer adsorption sites determined; *ab initio* calculations
630 1992	Monomer diffusion parameters; room-temperature STM
620 1994	Monte Carlo simulations of the growth of Si(001)
510 1995	**Review**: Monte Carlo simulations of the growth of Si(001)
631 1995	Adsorption sites of monomers determined; STM
632 1995	Effect of dimer buckling on monomer diffusion; *ab initio*
577 1995	Effect of dimer buckling on monomer diffusion; *ab initio*
633 1995	Effect of dimer buckling on monomer diffusion; *ab initio*
622 1996	Recent calculations of monomer adsorption sites; *ab initio*

Table 4.1. Si monomers on Si(001): selection of references.

a. *Monomers and their diffusion*

A "monomer" is a single adsorbed atom. We describe adsorption sites and diffusion paths of isolated Si monomers on a flat surface. Table 4.1 collects some relevant references. Interaction between monomers, with other mobile elements, and with surface steps is treated in Section 4.2.2.

Monomers are often called "adatoms". We will avoid this name because it is also used to describe certain atoms in stable structures, as in Si(111) 7×7, on Si(113), or in rebonded B-type steps on Si(001). In addition, some authors refer to any microscopic, mobile unit (monomer, dimer) adsorbed on the surface as to an adatom.

We begin with a monomer at its lowest-energy adsorption site. Next, we consider metastable sites and the energy barriers between them.

Si monomers migrate so easily that they cannot be detected by an ordinary STM: at room temperature all monomers are paired or adsorbed at step edges. In order to detect a monomer by STM, both measurement *and* deposition must be done at low temperature.

When Si atoms are deposited and imaged[631] at 160 K, they are immobilized on the flat surface in the so-called "twofold fault" or M sites. If we sketch the 2×1 dimer rows as: ≡ ≡ ≡ (dimers are represented by dashes, and dimer rows are their vertical files), then a monomer adsorbed at M can be sketched as: ≡•≡ ≡. It is bonded to two dimers of the same row and sits in the trough between the rows. This agrees with *ab initio* results.

A monomer trapped at M has relatively low energy. *Ab initio* DFT-LDA estimate of its formation energy[577] is $E_M \simeq 1.8$ eV. This energy is defined as the energy difference between Si in the bulk and Si at M. Its value corresponds to 34% of DFT-LDA cohesive energy of silicon.[c] Since E_c is the energy gained by creation of four crystalline bonds, E_M is nearly as low as expected for a *threefold* bonded atom. Yet a monomer at M makes only *two* bonds with the surface.

The energy at M is low thanks to rehybridization and the formation of a weak bond with subsurface atoms. The four valence electrons of the monomer are distributed over three states: two electrons are involved in two bonds and two electrons form a lone pair. These three occupied states use most of the s atomic orbital; one p orbital remains largely empty. The monomer orbitals which are unused in the backbonds are hyperconjugated with the substrate orbitals. This interaction resembles the formation of $p\pi$–$d\pi$ bonds between Si and N atoms in Si_3N_4 (Sect. 5.6.1).

Such effects are not captured by calculations with classical potentials (page 17). Classical theories can hardly model atomic-structure dependent quantum effects without sacrificing their major adventage, which is numerical and conceptual simplicity. Stillinger-Weber or Tersoff potentials yield for the M configuration spurious, strongly repulsive three-body interactions, though in reality the interaction with the third neighbour is weak and attractive. Classical adatoms avoid M sites, perceiving them as barriers rather than as adsorption sites. In the following analysis we consider only the results of *ab initio* calculations, as they provide a realistic description of the quantum system in any configuration.

Apart from the global minimum at M, there are other sites at which a monomer is locally stable. Their presence can be deduced from STM images.[631] When observed at 160 K, monomers are usually frozen at M sites and seen as well-localized spots. But ocassionally, maybe due to excitation from the tip, maybe due to interaction with other monomers (page 217), they become "blurred", apparently acquiring a metastable configuration.

Experiments can visualize the presence of metastable sites, but the only opportunity to study them is provided by theoretical methods. *Ab initio* calculations here give a consistent picture. Yet, a note of caution is needed. Due to practical limitations, contemporary *ab initio* studies are usually performed for systems of relatively small sizes.[d] Calculations

[c] E_M may be overestimated due to constrained relaxation of atoms and LDA inaccuracy.
[d] In *ab initio* studies of adsorption and diffusion of Si monomers, surface cells containing 16 to 24 surface Si atoms were used.[629,577,633,622]

intended to describe isolated monomers describe thus monomers which are so close to each other that they might interact by strain fields (page 217).

We now describe the computed energy surface of a Si monomer. *Ab initio* calculations agree that the best place for adsorption is at the M site: . Another good adsorption site is at the pedestal position H, , where the monomer makes complex bonds to four atoms of two dimers. A monomer at H has energy only about 0.1–0.2 eV higher than a monomer at M. On the way from M to H () there is a barrier $E_{MH} \simeq 0.6$–0.7 eV. Two H minima in the same dimer row are separated by a slightly lower barrier $E_{HH} \simeq 0.4$–0.6 eV. This barrier is located directly above the dimer bond: , that is, on the "short brigde" line.

Another local minimum, C, is on the "long bridge" (the line joining two atoms across the trough:), roughly mid way between the dimer atom and the center of the bridge, B (), that is, at: (). A monomer at C has energy ~0.5 eV higher than at M.

The saddle point on the MC path (→) is at least 0.8 eV above M. It is routinely computed above the saddle point on the HH path (0.6–0.7 eV). The path for diffusion *along* dimer rows is therefore from H to H′ (→), with stops at "parking lots" M ().

In order to migrate *across* the rows, the momomer must stride the trough. But the MM″ trip (→) cannot be done on a straight path, because a high barrier (above 1.2 eV) separates two neighboring M sites: . The detour leads through C and then through a saddle point at B (). It can be sketched as: → → → . The height of the saddle point at B is ~1.0 eV above the M minimum, so migration across the rows is more difficult than along them.

Dimer **buckling** () decreases the barrier on the MCBC″M″ path across the dimer rows from 1.0 eV to 0.75 eV, while the barrier for diffusion along the rows remains virtually unaffected (0.55 eV). This is the outcome of *ab initio* calculations[632,577] which employed a relatively big, 4×6 supercell (). An earlier study,[633] which used a smaller, 4×4 supercell (), found as the only effect a splitting of the C minima () into two classes: C_{up} (close to the "up" atom of the buckled dimer), and C_{down} (close to the "down" atom). The energy at C_{down} is lower than the energy at C_{up}, but both are still above the M and H energies.

This discrepancy, if not due to a technical problem such as an incomplete basis set or inadequate k-point sampling, may be caused by interaction of dimers with antiphase boundaries. These boundaries are places at which dimers neighboring in a row buckle in the same direction (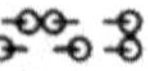, where the open circles mark the "up" atoms; the perfect configuration is ). Such boundaries are frequent on real surfaces (page 238). A monomer trapped at M pins the buckling of the dimers to which it is attached (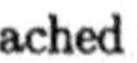): monomers occupy antiphase boundaries, as demonstrated by STM as well as by *ab initio* calculations. This effect couples the diffusion of monomers to the boundaries, causing a computational problem.

Nevertheless, *ab initio* estimates seems to be as accurate as experimental data. The latter come from interpretation[625,628] of observables such as island density. These observables can be interpreted in terms of diffusivities and activation energies when appropriate models are assumed. This procedure is, however, not necessarily unique. Consequently, the measured values depend on the adopted approach.[625,634]

Theoretical and experimenal diffusion barriers agree[627,629,625] within their error bars of ~0.2 eV (the accuracy is probably worse than that for diffusion across dimer rows). It appears that the barier is 0.5–0.7 eV along the rows. The barrier across the rows is higher by 0.2–0.3 eV.

This agreement notwithstanding, the accuracy of data on monomer diffusion is insufficient for reliable simulations of surface diffusion. Growth simulators must treat the diffusion barriers and pre-exponential factors as adjustable parameters.[620] These are *not* the only adjustable parameters needed;[635] consequently, the predictability is far from satisfactory.

b. *Ad-dimers and their diffusion*

When two Si monomers meet, they form an ad-dimer. In this Section we discuss atomic configurations of ad-dimers and ad-dimer diffusion on the flat surface. A selection of relevant references can be found in Table 4.2.

Ad-dimers are best studied by STM. They are immobile at room temperature; heating to 65°C is, however, sufficient to induce their migration.

Si ad-dimers appear large to STM. This is because they stand out from the surface and are approximately as sharp as the STM tip. The image of such an object is a convolution of the surface protrusion imaged by the tip and of the tip imaged by the surface protrusion. Though ad-dimers are easy to find, care must be taken that other objects are not confused with them. For example, occupied states of a MCM complex look almost identical to occupied states of an ad-dimer in its C configutation (page 219).

Ad-dimers on Si(001) have four basic geometries, which we label A

Ref year	Subject relevant to diffusion of Si ad-dimers
636 1991	"Protrusions" migrate on top of dimer rows; STM
626 1995	Rotation of ad-dimers; STM
637 1995	Rotation of ad-dimers; STM
631 1995	B-type ad-dimers are formed at room-temperature deposition; STM
626 1995	Ad-dimers at steps; classical potentials and STM
638 1996	Rotation of ad-dimers, energies of the four configurations; *ab initio*
149 1996	Diffusion parameters determined for B-state, D-state is stable; STM
622 1996	Diffusion of ad-dimers, energies of many configurations; *ab initio*
639 1996	B-type ad-dimers are formed at room-temperature deposition; STM

Table 4.2. Diffusion of Si ad-dimers: selection of references.

through D, following the nomenclature of Brocks and Kelly.[638] Ad-dimers of types A and B reside on top of dimer rows, while ad-dimers of types C and D sit in troughs. If the substrate dimers are represented by dashes and the ad-dimers by dumbells, these four configurations can be sketched as:

Substrate: A: B: C: D: .

There is an interesting controversy about the energetics of these structures. *Ab initio* calculations agree that the lowest-energy configuration is on top, while the configuration D is the least favorable one. When Si atoms are deposited at room temperature, ad-dimers indeed acquire the B geometry.[631,639] But annealing at ~125°C progressively transforms them[149] into the geometry D, the one with the *highest* computed energy!

It may be instructive to compare the computed energy differences:

	A	B	C	D	**range**	**periodicity**
Ref. 638	-0.01	0.00	0.30	1.10	1.11	4×4
Ref. 622	0.07	0.00	0.17	0.71	0.71	4×4

The energies[e] are given in eV per ad-dimer, whereby B is taken as the reference structure; "range" corresponds to the energy difference between the highest- and lowest-energy structure. Numerical conditions are similar in both calculations and the same surface cell, 4×4, was chosen. Discrepancies in relative energies are, however, significant, particularily for the problematic configuration D. If we assume, without going into uncertain details, that the difference in the energy range (0.4 eV) provides an estimate of the numerical error bar, we obtain the accuracy of ±0.2 eV. This is a relatively big error; usually, uncertainty of ~0.1 eV is claimed. But even this does not suffice to explain the disagreement with experiment.

The solution of this problem remains unknown. It is imaginable that

[e] Room-temperature kT (T=18°C) is 25 meV.

interaction between ad-dimers (the surface cells used in the calculations were small) destabilized the D state, or that the actual structure of the stable objects observed experimentally in troughs and considered to be D-type ad-dimers is different. It may contain a monomer, or even two dimers (page 5.2.3). Alternatively, ad-dimers could be trapped in the *metastable* state D if the D dimers and the on-top dimers (A and B) were connected by a "funnel" in the phase space. Broad reaction paths may lead from crests to troughs. But in the troughs they may narrow down, becoming easy to lose and difficult to find. The D state would then represent an entropy trap. A similar effect may be responsible for the apparent breaking of microscopic reversibility in H_2 adsorption and desorption on Si(001) 2×1 (page 269).

Ad-dimers are otherwise rather well understood. The in-trough configurations are immobile: when an ad-dimer falls down from the ridge (from A: , or from B:), it acquires the C state () for a while, and then becomes immobilized as D (). But on-top ad-dimers can move.

Their simplest motion is rotation A ↔ B (↔ . It can be seen in STM videos[626,637] even at room temperature. *Ab initio* dynamics[638] placed the upper bound for the energy barrier for this rotation at 0.9 eV, which is consistent with this motion being active at room temperature. From the statistics of A and B in STM, the energy difference between A and B was estimated to be 0.06±0.01 eV per ad-dimer,[626] with the B state being more favorable than A. Contemporary *ab initio* calculations may have problems to resolve such small energies, but the computed energy difference is indeed comparable to 0.06 eV.

B dimers migrate on top of dimer rows with activation energy of 1.0–1.2 eV. This agrees reasonably with the *ab initio* value[622] of 1.45 eV. *Ab initio* dynamics predicts that the ad-dimer atoms move together in a concerted way over barriers above the substrate dimer bonds. The same calculations yield a barrier of 1.8 eV for migration across the rows.

In summary, although the energetics of the four basic geometries of ad-dimers is unsolved, the key diffusion properties are known. Ad-dimer diffusion is easier along the dimer rows than across the rows. The path of easy diffusion, with a barrier of 1.0–1.2 eV, leads along the top of the rows. Ad-dimers may fall off this path to become trapped at D sites in troughs.

c. *Diffusion of dimer vacancies*

Dimer vacancies (DV) are also known as missing dimers or A-defects. If the 2×1 dimer rows are sketched as vertical files of dashes (), an isolated

Ref year	Subject relevant to diffusion of dimer vacancies
625 1991	Dimer vacancies can survive Si deposition; STM
640 1992	Dimer vacancies are mobile; STM
590 1992	At 450°C, vacancies move efficiently in both directions; STM
641 1993	Vacancy diffusion observed by STM
536 1993	Bond-breaking mechanism of vacancy diffusion; classical potentials
484 1993	Bond-breaking mechanism of vacancy diffusion; *ab initio*
603 1993	Concerted exchange diffusion of DV; classical potentials

Table 4.3. Diffusion of Si(001) dimer vacancies: selection of references.

dimer vacancy is represented by: . For a discussion of atomic and electronic structures of dimer vacancies and their fundamental complexes, the reader is referred to Section 3.3.1a. Table 4.3 collects some interesting, easily available publications relevant to diffusion of dimer vacancies.

An elegant STM study[641] has shown that the diffusion of dimer vacancies is anisotropic. Motion of vacancies was followed in real time at 500 and 525 K (227 and 252°C) so that their individual jumps were resolved. The vacancies migrate along dimer rows (from  to ) with an activation energy of 1.7 ± 0.4 eV and an attempt frequency of $10^{(16\pm4)}$ s^{-1}. If the frequency is 10^{13} s^{-1} (a typical value), this energy becomes ~1.4 eV.

The experiment traced vacancies in a selected dimer row, so the vacancy jumps *across* dimer rows (as from  to ) could not be distinguished from annihilation events. From the observed lifetime of vacancies in a dimer row one can estimate that the barrier in this direction is higher by at least 0.2 eV (assuming the same attempt rate). The apparent onset of efficient diffusion across dimer rows at temperatures between 400 and 450°C, seen in STM images of Xe sputtered surfaces[590] (page 235), suggests that this energy difference may be larger, maybe even 0.5-0.6 eV.

How does a vacancy migrate along a dimer row? We represent the dimer row with a sequence of thick vertical dumbells, and pairs of rebonded substrate atoms at the vacancy bottom by thin horizontal lines, so that a single dimer vacancy is: . One can now imagine the following series of atomic motions by which the vacancy proceeds into the next site:

Such diffusion is a series of bond breakings and reformings. Calculations which assumed various sequences of similar events ended up with rather high activation energies. An *ab initio* estimate[484] was 2.5 eV. A classical potential[536] also yielded more than 2.2 eV. These values may seem too high. But given that the "*ab initio*" activation energy was a rough static estimate based on approximate bond energies extracted from a few geometries, and that classical potentials are not reliable when hybridization is changed, this discrepancy between theory and experiment is not too conclusive. More investigation is needed to solve this problem.

The bond-breaking processes have an intriguing alternative,[603] akin to the "interstitial dimer" model[458,484] of DV (page 159). One of the edge dimers becomes depressed, pushing up the neighbouring atoms of the rebonded bottom and eventually taking over their bonds with the substrate and forcing the former bottom atoms to dimerize on top of the surface:

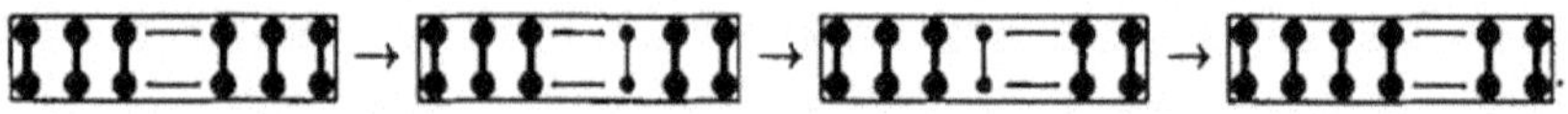

(Atoms of the depressed dimer are marked by small dots). This wavy motion was analysed by two model potentials: the classical Stillinger-Weber and a tight-binding potential, and the activation energy of 1.4–1.6 eV was obtained. This path is thus very plausible and should be studied *ab initio*.

In brief, dimer vacancies are mobile at temperatures in excess of about 200°C. Their diffusion is highly anisotropic (along dimer rows), and the activation energy in the direction of easy diffusion (along dimer rows) is 1.7 ± 0.4. The atomic mechanism of the diffusion needs clarification.

4.2.2. *Interactions between mobile objects on Si(001)*

This Section is devoted to interactions between monomers, ad-dimers, and dimer vacancies, including the formation of "diluted dimers" (page 218). Table 4.4 lists some bibliography. Simple sketches will show the surface viewed from the top: , where files of dashes are the dimer rows.

The simplest reactions on a surface on which Si atoms are being deposited is pairing of Si monomers into ad-dimers and capture of the monomers by immobile objects (surface steps and ad-dimer rows). The ad-dimers can annihilate with dimer vacancies, migrate to immobile objects, or nucleate new immobile objects by pairing and, subsequently, transforming into MCM complexes (). These complexes act as seeds of "**diluted dimer**" rows (), an initial stage of **epitaxial dimer** row growth (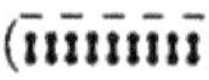). Dimer vacancies attract each other, do not react strongly with monomers, but can annihilate ad-dimers. The vacancies can thus assist epitaxy, hindering nucleation of two-dimensional islands, which would otherwise grow into three-dimensional mounds and cause surface roughening.

Ref	year	Subject relevant to interaction between mobile objects
642	1990	Diluted dimers (DD) observed for the first time; STM
625	1991	Dimer vacancies survive sub-monolayer Si deposition; STM
640	1991	Annihilation of DV's by Si deposition; RHEED oscillations
643	1992	DD's consistent with diffusion anisotropy; Monte Carlo simulation
644	1992	Ad-dimers can rotate on the surface; *ab initio* calculations
645	1993	"Surface dimerization" mechanism for Al/Si(001); *ab initio*
484	1993	Interaction between DV's; *ab initio* calculations
536	1993	DV's are poor traps for monomers; classical potentials
641	1993	Annihilation of DV's by ad-dimers; STM
623	1994	**Review**: Atomic processes in crystal growth
631	1995	Two monomers observed to form ad-dimer; STM
626	1995	Ad-dimers can rotate on the surface; STM, classical potentials
637	1995	Convincing arguments for DD atomic structure; STM
638	1996	DD formation explained by "surface polimerization"; *ab initio*
622	1996	Monomer–ad-dimer and ad-dimer–ad-dimer pairs; *ab initio*
542	1996	Diluted versus epitaxial dimer rows; *ab initio*
639	1996	Formation and growth of a DD row nucleus observed; STM
646	1996	Vertical electric field influence on monomer diffusion; *ab initio*
647	1997	Ad-dimer complexes, formation of epitaxial dimer strings; STM
648	1997	Monomers at monatomic steps; STM

Table 4.4. Interaction between mobile objects on clean Si(001): selection of references.

a. *Interaction between monomers*

An isolated monomer (page 200) spends most of its time trapped at M sites in troughs between dimer rows: , but prefers to travel on top of a row: 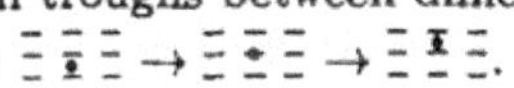. The barrier is 0.5–0.7 eV along the rows and 0.1–0.2 eV higher across the rows. This difference makes the diffusion at room temperature fast and practically one-dimensional.

Monomers feel each other. A hint that this affects diffusion barriers and paths comes from a low-temperature (160 K) STM study.[631] When two monomers separated by several dimer rows reach the same dimer *rank* (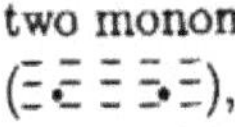), they may approach each other *across* the dimer rows, with a reduced barrier of approximately 0.5 eV. It is, however, possible that the diffusion is modified by the electric field of the tip[646] rather than by interaction between the monomers. The charge transferred from the STM tip to the monomer can also be of importance. But if this attraction is an intrinsic property of monomers, it may well be that it is mediated by buckling of dimers. As an argument against the latter one might quote that *ab initio* calculations[577] found the same diffusion barriers on differently buckled surfaces: 2×2 and $c(4\times2)$. Still, the theory might have thus far missed the right mode of monomer-monomer interaction (page 211).

An encounter between monomers produces[631] an ad-dimer of B type: . The *ab initio* binding energy[622] is $E = E(\mathrm{B}) - 2E(\mathrm{M}) = 0.76$ eV, in agreement with $E = 0.9 \pm 0.2$ eV deduced from simulations of homoepitaxial growth.[636] There is no additional barrier for ad-dimer formation;[622] the energy barrier for dissociation is estimated[622] as 1.33 eV.

b. *Monomer–ad-dimer and ad-dimer–ad-dimer interactions*

The main topic of this Section is the atomic structure, formation mechanism, and energetic stability of "diluted dimers", a particular configuration of ad-dimer rows. This involves a discussion of the interaction between two types of mobile objects: monomers (page 209) and ad-dimers (page 212). For a selection of references see Table 4.4 on page 217.

The Diluted Dimer Rows[642] (DDR) are metastable strings of ad-dimers with bonds parallel to dimer bonds. They are typical for low-temperature epitaxy. The ad-dimers sit in troughs between the substrate dimer rows, making a string *perpendicular* to the rows: . The perfect substrate is: , where each dimer bond is represented by a dash.

At first glance, it is natural that ad-dimers nucleate in troughs. Since the stable adsorption site of a *monomer* is in the trough ( ), isolated monomers should there spend most of their time. The initial structure of an ad-dimer could likely be a C, which is a pair of M monomers: . But when two real monomers meet, they end up as a B ad-dimer:[631] . It is not obvious how and why should B ad-dimers act as DDR seeds.

This problem was recently addressed by *ab initio* theory[638,622] and STM experiment.[639] It turned out that the seed of diluted dimer rows is not a single B ad-dimer, but more complex objects. Interestingly, the D state ( ) at which isolated ad-dimers appear to spend most of their time, is *not* in the proposed reaction path. It is imaginable that the D site is destabilized when ad-dimers approach one another (but see page 213).

To discuss the suggested path, we first recall that ad-dimers of type B can migrate along dimer rows, $\rightarrow$. This migration is enhanced when the moving ad-dimer is in a vicinity of another ad-dimer.[639] An isolated ad-dimer may relatively easily switch between A and B: $\leftrightarrow$.

If two B ad-dimers reach the same rank in neighboring rows ( ), the dimers below them open: . and the bonds between the undimerized atoms reform: , whereby each ad-dimer gains[622] $\sim$0.16 eV. This "BB twin" structure is stable.[639] It does not accept new B ad-dimers to form chains as a "BBB triplet" or a "BBBB quadruplet". Indeed, this would produce a narrow translational dimerization domain (Sect. 3.4.3), i.e., excessive stress. When a third B ad-dimer arrives as the next neighbour to the rank occupied by a BB twin, the three B ad-dimers switch frequently between B-BB () and BB-B () configurations.

A BB twin can convert into MCM, a DDR seed. One may speculate that the ad-dimers rotate to produce an AA pair, , which then decays into MCM: . *Ab initio* calculations indicate that an MCM complex has energy higher than that of a BB twin, by about 0.06 eV per ad-dimer, and lower than that of a CC pair (), by about 0.1 eV per ad-dimer.[622] These are small energy differences, maybe smaller than the accuracy of the calculation (page 213). For example, the close DB pair () was computed to have energy by 0.04 eV per ad-dimer lower than the BB twin, , but was not observed experimentally. We cannot discount the possibility that MCM geometry has nearly the same or lower energy than the BB-twin.

The relative stability and the role of MCM () in the growth of diluted dimer rows is related to the "surface polimerization" mechanism[645] in Al adsorption on Si(001). The ad-dimer destroys the π bonds of the dimers to which it is attached. This activates chemically the ends of the four affected substrate dimers. In the case of a C ad-dimer, the active ends are: , where the unsaturated, half-occupied "dangling orbitals" of the substrate dimers are marked with open circles. A CC pair has also four such active sites: , while in the MCM complex the active sites are terminated by monomers: . Since a monomer at M is in a *closed* electronic configuration,[f] such ends are less active chemically than the open ends. This stabilizes the MCM structure with respect to CC.

A MCM complex can readily act as a seed of a DDR. A diffusing monomer becomes attached to the monomer at one of the MCM ends, forming a MCC structure. This MCC string can then saturate its active end with another M monomer, and the whole process can repeat itself. A diluted dimer row was actually observed to grow from a MCM seed,[639] but the experimental conditions (room temperature STM) did not allow the monitoring of monomer-related processes. The reported mechanism was therefore related to ad-dimers and, consequently, was more complicated.

First, an ad-dimer of B type was found close to a MCM complex:

[f] Two electrons are used in two bonds, two fill a lone-pair orbital.

. Then the B ad-dimer rotated:  (this intermediate AMCM configuration was *deduced*, not observed). Finally, the AMCM complex restructured itself into MCCM: . This short diluted dimer string terminated with two M monomers was easily identified in STM images.

Migration of ad-dimers is thus an important component of the observed DDR formation process: an ad-dimer is captured by another ad-dimer or by an existing DDR. There may exist a similar mechanism which involves diffusing monomers. The only related study known to us is an *ab initio* calculation,[542] which used however relatively small supercells (4×4). An interesting observation[647] of C ad-dimer strings along [310] directions (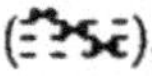), suggests that in vicinity of a C dimer, besides the main monomer adsorption site (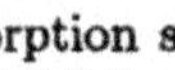) also metastable monomer adsorption sites exist, as: 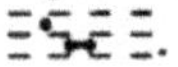. This is compatible with the active ends: 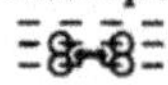.

No ad-dimer–monomer pairs have been reported so far. This may indicate that they are efficient monomer traps, immediately catching a monomer to become MCM. Another possibility is that that such pairs are not easily formed, that is, monomers either find each other more easily and build ad-dimers or become attached to surface steps. The *ab initio* estimate[622] is that the system gains about 0.2 eV when it changes from the "isolated ad-dimer – isolated monomer" state to MC (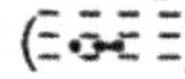). But trapping of monomer at S_B-r steps is more efficient: the energy gain is then five times larger, approximately 1 eV (Sect. 4.2.3a).

An STM image of a MCM complex can be easily mistaken for an image of an isolated C dimer: *ab initio* calculations[638] reveal that the M components are visible *only* in the images of empty states. The highest occupied MCM orbitals do not contain any significant contribution from the monomers. This is another example which shows that a reliable analysis of STM images requires pictures taken for both signs of sample-tip voltage. Inclusion of occupied as well as of unoccupied states becomes particularily important when new structures are being investigated.

These characteristic features of the MCM electronic structure, confirmed experimentally,[639] have a simple interpretation. The hybridization of an M monomer is close to sp^2. Two of the three sp^2 hybrids are involved in bonds with the substrate, and the third one becomes the location of a lone pair; the p_z-like state of the monomer is empty (Sect. 4.2.1a). The lone pair protrudes much less from the surface than dimer "dangling bonds", thus making it nearly invisible in STM images of occupied states.

The empty p_z state of the monomer sticks out and is easily detected.

Long diluted dimer strings: , are not stable. The energy of an ad-dimer in an epitaxial (or "dense") string: , is lower than in a DDR string (by 0.30 eV/dimer, according to an *ab initio* result[542]). Thus, long DDRs tend to break into pieces of dense strings. This decay is facilitated by the weakness of the substrate dimers which support the string:[638] the supporting dimers have bond length of 2.53 Å, while the adjacent "free" dimers of the substrate were calculated to have bond length of 2.21 Å. Such elongated bonds may open more easily upon the arrival of an additional adatom or ad-dimer. At any rate, it is an experimental fact[642] that diluted dimers disappear above 350°C.

Recent STM and *ab initio* studies[647] show that the epitaxial dimer rows are semiconducting: their occupied and empty states are separated by a gap. The calculations indicate that the lowest empty band of the row can be filled with electrons by doping the chains with alkali atoms.

c. *Interaction between dimer vacancies*

Interaction between dimer vacancies (DV) was treated in Sections 3.3.1a (formation of small vacancy clusters) and 3.4.1 (formation of vacancy lines). Here we will mostly summarize the essential points for completeness.

An isolated DV, , migrates along dimer rows (→) with a barrier of 1.7 ± 0.4 eV; the mean time between jumps is about 15 seconds at 500 K. The activation energy across rows (→) is higher by at least 0.2 eV. Two DVs in the same row attract due to the overlap of stress fields induced by the reforming of bonds at vacancy bottoms. This makes divacancies, 2DV (). *Ab initio* calculations estimate that 2DV has energy lower than two isolated DV by ~0.1 eV. As this gain is moderate, 2DVs may be mobile: → → . Attraction exists also across dimer rows (page 162). Eventually, DVs precipitate in clusters and "islands".

Dimer vacancies at high concentrations (but below ~30%) arrange in 2DV *lines* perpendicular to the dimer rows (). The average periodicity is $2\times n$, $8 < n < 10$. Surplus DVs are trapped at the line edges, as: , forming "split-off dimer" structures. Split-off dimers are also created when a migrating dimer vacancy encounters an *isolated* double dimer vacancy.

d. *Interaction of vacancies with ad-dimers and monomers*

STM videos recordered at ~230°C indicate that dimer vacancies (DV) are annihilated by the material moving on the surface (page 214). Model calculations suggest[536] that DVs are poor traps for monomers. At low deposition rates, the lifetime of a monomer on a DV is shorter than the time needed to capture the second monomer. At high deposition rates, ad-dimer nucleation and the growth of two-dimensional (2D) islands around them are more efficient than the filling of vacancies. The annihilation around 230°C is thus due to encounters with ad-dimers migrating across dimer rows.

This is plausible, because dimer vacancies survive[625] deposition of Si atoms even when done at 75°C. If vacancies would trap monomers, they would fill up. But ad-dimers must be nucleated before they can annihilate the vacancies. Thus, any process which prevents nucleation of ad-dimers (as adsorption at step or island edges) also saves the vacancies.

Since atomic configurations of the bottoms and edges of DV complexes are related to configurations of terraces and step edges, the interaction between ad-dimers and such complexes is likely to resemble the interaction with step edges. This subject is treated on page 234.

4.2.3. *Interactions between mobile objects and Si(001) steps*

We now consider processes related to surface morphology changes during growth and sublimation. This includes adsorption of adatoms (pages 224 and 228) and dimer vacancies (page 235) at surface steps. Since interaction of ad-dimers with steps has not been not studied directly, only some remarks in the Section on step fluctuations (page 228) are devoted to it. Geometry and energetics of steps are described in Section 3.3.2.

Since the atomic geometry at steps differs from the geometry of the perfect surface, monomers and ad-dimers may behave differently at step edges than on the flat surface. They may stick to the steps or may be reflected from them. Sticking to steps leads to smooth epitaxial growth by step flow. Reflection from steps may destabilize growth, yet it may have a stabilizing influence on surface roughness during evaporation (page 58).

An atom or a dimer arriving at a step may bounce if it encounteres a Schwoebel barrier (page 58), that is, an additional diffusion barrier making it more difficult to move over the step edge than to travel on the flat surface. Atoms arriving at the step edge from one side (for example, from the upper terrace) may be reflected by a Schwoebel barrier, while atoms arriving from the opposite site (in this case, from the lower terrace) may have a good chance to stick. The reason is that different atomic environments are experienced by atoms approaching the step from different directions.

Schwoebel barriers affect the surface morphology in growth and sublimation by linking the velocity of a step to the size of the associated terrace (Sect. 1.4.2b). High barriers on both terraces of a step decrease the probability that material will stick to the step and thus increase the probability that nucleation centers (ad-dimers and ad-dimer strings) will form on the terraces. The Schwoebel barriers on clean Si(001) have not been measured so far, although silicon ad-dimers were seen[229] reflecting from surface steps at 110°C (Fig. 4.2, page 234). There are experimental hints that attachment probabilities are similar for the material approaching steps from above and below,[230] that is, that the barriers are low. Significant barriers may possibly appear when an interface with another material is formed.

Ab initio calculations[576,577,632,633] suggest as well that Schwoebel barriers are weak on clean Si(001). Since the computed barriers are similar in height to the additional barriers needed to jump across dimer rows on the flat surface, they hardly affect growth. The only significant influence seems to be at low temperatures and high depostition rates: the Schwoebel barrier at S_B-r () weakens the sticking coefficient for the atoms arriving from the upper terrace, favoring growth by island nucleation. At higher temperatures and lower deposition rates, when the nucleation of ad-dimers and growth of islands is less probable than attachment to steps, the effect of these barriers is much less pronounced.

The process which affects the speed by which a step advances is the attachment of new material to flat segments of a step, that is, the creation of new protruding kinks at which most of incoming Si is later incorporated.[635] Steps of S_B type, particularily the rebonded S_B-r, (), are better sinks for incoming Si than S_A steps (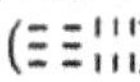); the latter advance primarily via kink flow (kinks on S_A steps are of S_B type). Diffusion along steps seems to be faster than on terraces. This facilitates sticking to steps by promoting nucleation of new ad-dimers from adatoms and/or ad-dimer pairs (Sect. 4.2.3b) from ad-dimers at the edges. Since S_B steps are much stickier than S_A steps, they move faster and, eventually, come close to or bump onto the latter, forming either close S_B-S_A pairs, or steps of biatomic height.

The effective sticking probability of silicon to Si(001) steps is much lower than one, at least at temperatures as high as 950°C. This is the conclusion of the recent LEEM study[230] which investigated the time evolution of the surface profile. The ratio of the transmission rate to the incorporation rate at 950°C was estimated in this work as approximately 40. In other words, the probability that silicon diffusing across a step will be in-

Ref year	Subject relevant to interaction of monomers with steps
227 1966	Schwoebel barriers
228 1991	S_B steps grow faster than S_A steps; STM
649 1991	Energy map of Si monomer over S_A and S_B; classical potentials
650 1991	Adsorption of photo-excited monomers at S_B; *ab initio* clusters
651 1992	Energy maps of Si monomer over steps; classical potentials
630 1992	Denuded zones. S_B are good monomer sinks, S_A are not; STM
652 1993	Energy map of Si monomer over S_A and S_B; classical potentials
633 1995	Energy map of Si monomer over S_A; non-selfconsistent *ab initio*
577 1995	Diffusion of Si monomers over S_A and S_B; *ab initio* calculations
632 1995	Diffusion of Si monomers over S_A and S_B; *ab initio* calculations
576 1995	Energy map of Si monomer over D_B; *ab initio* calculations
634 1995	Effective binding energy at steps; REM
635 1996	Growth anisotropy demonstrated; STM video

Table 4.5. Interaction between monomers and Si(001) steps: selection of references.

corporated is around 40 times smaller (at 950°C) than the probability that it will cross without interacting.

The sticking of material to step edges depletes the immediate neighborhood of steps from mobile silicon, hindering nucleation of two-dimensional islands in the vicinity of steps. The width of the resulting "denuded zones" provides information on surface diffusion and its anisotropy[653,630] and can be measured by STM[630] or REM.[654]

a. *Monomers in vicinity of steps*

Behavior of monomers at steps is an important ingredient of epitaxy (Sect. 4.4). Table 4.5 lists some papers which contribute to this subject.

Growth by step flow requires that individual monomers are temporarily trapped at the step edges. The permanent incorporation of a single atom into a step by changing the bonding topology of edge atoms is not easy because opening of a bond is an activated process, with a barrier of ~1 eV even if the bond is strained (page 228). A monomer must wait until another monomer comes (page 226), otherwise the step will not grow efficiently.

S_A steps (☰☰⁞⁞⁞), being only a weak perturbation of the surface geometry, are weak sinks for monomers. Binding energies of a monomer at S_A are similar to those on the perfect surface.[577,633] From the view point of a monomer, S_A edges resemble troughs between dimer rows on the flat surface. The *ab initio* global minimum[577,633] is at sites like: ☰☰⁞⁞⁞, which are nearly equivalent to the M site on the perfect surface: ☰☰☰.

When a monomer travelling on the lower terrace on top of dimer rows reaches S_A: , it arrives close to the "up" atom of an edge dimer (Fig. 3.21a, page 180). This site resembles the C_{up} site on the flat, buckled surface (pages 211 and 179). Such sites have high energies. The *ab initio* energy difference[577] between the C_{up} and the lower terrace ($\rightarrow$) is ~1.25 eV. The monomer is likely to be reflected unless it takes a detour through a C_{down}-like site: $\rightarrow$ $\rightarrow$ $\rightarrow$, with no barrier higher than that on a terrace.

Crossing the step to reach the upper-terrace pedestal site of S_A ( ) requires then 0.8 eV, which is roughly the same as the barrier for diffusion across dimer rows on the flat surface. Climbing from the lower to the upper terrace becomes energetically equivalent to a jump across a dimer row. Coming down S_A may be easier, but few atoms travel in this direction because the upper-terrace paths of easy diffusion are parallel to S_A edges.

In brief, S_A steps act chiefly as disruptions of the easy diffusion paths on lower terraces. They are weak sinks for monomers. A conclusion drawn from results of approximate (non-selfconsistent) *ab initio* results[633] was that the small variations of monomer binding energies at S_A steps with respect to the binding energies on a flat surface may cause a certain (unspecified) enhancement of monomer pairing rate at S_A step edges.

The situation at **rebonded** S_B-r steps ( ) is different.[577] There are several well-developed binding sites. Two of them are located on the "ledge" formed by the strained bonds of the rebonding atoms (and) and have energies −0.35 and −0.15 eV with respect to the global minimum on the flat surface. The configuration of the most favorable site, E: , with computed energy −1.0 eV, closely resembles the M site of the flat surface: . STM observations[228,635] confirm that S_B-r edges are much better sinks than S_A. The sticking coefficient at ~260°C was measured to be roughly 50 times larger for S_B than for S_A.

Monomers trapped at E ( ) have been recently observed by STM.[648] This is the only configuration at which they reside long enough to be seen at the temperatures used (between room temperature and 115°C). No monomers adsorbed at S_A or S_B-s could be found. This confirms the qualitative conclusions from *ab initio* calculations.

However, this very measurement may indicate that the *ab initio* binding energy at E is overestimated. The observed frequency of escapes from isolated E_0 traps ($\rightarrow$) suggests that the *barrier* on the easy diffusion path leading from an E_0 trap at a S_B-r-like end of an epitaxial ad-dimer row to a M site on the terrace, is ~1.0 eV. The *ab initio* calculation yields 1.0 eV as the direct binding energy, which is the *difference* between the energy at E at an S_B-r step and M at the terrace. The easy-diffusion barrier between M sites is 0.5–0.6 eV. Assuming that the binding at E and E_0 is comparable, the *ab initio* result suggests that the barrier from E_0 to M is around 1.5 eV, that is, 0.5 eV higher than in experiment. The reason for this discrepancy is unclear. Though it is in principle possible that monomers are bonded visibly less strongly at E_0 than at E, it is difficult to imagine why it should be so.

On the other hand, the energy needed to escape from E_0 might be delivered by the boundaries between buckling-phase domains of the terrace. Rapid flicking of surface dimers, which makes the buckled dimers appear symmetric in room-temperature STM images, is supposedly due to frequent passing of such antiphase boundary solitons (page 149). Each time a soliton hits one of the S_A edges of the ad-dimer string, it disappears, and its energy is dissipated. A single encounter with a soliton delivers thus about 0.1 eV to the ad-dimer string. Since a long ad-dimer epitaxial string crosses many substrate dimer rows which are "rails" for the solitons, large fluctuations of the string energy are possible. The energy collected from the solitons can then be either dissipated via emission of many phonons back to the surface and the bulk, or used in a single process to eject a monomer from the trap at the string end. This would decrease the lifetime t of the trapped monomer, resulting in an apparent decrease of the energy barrier[9] E_0.

Most monomers reach S_B-r from the upper terrace. A jump from to  requires 0.85 eV, but a jump  →  costs only 0.7 eV, slightly less than the 0.75 eV which the same calculation[577] yielded for the jump across rows on the flat surface, though more than the 0.55 eV found for the diffusion along the dimer rows. A monomer climbing to the ledge from the lower terrace must overcome a barrier higher than 1.0 eV. Such atoms (there are relatively few of them) are likely to be reflected.

Once on the ledge, a monomer can easily travel along the step.[577] A jump  → 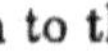 requires only 0.5 eV, even less than 0.55 eV obtained for diffusion along dimer rows. A monomer traveling on the ledge will finally fall down to the most favorable site E:  → 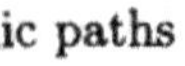.

One might expect that growth at S_B-r transforms a rebonded edge into a **simple** S_B-s step (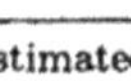). The atomic paths and energetics of this

[9] E_0 was estimated from $1/t = f \exp(-E_0/kT)$ at T=105°C, assuming $f = 10^{12-13}$ s^{-1}.

process have not been studied in detail, but one can imagine that a piece of S_B-s appears when a monomer traveling on the ledge meets an E monomer: 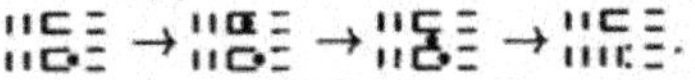. But the last step in this chain may involve a barrier. *Ab initio* calculations done for D_B-r steps[576] indicate that it is unlikely that a *single* atom opens a Si-Si ledge bond (a long horizontal dash in the sketch) and waits there for another monomer to complete the non-rebonded segment. There are two reasons for this. First, ~1 eV must be invested to break the bond. Second, the in-bond configuration is unstable (~0.5 eV above the M site). We speculate that at least a part of the energy needed to overcome the bond-opening barrier is delivered when two monomers traveling on the ledge meet: 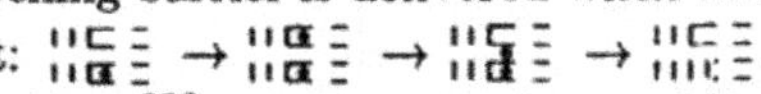. From *ab initio* results for ad-dimer energetics[622] we estimate that making a bond between two monomers can release approximately 1 eV of energy: 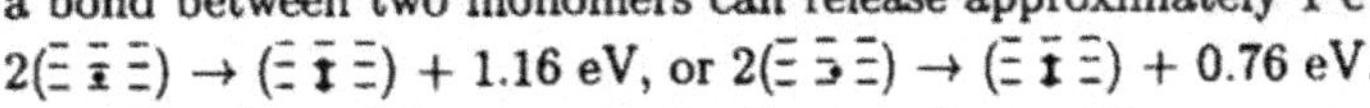.

The flow of S_B steps in the S_B-r → S_B-s → S_B-r sequence is further hindered by the poor chemical activity of S_B-s steps, which are weak sinks for monomers.[577] There are two local minima at the step edge (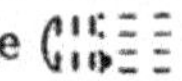 and 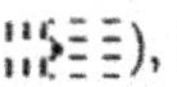), but the most favorable adsorption site is only about 0.1 eV deeper than the global minumum M on the flat surface. But S_B-s steps *are* in the growth path; STM images of samples on which Si has been deposited show approximately the same amount of S_B-r and S_B-s steps[h] (see, for example, Ref. 573). It cannot be excluded that dimers terminating some S_B-s fragments of the step are ejected, move along the step edge to other S_B-s sections, and stabilize them, turning them into S_B-r, or that vacancies are ejected from the step edge: 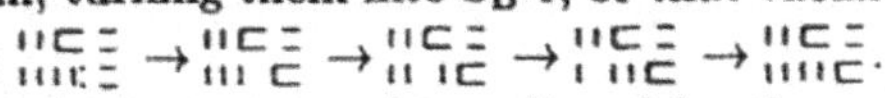.

Classical potential calculations, even when tailored for the use at surfaces, do not yield the correct adsorption sites at monatomic steps[655], just as they fail to reproduce the correct adsorption steps of the flat surface[643] (page 210). Without special corrections, classical predictions are wrong even qualitatively, yielding the non-rebonded steps (S_B-s, 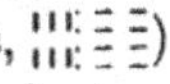) as good monomer sinks and rebonded steps as poor sinks.[649,651] A classical potential can reproduce the property of S_B-r as a good sink when it is modified such that it properly describes the elastic constants of the material, though it still yields S_B-s as a good sink.[656,655] It follows that the energetics of the binding sites is significantly affected by quantum

---

[h] S_B-s is avoided when steps do not grow but merely fluctuate (Section 4.2.3b).

mechanical effects and can be reliably *predicted* only by *ab initio* methods.

The relative speed of S_A and S_B steps depends on the relative efficency of kink flow (S_A steps) and direct step flow (S_B steps), which in turn is a function of sticking coefficients, diffusion barriers, *as well as* the concentration of free silicon on the terraces.[651] At a certain point the faster S_B-r steps will catch up with the slower S_A (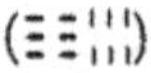) steps, and the growth will change from two-domain mode with monatomic step flow to two-domain growth mode with biatomic step flow (Section 4.4).

The flow of **biatomic** steps is an important element of Si(001) step-flow growth (Section 4.4). It is known that rebonded D_B-r steps (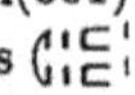) are formed in a natural way when S_B steps approach S_A steps. Unfortunately, monomer adsorption at, and diffusion over D_B steps has been hardly studied by *ab initio* methods. A Schwoebel barrier of ~0.2 eV for atoms arriving from the upper terrace at a D_B-r edge has been computed. Transformation of D_B-r into D_B-s by trapping and incorporation of material atomwise into the step ledges (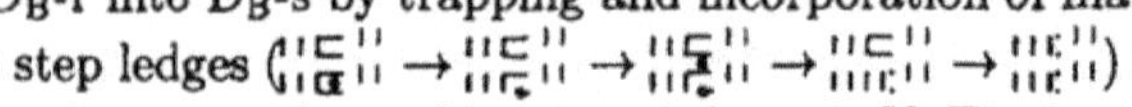) is rather difficult as it involves an additional barrier of about 1 eV. The barrier appears because in order to incorporate a monomer into the ledge, a Si-Si ledge bond (a long horizontal dash in the sketch) must be opened. Yet, these[576] are the only available *ab initio* results. Classical potential calculations for diffusion and adsorption at D_B steps have been done,[651] with the conclusion that these steps are poor sinks for monomers; given the failure of this approach for monomers at monatomic steps, this result needs verification.

b. *Ad-dimers, vacancies, and evolution of step edges*

We now describe the time evolution of step edges. We focus on the diffusion of material along monatomic steps (step fluctuation) and expand the discussion by some remarks on the adsorption of dimer vacancies and Si atoms at steps. See Table 4.6 for a list of related papers. Atomic geometries and step formation energies were covered in Section 3.3.2.

Steps **fluctuate** by diffusion of kinks, but are pinned (do not fluctuate) at the immediate neighborhood of defects resembling missing dimers, located on the lower terrace at sites adjacent to step edge.[657,658] Atoms leave S_B steps and are attached to them in packages of four (two dimers:

Once the correct geometry and energetics of the system is established, one can make an *a posteriori* fit and design a potential which will repeat this result. Such a potential may be very useful to examine more complex systems. The condition is that this increase in complexity must be due to effects of a well-known and well-parametrized classical origin.

Ref year	Subject relevant to evolution of Si(001) step edges
657 1992	First observation of kink flow; STM of S_B and S_A
590 1992	Preferential adsorption of DV at S_B steps; STM
655 1993	Bonding of Si monomers at kinks; classical potentials
659 1993	Fuzzy steps observed at 920°C; STM
574 1993	Fluctuations interpreted within "independent column" model; STM
660 1993	Scaling laws for various microscopic mechanisms of step fluctuation
620 1994	Monte Carlo simulation of epitaxial growth
578 1995	Steps grow by kink flow; STM
585 1995	Fundamental diffusion events in step fluctuations; STM
661 1995	Monte Carlo simulation of step flow
229 1997	Diffusing ad-dimers reflected from steps; STM atom tracking
662 1997	Step stiffness, capilary fluctuations

Table 4.6. Evolution of Si(001) step edges: selection of references.

=), forming or filling 2×2-sized dents in the edge, as  ↔  in a flat S_B segment (running horizontally in these sketches), or  ↔  at a kink.

The size of the package is due to the energy difference between rebonded S_B-r () and simple S_B-s (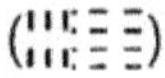) steps.[j] By emission or adsorption of a dimer, a S_B-r section would become a S_B-s section. The energy of S_B-s, however, is so high (Sect. 3.3.2) that this structure is avoided even at the price of emitting two dimers instead of one. Monte Carlo simulations can predict this behavior if they account for some of the substrate geometry, namely, for the creation and annihilation of broken bonds.[661] The two-dimer units are then reallocated along the edge. For example, the unit may move as a single file or become split into two separate dimers, but the details of the process are unknown. Alternatively, the fluctuations may occur by generation and annihilation of dimer vacancy pairs at the edges.

A kink is a natural emission site. Making a dent in a straight step ( → ) costs the energy needed to create two kinks, while a dent at a kink only shifts ( → ) or splits ( → ) an existing kink. The difference in the energy barrier for emission from straight segments and kinks corroborates this observation.[585,578] The activation energy of the emission process is consistently reported as around 1.2 ± 0.3 eV (Fig. 4.1).

The energy barriers and rate constants of processes involved in the fluctuations are determined from the step shapes recorded by STM at various substrate temperatures. This method poses a problem: pairs of opposite rearrangement events which happen between consecutive STM scans

[j] S_B steps are vertical in these two sketches; upper terrace dimers are thicker.

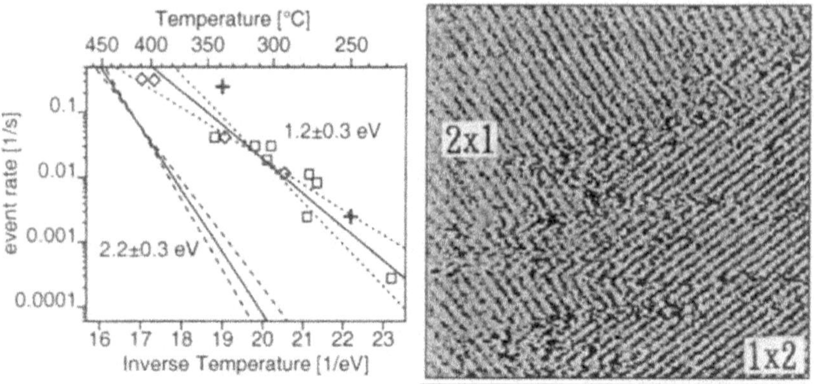

Fig. 4.1. Kink dynamics on S_B steps. **Left**: Compilation of STM data; detachment and attachment events per second per edge dimer. "Observed" event rates (+) include both attachments and detachments of basic mobile units (dimer files "="). Total event rates (□ and ◇) correct the observed rate by adding an estimated number of opposite events happening between scans. Data from: (+): Ref. 574, (◇): Ref. 578, (□): Ref. 585. Slopes 1.2±0.3 eV (fit) and 2.2±0.3 eV (shown for comparison, see text) are indicated. **Right**: STM image obtained at 920°C. Fuzzy step lines are visible. From Ref. 659.

are not recorded and must be accounted for by numerical models. Experimental kink energies (Sect. 3.3.2a) help to calibrate such models.[585]

How do S_A steps fluctuate? In principle in the same way as S_B steps,[658] only several times slower and with a markedly smaller amplitude (by the factor of five to seven at 450°C). This is consistent with the consensus that S_A steps are less active during epitaxial growth.[663] The emission and adsorption of material occurs in two-dimer packages,[657,658] as for S_B steps. The majority of events take place at kinks, as: ⁞⁞⁞⁞ ↔ ⁞⁞⁞.; the S_A step runs vertically in these two sketches.

Steps flucutate rapidly at temperatures above ~400°C (Fig. 4.1). "Rapidly" means here that the probability that a single file of dimers (⁞) changes its length during one second is close to unity. The shape of steps recorded at room temperature is thus expected to reflect a step configuration close to equilibrium at 350-450°C (page 183).

Let us now return to the **microscopic mechanism** of the fluctuations. The low activation energy (1.2 ± 0.3 eV, Fig. 4.1) indicates that the released material remains in contact with the step. But the character of the time dependence of the fluctuations[658] points to mechanisms which involve an exchange of material with terraces. Indeed, for temperatures at which steps reorganize, Si migrates on the surface in form of monomers, ad-dimers, or dimer vacancies (Sect. 4.2.1). We will now discuss this problem.

The event rate r is a sum of the attachment rate $r_a = k_a DC$ and the detachment rate $r_d = k_d \lambda$, where D is the diffusivity of the mobile material, C is its concentration, λ is the density of sites which can emit the mobile particles, and k_a and k_d are the reaction constants. If the system is in thermal equilibrium, we have $r_a = r_d$. In order to reduce the number of symbols, let us define the function $\text{act}(x)$ which returns the activation energy of its argument; naturally, $\text{act}(xy) = \text{act}(x) + \text{act}(y)$. The rate r of step fluctuation events is now activated with the energy

$$\text{act}(r) = \text{act}(r_a) = \text{act}(k_a) + \text{act}(D) + \text{act}(C) = 1.2 \pm 0.3 \text{ eV} \qquad (4.12)$$

(Fig. 4.1). Since in thermal equilibrium C becomes the equilibrium concentration, $\text{act}(C)$ is the formation energy of the mobile species.

First, we assume that the path between the detachment and attachment events involves diffusion on terraces. This means that the terrace acts as a reservoir which swallows each detached object and returns another one randomly somewhere at the step edge. One can estimate the lower bound for $\text{act}(r) = \text{act}(k_a DC)$ as $\text{act}(DC)$, taking the diffusivity D for the migration of material along dimer rows (the easy diffusion direction). It might be more realistic to approximate $\text{act}(k_a)$ by the Schwoebel barrier height (~0.2 eV), but since the heights and locations of Schwoebel barriers are uncertain, we set $\text{act}(k_a) = 0$. The value of $\text{act}(DC)$ is known (Sect. 4.2.1):

- $\text{act}(DC)$ is 1.5–2.7 eV for dimer vacancies;
- $\text{act}(DC)$ is 2.0–2.8 eV for monomers;
- $\text{act}(DC)$ is 3.0–4.4 eV for ad-dimers.

Since $\text{act}(r)$ is 0.9–1.5 eV, the dimer vacancies are the only resonable candidates. The characteristic activation energies of the vacancies would have to be: $\text{act}(D) = 1.3$ eV and $\text{act}(C) = 0.2$ eV. This implies that the barrier for dimer vacancy emission from steps does not exceed $\text{act}(r)$, that is, 1.5 eV.

The terrace-dominated mechanism is compatible with the functional dependence of the time evolution of step edges. When material is freely exchanged with terraces, the fluctuations evolve as $\langle h^2 \rangle \sim t^{1/2}$, where $\langle h^2 \rangle$ is defined as rms displacement of the step edge and t is the time.[664,660] (for more literature on the theory of interface fluctuations see Ref. 658). This functional dependence was indeed observed.[575,658]

Assume now that the detached material remains in contact with the

step, that is, that the diffusion takes place in a quasi one-dimensional channel along the edge. The low value of the activation energy for fluctuations suggests that every detachment event consists of two subsequent events:
1. A single dimer is ejected (S_B-r$\rightarrow S_B$-s+dimer, high activation energy).
2. The second dimer follows (S_B-s$\rightarrow S_B$-r+dimer, low activation energy).
The effective activation energy is determined by the first process.

We will now make a rough estimate of this energy, relying on the published *ab initio* data. The energy difference between S_B-r and S_B-s steps is of the order of 0.6 eV per edge dimer.[485] The formation of a ledge monomer (E site, , Sect. 4.2.3a) from kinks on surface steps costs[577] $\sim$0.8$\pm$0.3 eV/atom (page 207 and Sect. 4.2.3a). Two monomers release $\sim$0.8 eV when they form an ad-dimer on a terrace.[622] Let us assume that the energy gained by dimerization of two ledge monomers is also $\sim$0.8 eV. We obtain now act$(C) = 1.4 \pm 0.6$ eV/dimer for the energy difference between a kink with S_B-r edge and the same kink with S_B-s edge accompanied by a ledge ad-dimer (the estimate of the uncertainty includes no contribution from the dimerization energy). Comparing the lower bound, 0.8 eV/dimer, with the experimental activation energy of event rates, 1.2$\pm$0.3 eV, we arrive at act(D) = 0.7 eV as the upper bound for energy barriers on the reaction path. This is a reasonable value.

But the rate-limited, one-dimensional (1D) character of the diffusion process responsible for step fluctuations is in aparent conflict with the aforementioned measurements in which time evolution of step edges was followed.[575,658] Namely, a 1D process (material diffusing only along step edge) should lead to fluctuations $<h^2> \sim t^{1/4}$. This is qualitatively different from the observed $<h^2> \sim t^{1/2}$ dependence. We will, however, argue that $\sqrt{t}$ fluctuations may be caused also by diffusion on step ledges.

Formally, the $\sqrt{t}$ dependence of $<h^2>$ is due to random walk of kinks along the edge.[665] The theoretical description introduces a terrace solely as a source of uncorrelated Gaussian white noise $\eta(t)$ in the Langevin dynamics[666] of the step edge. This noise is added to the local attachment and detachment rates at each edge site i; the interface responds to $\eta_i(t)$ locally by ejecting or accepting two-dimer units with the speed proportional to the local deviation from the straight edge. The uncorrelated noise means that material is removed (or added) with the same probability at any site of the edge: $<\eta_i(t)\eta_j(t)> \sim \delta_{ij}$. For example, the kink configuration 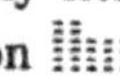 can be changed with the same probability into  or into 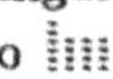. This is opposed to a correlated walk, in which no external reservuar is present, so that processes like  ↔  are forbidden: the amount of material must be conserved. In the absence of the randomizing reservoir, the fluctuation of step edges slows down, acquiring the $t^{1/4}$ dependence.

The bottom line is that if a strong enough, uncorrelated noise source $\eta(t)$ is present on the edge, the steps would fluctuate approximately as $\sqrt{t}$ also without exchanging *mobile* material with terraces. The uncorrelated noise may be due to emission and adsorption of vacancies. But it is not important for the vacancies to move, because the step edge itself is mobile. A vacancy may be created when only one dimer is emitted from the edge:  ↔  ↔ . A S_B-dv section is produced in this way, only to be covered by the moving kink. Such processes are less frequent than the usual double-dimer emission. However, they should be activated with a comparable energy, because they closely resemble the first, decisive step in the double-dimer emission. The main difference is that the second dimer does not follow, for example, because the arriving kink immobilizes it.

This model[301] is, naturally, very speculative as it relies on rough estimates of the involved energies. We cannot conclusively exclude that the steps fluctuate by exchanging mobile dimer vacancies with terraces. We cannot also exclude that the apparent disparity between the activation energies and the time evolution has yet another origin.

Before we leave this intriguing subject, let us recall that the activation energy for the exchange of material across terraces[494] in response to external strain (Sect. 3.2.2d) is 2.2 ± 0.3 eV. This value is markedly higher than the activation energy for diffusion along the edge (1.2 ± 0.3 eV, Fig. 4.1). It is not obvious how to explain this difference within the terrace-dominated model of step fluctuations. However, if the step fluctuations are dominated by diffusion along step edges, the problem disappears. Indeed, the rate r_e of material exchange across terraces must include the formation energy act(C) of the objects which are mobile *on the terrace*. Emission of monomers seems to be a plausible mechanism in this case. The value of act(CD), which gives the lower bound for act(r_e), is now act(CD)=2.4±0.4 eV (Sect. 4.2.1).

Finally, we note that the step flow velocity v due to electromigration (Sect. 4.2.3c) is activated with the energy act(v) = act$(k_a CD) \simeq 1.3$ eV (the paper reporting the corresponding experiment[654] quotes 1.6 eV, but this is probably a mistake, because the slope of the $v(T)$ Arrhenius plot in Fig. 4 of Ref. 654 is 1.3 eV). The value of act(v) coincides with the activa-

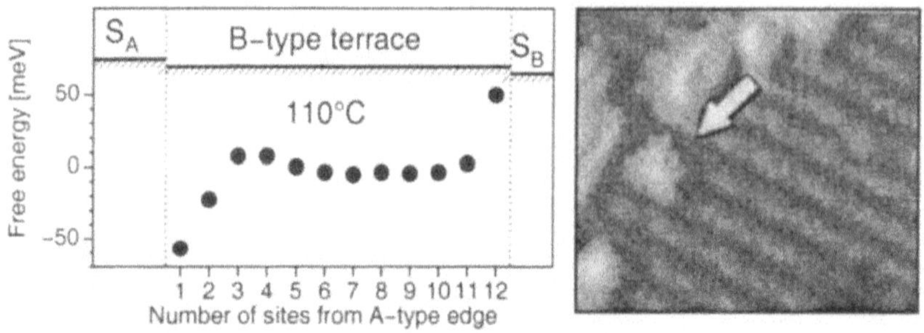

Fig. 4.2. Interaction between ad-dimers and Si(001) steps. **Left**: Relative free binding energy plotted against the position of the ad-dimer shown in the right panel. The type of the S_B step is unspecified. **Right**: Experimental configuration, STM image of empty states (compare Fig. 3.9b, page 151). An ad-dimer is trapped on a dimer row between S_A edge of an island and S_B edge of a dimer vacancy complex. From Ref. 229.

tion energy of step fluctuation activation energy. However, this coincidence may be misleading, because the experiment was performed on samples on which ~0.5 ML of silicon was pre-deposited. Therefore, there was a lot of potentially mobile material, initially agglomerated in tiny, loosely bonded clusters, such as diluted dimer rows (Sect. 4.2.2b). This means that act(C) is now the effective binding energy of monomers or ad-dimers in these clusters, rather than their formation energy. But if the electromigrating species are ad-dimers, then it is possible that act(C) $\simeq$ 0, because act(v) practically coincides with act(D) for ad-dimers (1.1±0.1 eV). Vanishing act(C) means that the deposited material is supersaturated in the mobile form.

An **ad-dimer** migrating on top of a terrace dimer row encounters an energy barrier at certain steps, as demonstrated in a beautiful STM study[229] in which an atom-tracking technique[149] was used to follow the motion of ad-dimers. An ad-dimer was trapped between S_A step of an island and an unspecified S_B step of a vacancy complex (Fig. 4.2). At the temperature of measurement, 110°C, both steps behave as rigid walls.

The free energy of an ad-dimer was found to increase by ~50 meV at the S_B edge (Fig. 4.2). The same dimer is weakly bonded (by ~50 meV) at the wall of the S_A step. The height of the Schwoebel barrier (page. 58) was not measured. Ad-dimer motion at 110°C is practically confined to a single dimer row, so a Schwoebel barrier comparable to that for migration across dimer rows (1.5 ± 0.2 eV) is compatible with the experiment.

The interaction potential (Fig. 4.2) is supposedly due to strain fields.

The character of the interaction is thus likely to depend on the step type and on the direction from which the ad-dimer arrives. It is unknown what happens to ad-dimers which do climb the step wall; they may become incorporated, or may simply continue their migration on the other side.

Dimer vacancies (DV) react preferrentially with S_B steps[590] (see also page 189). This seems to be at variance with the observation made at the same temperature (450°C) that the vacancies *pin* fluctuation of S_B steps,[657] that is, that Si from step edges does not want to fill them. A possible solution is that in spite of their identical appearance in STM images, these defects have different atomic structures. Since the defects studied in Ref. 590 were created by sputtering, they are missing dimers beyond any doubt. It follows that the objects which pin the step fluctuations are not DV or, if they *are*, that they are stabilized, for example, by adsorbates like hydrogen or water (page 3.3.1a). Another possibility is that the dimers diffusing on the ledge are reflected from DV edges, maybe as in Fig. 4.2.

Monte Carlo simulations of epitaxy suggest[620] that migrating vacancies may be an important factor opposing the unwanted nucleation of 2D islands during growth by annihilating island seeds, the ad-dimers.[k] This requires that new dimer vacancies are constantly created. Since S_B-s steps, present during epitaxy (Sect. 4.2.3a), are relatively unstable, we speculate that DV may be created with a lowered formation energy when S_B-s sections () ocassionally rearange into S_B-r + DV (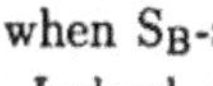). Indeed, dimer vacancy emission/adsorption may be responsible for step fluctuations.

c. *Electromigration*

A convenient way to heat a sample locked in an UHV chamber is to pass electric current through it. But this should be done with care, because DC currents affect the morphology of silicon surfaces. The reason is that the exchange of atoms between surface steps is influenced by the electric field.

There exists a technologically important effect akin to electromigration on silicon: the influence of DC current on diffusion along grain boundaries in metal lines which connect devices in integrated circuits[31]. However, it is not clear if there are enough parallels between these phenomena so that the knowledge acquired by electromigration studies on silicon surfaces can help to control electromigration in metal interconnects. Therefore, we view

[k] This work employed unphysical *single-atom* mobile surface vacancies instead of *dimer* vacancies. But its conclusions may be valid because they build on annihilation of ad-dimers by vacancies produced at S_B steps. Both processes are possible for DV.

Ref year	Subject relevant to electromigration
226 1951	Burton-Cabrera-Frank (BCF) theory of step/monomer interactions
667 1988	Electromigration on Si(001) reported for the first time
668 1991	Diffusion of monomers may be affected by electric field; theory
31 1991	**Book article:** Electromigration problems in CMOS interconnets
654 1992	Electromigration on Si(001); REM, micro-RHEED
669 1992	Domain conversion by electromigration on Si(001); theory
670 1992	Influence of lateral electric field on Si(001) steps; LCAO theory
659 1993	Electromigration on Si(111); STM
634 1995	Nearly single-domain Si(001) made by electromigration; REM
378 1996	Electromigration on mechanically stressed Si(001); STM
671 1996	Microscopic theory of electromigration on Si(111); *ab initio*

Table 4.7. Electromigration; selection of references.

this Section as complementary to our discussion of diffusion on Si(001). Table 4.7 lists some general references. Ref. 669 gives a phenomenological description of the electromigration in terms of the drift flux of mobile material, diffusion anisotropy, repulsive interaction between steps, and capture rates of Si at S_A and S_B from the upper and the lower terrace.

When the sample is resistively heated by passing DC current through it in the direction perpendicular to steps, and when the achieved temperature is high enough for exchange of material between the steps (~500°C), the domain structure of the surface is affected. The current flowing up the staircase moves silicon from S_B to S_A steps. The side view of the sample changes then from  to , where the dashed section is under a domain of type B (with dimers on the terrace above a step parallel to the edge). The current flowing down the staircase produces the opposite effect, changing 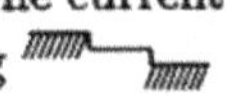 into . The transport stops when the repulsion between steps grows too strong. Mechanical stress can mask electrodiffusion: when the external strain reaches 0.05%, the influence of the current becomes negligible.[378] In the absence of stress, the surface can be made[634] nearly exclusively 2×1 or 1×2, that is, almost single-domain.

According to DFT cluster calculations, the electronic structure of steps depends on the lateral electric field.[670] The field applied in the step-down (step-up) direction shifts S_A states to lower (higher) energies. S_B steps behave in the opposite way. This is accompanied by changes Δn in the population of some orbitals; the magnitude of the effect was[670] $\Delta n \simeq 0.2e$, where e denotes an electron. It was suggested[670] that diffusion barriers and sticking coefficients at steps may be influenced as a consequence.

However, the electric field used in the calculations[670] (0.01 atomic units, or ~0.3 V/Å) corresponds to an extremely high macroscopic field (30 MV/cm). If we assume, for simplicity, that the voltage applied to a 1 cm long sample is 10 V, that most of the lateral field is for some reasons localized within 10 Å-wide regions near to steps, and that the average distance between steps is 1000 Å, we obtain the potential drop of only $\sim 10^{-4}$ V/step, that is, $\sim 10^{-5}$ V/Å. This is five orders of magnitude less than the electric field in the calculation. Assuming linear scaling of the charge transfer with the electric field, the orbital occupation changes are as small as $\sim 10^{-6}e$. If, in addition, we account for the fact that each mobile object interacts with steps many times before it sticks (the sticking probability of silicon at steps[230] is ~ 2% at 950°C, cf. Sect. 4.2.3), and that the mobile object responds to the occupation changes in several orbitals, the field causes the effective average difference in the orbital occupation of at most $\sim 10^{-3}e$. This should be compared to the effective charge of $\sim 0.1e$ estimated for the monomer electromigrating on the Si(111) surface.[671]

This analysis indicates that, if field-induced changes in the electronic structure of steps play a role in the electromigration process, the high field needed to induce such changes cannot be due directly to the voltage applied to the sample. Since we are not aware of any plausible indirect effects, we suppose that the role of the external electric field in the electromigration process is practically limited to providing a bias in the surface diffusion of positively charged, mobile material.[654,671] Note that the "wind force", that is, the force due to the impulse transfer from the charge carriers to the mobile silicon, is negligible on a semiconducting surface.[671]

4.3. Surface Vibrations and Phase Transitions

Two physically different classes of phase transitions exist on the clean Si(001) surface. First, vibrations of atoms affect the ordering of buckled dimers and, at temperatures close to the surface melting point, the dimerization itself. The corresponding phase transitions on flat surfaces are considered in Section 4.3.1. Second, lateral diffusion of silicon and interactions between steps mediated by strain fields on the surface and in the substrate cause transitions between arrangements of monatomic and biatomic steps. These transitions on stepped surfaces belong to Section 4.3.2.

Ref year	Subject relevant to phase transitions on flat Si(001)
389 1962	The first observation of $c(4\times2)$; LEED
672 1975	Ising model and random magnetic fields; theoretical concepts
360 1979	Buckling of dimers proposed; tight-binding theory
673 1983	Buckling mapped onto Ising hamiltonian; renormalization group
363 1985	Influence of defects on buckling observed; room-temperature STM
674 1985	Phase transition predicted at $T_c \approx 250$ K; Monte Carlo simulation
451 1987	Phase transition observed at $T_c \approx 200$ K; LEED
675 1988	**Review**: Ising model and random fields; theory
676 1989	Commensurate and incommensurate phases; mean field theory
372 1992	Flipping frequency of buckled dimers estimated; *ab initio*
677 1992	Rayleigh wave velocity; Brillouin light scattering
678 1993	LEED 2×1 pattern disappears at 1200°C
374 1994	Influence of defects on buckling substantized; low-temperature STM
514 1994	*Ab initio*-based Monte Carlo simulation
516 1994	Long- and short-range order; LEED streak patterns
679 1995	**Review**: Experimental aspects; LEED
680 1995	**Review**: Theoretical aspects; Monte Carlo simulation with defects
500 1995	Microscopic picture of dimer flipping; *ab initio* molecular dynamics
491 1995	Phonon dispersion; linear response from *ab initio* data
583 1996	Ratio of $c(4\times2)$ and 2×2 phases; variable-temperature STM
580 1996	$1/R^3$ terms in Ising hamiltonian; Monte Carlo simulation
20 1996	Si(001) surface melting; tight-binding molecular dynamics
382 1997	Soliton waves and flipping of buckled dimers; STM
681 1999	**Review**: DFT calculation of surface phonons

Table 4.8. Phase transitions on flat Si(001): selection of references.

4.3.1. *Surface transitions on the flat Si(001)*

By a "flat surface" we understand here a wide terrace unaffected by steps on its boundary. Of two phase transitions on flat Si(001), the best studied is the order-disorder transition between the $c(4\times2)$ arrangement of buckled dimers and their disordered phase (denoted 2×1 after the signature of its LEED pattern). It takes place at about 200 K, is of second order, and is sensitive to the presence of surface defects. Less is known about the other transition, between the dimerized and (allegedly) undimerized surface. The 2×1 LEED patern, indicative of dimerization, disappears around 1200°C, but it is not completely clear what really happens at this temperature (page 250). Tight-binding molecular dynamics suggests surface melting with the dimerization of the topmost atoms largely preserved, though highly disordered.[20] Table 4.8 collects some references relevant to these topics.

We begin with a summary of the salient features of the reconstructions involved in the order-disorder transition. Next, we outline a theoretical

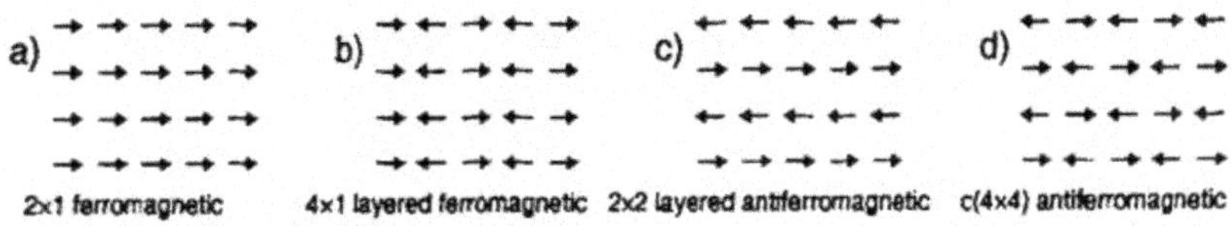

Fig. 4.3. Spin representation of Si(001) buckling phases. Each buckled dimer is mapped to a "spin $\frac{1}{2}$" orientated along the dimer bond, so that the spins indicate the two possible orientations of a buckled dimer. (**a**) 2×1, ordered "ferromagnetic"; (**b**) 4×1, layered "antiferromagnetic"; (**c**) 2×2, layered "antiferromagnetic"; (**d**) $c(4\times2)$, "antiferromagnetic". The 2×1 and 4×1 structures do not occur in nature. They are used to extract the nearest-neighbour integrals V, D, and H from their computed energies. See also Fig. 2.5 on page 93 and Table 4.9.

method used to study phase transitions. Finally, we present the available experimental and theoretical data and examine their mutual consistence.

a. *Structures, vibrations, and the Ising hamiltonian*

The energetically stable structure of clean Si(001) is the $c(4\times2)$ arrangement of buckled dimers: , where the dashes represent dimers viewed from the top and the circles mark the "up" atoms (Fig. 2.5e on page 93 and Fig. 4.3d). It has nearly the same surface energy as the Si(001) 2×2 (, Fig. 4.3c). The common feature of $c(4\times2)$ and 2×2 is that the direction of buckling alternates along each dimer row (each file of dashes). The main difference between them is the phase of buckling in the neighboring rows.

Arrangements of buckled dimers are often described as "ferromagnetic" and "antiferromagnetic". There is a good reason for that. Buckled dimers can be mapped onto spins ($=\frac{1}{2}$, $=-\frac{1}{2}$, see Fig. 4.3) and their interaction can be represented by **Ising spin hamiltonian**. This is useful in studies of phase transitions. Such studies can verify atomistic theories by checking their consistence with the observed temperature effects.

The input data is provided by estimated dimer-dimer interactions. Computed differences between energies of various buckling patterns are translated into exchange integrals in an effective spin hamiltonian $\mathcal{H}$:

$$\mathcal{H} = -\frac{V}{2}\sum_{i,j}\sigma_{ij}\sigma_{i\,j\pm1} - \frac{H}{2}\sum_{i,j}\sigma_{ij}\sigma_{i\pm1\,j} - \frac{D}{2}\sum_{i,j}\sigma_{ij}\sigma_{i\pm1\,j\pm1} + \mathcal{H}', \quad (4.13)$$

where V, H, and D are the nearest-neighbor integrals (Fig. 4.3a), and $\mathcal{H}'$ collects all higher-order coupling terms. The operator σ_{ij} acts on j-th dimer in i-th row: $\sigma(\;) = \sigma(\rightarrow) = 1$, $\sigma(\;) = \sigma(\leftarrow) = -1$. Once

V, H, and D are computed, the system may be solved employing the renormalization group theory, the mean field approximation, or using a purely numerical method, such as a Monte Carlo simulation.[l] We focus on Monte Carlo results because they are the most illustrative and because in such simulations it is straightforward to include surface defects.

We set $\mathcal{H}' = 0$, assuming that the energetics is determined by the nearest-neighbor and next-nearest neighbor (NNN) interactions. In this NNN approximation, the dimer (i, j) occupying the j-th place in the i-th row of dimers interacts only with dimers $(i \pm 1,\ j \pm 1)$. The energy per dimer of the ordered 2×1, 2×2, and $c(4\times2)$ phases and of the disordered 2×1 phase can be now expressed as (cf. Table 4.9):

$$\begin{aligned} E_{2\times1} &= -V - H - 2D, \\ E_{2\times2} &= +V - H + 2D, \\ E_{c(4\times2)} &= +V + H - 2D, \\ E_d &= 0 \ \text{(disordered phase)}. \end{aligned}$$

Coupling to more distant neighbors is treated in Refs. 673, 676, and 580. No significant change in the thermodynamic properties is expected when the set of interaction parameters is extended to cover more neighbors. Only if the contributions from third and fourth neighbors were comparable with the diagonal first-neighbor terms V and H, a new physical effect might be observed. In this case, the low-temperature ordered phase and the high-temperature disordered phase could be separated by an "incommensurate" phase, with the buckling direction changing quasi-randomly. The term "incommensurate" comes from mean field theory: the expected value of the buckling direction, $<\sigma_{ij}>$, is formally periodic, but the periodicity is incommensurate with the surface lattice. This third phase is not known for Si(001), though it may be present[504,676] on Ge(001).

Inclusion of higher interaction terms is not practical for our purpose. The energy differences (Fig. 4.3) which determine the exchange integrals are small. An extremely careful calculation is needed to obtain even the correct *sign* of the energy difference between the 2×2 and $c(4\times2)$ structures. Fitting with higher-order terms might thus be physically meaningless.

[l] In a single numerical Monte Carlo step, each dimer is flipped with a probability determined by the resulting change of energy. When the energy change $\Delta E < 0$, the flip is accepted. In the other case, the flip is accepted with probability proportional to $\exp(-\Delta E/kT)$. This is the so-called Metropolis algorithm. See also page 18.

system	q_x, q_y	$\Delta E(V,H,D)$	ΔE, meV
2×1	0, 0	$-2V-2H$	90.6
4×1	π,0	$-2V+2D$	117.6
2×2	0, π	$-2H+4D$	1.2
c(4×2)	π,π	0	0.0

$$\mathcal{H} = -\tfrac{V}{2}\Sigma\left(\substack{\rightarrow\\ \rightarrow}\right) - \tfrac{H}{2}\Sigma\left(\rightarrow\rightarrow\right) - \tfrac{D}{2}\Sigma\left({}_{\rightarrow}{}^{\rightarrow}\right)$$

Parameter	Value, K	Value, meV
V	-602	-51.9
H	77	6.6
D	42	3.6

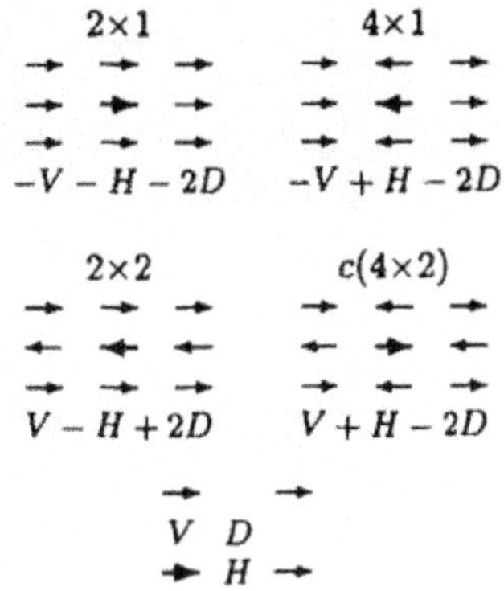

Table 4.9. *Ab initio* data[514] on phase transitions on flat Si(001). The upper part shows change of buckling phase (wave vector) across dimer rows (q_x) and along dimer rows (q_y), and total energy differences (ΔE) per dimer for various reconstructions with respect to c(4×2). The lower part lists the resulting exchange integrals V, H, and D for the effective Ising spin hamiltonian $\mathcal{H}$ displayed symbolically above the table. The reconstructions, their energies expressed in terms of the exchange integrals, and the definitions of the integrals are shown schematically on the right. Compare Fig. 4.3 and Eq. 4.13.

While the NNN approximation does not seem critical for Si(001), the procedure of mapping onto the spin hamiltonian contains a significant simplification: $\mathcal{H}$ is static. It does not account for coupling between **thermal vibrations** of dimers and the exchange integrals. In principle, dynamical effects could be modeled by molecular dynamics. Yet, such a simulation would have to describe the motion of around a hundred thousand atoms. Reliable *ab initio* molecular dynamics is feasible today for systems smaller than a thousand atoms. Even if the performance of numerical techniques continued to double every two years,[493] we would have to wait nearly two decades for the first result. In the meantime, a reliable but more feasible method is likely to be designed. It may integrate an *ab initio*-based description of local physics and a less rigorous treatment of long-range interactions. Verified in studies of surface phase transitions and then applied to other systems, a method like that might support simulations for technology.[28]

Surface vibrations can nowadays be computed from dynamical matrices obtained from *ab initio* calculations in small cells.[491] An element ij of this matrix describes the forces acting on atom i in response to small displacements of atom j. Once the matrix is found (which requires many well-converged simulation runs), the phonon band structure can be calculated.

Such calculations[491] have found a low frequency dimer twisting mode which resembles the dimer twist once proposed[425] to explain LEED intensities of Si(001) 2×1 (page 104). It is interesting to see that this idea, fervently defended by its authors but then proved wrong, has eventually re-emerged, though in a different form. The same twisting mode was found in a full-blown *ab initio* molecular dynamics run employing 4×2 surface supercells.[500] It follows that the dynamics of dimers is indeed quite complicated. But the influence of such effects on the effective interaction parameters in the Ising hamiltonian has not been studied.

Ab initio **interaction parameters** V, H, and D were extracted from static calculations[514] (Table 4.9). For convenience, we express these parameters in units of temperature (1 meV $\simeq$ 11.6 K). The interaction between dimers which are neighbors in the same row ("vertical") is strong and antiferromagnetic: $V = -2 <^{(i,j+1)\rightarrow}_{(i,j)\rightarrow}| \mathcal{H} |^{(i,j+1)\rightarrow}_{(i,j)\rightarrow}> = -602$ K. The interactions across rows are weak and ferromagnetic: $H = 77$ K (horizontal) and $D = 42$ K (diagonal). The vertical term V firmly keeps the buckling in an antiferromagnetic state. In conjunction with the antiferromagnetic V, the horizontal term H favors the layered antiferromagnetic structure, 2×2, in which the dimers in neighboring rows buckle in phase. The diagonal term opposes this tendency: since the buckling alternates along each row, the ferromagnetic buckling of the diagonal neighbour corresponds to the $c(4\times2)$ structure. The two diagonal members win over the single horizontal member, and the effective interaction between the antiferromagnetic rows ends up as antiferromagnetic, though very weak: $H - 2D = -7$ K.

b. *Stability of phases*

Since the characteristic temperatures for the interactions along and across the rows, $T_y = -V$ and $T_x = 2D - H$, differ by two orders of magnitude, one might naively expect that there are three distinct phases:

1. The ordered $c(4\times2)$ phase below $T_x = 2D - H = 7$ K;
2. A striped phase ordered along the rows and highly disordered across the rows, for $2D - H < T < T_y = -V \simeq 600$ K;
3. A highly disordered phase above $T \simeq T_y \simeq 600$ K.

Total disorder is expected above the temperature T_c at which the free energy difference between the disordered phase and the $c(4\times2)$ phase vanishes. In the mean field approximation, $T_c = 2(2D - H - V) \simeq 1200$ K. This means that some order should be there up to nearly 1000°C.

Yet, the $c(4\times2)$ phase is stable at temperatures much higher than

7 K. The $c(4\times2)$ LEED pattern is clearly visible below 200 K, and Monte Carlo simulations with the parameters of Table 4.9 are consistent with this observation. What is the effect that keeps the $c(4\times2)$ phase above T_x?

It is the strength of the intra-row interaction, V, which not only enforces the antiferromagnetic order within each row, but it also indirectly helps to preserve the order between the dimer rows. Flipping a *single* dimer embedded in the $c(4\times2)$ matrix costs much more that the 7 K of the row-row coupling. The energetic cost of such an orientation defect is in fact $E_o = 2(2D - H - V) = T_c = 1218$ K and is dominated by $-2V = 1206$ K from the antiferromagnetic interaction with the two neighbors in the same row. This high cost per flipped dimer can be reduced when a whole portion of the row flips together, producing a rod-like 2×2 domain. The rod can grow easily, because a new flip costs only 14 K, which is the energy difference between the $c(4\times2)$ and 2×2 reconstruction: the number of unfavorable vertical interactions within the row does not change when the rod grows. The flip energy accumulates in form of the interface energy and finally limits the rod length. The fundamental excitation unit is *not* a single orientation defect but a rod of flipped dimers, with a length of the order of the correlation length along the dimer rows.[481,382] Thus, the temperature at which the $c(4\times2)$ structure is destroyed is not $T_1 = 2D - H = 7$ K but rather $T_\xi = 2\xi(T_\xi)T_1$, where $\xi(T_\xi)$ is the correlation length at the temperature T_ξ. In addition, the interface energy makes the rods stick together, promoting formation of patches of $c(4\times2)$ *antiphase*, which we note as $c(4\times2)^-$. The dimers in each antiphase domain buckle as $c(4\times2)$, but the phase of their buckling is opposite to that of the matrix, $c(4\times2)^+$.

We thus expect that the fundamental excitations are antiphase boundaries between $c(4\times2)^+$ and $c(4\times2)^-$. Since the energy of such an excitation is chiefly controlled by the value of V, it is plausible that the $c(4\times2)$ phase is stable even up to a few hundred K. How does this compare with the results of simulations and, even more interestingly, with experiment?

Monte Carlo simulations[514] with the *ab initio* parameters listed in Table 4.9 locate the order-disorder transition at $T_{o-d} \simeq 320$ K. At this temperature, the long-range[m] order parameter drops rapidly. Since the transition temperature is largely controlled by the strong V interaction,

[m] "Long-range" means here "of the size of the simulated system". The calculation considered 10000 dimers arranged in 20 dimer rows, with periodic boundary conditions. An extremely long row is in this context a row of 500 dimers, with length of about 2000Å. A typical coherence length of the electron beam in LEED is several hundred Å.

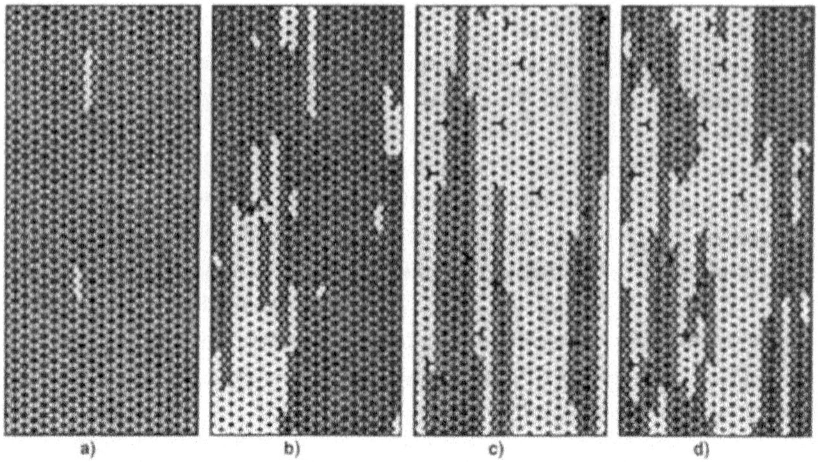

Fig. 4.4. Order-disorder transition on flat Si(001). Snapshots of Monte Carlo simulations. "Up" atoms are indicated as spheres. The atoms of the majority domain are grey, those of the minority domain are white. The transition temperature computed with *ab initio* parameters is $T_{o-d} = 320$ K. **Left panels**: undefected surface. **(a)** $T = 0.78T_{o-d}$. **(b)** $T = 1.03T_{o-d}$, **Right panels**: defected surface. 1% of the surface covered by C-like defects. **(c)** $T = 0.78T_{o-d}$. **(d)** $T = 1.03T_{o-d}$, From Ref. 680.

it is only slightly sensitive to small relative changes $\Delta E/E$ in the energy difference $E = 4D - 2H$ between the $c(4\times2)$ and 2×2 structures. For instance, when E was assumed to be 0.5 meV (3.0 meV) instead of 1.2 meV, the critical temperature became 280 K (400 K) instead of 320 K.

As evident in the snapshots from the simulations (Fig. 4.4b and 4.4a), the disorder perpendicular and parallel to the dimer rows starts at the same temperature. This is consistent with our expectation that the fundamental excitations of the $c(4\times2)$ matrix are antiphase boundaries between islands of $c(4\times2)^-$ embedded in the $c(4\times2)^+$ matrix. Simulations show as well that the dimers flip most easily at antiphase boundaries. Indeed, flipping a dimer in an orientational defect at a boundary (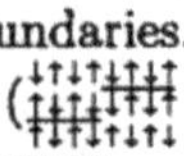 to 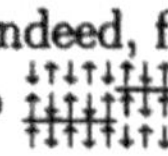) merely shifts the defect; the boundary length does not vary.

c. *Influence of surface point defects*

LEED observes a sharp $c(4\times2)$ diffraction pattern up to about 200 K. Close to this temperature, the quarter-order spots become diffuse and gradually disappear. This is consistent with the theoretical result that the disorder takes off simultaneously in both directions. The simulated transition temperature of 320 K is higher than the experimental value; yet, given the approximate form of the spin hamiltonian and a purely *ab initio* input, this is quite a satisfactory agreement. But another observation may be alarming: while simulations predict a sharp drop of the long-range order at the transition temperature, LEED experiences a gradual weakening of the $c(4\times2)$ reflexes. The first observation of the $c(4\times2)$ structure was done[388] by *room temperature* LEED, about 100 K above the transition point!

This apparent discrepancy can be attributed to the presence of surface defects, not included in the simulations but unavoidable on real surfaces. A typical concentration of surface point defects (Section 3.3.1) is of the order of a few percent, that is, a few dimer sites per hundred are occupied by tiny, dimer-sized flaws in the surface reconstruction.

That defects influence the buckling of dimers has been clear since the publication of the first STM images of Si(001); see Fig. 2.8, page 108. Some of these defects (dimer vacancies, Section 3.3.1a) are intrinsically symmetric and their effect on buckling is probably limited to local changes of buckling energies and barriers for flipping. Since the vacancies may appear in several atomic configurations, their influence on buckling may vary. The other abundant defect, the C-defect (Section 3.3.1b), is asymmetric and can pin the direction of buckling. Also steps (Section 3.3.2), present on all real surfaces, affect the buckling, though in a different manner. Monatomic steps with dimer rows on the upper terrace running paralell to the step edge (S_A steps, page 179) support the $c(4\times2)$ structure. Monatomic steps with dimer rows on the upper terrace running perpendicular to the step edge (S_B steps, page 181) may favor 2×2 over $c(4\times2)$. A two-domain Si(001) surface contains necessarily steps of S_A and S_B type.

In Monte Carlo simulations[514,680] defects were modeled by sites which fix the buckling direction of the first-neighbor dimers in the same row. It turned out that as few as 1% defect sites affect the transition drastically. Snapshots of the defective surface taken slightly above the order-disorder transition temperature T_{o-d} (Fig. 4.4d) and below T_{o-d} (Fig. 4.4c) show that the defects are very efficient seeds of antiphase domains. With defects present, there is little difference in the surface state above and below the transition temperature. This is substantiated in the diagram in Fig. 4.5a, which plots the long-range-order parameter against the temperature: in comparison with the perfect surface, the long-range order of the defected surface increases only slightly with decreasing temperature.

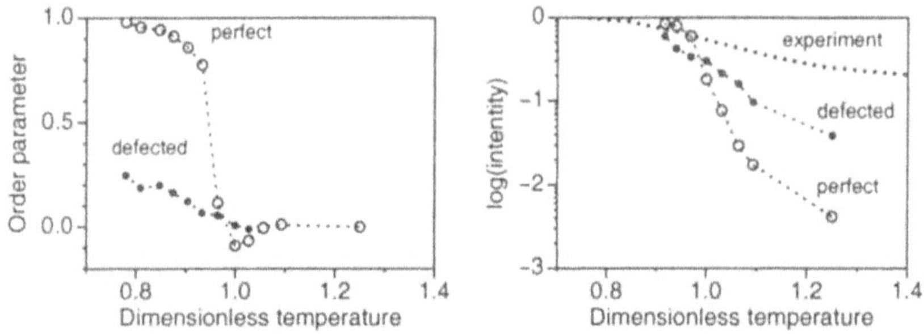

Fig. 4.5. Order-disorder transition on flat Si(001) for perfect (open circles) and 1% defected (full circles) surfaces. The dimensionless temperature is measured in units of T_{o-d} = 320 K (theoretical results[680]) and T_{o-d} = 200 K (experimental data[451]). (a) Temperature dependence of simulated $c(4\times2)$ long-range-order parameter. "Long-range" means here "within 2000 Å along dimer rows and 150 Å across dimer rows". (b) Temperature dependence of simulated (circles) and measured (dashed line) LEED intensity for a quarter-order spot, normalized at 0.75 T_{o-d}.

The influence of defects on the long-range order can be investigated rigorously.[675] One considers the long-range magnetic order in a two-dimensional Ising spin system when a random magnetic field, equivalent to surface defects, is applied. The random field destroys the long-range order at any temperature. The long-range order in Monte Carlo simulations is simply due to the limited size of the model system. Does this mean that *any* amount of defects would destroy the LEED $c(4\times2)$ pattern?!

Not at all. LEED intensity depends on *short*-range order within the coherence length of the incident beam. The area of the surface visible to LEED is of the same magnitude as the area included in the Monte Carlo simulations. Thus, the temperature dependence of the LEED intensity can be estimated from the temperature dependence of the short-range order in the simulations. The result is shown in Fig. 4.5b: the influence of defects is to smear out the transition region, just as observed experimentally.

The importance of surface defects was also highlighted by an investigation of temperature dependence of streak patterns in LEED, that is, by an analysis of the shape of the diffraction reflexes (Fig. 4.6). Disorder broadens the peaks by adding a Lorentzian tail to the Gaussian profile of the spot. Asymmetry in correlation lengths makes the Lorentzian tail widths dependent on direction, shaping the reflexes into streaks. It was found that half-order spots form clear streaks even at 350 K, well above the transi-

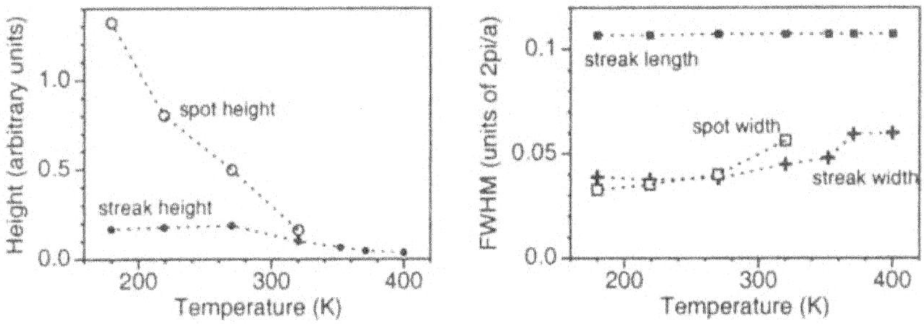

Fig. 4.6. Temperature dependence of the structure of a quarter-order LEED reflex. The reciprocal-space distribution of reflex intensity is decomposed into a Gaussian peak (spot profile) and a Lorentzian tail (streak profile). The spot height (left panel, hollow circles) decreases gradually with increasing temperature T, while the spot width (right panel, hollow squares) increases slightly. The streak height (left panel, heavy circles) decreases with T. The streak width (right panel, crosses), related to disorder along dimer rows increases with T, roughly following the spot width. The streak width (right panel, heavy squares), related to disorder across dimer rows, remains constant. From Ref. 516.

tion temperature. This indicates that some long-range order (within the coherence length) exists along the dimer rows even above the transition temperature, but the ordering across the rows is mostly destroyed. Such an effect is consistent with simulations (Fig. 4.4) for a defective, but also for a perfect, surface. Where then is the promised influence of defects?

The finding which requires that defects are invoked is related to the temperature dependence of the LEED reflexes (Fig. 4.6). Though the quarter-order reflexes (Gaussian spots and Lorentzian tails) are weakened by increasing temperature and the disorder along the dimer rows increases, the disorder across the dimer rows remains virtually unchanged. This cannot be explained by a perfect surface, at least not within a simple Ising model. But pinning centers can easily account for it. The strong coupling V in the direction of dimer rows extends the randomizing effect of defects over appreciable vertical distances (Figs. 4.4d and 4.4c). Thus, the surface observed across the rows appears as composed of randomly arranged rods with phases determined by the randomly distributed defects. In other words, the "horizontal" streak width is mostly determined by the density of surface defects, and hardly changes with temperature. On the other hand, the length of the vertical rods, that is, the correlation length along the dimers, is mostly determined by the strength of the exchange integral V, and as such it shows the usual temperature dependence.

d. *Influence of surface steps*

The influence of steps on phase transitions has been less investigated. The only theoretical work which considered the steps, a Monte Carlo study,[580] focused instead on the effect of inclusion of long-range dipole-dipole interactions between the dimers. The steps were included only as free boundary conditions, with no explicitely pre-defined preferential influence on the type of reconstruction, that is, $c(4\times2)$ versus 2×2. Among experimental studies, one should mention in this context the work of Ref. 583. It was observed that dimers between closely spaced steps of S_B type (metastable $2\times n$ structure, see page 194) appear symmetric in STM images taken at 190 K. This is unusual, since the temperature of 190 K is sufficiently low to slow down the flipping rate such that STM images of 2×1 surfaces show buckled dimers. How does this relate to the subject of this Section?

This unusual suppression of buckling in STM images can be understood on the basis of what we have already said. First, the dimer rows on the $2\times n$ surface are very short. They contain only about six dimers. This must drastically reduce the critical temperature T_{o-d} for the order-disorder transition: all of the dimers in an entire row can be easily flipped, with an energy cost of only $6(4D-2H) \simeq 84$ K. Second, in consequence of this lowering of T_{o-d}, antiphase domain boundaries become very frequent at 190 K. Third, antiphase domains are the places where, according to Monte Carlo simulations (and to common sense) the dimers flip most easily.

Thus, the shortening of dimer rows by the vacancy rows on Si(001) $2\times n$ may be viewed as the natural cause of the symmetrical appearance of the dimers in low-temperature STM images. Yet, this reasoning is only a speculation. In particular, it should be noted that Monte Carlo simulations did not take into account the real barrier for flipping; the only parameter was the energy difference between the initial and final configurations. Second, the $2\times n$ structure in the above experiment was apparently induced by unintentional contamination with a trace amount of nickel. It cannot be excluded that the presence of Ni has a dynamic effect on the flipping rate, which is quite independent of the effects described above.

In conclusion, we have seen that the second-order phase transition between the ordered $c(4\times2)$ and the disordered "2×1" arrangements of buckled dimers can be understood on the basis of a simple Ising model. The interaction parameters can be obtained with reasonable accuracy from *ab initio* calculations. The transition is influenced by the surface defects

Ref year	Subject relevant to phase transitions on vicinal Si(001)
682 1975	**Book article**: Statistical thermodynamics of clean surfaces
683 1980	Fundamental concepts for step-step interactions
367 1988	External strain affects ratio of 2×1 and 1×2 areas; LEED
56 1988	Interaction between monatomic steps; elasticity theory
684 1990	Phase diagram; elasticity theory
568 1990	Phase diagram; elasticity theory and classical potentials
134 1990	**Review**
277 1991	Phase diagram, (110), (210), (100) zones; SP-LEED
494 1991	Surface stress anisotropy measured; LEED and elasticity theory
685 1991	Phase separation on vicinal surface; STM
559 1991	Mixed phase; elasticity theory and classical potentials
520 1991	**Review**: Thermal stress and domain occupation; LEEM
558 1991	Phase diagram, [110] zone, angle dependence (500 and 800°C); μRHEED
561 1991	Phase diagram, [110] zone, angle and temperature dependence; LEED
562 1992	Wavy monatomic steps on flat surfaces; elasticity theory
563 1992	Wavy monatomic steps on flat surfaces; LEEM
686 1992	Phase diagram, interaction potentials; LEED and elasticity theory
517 1992	Interaction parameters; clasical potentials and elasticity theory
678 1993	LEED 2×1 pattern disappears at 1200°C
519 1993	Interaction parameters; STM and elasticity theory
518 1993	Interaction parameters; STM and elasticity theory
687 1994	Limits for vicinality, azimuths other than [110]; STM
557 1994	**Review**: Thermodynamics of step-step interactions
509 1994	**Review**: Strain effects
688 1995	Miscut dependence of step topography (influence of adsorbates)

Table 4.10. Phase transitions on vicinal Si(001): selection of references.

which induce a significant disorder even at low temperatures, breaking the surface into domains separated by antiphase boundaries.

4.3.2. *Vicinal Si(001): phase transitions, interaction of steps*

Phase transitions of vicinal Si(001) are caused by interaction between surface steps. The phase diagram describes the dependence of type and distribution of steps on the miscut angle θ, temperature T, and strain ϵ. Its relative complexity is a consequence of interactions between three types of steps: the monatomic S_A and S_B steps and the biatomic D_B steps. We focus on the dependence of step morphology on the miscut angle (for miscuts in the [110] azimuth) and on temperature. A schematic representation of the step phases was given in Fig. 3.16, page 172. Table 4.10 lists some references relevant to step interactions and phase transitions on vicinal Si(001).

Vicinal Si(001) has three basic phases: biatomic steps, monatomic steps, and "wavy" monatomic steps (Fig. 4.7). The θ–T diagram is bounded

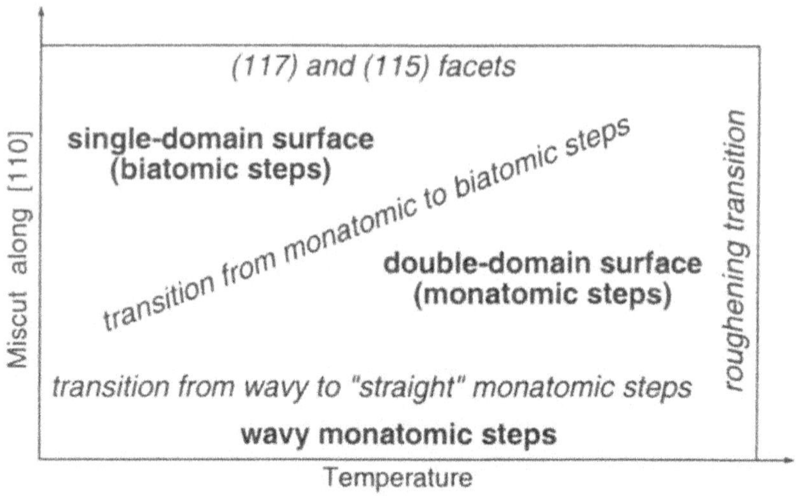

Fig. 4.7. Schematic phase diagram of vicinal Si(001). See references in Table 4.10.

by faceting into Si(115) and Si(117) at miscuts above ~9° and by a roughening transition at $T_R \simeq 1200°$ C. Let us cite from Ref. 277:

> In the (110)-zone, single layer steps are found with a tendency for step pairing with narrower A- and wider B-terraces. Beyond $\theta = 1.5°$, the A-terrace width locally becomes zero so that paired single steps and true D_B-type double steps coexist, probably stabilized by kinks. Kinked D_B-steps cause a correlated lateral shift of dimer rows across steps. Only beyond $\theta \simeq 8°$, the clean surface is completely D_B-stepped. The pairing mechanism is based on the different contribution of S_A- and S_B-type single steps to surface strain. A mean terrace inclination decreasing with increasing terrace width is observed which is compatible with a constant edge depression. For $\theta > 9°$, the special stability of the (117) orientation causes separation into (117) facets ($\theta = 11.4°$) and less tilted stepped domains. The step topography is similar in the (210)-zone. The steps there can be considered as strongly kinked (110)-type steps. In the (100)-zone, only single steps without pairing are observed up to $\theta = 6°$ degrees which agrees with the equality of both sublattices with respect to the edge direction.
>
> [After Schröder-Bergen and Ranke[277]]

It has been demonstrated[575] that around T_R steps cease to exist: their free energy (Fig. 3.23, page 184) and the line tension vanish. The

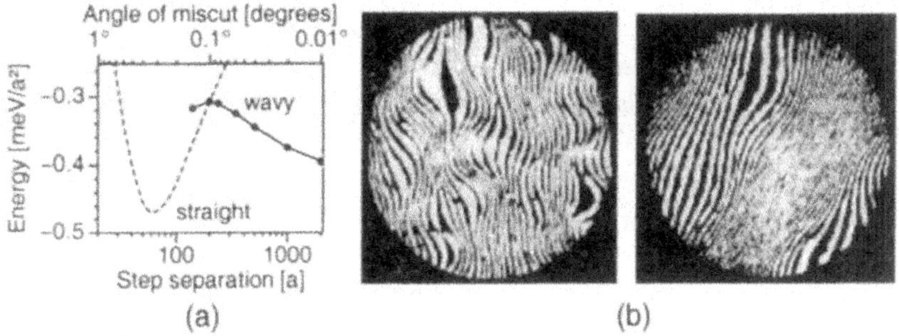

Fig. 4.8. Wavy steps on flat Si(001). **(a)** Calculated step energy per area $A = a_s^2 = 3.84 \times 3.84$ Å^2 versus miscut angle θ. The dashed line corresponds to straight steps, the dots to wavy steps, the solid line is to guide the eye. From Ref. 562. **(b)** LEEM images of Si(001) with $\theta \simeq 0.1°$. **Left**: wavy steps. **Right**: a faceted area. From Ref. 563.

roughening[563,678,689] which the surface undergoes at T_R can be therefore interpreted as a phase transition[575] rather than as a kinetic effect[563,678] (but see also page 261). Results of recent tight-binding molecular dynamics simulation[20] allow to interpret this transition as being due to melting of the two topmost atomic layers. According to the simulation, the majority of surface atoms is still dimerized, but the structure is heavily disordered and the arrangement of dimer bonds changes dynamically. In other words, the entropy makes 2×1 dimer rows unstable. Indeed, the LEED 2×1 pattern disappears at this temperature.[678]

We will limit our analysis to temperatures below T_R and to miscut angles below the critical angle for faceting. We begin with the role of step energies in the phase structure of vicinal Si(001) surfaces.

Some data suggest that D_B steps are stable against splitting into S_A and S_B (Sect. 3.3.2b). One might therefore suppose that the monatomic steps on surfaces with small miscuts reflect kinetic limitations due to excessive mass transport along the large distances needed to transform a freshly prepared, mono-stepped surface into a bi-stepped surface. But the elastic interaction between monatomic steps[684,568] stabilizes the mono-stepped surface for small miscuts at $T = 0$ K (Fig. 3.26, page 190). It is driven by the anisotropy of the surface stress tensor (Sect. 3.2.2d). At finite temperatures, entropy contributes to a further stabilization of the phase with monatomic steps, so that the critical miscut angle Θ_c above which the biatomic steps are favored is an increasing function of T (Fig. 4.7).

Another interesting phase,[563] **wavy monatomic steps** on very flat surfaces (Fig. 4.7), is due to the same stress-driven interaction.[562] In contrast to steps on vicinal surfaces, which appear to be straight when viewed in the micrometer scale, the steps on a flat surface become undulated, with wave lengths and amplitudes in the micrometer range. Surfaces with very small miscuts are faceted into flat regions with wavy monatomic steps and tilted regions with "straight" monatomic steps (Fig. 4.8).

The first-order transition between the **mono-stepped** and the **bi-stepped** surface (Fig. 4.7) was suggested to occur in a sharp line $\theta_c(T)$ separating monatomic and biatomic phases.[690] Alternatively, it may extend over a transition region[691] in which the mono- and bi-stepped areas are either mixed or form facets with different orientations.[685]

Which theory is the correct one? It is now known that this transition does *not* occur on a sharp line. With increasing miscut angle θ, the surface area fraction f covered by one of the two orientational domains (for example, 1×2) does not change sharply, but decreases gradually over a broad range of miscut angles[686,687] This agrees with the mixed-region theory, but also with the phase coexistence theory.

In the transition region, STM images usually show that D_B steps split into S_A and S_B parts, forming a mixed phase. These images cannot exclude that the mixed phase is metastable, enforced by diffusion kinetics. At least one STM study[685] indicates a phase separation (phase coexistence). The general conclusion from a simple elastic model[559] is that the transition region constitutes a mixed phase in which monatomic and biatomic steps alternate. This avoids kinetic bottlenecks due to the huge surface mass transport needed for separation of monatomic and biatomic steps. Such mixing is in agreement not only with elasticity theory[559] but also with results of atomistic calculations using empirical classical potentials.[559]

The asymmetry in the domain occupation ($f \neq 0.5$) is not solely due to biatomic steps. Calculations with classical potentials indicate[559] that even at angles for which the steps are exlusively monatomic the domains are not equally occupied. The energy of a symmartic ($f = 0.5$) surface can be lowered by moving S_A steps into the regions of increasing displacements induced by the reforming of the bond ("rebonding") at S_B-r steps. Some occupation assymetry has been measured for all miscuts.[686] Another important factor is stress. Thermal stress induced by subsequent heating and fast cooling cycles as well as stress around centers which pin (halt) step motion (such as SiC precipitates) can turn a well-oriented, two-domain surface

($f \simeq 0.5$) into virtually a single-domain ($f \simeq 1$) surface.[520]

We now consider the **interactions** responsible for this rather complex behavior. The study of step interactions and their relationship to growth mechanisms has its foundation in the studies of growth spirals on crystals.[134] Several general principles have been established:

- steps interact with elastic repulsion;
- step interaction controls the equlibrium surface topography;
- step interaction can be overshadowed by kinetic effects.

Step-step interactions and the resulting surface morphology depend on:

- tilt angle, in experiment usually about $[1\bar{1}0]$;
- temperature (through the density of thermally excited kinks);
- external strain;
- kinetics of surface diffusion, cooling rates;
- kinetics of sublimation or growth.

An essential feature of Si(001) is that its surface stress is tensile in the direction of dimer bonds, as ←⋮⋮⋮⋮→, and compressive in the direction perpendicular to them, as ⋮⋮⋮⋮⇅ (page 138). This opens a possibility for stress relief by alternating orientational domains of dimerization. Still, this type of relaxation has its energetic costs. First, the domains are neccesarily separated by monatomic steps, so some energy must be invested in order to make the steps (page 183). Second, relaxation of the surface layer strains subsurface layers, producing stress domains below the surface. Such domains may extend vertically to about 100 monolayers below the surface.[568] But it can be shown that energy is gained if the separation between steps is large enough. This is an example of a general rule: it follows from elasticity theory that *surfaces which reconstruct with broken orientational symmetry and exhibit an anisotropic intrinsic stress tensor are unstable to the formation of elastic-stress domains.*[56] A perfectly oriented surface can thus lower its elastic energy by giving up its atomic flatness.

In other words, steps on a surface like Si(001) interact through the surface stress tensor σ. We will now quantify this interaction. Assume that the surface is broken into strip-like, parallel minority (1×2) and majority (2×1) domains of lengths r_1 and r_2, respectively. The details of the reconstruction at the step edges at the domain boundaries are unimportant for the long-range effects which we are considering. Elasticity continuum theory applied to such a system reveals[56,559] that the energy $aE_\sigma(r_1, r_2)$ gained by relaxation of the surface area $A = (r_1 + r_2)a$ varies with domain length (that is, step separations) r_1 and r_2 as:

$$E_\sigma(r_1, r_2) = -2\lambda_\sigma \ln\left(\frac{r_1 + r_2}{2\pi a_s} \sin\frac{\pi r_1}{r_1 + r_2}\right), \qquad (4.14)$$

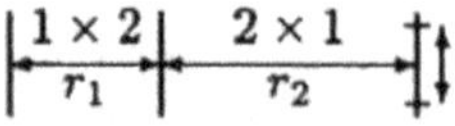

where $a_s = 3.84$ Å, the surface lattice constant, enters as a cutoff length below which all relxation effects are included in the step formation energy and not in the stress relaxation. Note that the distance $r = r_1 + r_2$ is a constant for misorientation steps. The energy λ_σ is proportional to the square of the stress anisotropy $\Delta\sigma$. The value of λ_σ was determined as:

- $\lambda_\sigma = \frac{(\Delta\sigma)^2}{2\pi}\frac{1-\nu}{\mu} = 5.5 \pm 1$ meV/a_s (LEED $\Delta\sigma$);
- $\lambda_\sigma = 8 \pm 4$ meV/a_s (STM, Zandvliet and Elswijk[519]);
- $\lambda_\sigma = 2.4$ meV/a_s (STM, Swartzentruber *et al.*[518]);
- $\lambda_\sigma = 9$ meV/a_s (classical potential calculations, Poon *et al.*[517]).

The value labelled "LEED $\Delta\sigma$" is obtained from stress anisotropy[494] and elasticity theory, assuming that bulk values of Poisson ratio ν and bulk modulus μ are valid near the surface. The STM results assume that $\Delta\sigma/\lambda_\sigma = 49 \pm 7$ Å^{-1}, as measured directly from the initial slope of the dependence of the domain population on the strain[494] (Sect. 3.2.2d).

In brief, a monatomic step sits between neighboring steps in the potential $E_\sigma(r_1, r_2)$ of long-range strain fields arising from anisotropy of the surface stress tensor. Each step is being acted upon by a "force monopole" proportional to the surface stress anisotropy $\Delta\sigma$. This is similar to the situation of a membrane separating two gases of different pressures, where the pressure difference is an analogue of the effect of the stress anisotropy.

A complete description of step interactions must include additional terms. If λ_{SA} (λ_{SB}) stand for the local energetic costs to create a *unit* length a of S_A (S_B) step,[n] the net energy E_{step} of the surface area A is:

$$E_{step} = \lambda_{SA} + \lambda_{SB} + E_\sigma + E_d + E_{\sigma d}. \qquad (4.15)$$

E_d represents the elastic interactions between local, subsurface strain fields produced by the relaxation at the step edges. It may be viewed as interaction between "force dipoles" located at the edges:[683] E_d is proportional to $1/r^2$, where r is the separation between steps. $E_{\sigma d}$ describes the coupling between the force monopole and the force dipole terms.[559] In our case of

---

[n] "Local" means that the steps should be viewed as decoupled from the surrounding elastic fields. Generally speaking, the formation energies λ_i contain only the atomic relaxation effects within a certain cutoff distance, usually taken as a_s (cf. Eq. 4.14).

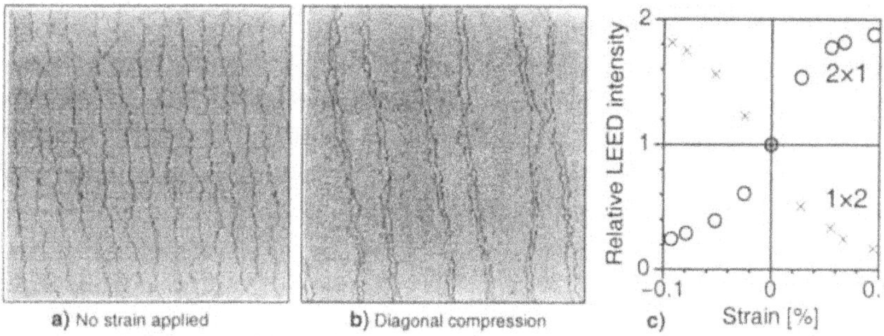

Fig. 4.9. Influence of strain on orientational domains of Si(001). Surface misoriented $\sim 0.1°$ from (001) toward [010]; dimer bonds run at $\pm 45°$ to the average direction of steps (dark lines in STM), that is, all steps are equivalent, having the same amount of S_A and S_B sections. **(a)** No strain, and **(b)** -0.09% strain (diagonal compression from upper left to lower right); adapted from Ref. 569. **(c)** Relative LEED intensities of 2×1 and 1×2 spots as a function of strain; from Ref. 367.

an ordered array of monatomic steps, the expressions for E_d and $E_{\sigma d}$ are:

$$E_d = \lambda_d\left(\frac{a_s}{r_1 + r_2}\right)^2, \tag{4.16}$$

$$E_{\sigma d} = -\sqrt{3\lambda_d\lambda_\sigma}\frac{2a_s}{r_1 + r_2}\tan\frac{\pi(r_1 - r_2)}{2(r_1 + r_2)}, \tag{4.17}$$

where r_1, r_2, and a_s are as defined for equation 4.14. According to Zandvliet and Elswijk,[519] the ratio $\lambda_d/\lambda_\sigma = 125$. This value is comparable to the ratio of about 80 obtained[517] in classical potential calculations.

Interactions between *biatomic* steps do not include the E_σ term. Since the bi-stepped surface is primitive, no force monopoles exist at D_B edges, and λ_σ vanishes. The interaction is thus governed by the force dipole term λ_d. The value of λ_d for biatomic steps has not been measured. According to calculations[517] with classical potentials, the dipole interaction between biatomic steps is about twice as strong as between monatomic steps.

Given that the energy of monatomic steps depends sensitively on the surface stress tensor, it is natural that surface stress is another variable which controls the phase structure of vicinal Si(001). Indeed, the minority domain fraction f is a sensitive function of external strain ϵ (Fig. 4.9). Marchenko[515] and Alerhand *et al.*[56] developed a theoretical method describing the population changes of orientational domains as a function of ϵ. According to their analysis, in the presence of external strain the expression 4.15 for step energy E_{step} should be extended by an additional term

$-\frac{1}{2}p\epsilon\Delta\sigma$, where p is the domain occupation anisotropy $p = 1 - 2f$, and $\Delta\sigma$ stands for the surface stress anisotropy. When the miscut angle increases while the strain ϵ is kept constant, the occupation anisotropy p decreases. Note that external strain may be intentional, for instance, applied by bending the sample, or unintentional, as caused by surface contaminations or deformations due to thermal treatments (thermal stress).

The dependence of p on ϵ is the basis for the measurement of $\Delta\sigma$. The measured $\Delta\sigma = 70 \pm 10$ meV/Å^2 coincides with theoretical *ab initio* estimates when the buckling of dimers is taken into account (page 138).

4.4. Deposition and Sublimation of Si

We now consider the evolution of surface morphology caused by deposition and desorption of Si: step flow and bunching, formation of facets, islands, hillocks, and pits. A selection of references is listed in Table 4.11.

High **deposition** rates lead to accumulation of adsorbed Si on terraces, because new atoms are adsorbed faster than the old atoms can reach the step edges. Si deposited at low temperatures also gathers on terraces, because the diffusion is slow. Therefore, high deposition rates and low deposition temperatures promote nucleation and growth of islands. In its extreme, this leads to buildup of three-dimensional islands and mounds; the surface roughens and the crystallographic quality of the grown film becomes low.[233,692] The typical situation is that of a mixture between step flow (adsorption at misorientation steps) and island growth (nucleation of 2D islands and adsorption at steps on the island edges).

The best growth mode is by step flow, without islanding. Islands may cause roughening. For example, when two islands meet, there is a 50% chance that a translational domain boundary of dimerization will be formed (Sect.3.4.3). Such boundaries trap adsorbed monomers and thus nucleate growth of a new layer even before the previous layer is complete.

When adsorption at steps (miscut steps or island edges) dominates, steps of monatomic height tend to coalescence into biatomic steps. A surface that initially has two domains may become virtually a single-domain surface after deposition of only 0.5 monolayer of atoms at ~500°C.[693] This is because S_B steps are better sinks for monomers than S_A steps (Sect. 4.2.3) and advance faster. Anisotropy of diffusion plays little role

Ref	year	Subject relevant to Si(001) step flow and bunching
226	1951	Burton-Cabrera-Frank (BCF) model of epitaxial growth
694	1985	Work function, E_F, photoionization threshold, various morphologies
695	1989	Monatomic → biatomic transition in growth; RHEED, Monte Carlo
663	1989	Stability of well-oriented D_B-stepped Si(001); STM after MBE
696	1989	Atomic steps during sublimation; LEEM
697	1989	Ripples on Si/Si(001); MBE
698	1990	Epitaxial growth; continuum model
570	1991	"Kissing sites"; STM
228	1991	"Denuded zones"; STM
520	1991	**Review**: Step flow and bunching during sublimation; LEEM
636	1991	Simulation of MBE growth; Monte Carlo based on STM data
651	1992	Equations governing growth by step flow; microscopic theory
590	1992	Stability of well-oriented D_A-stepped Si(001); STM after sputtering
573	1992	S_B-r and S_B-s comparably abundant after low-temperature epitaxy
699	1993	Step bunching produces {113} and {111} facets; SEM
700	1994	Surface diffusion, bunching and fluctuation of steps; Monte Carlo
701	1994	Monatomic → biatomic transition in growth; analytical model
702	1994	"Lochkeim" formation, nucleation rate; STM, BCF theory
623	1994	**Review**: Atomic processes in crystal growth
557	1994	**Review**: Morphology described with steps as basic units, thermodynamics
703	1995	Theory of step bunching + references to simple models
576	1995	Atomistic theory of biatomic step bunching; *ab initio*
233	1995	**Review**: Epitaxial growth by MBE at low temperatures
138	1995	**Review**: Si(001) vicinals in deposition and sublimation; REM
578	1995	Steps grow by kink flow; STM
692	1996	Roughening in low-temperature epitaxy, single- and double-domain; AFM
635	1996	Mechanism of step flow growth illustrated; STM video
704	1996	Ostwald ripening of Si(001) islands
705	1996	Origin of ripples on Si/Si(001); MBE, AFM, Monte Carlo
230	1997	Probability of Si incorporation at Si(001) steps; LEEM

Table 4.11. Step flow and step bunching; selection of references.

in this phenomenon.[701] Such **step bunching**, that is, creation of higher steps by lower steps coming together (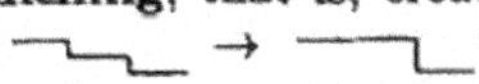), can be also caused by step flow instabilities due to Schwoebel barriers (page 58), electric fields (Sect. 4.2.3c), translational domains of dimerization ("kissing sites", page 185), or defects which stop the motion of a step.

Kinetic flow of biatomic steps can be achieved on surfaces on which S_A steps alternate with S_B steps, that is, in samples misoriented towards {110} directions. The double-step flow is stabilized by the flux. When no material is being deposited, biatomic steps are stable only on surfaces misoriented by several degrees (Sect. 4.3.2; Fig. 4.7 on page 250). With a given flux, higher temperature allows the growth to relax back to equilibrium single steps. In general, high temperatures destabilize biatomic steps (Fig. 4.7). Even on surfaces misoriented by as much as 6°, biatomic steps begin to dissolve into monatomic steps at 800-900°C.[561]

Interestingly, bilayer growth on a *well-oriented* Si(001) surface has been reported as well.[695] It was deduced from RHEED intensity oscillations[o] during deposition at substrate temperatures above ~500°C. The growth mode could be varied between single- and double-step flow by adjusting the temperature. However, the interpretation of this result is not straightforward because the deposition was initiated on a single domain surface obtained from the two domain surface through long annealing at 1000°C. This indicates that the growth might have been affected by thermal stresses (page 252 and Sect. 4.3.2) or by electromigration (Sect. 4.2.3c).

A well-oriented single-domain surface produced by epitaxy could be an interesting substrate for growth of GaAs films (page 188). But annealing at 450°C causes the metastable, single-domain surface with D_B steps to revert into the usual two-domain surface with S_A and S_B steps.[663] Typical temperatures in epitaxial growth are higher than that. It was therefore suggested[590] that a well-oriented surface with D_A steps (page 189) is used instead. Such surfaces can be made, for example, by Xe sputtering, and are stable up to 750°C. However, the high cost of making GaAs optical devices on-chip with silicon CMOS structures may render such systems too expensive for commercialization.

But let us return to physics. We noted (Sect. 3.3.2b) that rebonded biatomic steps, D_B-r (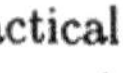), have a lower formation energy than non-rebonded D_B-s structures (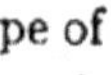). It should be stressed that this assumed that the steps are in equilibrium with the *bulk*. During growth, the relevant chemical potential of silicon is that of the incoming material, that is, higher than that of the bulk[p] According to *ab initio* results,[576] high chemical potential, such as when the vapor pressure is high, may stabilize the D_B-s steps. Does this have a practical consequence? This is an open question. In any case, the preferred type of the step may be important for the atomic mechanism of step flow, determining the sticking coefficients and the diffu-

[o] When a layer is growing, the surface is usually rough and RHEED intensity is low. When the layer is complete, the surface smoothes out and the RHEED intensity is high. The frequency of the oscillations measure thus the effective height of steps during growth.

[p] This is important because the areal density of Si atoms on a surface covered with D_B-r steps is different (higher) than on a surface covered with D_B-s steps. The chemical potential of incoming Si does not differentiate between S_B-r and S_B-s, which have both the same density of Si atoms.

sion bias (Schwoebel barriers, page 58). The structures formed when two steps meet depend on whether their edges were rebonded or not.[576]

An interesting morphological structure of Si(001) which may be considered as related to step bunching are the **ripples** appearing when quite thick (~1 μm) homoepitaxial layers are grown by MBE. The ripples resemble the pattern left on beach sand by water and can be observed on a well-oriented surface slightly tilted in a low-symmetry direction. Their height is around a few nm and the average distance between the mounds is in the micrometer range. The surface on the mounds remains therefore well oriented, deviating from [001] by a fraction of degree. The period and orientation of the ripples depend on the misorientation of the surface.[697] This structure is caused by growth kinetics and disappears when the surface is annealed at the growth temperature.

In order to explain the ripples, the dependence of step structures on the chemical potential of the deposited mobile material (that is, on the local surface concentration of the mobile silicon) was invoked.[697] Alternatively, one may view the ripple formation as a specific case of step edge instabilities which may be caused by specific growth conditions.[698] Following the latter idea, a recent study[705] has linked the ripples to **kink bunching**. Indeed, growth of steps proceeds mostly at kinks. Formation of new kink sites is energetically unfavorable (Sect. 4.2.3b). When the number of attachment events exceeds the number of events in which material moves along the edge, numerous kinks are likely to become smoothed out. This produces straight, long S_A segments. Since ideal, straight steps on [100]-miscut surfaces have the same amount of S_A and S_B segments, the smoothing corresponds to local variation of step-step distances. The process is statistical and the standard deviation is an increasing function of growth time, leading eventually to a mesoscopic ripple structure.

How does silicon arrive at steps: as monomers or as ad-dimers? Ad-dimers are certainly more stable on terraces than monomers, though also less mobile. The barrier for ad-dimer diffusion is by ~0.4±0.2 eV higher than that encountered by migrating monomers (page 207). Both barriers are, however, low enough for an efficient diffusion at temperatures typical for epitaxial growth. The energy of an ad-dimer is by ~0.4 eV/atom lower than the energy of a monomer (page 207). When two monomers meet, they form an ad-dimer in a barierless reaction; dissociation of an ad-dimer into monomers is hindered by a barrier of 1.3 eV (Sect. 4.2.2a). The energy difference between a stable monomer and a stable ad-dimer is high enough that in thermal equilibrium practically all "adatoms" (free Si on terraces) exist as ad-dimers. Even at 1000°C only a few procent or less of the adatoms are monomerized. Equilibrium diffusion on terraces should be therefore dominated by ad-dimers under all conditions of practical importance.

Naturally, deposited Si atoms may not always have enough time to equilibrate on terraces. In other words, they may become trapped at surface steps before they dimerize on terraces. Some data indicate that monomer-dominated growth can be observed. To wit, a "solid wall" Schwoebel barrier has been demonstrated at 110°C for ad-dimers approaching S_B steps from the upper terrace (page 234). But studies of denuded zones around steps on surfaces misoriented in such a way that monatomic steps run at ±45° to dimer bonds (each step contains the same amount of S_A and S_B segments) indicate[228] that there are no significant Schwoebel barriers for atoms attaching to S_B, at least at 290°C.

In addidtion, the activation energy for surface diffusion measured in the denuded-zone experiment[630] was 0.67±0.08 eV, which is a value typical for monomer diffusion but too low for ad-dimer diffusion (Sect. 4.2.1). These results allows one to argue that new material is delivered to steps in the form of monomers, at least when the temperature and the impingent flux are not too high. The monomers pair up at step edges and at the same time become incorporated into the step (page 226).

On the other hand, it is possible to *deposit* Si dimers. This can be done by CVD (page 275) from Si_2H_6. When the temperature is high enough for H to desorb from the surface (~550°C or more), CVD from SiH_4 corresponds to deposition of Si monomers while CVD from Si_2H_6 corresponds to deposition of dimers. A recent LEEM study[706] reported that SiH_4 and Si_2H_6 CVD at 650°C proceed practically in the same way. This result can be interpreted as an evidence for dimerization of monomers on the terraces and for growth dominated by diffusion and attachment of ad-dimers.

However, it may be that the deposited "dimers" dissociate during the adsorption process, for example, like: 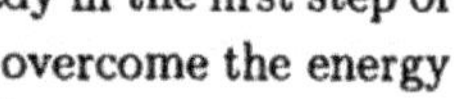. Note that, effectively, a new Si–Si bond is created already in the first step of this process. The resulting energy gain may suffice to overcome the energy barrier in the second step, 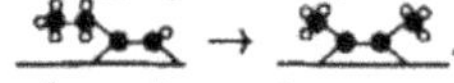. Its final configuration is in fact a transition state in the adsorption process completed by entropic dissociation of the reaction products.

So far we have been discussing monatomic steps and their bunching into regularly flowing biatomic steps. If a smooth flow of biatomic steps becomes hindered by an obstacle, such as a tiny precipitate of SiC, the moving steps pile up, forming large bunches which eventually develop into **facets** or **hillocks**. The facets have often a well-defined crystallographic orientation,[699] usually (111) or (113). When D_B-s steps meet, a (111) facet is produced. When D_B-r steps meet, a (113) facet grows. Kinetic pathways for these processes have been recently investigated *ab initio*.[576] Growth conditions which can cause instabilities in step flow and lead to step bunching on clean surfaces are discussed in Ref. 698.

Finally, let us describe the surface evolution during the process of **desorption**.[696,138] Reversing the growth process, atoms are released from steps and diffuse onto terraces, from which they can detach then into the vacuum. This leads to sublimation by step flow.

Sublimation can cause kinetic roughening of the surface. When the retreating steps are pinned at defects, hillocks grow. At high sublimation rates, dimer vacancies can be created on terraces and the vacancies can coalesce into vacancy islands, the so-called "Lochkeime" (hole nuclei). The holes are of monolayer depth. Since their growth is hindered by adsorption of silicon emitted from misorientation steps, the temperature at which they appear increases monotonically with the miscut angle, in other words, with decreasing separation between the steps. At 1170°C, that is, close to the Si(001) roughening transition (~1200°C, Sect. 4.3.2), the rate r of Si sublimation from Si(001) is[575] about 0.12 ML/s and decreases to below 0.02 ML/s at 1100°C. The sublimation rate can be written as

$$r = fe^{-(E_a+E_v)/kT}, \tag{4.18}$$

where E_v is the desorption barrier for the mobile material, E_a is its formation energy, $E_a + E_v = 4.3 \pm 0.3$ eV, and f is the attempt frequency of 10^{13}–10^{14}/s (see Ref. 702 for references and discussion).

Beside steps, **dislocation** exit points are in principle other weak spots from which atoms can easily escape. This may lead to formation of thermal pits, similar to etch pits. The walls of the pits may have a well-defined orientation (faceted pits), but rapid sublimation produces cone-shaped pits. Such pits have been observed on Si(111).

The problems of chemistry and biology can be greatly helped if our ability to see what we are doing, and to do things on an atomic level, is ultimately developed

– R. Feynmann

CHAPTER 5

Adsorption on Silicon Surfaces

J. Dąbrowski

Interaction of foreign atoms and molecules with Si surfaces. Adsorption sites, the nature of bonding, surface diffusion.

5.1. Introduction

Adsorption of atoms and molecules on silicon is a subject of countless research programs. Here we emphasize Si(001) and its interactions with hydrogen, halogens, oxygen, nitrogen, their compounds such as H_2O and NH_3, and some transition metals. This choice of topics is a compromise between their scientific importance, technological relevance, our wish to present a selection of inter-related issues, and our own expertise. Due to space and time constraints, we left out elements of groups V (other than N) and III, even though In, Sb, and particularly B, P, and As are essential to CMOS technology (Sect. 1.2.2) simply because they are the dopants.[a] The alkalis, rare earths, and rare gases are omitted as well. Our discussion of transition metals is focused on Ni. Although the technological importance of nickel silicides is at the moment negligible (even though a low-temperature NiSi technology has been demonstrated[707]), contamination of Si surfaces with Ni is common under laboratory conditions and may lead to effects which are easily misinterpreted and attributed to the clean surface. Moreover,

[a] See Sect. 6.5.5c for interactions between dopants and SiO_2/Si(001). References to recent work on adsorption of dopants on Si can be found, e.g., in Ref. 215, 708, and 709.

$NiSi_2$/Si interfaces resemble $CoSi_2$/Si interfaces, where $CoSi_2$ has already been implemented in some industrial technologies.[710]

Our discourse evolves around the issues which we introduced in the previous Chapters: surface defects, buckling of dimers, surface diffusion, step flow. For instance, dimer buckling may assist sticking of water on the clean Si(001) surface (the sticking probability is one). We also lay the ground for analysis of the oxidation of silicon (Chapter 6). For example, pyrogenic steam is used as a robust oxidant in production lines, and hydrogen not only passivates troublesome defects at SiO_2/Si(001) interfaces, but is also responsible for weak spots in the gate oxide which may kill an overworked transistor, and with it the whole chip.

The area of surface chemistry is full of interesting, important, and unsolved problems. It is virtually unexplored by atomistic calulations, and *ab initio* studies are particularily rare. Where *ab initio* results do exist, they usually provide real help in the interpretation of experiments. On the other hand, there is a clear disagreement (Sect. 5.2.2) between H_2 desorption energy barriers computed by two different classes of *ab initio* methods: DFT- and HF-based (Sect. 1.3.1a). This issue is critical for the credibility of *ab initio* techniques and must be settled by a careful, unbiased study combining a well-designed experiment and calculations of both types.

This Chapter begins with the presentation of the simplest adsorbate: hydrogen. As a matter of fact, interactions of H with Si will turn out to be astonishingly complex. Subsequently, we consider somewhat related species, halogens. They are the active components in commercial etching systems. Oxygen, our next subject, is an important topic of this Chapter, chiefly because adsorption of O is the first step of oxide formation. Water is then discussed for its role in wet oxidation procedures. N and nitridation found their place because it is hoped that Si_3N_4 may one day substitute SiO_2 as a gate dielectric; even if it does not, nitridation of gate oxides is likely to become a major approach to improve SiO_2/Si(001) interfaces and/or control the thicknesses of extremely thin oxide film. Finally, we devote a few pages to transition metals, mostly nickel.

We support our discussion by sketches like . Full circles in this sketch of the clean Si(001) 2×1 are the Si dimer atoms viewed from the perspective of the [110] direction. Each dimer atom is bonded to the substrate by two backbonds. Only one backbond per atom is visible from this perspective (Fig. 3.1, page 130). The long line at the bottom represents the Si(001) substrate. Electrons in the "dangling" bonds are symbolized by small dots. Ocassionally, the top view of the surface will be drawn, with dashes indicating surface dimers. The perfect 2×1 surface as seen in room temperature STM images is thus sketched as . Some important features, as "up" atoms of buckled dimers () or adsorbate atoms A () will be indicated by circles. For instance, the $c(4\times2)$ arrangement of buckled dimers seen in low-temperature STM images is sketched as .

5.2. Hydrogen: Passivation and Etching

Hydrogen on Si surfaces is a popular research subject. In the last few years, it has evolved into a model system for studying kinetics and dynamics of adsorbate interactions with semiconductor surfaces.[141] Table 5.1 lists only some selected publications. Due to limited space, we restricted this list to reviews, subjectively chosen early milestones and most recent articles, and several papers covering topics complementary to those discussed here.

Si surfaces are hydrogenated by an aqueous HF solution or by contact with gaseous hydrogen. Since H_2 does not dissociate spontaneously on undefected low-index Si, it is common to place above the surface a hot (~1500–1800°C) tungsten filament[711] (wire) which breaks H_2 into atomic H. Wet hydrogenation is typically done by a dip into a dilute (1–10%) solution of HF in ultrapure H_2O or into buffered HF, that is, $HF{:}NH_4F{:}H_2O$.

H and Si interact during technological processes. Chemicals containing HF etch the native oxide (Section 6.3.1). H passivates the broken bonds produced during oxidation. Hydrogen passivation of the surface is a beneficial factor during CVD epitaxy. H influences electrical properties of amorphous Si, which is important in solar devices (references in Ref. 265).

Ab initio calculations for H on Si require various degrees of sophistication, depending on the issue addressed. When energies of different configurations of H attached to the surface through bonds of similar character are compared, then the standard DFT approach with the local density approximation (Sect. 1.3.1a) may be sufficient.[b] Also zero-point vibrations can be ignored in this case.[712] The situation gains complexity when adsorption and desorption energies and reaction paths are considered. Approaches beyond the simple LDA and inclusion of spin effects may then be imperative; see,

[b] But see comments and references in Ref. 712, and Ref. 81.

Ref year	Subject relevant to hydrogen on silicon
713 1976	First demonstration of Si(001) 1×1–H dyhydride; UPS, LEED
714 1978	First self-consistent calculation for H/Si(001) 2×1
345 1986	Hydrogen on Si(110); LEED, EELS
265 1988	**Book article**: Kinetics, structures, energy loss spectra
715 1989	First-order kinetics of H_2 desorption from Si(001); isothermal desorption
716 1990	Real-time STM of H adsorption on Si(111); activity of adsorption sites
717 1990	Structure of Si(001) 3×1–H, etching, saturation; STM
525 1991	**Review**: Electronic structure of H/Si(001) 2×1
718 1991	Si(001) 1×1–H is disordered, H saturation coverages; TPD
708 1994	**Book chapter**: Kinetics, structures, phase diagrams
719 1994	H on Si(112) and Si(113) step edges; infrared absorption
720 1995	**Review**: Dynamics of H adsorption/desorption, Si(001) and Si(111)
721 1995	Sticking coefficient, Si-H bond energy, activation energies; D_2 beams
722 1995	**Review**; optical vibrational spectroscopy
723 1995	Adsorption/desorption; *ab initio*, LDA, gradient correction
724 1995	Adsorption/desorption; *ab initio*, LDA, gradient correction
144 1996	**Review**; STM
542 1996	Hydrogen and C-defects; *ab initio*
725 1996	Electron-stimulated desorption of hydrogen
726 1996	STM "litography" by local removal of H resist from Si(001)
727 1996	Anisotropy of vibrational polarizations
728 1996	Core level shifts for Si(001) 2×1–H; *ab initio*
141 1996	**Review**: Adsorption/desorption, diffusion, bond strength, also Si(111)
326 1996	**Review** on Si(113): Hydrogen on Si(113)
729 1996	Adsorption/desorption kinetics of H_2, Si(001); dynamical model
79 1997	**Review**: Dynamics of Si(001)–H adsorption/desorption; *ab initio* cluster
80 1998	**Review**: Dynamics of Si(001)–H adsorption/desorption; *ab initio* DFT
318 1999	Monohydride on a high-index surface: Si(114)
730 1999	H_2 adsorption on vicinal Si(001); SHG
582 1999	H_2 adsorption on vicinal Si(001); *ab initio* DFT
153 1999	**Review** on molecules on semiconductors: SiH_4, Si_2H_6 on Si(001)

Table 5.1. Hydrogen on silicon; selected references.

for example, Refs. 96, 724, 731, and the remarks on page 274.

5.2.1. *Hydrogen adsorption, etching, and structures on Si(001)*

The process of dissociative adsorption of a hydrogen molecule on a clean silicon surface is exothermic with energy gain E_a of roughly 2 eV. For example, the H_2 adsorption energy on Si(001) 2×1 was obtained as approximately 1.9 eV from DFT-LDA calculations.[582] If we assume the H_2 bond strength of 4.5 eV (Table 1.4, page 57), then the adsorption energy $E_a \simeq 2$ eV corresponds to the Si–H bond strength of ~3.1 eV. The experimental estimate[721] is ~3.4 eV. This value is by ~0.6 eV less than the Si–H bond strength in SiH_4 (Table 1.4).

In spite of this energy gain, the probability of H adsorption from molecular ambients is negligible on flat surfaces. The **sticking coefficient** of H_2 is low, possibly because the surface degrees of freedom play an important role in the reaction and the light H_2 molecule does not couple to them efficiently enough.[721] This interpretation is consistent with the fact that the sticking coefficient *increases* with increasing temperature: such an increase indicates that the presence of surface phonons is essential for efficient sticking.[732,721] Interestingly, the energy barrier on the adsorption path, estimated[721] to be ~1 eV, seems to be missing in the desorption path (Sect. 5.2.2). The reason for this apparent breaking of the microscopic reversibility is still debated, but in general it is probably related to the same "funnel" in the configuration space which limits the sticking coefficient.

The sticking coefficient of H_2 it is greatly enhanced on *misoriented* Si(001) and Si(111). This enhancement suggests that surface steps play an active role in the adsorption process.[733] Indeed, the results of recent SHG measurements[730] and *ab initio* DFT calculations[582] consistently point towards selective adsorption of H_2 at biatomic (D_B) steps. After the molecules are dissociatively adsorbed on steps, the H atoms diffuse away from the edges. The corresponding activation energy of 1.7±0.15 eV[730] is consistent with the **diffusion barrier** of 1.68±0.15 eV deduced from an STM study[734] for H atoms migrating along the dimer rows on Si(001) 2×1. The latter value assumes the attempt frequency of 10^{13} s^{-1} (page 208). Theory arrived at similar energy barriers for diffusion. Calculations using model potentials based on *ab initio* CI cluster data[118] yielded the diffusion barrier of 1.65 eV along the dimer rows and 2.7 eV across them (model potentials were used in order to overcome the difficulties due to small cluster size in CI calculations). Although DFT-LDA slab calculations[735,734] revealed smaller barriers (~1.3 eV along and ~1.8 eV across the rows), GGA corrections to DFT-LDA data increased[734] the barrier along the rows to 1.51 eV; the correction for diffusion across the rows was not computed.

H/Si(001) surfaces obtained by **dry hydrogenation** have 2×1, 3×1, or apparent 1×1 periodicities. In the 2×1 phase, H saturates Si dimers. We will sketch them as  or simply as 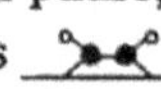. This is the monohydride phase: each surface Si atom has a single H attached (open circles). The dimers are symmetric. The 2×1 phase forms when H is adsorbed at 350°C or at higher temperatures. It persists up to ~550°C, until H desorbs.

Besides merely passivating the surface, hydrogen atoms can attack

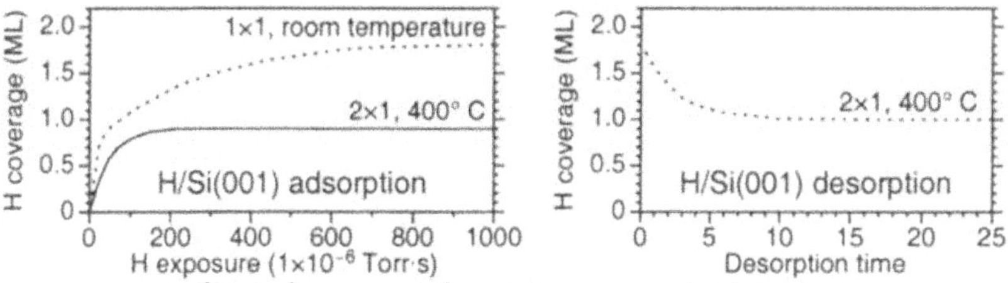

Fig. 5.1. **Left**: adsorption of H on Si(001) at room temperature (broken line) and at 400°C (solid line). **Right**: desorption at 400 °C. From Ref. 736.

Si–Si bonds.[737] As evidenced by *ab initio* calculations,[738] a positively charged H atom in the bulk silicon is stable in a Si–Si bond. Also a neutral H atom in the bulk acquires this position. Only H^- prefers to stay outside the bonds. When surface dimer bonds are attacked, they tend to open. This leads ultimately to the 1×1 and 3×1 structures. In the 3×1 phase, each third Si atom is undimerized and saturated with two H atoms (dihydride, ), so that di- and monohydrides alternate: 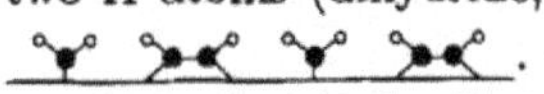. The 3×1 structure forms when H is adsorbed at ~130°C. But if the H supply is not cut off after the 3×1 is completed, hydrogenation continues and patches of 1×1 dihydride begin to grow.[717]

One might suppose that the 1×1-H phase is a pure dihydride, that is, it constitutes a truncated-bulk silicon crystal (Fig. 3.1e, page 130) with all broken bonds saturated by H. But in reality, this surface is a disordered mixture of mono-, di-, even trihydride, whereby the composition and the surface morphology depend on the preparation details. The 1×1 surface is full of etch pits and apparent etch products.[739] Flat regions with truly 1×1 periodicity exist, but only locally. Why is this the case?

The answer: because H can intervene in Si–Si bonds and because dihydrides are stable below 350°C. The former enables etching, the latter secures a path to higher hydrides. When dihydrides are there, etching can proceed on the whole surface[740,741] and not only at defect sites.[742] The disordered 1×1 structure needs low temperatures, that is, etching conditions. The well-ordered 2×1 and 3×1 phases require that there is no etching.

Clearly, no etching occurs above the dihydride stability limit. The ordered 2×1 phase is then formed (Fig. 5.1, left). The 3×1 phase, which exists at temperatures at which dihydrides are stable, is reasonably ordered because at ~120°C hydrogen atoms segregate to the surface.[743]

Nevertheless, prolongated hydrogenation corrodes the 3×1 surface. This is because the 3×1 is then converted locally to 1×1 dyhidrides. Etching of dihydrides ordered in the 1×1 array is facilitated by a high areal density of H atoms in this phase (Fig. 5.2a). These atoms are so closely packed that they repell each other. The H–Si–H bond angle, which should be tetrahedral (109.5°, as in the 3×1), decreases[744] to 102°. *Ab initio* calculations show that a canted 1×1 phase, with tilted H and surface Si atoms (Fig. 5.2b), has a lower energy than the symmetric phase.[744] This distorsion reduces the repulsion between hydrogen atoms. It also weakens Si backbonds, so that the formation energy of a trihydride (SiH_3, Fig. 5.2c) computed *ab initio*[742] becomes negative. Etching is now easier than in the 3×1.

Hydrogen etching works efficiently only below the temperature of 3×1 formation.[737] The trihydrides which are present on the surface hydrogenated below ~100°C combine with hydrogen during annealing around 375°C and evaporate as volatile SiH_4; at higher annealing temperatures also Si_2H_6 and other volatile Si_xH_y species are produced and desorbed.[737]

Room-temperature etching is rather slow. Cooling down to −170°C stabilizes the intermediate hydrides and the reaction procees efficiently,[741] with continuous emission of volatile SiH_4. The geometry of Si(111) makes dihydrides even more unstable than on Si(001), which results in slower etching rates of Si(111) than of Si(001) under the same ambient conditions.[606]

Wet hydrogenation, done by etching in HF, has a more complex chemistry because the surface is usually covered by a "native" oxide layer produced, for example, by standard wafer cleaning (Sect. 6.3.1). This oxide is stripped by the acid (page 279). Ideally, the substrate should be merely terminated with H. In practice it is etched as well, mostly by remnant OH^- ions[745]. They attack Si anisotropically, along (111) planes. Consequently, weakly basic (pH~8) solutions, such as obtained by ammonia buffering of HF, improve the homogenity of H/Si(111) by removing asperities.[746] A similar effect can be achieved by boiling in ultrapure water after HF etch.[747]

H/Si(001) surfaces prepared in HF are microscopically rough.[748] The same instability of dihydrides which is responsible for the roughness of H/Si(001) 1×1 after dry hydrogenation promotes roughening of wet-hydrogenated surfaces. Results indicating that atomically flat Si(001) can be obtained by a careful HF:HCl treatment in very acidic sulutions have been recently presented.[257] However, HF etch is followed by a rinse in water to remove loosely bonded HF molecules and remnant F atoms which

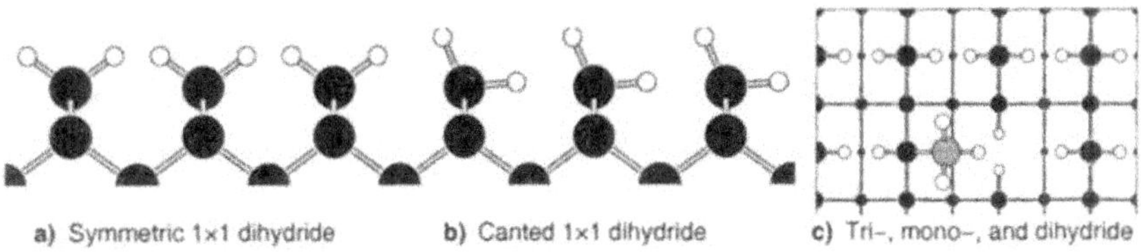

Fig. 5.2. Schematic ball-and-stick models of Si(001) hydrides. Hydrogen atoms are small and white. **Left**: Side view. **(a)** Symmetric and **(b)** canted Si(001) 1×1 dihydride. **Right**: Top view; atoms in deeper layers are smaller. **(c)** A trihydride cut off by H atoms from a symmetric dihydride surface. The Si atom of the trihydride is gray.

terminate some of the surface bonds (page 279). The OH radicals from water can then enhance the atomic-scale roughness.[257]

H-passivated Si(001) does not oxidize immediately in air.[749] This is a highly desirable property, because conventional CMOS manufacturing involves temporary storage of wafers after HF cleaning. Unfortunately, the chemical inertness of HF-cleaned wafers is significantly reduced when the substrate is heavily doped.[750] In any case, the hydrogen-saturated surface is hydrophobic (that is, it is not wetted by water). The angle of contact between water droplets and the surface depends on the pH of the droplet. It decreases with increasing pH, indicating acidic nature of the surface.[750]

A surface hydrogenated by an HF dip can be imaged with atomic resolution in air by STM.[751] The micrographs show the square 1×1 lattice of the dihydride. But the passivation may be not perfect: higher-order superlatticies, such as $c(2\times1)$, have also been seen in this air-STM study.

5.2.2. *Desorption of hydrogen and buckling of dimers*

Hydrogen desorbs from Si associatively: two H atoms come together and escape as an H_2 molecule. The measured[715,752,753,754] activation energies for H_2 desorption from Si(001) vary from 2.0 eV to 2.5 eV (see Ref. 723 for more references). This seemingly straightforward process is in fact little understood, at least for the Si(001) surface. There are no less than two intriguing problems related to it. First, although physical processes are generally expected to be reversible in the microscopic scale, this principle of detailed balance appears to be broken[755] for H_2 desorbing from Si(001). Second, even though *two* adsorbed H atoms must meet in order to desorb, the desorption rate r_d depends *linearly*[715] on the H coverage Θ_H of Si(001) (first-order kinetics). In both cases explanations have been given, but the actual mechanisms are controversial. Naturally, this interesting behavior of H_2 motivated numerous studies, experimental as well as theoretical.[c]

Since H desorption proceeds through emission of H_2, it should be the **reversal** of H_2 dissociative adsorption. Let us sketch the energy changes on this path as: 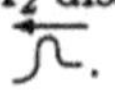. The arrow indicates here the direction of the adsorption process, which begins as  and ends as . Naturally, the energy barrier encountered during the process of desorption,  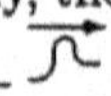 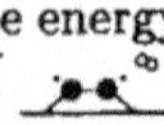, should affect the energy distribution of the desorbing molecules. However, the molecules escape with nearly thermal energies, that is, with no significant excess energy that they should have collected while rolling down the barrier.[755] In other words, the energy barrier[141] which hinders H_2 adsorption seems to be absent in the desorption path.

This effect is most probably associated with strong coupling of the reaction coordinates to the surface degrees of freedom.[732] In other words, the reaction path that joins the free and dissociatively adsorbed molecule resembles a "funnel" with a wide opening on the surface side (where substrate phonons can easily interact with the adsorbate vibrations) and a narrow opening on the vacuum side (where the interaction with the surface phonons is disrupted due to the mass mismatch between hydrogen and silicon). There are two consequences of such a coupling. First, the most probable adsorption and desorption processes may follow different routes. This is *not* a violation of the microscopic reversibility, but simply an entropic effect. Second, at least a part of the energy collected by the desorbing molecule while coming down the adsorption barrier (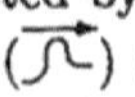) may be returned to the crystal.[756] The idea of the coupling to surface degrees of freedom[755,732] is corroborated by the results of *ab initio* DFT calculations[723,724,731] which we discuss below. Similar entropic effects may be responsible for trapping of ad-dimers at energetically unfavorable sites (page 213).

But why is the **desorption kinetics** linear? An early STM study[757] provided an influential hint: H_2 desorption from Si(001) 2×1-H apears to be accompanied by pairing of H on the same dimer ( → 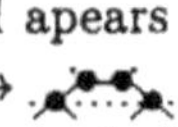 + H_2). Numerous isolated clean dimers are seen in STM images acquired after partial desorption of H_2 from nearly saturated Si(001) 2×1-H (Θ_H close to 1 ML). If dimer bonds are represented by dashes and H atoms by dots, top view of a 2×1 monohydride surface from which a single H_2 molecule

---

[c] For brief overviews, see Refs. 753 and 758. Comprehensive reviews representing opposite opinions are given in Refs. 79 and 80.

desorbed is:  (or , if we zoom in). Given that two H atoms prefer to passivate dangling bonds of a single dimer () rather than single dangling bonds of separate dimers (), the tendency of H_2 to escape from  would be a natural explanation of the linear kinetics.

To wit, since a great majority of H atoms is **pre-paired**, at least at $\Theta_H \gtrsim 0.01$ ML, the concentration Θ_2 of desorption centers occupied with two atoms () is proportional to Θ_H, leading to first-order kinetics ($r_d \sim \Theta_H$). Second-order kinetics ($r_d \sim {\Theta_H}^2$) in conjunction with this desorption mechanism ( $\rightarrow$  + H_2) would indicate that $\Theta_2 \sim {\Theta_H}^2$, that is, that a significant fraction of H is unpaired (). This can happen at low Θ_H or when the reaction constant K of the pairing is small: $\Theta_2 = K{\Theta_1}^2$, where $\Theta_1 = \Theta_H - 2\Theta_2$ is the concentration of unpaired H.

H atoms have good reasons to fully occupy dimers. First, the electrons in the "dangling" bonds of a clean dimer (the small dots in ) form a weak π bond. A single H atom adsorbed on a dimer () disrupts this bond, leaving one of the dimer Si atoms with a truly unsaturated, dangling orbital. In addition, adsorption of an H atom on a dangling bond reduces the buckling () of the affected dimer noticeably (Fig. 5.6, page 302). The consequence of reduced buckling is that the elastic strain due to dimerization is less efficiently relaxed (Sect. 3.2.2d). Moreover, with one side of the dimer chemically passivated it is more difficult to decrease the energy of dimer electrons by rehybridization of their orbitals (Sect. 3.2.2b). The net effect is that the energy is lowered when two H atoms occupy the same dimer instead of two: 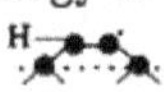 +  $\rightarrow$  +  + energy gain ΔE. Given the value of ΔE, one can use statistical mechanics[759] to estimate the fraction of H atoms which occupy single dimers and contribute therefore to the second-order term in the desorption rate. Conversely, one can use this type of analysis to estimate ΔE from the measured $r_d(\Theta_H)$. In practice, $r_d(\Theta_H)$ becomes non-linear[753] below $\Theta_H \simeq 0.01$ ML.

The value of the pairing energy ΔE has been obtained from *ab initio* cluster[760,761,762] (0.1–0.2 eV) and DFT slab[547,763] (0.25–0.5 eV) calculations and deduced from experimental STS[757] (0.8 eV) and TPD[715,752,753,759] (0.25–0.3 eV) data. The reason of the discrepancy between the measured values of ΔE is unclear. But the important difference between STS and TPD data set is that the latter corresponds to low coverages ($\Theta_H \sim 0.01$ ML or lower), while the former is collected at a submonolayer coverage.

It was suggested[753] that he inconsistency between the values of ΔE

decuded from TPD and STS stems from difficulties in the analysis of STS data. In later work[547] it was proposed that a contribution to the statistics of the pairing reaction may come from correlation between neighboring dimers. As emphasized in Ref. 753, the usual analysis of the pairing reaction assumes that the hydrogen occupation of neighboring dimers is uncorrelated. The *ab initio* study of Ref. 547 found that H atoms occupying neighboring dimers may become weakly bonded in the so-called *cis* configuration: 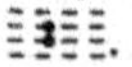. Moreover, there is an attractive force which tends to arrange monohydride dimers () into strings extending along the dimer rows (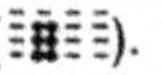). This force contributes to the effective pairing energy observed at higher coverages. Such strings have been seen in STM images.[757]

The detailed **desorption mechanism** remains a mystery. Several schemes are plausible. First, the desorption can be initiated on a monohydride dimer by the pre-paired H atoms either directly (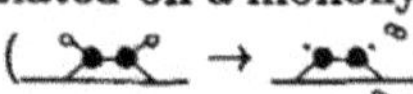) or through the so-called isomerization path (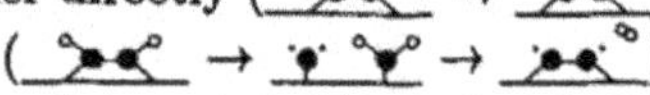). In the latter case, H_2 desorbs from a temporarily created dihydride (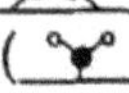). Second, H atoms can be collected from monohydride dimers by diffusing Si monomers, so that a mobile dihydride is created from which H_2 subsequently desorbs (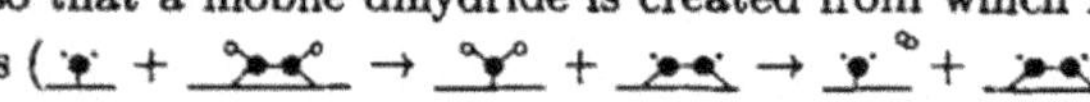). Third, hydrogen atoms may "evaporate" from monohydride dimers onto the surface (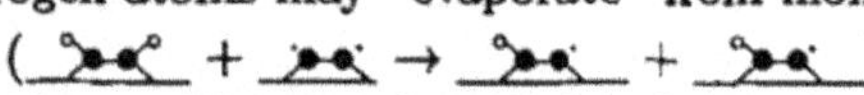), diffuse separately to a defect site where a Si atom is loosely bonded, convert this Si atom first into monohydride and then into dihydride, and eventually desorb from the dihydride, restoring the defect site to the initial state.

The most popular mechanism is probably the direct desorption from monohydride dimers:[752] 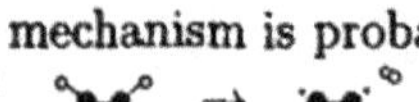. It is partially supported by a bulk of *ab initio* DFT calculations (see Refs. 80 or 582 for references and discussion and page 20 for a brief description of DFT). The DFT results show and quantify the intimate coupling between the surface degree of freedom (dimer buckling: , Sect. 3.2.2b) and the height of the energy barrier. The substrate rearangement energy at the barrier configuration (transition state configuration) is about 0.15 eV. The barrier towards dissociative *adsorption* is distinctly lowered when the buckling angle decreases. The lowest barrier is found for the asymmetric transition state (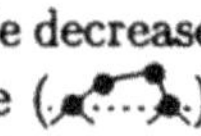) at the buckling angle of ~13° (the buckling angle is ~20° on the clean surface). This barrier amounts to ~0.4 eV in fully consistent GGA calculations[731]. The barrier associated with the symmetric transition state () is somewhat higher than that for the 13° buckled transition state.

Taken together with the computed H_2 adsorption energy of ~1.9 eV, the DFT desorption energy for the direct process is ~2.4 eV, well within the experimental range. On the other hand, all-electron quantum chemistry calculations (page 20 and Sect. 1.3.1b) arrived at much higher energy barriers (for example, Ref. 758 reported barriers of nearly 4 eV), so that these authors consider the direct desorption from monohydrides as a rare rather than dominating event. This issue is still not settled and calls for a careful comparative study. We will briefly return to it later in this Section.

The alternative suggestions employ dihydrides as the centers from which H_2 molecules are released. A hint that the desorption may proceed through isomerization (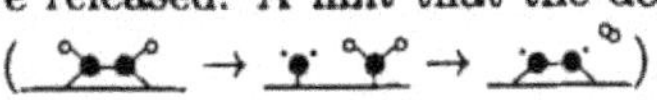) comes from studies of Si(001) oxidation by H_2O (page 305). Namely, Si–H bonds are stabilized against H_2 desorption when an intervening oxygen strenghtens the dimer bond (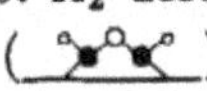), so that the isomerization is made difficult and H desorbs only when SiO begins to desorb. It is, however, not clear whether both ends of the oxidized dimer are indeed saturated by hydrogen, or the H atoms are isolated (). Moreover, the oxygen atom intervening in the dimer bond may hinder also the direct desorption process.

The isomerization path has been rejected on the basis of quantum chemistry results obtained for small cluster models:[758] the amount of dihydrides created by isomerization was computed to be negligible. No corresponding DFT slab calculations exsist up to now. The defect-mediated process[764] may be at variance with experimental evidence.[732] The dihydride mechanism which seems to attract most attention is the desorption assisted by diffusing Si monomes[765] (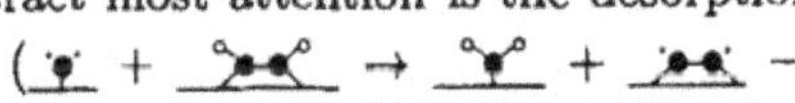
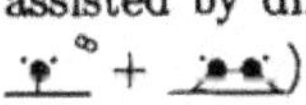). Succesfull Monte Carlo simulations of desorption kinetics, based on *ab initio* data, have been recently performed.[766] Although the "diffusing monomer" was assumed to have an unrealistic atomic structure, this should not lead to problems because the key details are related to the interactions between the monomer and two H atoms and not between the monomer and the Si matrix.

However, the monomer-mediated mechanism requires that the formation energy of monomers is low. Indeed, since desorption occurs from hydrogenated monomers, the desorption rate should depend on the concentration of monomers, that is, the formation energy of monomers should be added to the activation energy of desorption. Since the monomer formation energy is ~1.8±0.3 eV (DFT-LDA estimate, page 207) and the energy barrier for H_2 desorption from an isolated dihydride was computed[758] as ~2.5 eV, the activation energy of the monomer-assisted desorption should be at least 4 eV. This is much more than the experimental 2.0-2.5 eV.

So let us revert to direct desorption from monohydride. The CI cluster calculations of Ref. 758 did not reproduce the low energy barrier which was obtained in slab DFT studies.[723,724] But at the same time, the CI study did not pass the key reliability test: it did not reproduce the buckling of dimers (Chapter 2 and Section 3.2.2). This may exemplify a major failure of physically advanced but numerically heavy *ab initio* CI techniques. They seem to suffer from precisely the same difficulty which plagued early *ab initio* DFT calculations: practical limits in computer resources enforce the use of small clusters and small basis sets (Sect. 1.3.1b). The sophistication of the theoretical aparatus may be still played down by unphysical constrains imposed by the numerical algorithms and computers.

In addition to the geometry constraints due to small cluster sizes and problematic boundary conditions on cluster edges, the high energy barrier computed by CI for simple pairing may be caused by an overestimate of the strength of the intra-dimer π bond. The same overestimate prevents the dimer from buckling (Section 2.2.5a). However, since both desorption paths lead through many electronic configurations, it is imaginable that this error cancels out and that the final CI result is correct. CI may be actually *necessary* for a correct description of the energy differences between various bonding configurations of hydrogen.

Although we tend to prefer the direct monohydride desorption model over the other models, we think that the problem has not been solved. In particular, DFT slab calculations for dihydride desorption are still missing.

Another interesting observation[720,80] is that the H_2 molecules desorbing from Si(111) and Si(001) are internally excited in a similar way: they are vibrationally hot ($E_{\text{vib}} > kT$) and rotationally cold ($E_{\text{rot}} > kT$). Given that the atomic structures of these surfaces are very different, this is an unexpected result. In particular, there are no monohydride dimers on Si(111); the structure of hydrogenated Si(111) is dominated by Si "adatoms" which are threefold back-bonded to the substrate and terminated by hydrogen. But this result may be in fact compatible with the desorption from monohydride dimers. Since the direct desorption path from the dimers proceeds through an asymmetric transition state, by the end of the desorption event both H atoms interact with only one Si dangling bond (). This makes Si(111) and Si(001) desorption similar at least during their final steps.

Finally, we recall that there is a high probability (~50%) that instead of making a bond with silicon, an approaching H atom will pick up an already adsorbed H atom to form an inert H_2 molecule.[715,767,768] This reaction (**Eley-Rideal** mechanism, Sect. 1.4.2c) is driven by differences in bond energies in H_2 molecules (4.5 eV) and in Si(001) monohydride (more than 1 eV, according to recent *ab initio* calculations[769,582]). A detailed theoretical, *ab initio*-based study of the dynamics of such hydrogen abstraction from Si(001)-2×1 by hot gaseous H is reported in Ref. 770 and 771.

The Eley-Rideal reaction is the reason why the H coverage in the 2×1 phase produced by H adsorption at 400°C saturates below unity (Fig. 5.1, left). A similar effect hinders complete H passivation of broken bonds near SiO_2/Si(001) interfaces. A coverage of the Si(001) by a full monolayer of H can be obtained by H desorption from 1×1 surface (Fig. 5.1, right). The 2×1 phase is then produced, illustrating the fact that dihydride does react with H to form volatile Si-containing species; the desorbed material consists of H_2 molecules. But the 2×1 surface obtained in this way is not flat.[606]

5.2.3. *Hydrogen and epitaxial growth*

Desorption of hydrogen from the surface sets a major limit[772] in the Chemical Vapor Deposition (CVD) growth from SiH_4. In CVD, the silicon is often delivered to the surface in the form of silane (SiH_4) or disilane (Si_2H_6) vapor. If the deposition is so fast or the substrate temperature so low that some H atoms are buried below the surface, an amorphous SiH_x film grows. In principle, the lowest temperature T_e by which the film is crystalline (epitaxial temperature) should be that at which monohydride becomes unstable, that is, ~550°C. However, a monohydride surface layer is not a problem and it is sufficient[743] that H segregates to the surface (above 120°C, Ref. 743), and that no dyhidride is present (above 350°C). Such extremely low-temperature films, though crystalline, contain many defects. In practice, T_e can be reduced to ~425°C if the deposition is very slow.[773]

That T_e is as high as 425°C may be due to reduced diffusivity of silicon on hydrogenized surfaces. Even small amounts of adsorbed hydrogen slows down surface diffusion. Already 0.04 ML of H coverage doubles the effective activation energy for Si diffusion, as indicated by Monte Carlo simulations based on STM data.[549] This may be because Si monomers must take a detour each time they meet a monohydride dimer. The interaction of Si monomers with the isolated monohydride dimers may be in fact more complicated, because *ab initio* results[774,775] suggest that the presence of hydrogen should increase the barrier to ~1.0 eV, while the experiment suggests[549] ~1.3 eV. We speculate that the presence of "roadblocks" increases the probability of ad-dimer formation, so that the dominating mobile species change from monomers to ad-dimers (cf. also page 260).

Diffusion of silicon on a saturated monohydride surface is possible, but the barriers are higher than on the clean surface.[774,775] The migrating silicon is then bonded with the substrate silicon and covered with hydrogen. The monomer may become temporarily captured on top of a dimer from which it stole its two terminating H atoms to saturate its own dangling bonds.[774] Silicon diffusion on Si(001) 1×1-H is even more difficult.[774,775] This *ab initio* observation is consistent with the experimental fact that the presence of dihydride destroys epitaxial growth.[743]

In general, hydrogen is beneficial for epitaxy (see references in Refs. 549 and 743, and Ref. 772). By passivating the surface, hydrogen protects it from contaminants. At the same time, a passivated surface is no longer a place where a dopant atom could localize its extra electron or hole on a dangling bond or where a Ge atom could profit from the fact that its dangling bond has a lower energy than that of a Si atom. Thus, hydrogen reduces surface segregation of dopants during the deposition of doped layers and facilitates intermixing of Si and Ge during the deposition of SiGe layers. Finally, since hydrogen saturates the orbitals of deposited Si atoms which are unused in bonds with the crystal, the process of incorporation of new material into the crystal structure, quite complex when atomic Si is deposited (Section 4.4), can be now simplified.

Studies of SiH_4 adsorption on the clean Si(001) 2×1 at low temperatures provide information about the decomposition paths of silane.[d] But STM images are not easily interpreted without support from *ab initio* calculations of atomic and electronic structures. For example, clover-like objects of the size of approximately two Si(001) dimers form when SiH_4 is deposited between 280 and 420°C. A configuration consisting of four SiH_2 groups, decomposed from SiH_4 molecules and attached to four substrate atoms, was proposed to explain them.[776,777]

Maybe a more realistic structure of the "clover" was proposed in Ref. 778: two silicon ad-dimers form a tetramer located above the trough

[d] See, for example, Refs. 779, 776, 780, and references therein.

between dimer rows: 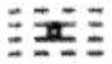 (or 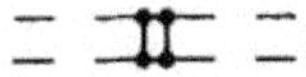, if enlarged). This model is based on *ab initio* DFT-LDA calculations. The computed energy of the tetramer was by 0.18 eV lower than the energy of two isolated ad-dimers and the characteristic features of STM images were reproduced. Note that the ad-dimers in this structure resemble to a some extent the ad-dimers in the dimer-chain model of Fig. 2.6e (page 94). In both cases, ad-dimers occupying the same trough between two dimer rows are mutually bonded, forming pairs (as in the "clover") or chains (as in Fig. 2.6e).

5.3. Halogens: Etching

Halogen atoms (F, Cl, Br, I), particularily F and Cl, are involved in many technological processes: dry and wet etching of Si and SiO_2, Chemical Vapor Deposition (CVD), dry and wet thermal oxidation. Plasma etching with fluorine-containing precursors is commonly used.[245,781,782] The advent of shallow trench isolation techniques (page 378) due to aggressive miniaturization of CMOS devices (page 11) brought etching into focus (in the context of trench etch). New recipes, usually designed intuitively, are constantly being examined. Studies of etching with halogen atoms and molecules are hoped to make this development more efficient. Naturally, investigations which consider the pecularities of the CMOS process will have the highest impact. It was found, for example, that it is beneficial for the gate oxides if highly electronegative substances such as SF_6 are avoided.[783] Polysilicon etch with HBr + O_2 was found to be mild to the oxides.[784]

We can only touch this vividly evolving field. We focus on those aspects of etching by Cl_2, Br_2, and I_2, which are linked to our discussion of dimer vacancies (Sect. 3.3.1a) and their interactions (Sect. 3.4.1). Remarks on F will be made as well, also in the context of HF cleaning. Table 5.2 selects papers related to halogen interactions with Si surfaces.

Similarly to hydrogen, a halogen atom is missing one electron for completion of its valence shell. The behaviour of halogens on silicon is thus in many aspects similar to that of hydrogen (Section 5.2) and OH (Section 5.5). Halogens attach themselves easily to surface dangling bonds, building monohalides (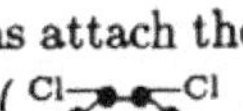 or, in general, ) and dihalides (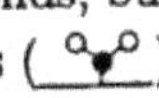). For example, STM images of a 3×1 reconstruction analogous to Si(001) 3×1-H (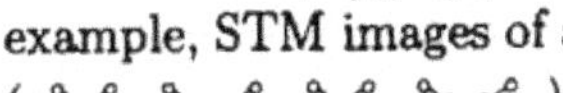) have been recorded for adsorbed bromine;[785] the

Ref year	Subject relevant to halogens on silicon
786 1986	Etch rates in CF_4/O_2 plasma, dependence on doping
787 1986	Cl etch products; TPD
788 1988	F-Si interactions, F/Si(111), bulk Si; *ab initio*
789 1988	**Book article**: F and Cl dry etching
252 1988	F termination of Si(001) during HF etch; XPS
790 1989	F, adsorption, desorption, temperature dependence; XPS, MB
791 1989	Cl, Br, I etch products; TPD
792 1981	Si(001), Si(111), Si(110) F termination, oxidation after HF etch
245 1992	**Review**: Etching
793 1993	Br, layer-by-layer etching; STM
785 1993	Si(001) 3×1-Br; STM
794 1993	Manipulation of Cl atoms on Si(001) by STM
795 1994	Br, dissociative chemisorption of B_2, Si(001) step evolution; STM
600 1994	Br, temperature dependence, surface morphology; STM
708 1994	**Book chapters**: Surface reconstructions induced by halogens
796 1994	Cl, etching under various conditions; molecular beams, TPD
797 1995	Cl, parameters controlling etching; STM
798 1995	Cl, growth of missing dimer rows; STM, Monte Carlo
601 1995	I, structures, uptake, etching; STM, SCLS
799 1996	F and Cl, adsorption sites, bonding geometries; ESD
800 1996	Br, adsorption sites, temperature dependence; nuclear methods
75 1996	F, dynamics of surface reactions; classical potential from *ab initio*
801 1997	F_2 treatment of H/Si(001); infrared reflection-adsorption spectroscopy
802 1997	Effect of Cl etching in CVD; thermodynamical analysis
803 1997	Surface diffusion of Cl, CVD equipment simulation; *ab initio* cluster
804 1997	Cl, barrier for $SiCl_2$ formation; *ab initio* slab
805 1999	**Review**: Plasma processing for CMOS

Table 5.2. Halogens on silicon; selected references. The surface orientation is Si(001), unless explicitly stated otherwise. MB stands for molecular beams.

main difference to H is that volatile $SiBr_2$ desorbed, leaving 3×1 rows of missing atoms, separated by dimers: . Cl atoms have been observed[794] to jump between two dangling bonds of a dimer, as the OH moiety does (Sect. 5.5.2). This experiment detected also a metastable adsorption site above the dimer bond (dimer bridge site), just as predicted[712] for OH. Such metastability was computed for chlorine[806] as well as for bromine.[807] A study using the radioactive ^{77}Br isotope[e] found some evidence for bromine atoms occupying these dimer bridge sites.[800]

5.3.1. *Fluorine and hydrofluoric acid*

Fluorine (F, electronic configuration [He]$2s^22p^5$) reacts strongly with Si. The Si–F bond is the strongest bond that Si makes with any element (Table

[e] Radioactive isotope techniques afford analysis of small coverages, such as 0.001 ML.

1.4, page 57. Hydrofluoric acid (HF) is the only acid to which Si yields, and F_2 is the only halogen molecule which spontaneously etches Si surfaces.

Fluorine atoms are the smallest among the halogens and the most electronegative among all elements. They easily penetrate the substrate and attack Si–Si bonds "from the inside". *Ab initio* cluster calculations[808] estimated that atomic F needs ~1 eV to enter the first subsurface layer of Si(111). Fluorine uptake does not saturate with exposure to F_2. Nevertheless, the adsorption is markedly faster in the initial phase (up to approximately 1.5 ML) than at higher coverages and a kind of "surface passivation" of room-temperature surfaces can be observed.[245] Atomic fluorine combines with Si even more efficiently. When three monolayers of F have been adsorbed, the adsorption rate is still half its initial value.

The initial sticking probability[790] of **atomic** (as in plasma etching) or **molecular** (F_2) fluorine is close to 0.5 in a wide temperature range around −150 to 330°C. Calculations with classical potentials based on *ab initio* cluster CI data[809] indicate that all incident F_2 molecules react with the clean surface, whereby either both or only one of the two F atoms may become adsorbed. When the surface is partially fluorinated, the reaction probability was computed to drop, in agreement with measurements. Experimentally, etching with fluorine generates volatile SiF_2 and SiF_4 molecules. Silicon tetrafluoride desorbs already below ~180°C and is the major etch product for F coverages above $\Theta_F \simeq 2$ ML. Silicon difluoride is the major etch product for lower coverages; it desorbs around 530°C.[790]

It is possible to terminate some dangling bonds of the Si(001) dimers with F atoms, for example, by exposing at room temperature the surface briefly to XeF_2 vapor.[799] XeF_2 is a clean source of higly reactive fluorine; the role of Xe in the reaction is unclear. Since the adsorption of F_2 is strongly exothermic,[f] one of the F atoms is likely to be ejected away from the surface; this results in only one Si–F bond per dimer (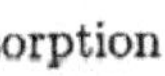). Higher XeF_2 exposures produce[810] first a SiF_3-rich layer and then volatile SiF_4.

Hydrofluoric acid (HF) is used to remove native oxide and microparticles from Si surfaces (Sect. 6.3.1). It etches the oxide more efficiently than it etches Si (although the difference is less pronounced when the silicon substrate is heavily doped[750]) This provides a practical cleaning method of a growing importance in CMOS technology[235] and – to some extent – enables investigations of the substrate directly below the SiO_2/Si

[f] The F–F bond energy is 1.6 eV, the Si–F bond energy is 5.6 eV (Table 1.4, page 57).

interface. The roughness of the interface can be studied in this way.[811,812]

HF etching opens all dimer bonds of Si(001) and the dominant surface species are dihydrides, 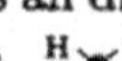. The fluorine atoms saturate some of the dangling bonds of triply backbonded Si atoms at defect sites and on step edges.[792] Indication for many subsurface SiF bonds was found as well.[813] This passivation of the dangling bonds protects the surface temporarily against oxidation[792,749] but does not provide a good protection against hydrocarbons.[255,252,253]

How can HF passivate the surface with hydrogen despite that SiF bonds are stronger than SiH bonds (Table 1.4, page 57)? It seems that this is mostly a kinetic effect. Let us follow the etching process step by step. HF dissolves SiO_2 by opening Si–O–Si (siloxane) bridges and, subsequently, by substituting OH radicals by fluorine atoms:

$$\text{Si–O–Si} + \text{HF} \rightarrow (\text{Si–OH F–Si}), \quad (5.19)$$

$$(\text{Si–OH F–Si}) + \text{HF} \rightarrow (\text{Si–F F–Si}) + H_2O. \quad (5.20)$$

It is plausible that the siloxane opening proceeds in two steps. First, a proton attaches itself to the oxygen atom: Si–(OH^+)–Si. Second, the proton is neutralized by an electron from the neighboring Si atom or from F^-, one of the Si–O bonds opens, and the exposed Si atom becomes terminated by fluorine. When all oxygen bridges of a Si atom are ruptured, the atom is liberated to the solvent. It follows that immediately after the oxide film is removed, the silicon surface is terminated by F atoms. Any remnant OH should be removed by HF, at least, when the pH is sufficiently low.

However, the etching does not stop at this point. This is because[814] the backbonds of a fluorinated surface Si are polarized (F is much more electronegative than Si), and so are HF molecules (F is much more electronegative than H). Note that since the HF dissociation in water is low,[750] there are many HF molecules available in the solution. A polarized backbond attracts and pre-orients an HF molecule such that the F end of the molecule is closer to the fluorinated Si atom on the surface and the H end is closer to the subsurface Si atom. The Si–Si backbond is now attacked:

$$\text{Si–Si–F} + \text{HF} \rightarrow (\text{Si–H F–Si–F}). \quad (5.21)$$

Ultimately, a silicon fluoride molecule is detached and the surface becomes saturated by H. Since the electronegativity difference between Si and H is much smaller than between Si and F, the backbonds on a hydrogenated surface are less frequently attacked by HF molecules than the backbonds on a fluorinated surface. In other words, the etching process slows down after all atoms that were bonded to oxygen are removed by fluorine.

An HF etch leaves therefore the surface terminated predominantly by dihydride species.[815,816] This picture of the etching process is substantiated by results of *ab initio* calculations done for small clusters of atoms.[817]

But why does hydrogen remain on the surface instead of being substituted by fluorine in the reaction: Si–H + HF → Si–F + H_2? Indeed, an estimate from bond energies gives for this process a gain of 1.2 eV. However, the calculations of Ref. 817 indicate that this reaction has a higher energy barrier than the barrier encountered during the attack of HF on the polarized backbond of a fluorinated surface atom. This higher barrier for H substitution slows down the etching, so that the thermal equilibrium corresponds to the surface passivated chiefly by hydrogen.

Another effect which occurs durnig a wet HF etch is the interaction between the Si–Si backbonds and water: a backbond may be attacked by a H_2O molecule or by a proton and an OH^- radical (OH etching, page 306). In consequence, the kinetics and relative efficiency of oxidation and etching as well as the location of the preferably attacked backbonds (on the flat surface, at defects, at step edges) depend on the relative concentration of the moieties which take part in the reactions: HF, H^+, OH^-. Thus, the pH of the solution becomes an important factor in the design of the etch. The pH can be varied over a wide range[818] (between 0 and 12) independently on the HF content by adding NH_4OH or HCl to the solution. A common way[818] to maintain the pH constant at approximately 5.0 is to use ammonium fluoride as a **buffering** agent (7:1 NH_4F:HF).

Most of the surface F atoms become replaced by OH groups in the reaction Si–F + H_2O → Si–OH + HF when the surface is **rinsed** in ultrapure water[745] to remove losely bonded reaction products and molecules such as physisorbed HF.[746] But even after the rinse some residual F (up to a few percents of a monolayer[792]) remains at defect sites. These atoms can be detected on the surface by XPS directly after the rinse,[745] or by SIMS at the SiO_2/Si(001) interface after oxidation in a conventional furnace.[819]

What happens if the HF-treated wafer is not rinsed but merely dried in a spin drier? The amount of surface fluorine is then much higher. Several times more fluorine atoms survive now the high-temperature oxidation process and segregate to the oxidizing SiO_2/Si(001) boundary.[819] This segregated fluorine facilitate the stress relaxation during oxidation and improve the dielectric strength of the oxide film (Sect. 6.5.5b). On the other hand, the presence of fluorine in the oxide assists the penetration of boron (Sect. 6.5.5c) and is thus, in general, unwanted in ultrathin gate oxides.

Boiling in water seems to be more efficient in removing F than a room-temperature rinse. But the OH groups saturating the dangling bonds of many of the Si atoms on the surface treated with water may be troublesome if the substrate is prepared for epitaxy:[254] the oxygen from OH attacks Si during the contact with water and also later, during heating (Sect. 5.5.3). The resulting oxide islands disrupt the epitaxial growth.

5.3.2. *Chlorine, bromine and iodine*

Chlorine (Cl, [Ne]$3s^2 3p^5$) is present as silicon tetrachloride ($SiCl_4$), dichlorosilane (SiH_2Cl_2), or trichlorosilane (SiH_1Cl_3) in gas mixtures used in chemical vapor deposition (CVD) of Si, SiGe, Si_3N_4, and SiO_2 (see Ref. 803 and 820 for references). Added to the oxidizing ambient (for example, as HCl), Cl binds and removes impurities as well as excess Si, helping to keep the mobilie oxide charge and the emission of Si interstitials at acceptable levels (Sect. 6.3.2). Cl etching of the surface is an important factor in CVD,[821,802] and occurs, for instance, in such bidirectional deposition/etching reactions as $SiCl_4(gas) + 2H_2(gas) \rightleftharpoons Si(solid) + 4HCl(gas)$, or $SiCl_4(gas) + Si(solid) \rightleftharpoons 2SiCl_2(gas)$. *Ab initio* theory can address the atomic and electronic structure of adsorbed Cl[822,823] and elucidate the details of surface reactions, validating models for practical simulations.[803]

Atomic Cl etches Si(001) even below room temperature,[796] building $SiCl_4$ molecules which desorb around −120°C. When the temperature of the substrate exposed to atomic Cl is between 330 and 580°C, the dominant product changes to $SiCl_2$. Applied as Cl_2, chlorine etches mildly. Cl_2 must dissociate on the surface before the etching begins. Cl atoms are trapped by the dimers (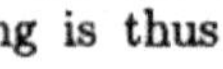) and the path to volatile $SiCl_x$ is severed.

Cl_2 etching is thus minimal at room temperature. But when the temperature is high enough for the adsorbed Cl to break dimers: 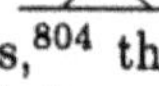 → 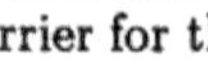, it becomes efficient. According to *ab initio* estimates,[804] the barrier for this bond breaking is close to 2 eV; the configuration  is metastable, with the energy higher than that of the monochloride (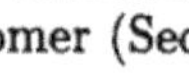) by ~1.3 eV. The Si partner, , diffuses away as a mobile monomer (Section 4.2.1a), while the dichloride () desorbs later as $SiCl_2$. Recent STM measurements[824] confirmed the isomerization mechanism.

This process, conceptually analogous to oxygen etching with silicon ejection (Section 5.4.3a), generates dimer vacancies (, where  represents the top view of the ideally dimerized surface). The unreacted material is collected in "epitaxial" Si ad-dimer strings (the thick lines in , Sect. 4.2.2b) on top of the terraces.[797] The vacancies are arranged in clusters expanding as dimer-wide trenches with S_A-type edges () and with ocassional branches across dimer rows[608] (). The temperature at which the S_A-edged trenches form (580°C) is close to 600°C, that is, to the temperature at which *random* dimer vacancies tend to make divacancy trenches with S_B-type edges (, Section 3.4.1). Therefore, this preference for S_A-type etch pits indicates that etching occurs mostly at steps. This is consistent with the enhancement of the Cl_2 etch rate observed on defected surfaces.[796] The shape of the trenches is thus likely to reflect the relative etch probabilities of S_B and S_A step edges.

An interesting consequence of the isomerization reaction path[824] is that steady-state etching can become self-limited at high chlorine fluxes, because high Cl coverages hinder the diffusion of Si monomers, severing the reaction path in this way. The same effect contributes to slow etching rates by Cl_2 at low temperatures. And how does the formation of missing dimers affect the interaction of the surface with chlorine? It seems that some Cl atoms can be bonded in adsorption sites associated with surface vacancies[825,826] and give rise to an additional low-temperature $SiCl_2$ desorption peak. When the experiment is done on a surface originally saturated with chlorine (at 100 K), this desorption peak occurs at ~280°C (the main desorption peak is around 700°C). Some influence of the defect structure and, in general, of the surface damage on the desorption process can be observed as slight changes in the shape and position of the desorption peak, which depend on the surface preparation.[825,826]

Bromine (Br, $[Ar]4s^2 4p^5$) interacts with silicon less strongly than than fluorine and chlorine. This is exemplified by the desorption products from Si(001) surfaces exposed to bromine atoms.[791] When the substrate temperature is as low as 100 K (-173°C) at the moment of adsorption, heating to 200 K (-73°C) releases Br_2 molecules and B atoms from the surface. The main product of the etching reaction is $SiBr_2$, which desorbs around 250°C. Desorption at ~500 °C may involve $SiBr_4$. Bromine atoms, $SiBr_2$, and $SiBr_4$ evaporate from the surface around these characteristic temperatures also after room-temperature exposure to bromine.[791]

Generally, bromine etches silicon in a similar way as does chlorine. Its pecularity is the 3×1 structure with missing rows described above (page 278). It can be created[785,824] when the sample is quenched to room temperature after covering the surface at 630°C with ~1 ML of bromine from Br_2 with sufficiently high fluence. Too low flux of bromine results in a different steady-state structure, which is just the clean 2×1 with many etch pits and random bromine atoms decorating some of the dimers.

STM images of Si(001) exposed to small amounts of Br_2 at room temperature show other exciting features. Besides occasional dimer vacancies created by etching, single dimers saturated by bromine () are seen. Occupied states of these monobromide sites are imaged similarly to dimer vacancies, but they are not so "deep" (that is, they appear brighter than dimer vacancies, but darker than clean dimers). The most interesting observation is that also single atoms appear to be missing.[795] These "atomic vacancies" are probably indeed vacant Si sites, the result of the familiar process:  →  → , whereby a mobile Si monomer () is ejected. The dibromide (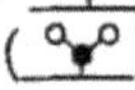) is more stable, but after a while it is gone as well, so that the site changes into a dimer vacancy.

Iodine (I, [Kr]$5s^2 5p^5$), the heaviest of the stable halogens,[g] is also the least reactive. The major desorption products[791] are I_2 and I, detected at the desorption peak temperatures around -20°C, 250°C, and 450°C. The apparent etching product is SiI, which desorbs at approximately -70°C and 150°C. As usual, the exact desorption temperatures depend on the adsorption conditions (temperature, coverage), cf. Fig. 5.4 on page 295.

Though I_2 does etch Si(001), ejecting Si monomers just like the other halogens do, its action is slower than that of the lighter halogens. The exposure soon produces a steady state with the regrown dimer chains and the substrate covered by 2×1 monoiodide (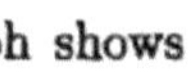). An STM image taken after room-temperature exposure to I_2 reveals the same 2×1-I structure.[601] A careful examination of this micrograph shows that the order is actually $p(2\times2)$: 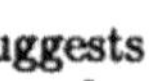, where the circles mark brighter spots in the image of occupied states. Since the coverage is close to full monolayer, this 2×2 arrangement suggests that the iodine atoms in the monoiodide are "canted" (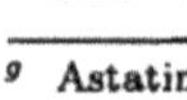 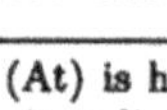, side view), evidently due to the same steric repulsion which causes canting of dihydride on Si(001) 1×1-H (Section 5.2). It is

[g] Astatine (At) is heavier than iodine but, as its name indicates (from Greek *astatos*, unstable), it is radioactive. Its longest-lived isotope has a half-life time of 8.3 hours.

difficult to say to what extent is this canting accompanied by buckling of the underlying dimers ().

Iodine coverages of around 0.5 ML induce a $c(4\times2)$ structure assigned to monoiodide dimers () alternating with clean dimers (), as in:  where the iodine atoms are indicated by circles. Coverages markedly below 0.5 ML contain randomly distributed, isolated monoiodide dimers: 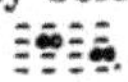. As opposed to dimers saturated by other halogens, monoiodide dimers are brighter than clean dimers. The contrast depends on the bias voltage.[601] The STM tip can move both iodine atoms from one dimer to another, but no single-atom events have been observed.[601]

Besides the saturated monoiodide dimers (the so-called *trans* configuration of iodine atoms), STM images reveal occasional occurences of the *cis* configuration: 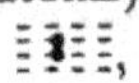, in which two half-iodized dimers () occupy neighboring sites in the same row. The *cis* structure becomes more abundant as the coverage decreases, in analogy to the theoretical predictions for the *cis* hydrogen (page 272). The assignment of this structure to *cis* iodine[601] is, however, tentative. One would expect that the unsaturated dangling bonds of dimers appear as bright spots in STM images; yet they are invisible. In addition, *cis* hydrogen was computed to be less favorable energetically than *trans* hydrogen.[547]

We speculate[301] that the *cis* iodine is an etch feature and represents a diiodide pair: 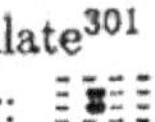 (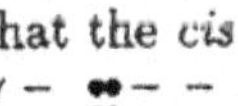 in magnification). The pair is possibly the product of two isomerization reactions of the type known for bromine: 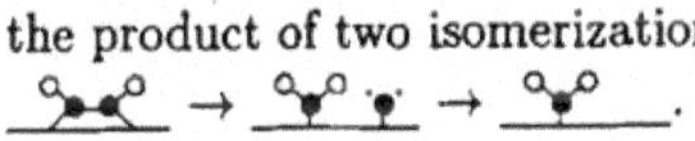. As a matter of fact, similar *cis* objects can be found in STM images of surfaces exposed to bromine.[795] Our interpretation is that this reaction occurs preferentially at neighboring monohalide dimers because the emission of a Si monomer produced in the course of isomerization leaves two subsurface Si atoms with a broken bond each (cf. Fig. 3.13b, page 164). These bonds can be reformed as at the bottom of a single dimer vacancy (Fig. 3.13a). When the reaction takes place at neighboring dimers, the bonds can reform as at the bottom of a double dimer vacancy (Fig. 3.13c). In our model, the correlation between isomerization events accompanied by monomer emission is therefore related to the tendency of dimer vacancies to combine in pairs (Sect. 3.3.1a). The gradual reduction of the abundance of *cis* structures with the increasing surface concentration of iodine can be understood as the effect fully analogous to the suppression of chlorine etching at high Cl coverages[824] (page 283).

Ref year	Subject relevant to O/Si(001)
827 1985	Vibrational frequences, core level shifts; EELS, XPS
694 1985	Work function, E_F, photoionization threshold (various orientations)
828 1987	Adsorption and desorption kinetics, sticking coefficient, many references
265 1988	**Review**: O adsorption on Si(001), atomistic aspects
829 1988	Oxidation states on Si(001), O_2 at 750°C; SCLS
252 1988	Buildup of hydrocarbon contamination in air on surface oxide; XPS
830 1990	Oxygen adsorption sites; *ab initio* theory
716 1990	Oxygen adsorption sites on Si(111) 7×7, including steps; STM
831 1991	**Fundamentals**: O adsorption on Si(001), atomistic aspects
832 1992	Initial stages of adsorption; STM
833 1992	Oxygen adsorption sites; *ab initio* simulation of STM
530 1992	Silicon ejection, oxidation of defects and steps; STM
834 1992	SiO desorption, temperature, coverage dependence; TPD, isotopes
835 1993	**Review**: Point defects and charge traps in SiO_2 and SiO_2/Si
836 1993	**Review**: Molecular and atomic O on Si(001) and Si(111)
837 1994	Oxide nucleation and surface roughening; STM
838 1994	Ordered structures during thermal oxide formation; STM
839 1994	Phase transition theory of submonlayer oxidation; real-time XPS
708 1994	**Book chapter**: Kinetics, structures, phase diagrams
840 1995	Active and passive oxidation, surface roughness; STM
841 1995	Ordering of thermally grown oxide films; X-ray diffraction
842 1996	Importance of initial surface quality; TEM, STM, AFM
843 1996	Oxygen adsorption sites; *ab initio* theory
337 1996	Oxidation states of Si(113), O_2 at 750°C; SCLS

Table 5.3. O/Si(001): selection of references

As expected, I_2 etching performed at the substrate temperature high enough to desorb SiI_2 efficiently and allow for an efficient equilibration of the etch pits, leads to the $2\times n$ dimer vacancy lines (Section 3.4.1) running across substrate dimer rows. Such a quasi-ordered structure has been indeed observed[601] after I_2 etching at 630°C.

5.4. Oxygen: Isolation and Protection

Below we describe atomic-scale details of the initial stage of interaction between oxygen atoms and silicon surfaces: adsorption sites, the reaction products, and the formation of the first oxide layer. The focus is on the Si(001) surface. A selection of references is given in Table 5.3.

The practical quality, or "integrity", of the gate oxide (Section 1.2.2) is sensitive to the initial state of the surface. For example, consider how the roughness of the interface between silicon and a thermal oxide film[h] depends on the film thickness t_{ox} and on the method chosen to clean the surface.[844] Oxides grown at 800°C on commercial wafers cleaned by a standard RCA etch (Section 6.3.1) have the maximum interface roughness for film thicknesses $t_{ox} \simeq 40$ Å, or 4 nm. This is close to t_{ox} which is typical for modern gate oxides. But when the surface is cleaned by heating in ultrahigh vacuum (UHV), 40 Å thick oxides have a roughness of only 1 Å rms.[844] The difference is apparent. The initial surfaces differ dramatically as well. RCA cleaning leaves the surface covered by a few atomic layers of a "native" oxide film, while the UHV cleaning produces an atomically clean and atomically flat surface. The difference in price is obvious as well: RCA is simple and cheap, while UHV processing is complicated and expensive...

Investigations of early oxidation stages of silicon surfaces elucidate the mechanisms by which oxygen combines with crystalline silicon. Besides being an excellent scientific exercise, such studies contribute to our knowledge on the formation of SiO_2 films and SiO_2/Si(001) interfaces. This knowledge may add a new dimension to the control of the interface quality.

Oxygen is usually adsorbed from O_2 gas, whereby the reaction proceeds through molecular precursors. Atomic oxygen can be adsorbed from N_2O atmosphere (Sect 5.6.2). Water and steam are important oxidants (Sect. 6.3.2). Accidental exposure to oxidizing ambients may cause problems. Trace amounts of oxygen etch and roughen unprotected surfaces (Sect. 5.4.4). Surface contamination with oxygen deteriorates silicon homoepitaxy, by MBE[530] as well as by CVD.[772] A practical property of Ti silicidation (Section 1.2.2) is that Ti reduces the Si surface, making tiny oxidized areas not troublesome during the process of contact formation.

A simple general principle accounts for the site-selectivity in the initial stage of silicon oxidation: sites with the highest DOS at the Fermi energy are the most reactive.[530] This is because electrons at the Fermi level are energetically most distant to the lowest empty states of oxygen, and the energy gain from electron transfer between silicon and oxygen is then maximized. For Si(001) this means that C-defect sites (page 165) are the most active. For the Si(111) 7×7 surface (page 69) this means that corner-adatom sites at the faulted-half of the unit cell are oxidized first.[i]

Oxygen adsorption strongly changes the Fermi energy E_F and the

[h] Thermal oxidation is the standard method used to grow MOS gate oxides.

[i] For example, see references in Ref. 530 and 845.

photoionization threshold. On clean surfaces, the distance between the valence band maximum and the Fermi energy varies strongly with the surface orientation (in other words, the surface-induced band bending varies). After oxygen exposure, E_F becomes pinned at a nearly constant position ~0.1 eV above the midgap, probably by defect-related surface states.[694]

5.4.1. *Oxygen and its bonds with silicon*

Oxygen is one of the most electronegative elements in the periodic table, preceded only by fluorine. It belongs to the group VIa and its electron configuration is $1s^2 2s^2 2p^4$. Oxygen readily makes single and double bonds with other *sp* atoms; in spite of the high electronegativity of O, these bonds are only partially ionic, with a significant covalent component.

As an example, consider a Si=O bond, such as in the so-called "silanone complex". This complex is a Si surface atom with two backbonds and an oxygen atom on top: 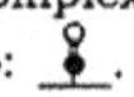. The O atom is here the open circle, and the double bond is indicated by a double line. If viewed from the direction of the arrow: → 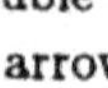, the silanone complex would look like: 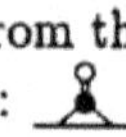.

The double Si=O bond is formed by four orbitals. The Si atom contributes its "dangling bond" sp_z hybride, the "bridge bond" p_x orbital (Fig. 3.2, page 132), and two electrons. The oxygen atom contributes p_x and p_z orbitals and two electrons. Due to the high electronegativity of oxygen, the resulting σ and π bonds have more weight on O than on Si.

A single Si–O σ bond is formed, for example, when an oxygen atom attaches itself to a Si dimer: 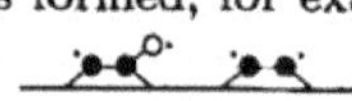. This leaves an unsaturated electron on the oxygen. Since the other atom of the dimer has an unused electron in its dangling bond, and since oxygen is more electronegative than Si, the oxygen will accept this electron: 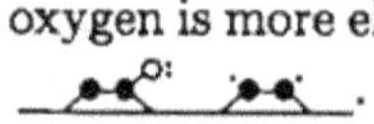. Naturally, in some geometries with single Si–O bonds no electron transfer is needed. This happens when the oxygen atom intervenes in a Si–Si bond. On Si(001), the object of such an attack can be the dimer bond (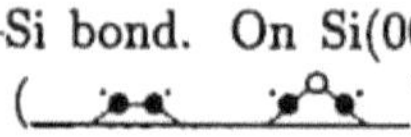) or a backbond (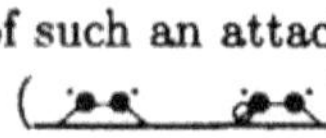). On the Si(113) surface, the dangling bonds of adatoms on the clean Si(113) are occupied with two electrons (that is, the adatoms are negatively charged; Section 1.5.4), so the extra electron is available on the spot when an oxygen atom attaches itself to an adatom.

Configurations with no additional charge transfer, as in the Si(113) example, are expected to have a lower energy than configurations which invoke an additional local charging of silicon by transfer of electrons from other regions of the surface (as in:). Two single σ bonds should be more favorable than a double bond with σ and π components.

Substances based on single O–O bonds are known as peroxides. An example is hydrogen peroxide, H_2O_2: H–O–O–H. Oxygen molecules chemisorbed on Si(001) are thus termed peroxide bridge () and peroxy radical (). The latter has an chemically active oxygen orbital. We indicated this orbital by a dot. In reality, it is to be expected that this orbital has accepted an electron: 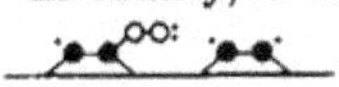

5.4.2. *Adsorption of O_2 on Si(001)*

When O_2 molecules are brought in contact with the clean Si(001) surface, they are initially physisorbed, that is, the molecular bonds remain unchanged. This physisorbed state is observable only below the liquid nitrogen temperature.[846] If the temperature remains below 50 K, the initial sticking probalility (initial sticking coefficient, S_0), is close to 1. At higher temperatures, all molecules either overcome the chemisorption barrier and transform into a chemisorbed precursor molecular state, or desorb. The height of the involved barriers is probably around 0.1 eV.

Since O_2 can desorb, the sticking coefficient S_0 should drop as the temperature T increases. Indeed, S_0 exceeds 0.015 at kriogenic temperatures but at room temperature it is 0.01–0.005 (depending on the oxidation procedure). The oxidation kinetics suggest that this drop is rather due to decreased probability of physisorption than to desorption of physisorbed molecules.[828] Around 330°C, S_0 begins to increase with temperature.[828,836]

Chemisorbed O_2 molecules appear to STM as protrusions located between dimer rows. Thus, there are two most natural candidates for their atomic configurations: peroxy radical (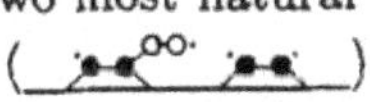) and peroxide bridge (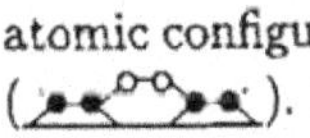). In the above sketches, oxygen is indicated by empty circles.

Oxygen adsorbed in a molecular form is not stable. It dissociates, with activation energy of about 1.3 eV/atom (single O–O bond energy[225] is 1.44 eV) and with attempt frequency around 10^{16}/s. This means that the room temperature lifetime of a chemisorbed O_2 molecule is roughly a day. From results of *ab initio* calculations[830,843] one can estimate that during the whole chemisorption process as much as 6 eV/molecule may be released. This is more than is needed to break a bulk Si-Si bond (2.2 eV).

This dissociation energy is released in at least two steps. A probable scenario is as follows. First, an electron from the surface reaches the highly electronegative O_2 and is trapped in the antibonding π^* molecular orbital.[830] The double O=O bond of the physisorbed O_2 changes in this way into a weaker, single O–O bond. Such a peroxide bond is also characteristic of a chemisorbed molecule; as a matter of fact, it cannot be excluded that the molecule chemisorbs before it captures the negative charge. In the second step, the single O–O bond is broken by thermal vibrations (it is weaker than a Si–O or a Si–Si bond) and the molecule dissociates.

5.4.3. *Adsorption sites of atomic oxygen on Si(001)*

The stable adsorption site of an isolated oxygen atom on the clean Si(001) 2×1 surface is the dimer backbond site: . Oxygen atoms in dimer bonds () have somewhat higher energy. An isolated chemisorbed O atom can be excited into the atomic configuration of a silanone complex (). Such complexes are mobile and contribute to the process of phase separation between oxidized areas of the surface and areas free of oxide. A silanone can desorb as an SiO molecule (Section 5.4.4), which leads to etching if the temperature is too high and the oxidant pressure is too low.

a. *Silanone complexes and silicon ejection*

Microscopic details of O_2 decomposition on Si(001) are unclear. A peroxide bridge O_2 may move to an on-dimer position, open the underlying Si dimer, dissociate, and produce a pair of surface silanone complexes (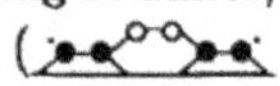 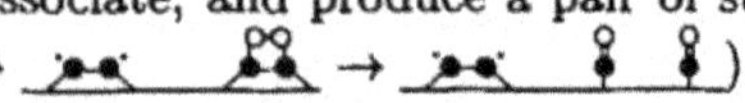). We now consider consequences of this reaction. A path leading *directly* to stable adsorption sites is then discussed.

Each **silanone complex** () resembles a Si monomer (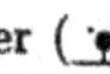). Two backbonds connect the Si atom to the substrate. This consumes two of the four electrons of the atom. The remaining electrons are located in a low-energy state (pages 209 and 288). The silanone complex is a natural candidate for the precursor of volatile SiO. Desorption of a SiO molecule requires[828,831,834] only 3.5±0.3 eV, less than the ~4.6 eV needed to break two Si–Si bonds. This is not astonishing because a SiO molecule is quite stable with its triple bond (:Si:::O:) resembling that of carbon monoxide, CO. But the attempt frequency for SiO desorption[831] is high, 10^{19} s^{-1}.

Given the similarity between a monomer and a silanone, one would expect silanone complexes to migrate easily on the surface, just as Si monomers do. Surprisingly then, the activation energy of oxygen surface diffusion on Si(001) is[661] 2.4 eV, nearly four times the activation energy of Si monomer diffusion, and by 0.6 eV higher than the energy barrier for migration of oxygen in bulk silicon! The corresponding attempt frequency is[661] around $10^{16}s^{-1}$, three orders of magnitude higher than the value usually accepted for diffusion of Si monomers.

Before we address this inconsistency, we explore the fact that the reaction 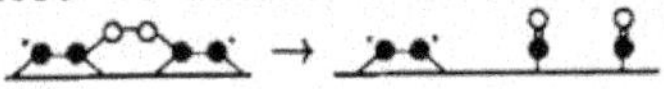 produces two mobile silanone complexes. They are likely to diffuse away, leaving behind them a dimer vacancy (, where  symbolizes the top view of a perfectly dimerized surface and each dash indicates a Si dimer). The ejected material may diffuse to steps. Alternatively, the mobile elements may combine into ad-dimers or two-dimensional islands, similar to those produced during Si homoepitaxy. Such vacancies, islands and ad-dimer-like objects generated during chemisorption of O_2 on Si(001) have been observed[832,845] by STM.

There are other possible sources of the ejected material and other possible reaction products. Si *monomers* () may be released when a single Si atom is surrounded by three O atoms.[845] Indeed, a lowered formation energy of Si interstitial was computed when the initial atomic configuration had Si atoms with three Si–O bonds and one strained Si–Si bond.[847] Similar *ab initio* calculations have shown that Si is ejected when several subsequent surface atoms in one $[1\bar{1}0]$ row are oxidized.[848] One can imagine that O_2 chemisorption energy is used to eject Si atoms. Finally, Si monomers can be generated together with silanone complexes in a number of reactions involving in-bond oxygen, such as the reactions described below. Examples are: dissociation of a peroxide bridge or decay of a migrating silanone.

b. *Oxygen atoms in Si–Si bonds*

Regardless of the detailed mechanism of Si ejection, it is clear that the chemisorption does not end at a silanone pair (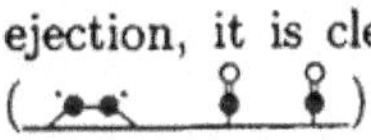). The atoms of a broken O_2 molecule go into silicon **dimer backbonds** (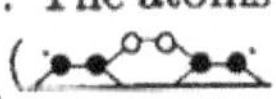 → 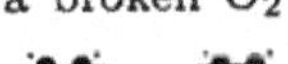). But a reaction like 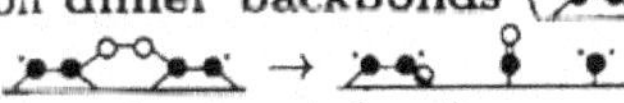 is also imaginable. Since both the silicon monomer () and the silanone complex () are highly mobile, a process of the latter type would generate dimer vacancies and, possibly, ad-dimers and two-dimensional islands.

The in-backbond structure, indicated by numerous experiments, is consistent with *ab initio* results.[843,848] Experimental STM images agree with the images simulated *ab initio* for a dimer with an O atom in a backbond. After dissociative chemisorption of O_2, some dimers look as if strongly buckled. This is exactly what the calculations predict.[843]

Another plausible adsorption site is the **dimer bridge**, with the oxygen inserted into a dimer bond[849,832] (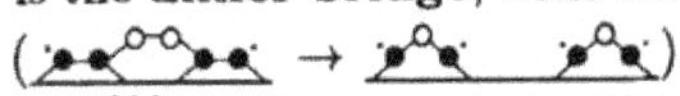). *Ab initio* theory initially supported this model[830] as the most favorable energetically, but the backbond site was not considered in these calculations. STM images simulated later for the dimer bridge structure[833] failed to reproduce the experimental data. Oxygen atoms in the dimer-bond position should be invisible in STM, and the Si atoms of the dimer should be somewhat brighter. The on-dimer protrusions observed by STM are usually much brighter than the surrounding substrate dimers and are better interpreted as ad-dimers produced in a silicon ejection process (page 291).

The oxygen atoms in Si–Si bonds are not visible in STM images because Si–O bonds are strong. The bonding orbitals are hidden deep in the valence band and the antibonding orbitals lie high in the conduction band. Moreover, the orbitals of oxygen atoms, being strongly localized, do not extend in space far enough to be detected by the STM tip. In further stages of oxidation, oxidized areas appear in the images as dark patches. This reflects the wide energy gap (~9 eV) of silicon dioxide.

Since Si–O bonds are much stronger than Si–Si and O-O bonds,[850] the dissociative chemisorption releases a lot of energy. As a consequence, O atoms may be transported to various locations. Indeed, even when the oxygen coverage is very small (less than 1% of a monolayer), core level peaks attributable to all four oxidation states of silicon (page 399) can be measured,[851] including fourfold-oxidized silicon atoms, as in SiO_2.

An interesting observation is that intensities of the core level peaks from the single-, double-, and triple-oxidation states are comparable. The presence of many triply-oxidized atoms already at low coverages shows that O atoms migrate and that they tend to come together.[852]

We can now return to the problem of the high activation energy and the attempt frequency of oxygen surface diffusion (page 290). An oxygen atom can cover large distances travelling in a silanone complex. If the atom was trapped in a Si–Si bond, the activation energy for diffusion, E, is dominated by the barrier for excitation (such as  → 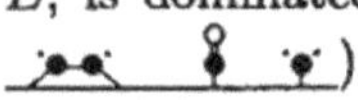). The length of a single "hop" is determined largely by the probability of reverting into an in-bond state. In a kinetic model the diffusion rate is $f \exp(-E/kT)$, so this high mobility translates into a high attempt frequency f (page 208). Since the trapping process must be activated, the covered distance should increase with decreasing temperature. But we are not aware of any experimental verification of this prediction.

For completeness we note that when the oxygen migrating in a silanone complex () falls into a Si–Si bond, a free Si monomer appears:

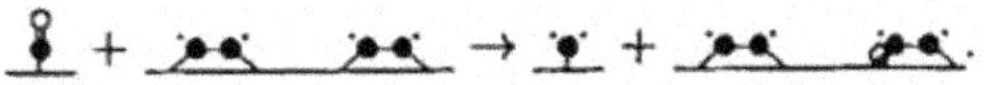. A monomer is also generated when an in-bond O atom is excited (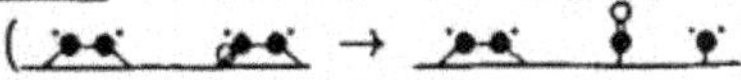). Therefore, it is possible that the ad-dimers and two-dimensional islands generated during oxidation contain only Si atoms which arrive as Si monomers, no matter what the mechanism of the Si ejection is like.

c. *Surface defects and initial stages of oxidation*

The most reactive sites are those with the highest density of states at the Fermi energy E_F (Sect. 1.4.2b). A perfectly dimerized Si(001) is semiconducting, so few perfect dimers are affected when the first O_2 molecules arrive. The oxidation starts at C-type defects (page 165), the structural flaws of clean Si(001) which have states at E_F. Dimer vacancies do not introduce such states and, accordingly, they do not oxidize so fast. But S_B-r step edges (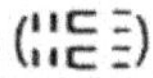) readily accept oxygen.[530] This is consistent with the observation of local electronic states in the forbidden gap at S_B-r (page 187).

This site selectivity for initial oxidation is clearly visible in STM images.[530] Also optical measurements[853] indicate that steps are particularly affected when oxidation begins. However, enhanced relative reactivity of steps appears to be a function of preparation conditions. For example, oxidation conducted at 600°C and 1.33×10^{-6} mbar (entirely in the passive oxidation regime, Section 5.4.4 and Fig. 5.3) starts from both step edges and terraces.[838] A possible explanation is that the site selectivity of the sticking coefficient falls off with increasing temperature because the site-dependent kinetic factors (energy barriers) controlling the adsorption process become less important at higher temperatures.

Intuitively, one expects that the degree of order at the SiO_2/Si interface is affected by the degree of order in the first oxidized layer. Though even a perfect surface oxide layer cannot guarantee an ordered interface, an atomically disordered first layer can hardly lead to the formation of atomically ordered interface during later stages of oxidation. This is corroborated, albeit indirectly, by the observation[854] that SiO_2 microcrystalites are formed at steps of vicinal Si(001). The first oxide layer relaxes most easily in the vicinity of surface steps, and it is there where one would expect that the most ordered structures nucleate. Again, this may depend on the growth conditions; the STM study cited above[838] found that at 600°C and 1.33×10^{-6} mbar of O_2 there is a substantial degree of order along the steps as well as on terraces. No significant roughening due to silicon ejection was observed in this case, which indicates that the production of the mobile

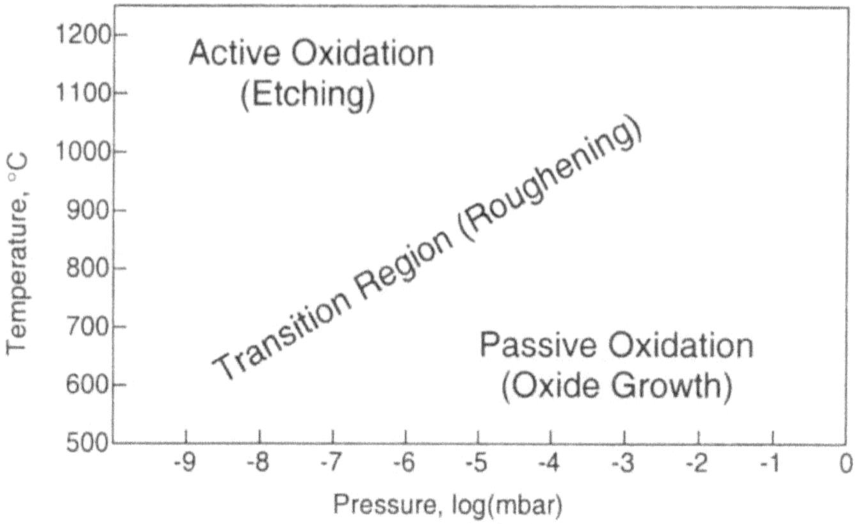

Fig. 5.3. Silicon surface oxidation regimes. For more data and discussion, see Refs. 388, 852, 839, and 840.

silanone complexes is suppressed under these conditions.

5.4.4. *Active etching and passive oxidation*

Depending on the temperature and oxygen pressure, there are two distinct oxidation regimes, "active" and "passive". [836]The **active** regime corresponds to high temperatures (above ~600°C) and low oxygen flux. In this case, no stable oxide is formed. Si and O atoms form volatile silicon monoxide (SiO), which escapes to the ambient (Sect. 5.4.3a). The result is that oxygen etches the substrate. An oxide layer can be formed only when the temperature is sufficiently low and the flux is sufficiently high. These conditions allow nucleation and growth of stable oxide structures. This is the **passive** oxidation regime. For low O coverages, oxidation may be described[839] by a phenomenological first-order phase transition theory. As noted in Ref. 839, the features typical for a system which undergoes such a transition have been established for the initial oxidation stage:

- There is a critical oxygen exposure, coverage, and pressure;
- Oxide islands distinct from an adsorbed phase are formed;
- The adsorbed phase and the oxide phase coexist during oxidation.

Oxidation in the transition region (Fig 5.3) roughens the surface.[837]

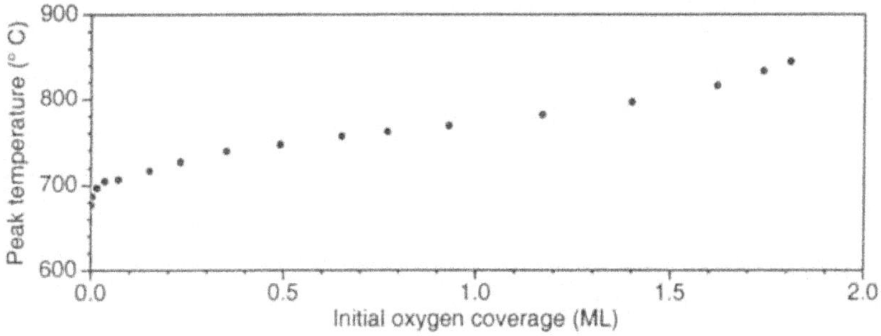

Fig. 5.4. Dependence of the peak temperature of SiO desorption from Si(001) on the initial oxygen coverage. Data from Ref. 834. The change of the peak position is due to the change in the attempt frequency; the activation energy remains constant at 3.5±0.3 eV. Note that this is not necessarily a "general curve": the local structure of the surface oxide and, consequently, the local desorption temperature, may depend on the microscopic quality of the surface before oxidation.[855]

The areas where stable oxides have nucleated are protected from etching, but the rest of the surface is not. So they evolve into hillocks surrounded by valleys several monolayers deep, from which the material has been etched away.[856] The roughening in the transition region is thus related to surface diffusion and can be suppressed when the diffusion is suppressed. This seems to happen during oxidation of the Si(113) surface.[332]

Many technological processes include transient, low-temperature exposure to background levels of O_2 and H_2O, which may lead to roughening of the surface via active oxidation. A thin oxide layer may be used to protect the surface from the detrimental influence of trace (such as 100 parts per billion) oxygen concentrations.[857]

Since the volatile species is SiO and the desorption may have to be initiated by the excitation of an in-bond oxygen (such as  ) to a silanone (), the active oxidation dominates (1) when the temperature is high enough for silanone to detach as SiO, and (2) when the oxygen supply is low. Configurations with only one bond of a Si dimer atom attacked by oxygen (Si–**Si**–O–Si), transform into silanones more easily than configurations of Si–O–**Si**–O–Si type, with both bonds of the same Si atom containing oxygen. The oxide film can nucleate and grow only if there is enough oxygen to assure that Si–O–**Si**–O–Si and higher oxidation states are frequently formed, or if the temperature is low enough to forbid desorption of SiO altogether. As expected for such a mechanism, the peak temperature of SiO desorption depends strongly on O coverage[834] (Fig. 5.4). The activation energy barely changes with the coverage; the changes in the desorption rate are dominated by the effective attempt frequency.

Desorption of SiO produces[832,856] dimer vacancies (Sect. 5.4.3a). The vacancies diffuse already above 200° C, so they can migrate to surface steps (page 235), which causes etching by step flow. Dimer vacancies react easier with S_B-type steps (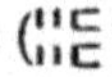 or 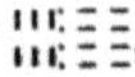, see Sect. 3.3.2 for notation) than with S_A steps (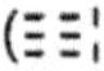), thus S_B steps retreat faster than S_A steps, and the surface becomes predominantly single-domain.[840] Pinning of step flow at oxidized areas prevents formation of a purely A-type surface. If an exclusively A-domain surface is desired, it can be made by Xe sputtering (page 189).

High concentrations of dimer vacancies (such as 10% of dimer sites) hinders their diffusion. They attract each other and arrange in lines. In this way, active oxidation can produce metastable $2\times n$ structures (page 194).

5.5. Water: UHV Pest and a Renowned Oxidant

Water[858] is able to oxidize silicon, acting both as a reactant and as a catalyst. This makes H_2O a key component of several oxidation schemes (Chapter 6). Since surface oxides affect chemical vapor deposition (CVD), water in the CVD gas mixture or on the surface may influence the CVD process.[802] Moreover, water is a major component of the residual gas in a typical Ultra High Vacuum (UHV) setup.

This Section deals mainly with $H_2O/Si(001)$. Table 5.4 lists selected references, including some work relevant to other orientations of Si. Growth of $SiO_2/Si(001)$ in water-containing ambients is considered in Chapter 6.

Water attaches itself to Si(001) very efficiently. Its sticking coefficient of 1 prevails virtually up to the saturation coverage,[859,338] which is 0.5 monolayer.[j] This means that at room temperature, one hour exposure to UHV with 10^{-11} mbar of residual H_2O pressure covers 2% of the surface with water. Since a typical UHV equipment provides 10^{-10}–10^{-11} mbar, water contamination of Si(001) surfaces may become a problem.

At low coverages, some water molecules are possibly attached to Si(001) as tiny needles interconnected by the same kind of bonds as in ordinary ice. These molecules appear as depressions in STM images (page 303).

[j] Naturally, a multilayer film of ice can be grown at low temperatures.

Ref year	Subject relevant to the H_2O/Si system
860 1971	Dissociation of H_2O into H and OH moieties proposed
861 1975	Molecular H_2O, room temperature; EELS
862 1982	Only Si-H and Si-OH vibrations found; EELS
863 1983	Molecular H_2O is unstable, electronic structure; semiempirical
864 1983	Molecular H_2O, room temperature; arguments from UPS, PDS
865 1984	H_2O dissociates, 80–500 K; strong arguments from IRS
866 1985	$[1\bar{1}0]$ azimuth, sticking, role of defects, 350-870 K; AES, SCLS, XPS
858 1987	**Review**: Adsorption of water, fundamental aspects
867 1989	Moisture in RT air needed to oxidize Si(001); XPS
868 1991	H_2O/Si(111) 7×7, local chemical activity; STM
537 1993	Low coverage H_2O, molecules seen between dimer rows; STM
538 1993	High coverage, islands of dissociated H_2O; STM during exposure
749 1994	Oxidation of H/Si(001) and H/Si(111) from air and water; IRS
869 1995	Si(001) and Si(111), 90 K and RT, Si–O–Si found; SCLS, XPS
712 1995	Binding and diffusion of H and OH on dimers; *ab initio* DFT slab
728 1996	H and OH on dimers, SCLS and geometry; *ab initio* DFT slab
144 1996	**Review**; STM
870 1997	Molecular and dissociative adsorption; *ab initio* DFT cluster
338 1997	Adsorption kinetics on Si(001), (115), (5 5 12), (112); UPS, LEED
871 1998	Dissociative adsorption, bond lenghs/angles; photoelectron diffraction
872 1998	Dissociation, long exposure, mechanism of wet oxidation; IRAS

Table 5.4. H_2O on Si; selected references. The surface is Si(001), unless stated otherwise.

Many of "missing dimers" (Sect. 3.3.1a) or "C-type" defects (Sect. 3.3.1b) observed on "clean" Si(001) may be such ice pieces.

If chemisorbed H_2O molecules exist, they are nevertheless unstable. When the coverage increases, they break into O and OH moieties which then saturate dimer dangling bonds (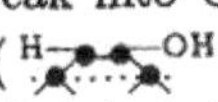). Some of the O atoms may become liberated and attack Si–Si bonds, initiating oxidation. Heating removes H altogether, leaving the surface partially oxidized.

The above scenario appears to be a widely accepted one, yet it is not obvious that it is entirely correct. In particular, it is not clear whether molecularly adsorbed water really exists. It took, however, many years of difficult research and controversies (Table 5.4) to establish this uncertain consensus. The role of STM studies was instrumental in this process, though STM is not a device suited for an efficient chemical analysis.

It is well known that even a small amount of water speeds up oxidation. There are indications that very dry air does not oxidize H-passivated Si surfaces at room temperature.[749,867] It is moisture that catalyses the growth of such native oxides. Similarly, "wet" thermal oxidation procedures are much faster than "dry" thermal oxidation schemes (Chapter 6).

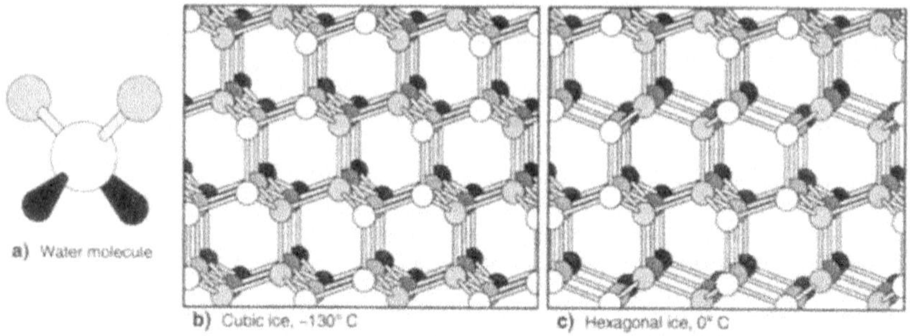

Fig. 5.5. Water molecule, hydrogen bonds, and cubic ice. **(a)** H_2O molecule. Oxygen atoms are big and white, hydrogen atoms are small and grey, *fully occupied* lone pairs of oxygen are indicated by black clubs. The O–H bond length is 0.957 Å, the H–O–H angle is 104.5°. **(b)** Low temperature form of ice, formed by crystallization of water around −130°C: a cubic crystal with diamond-like lattice. Oxygen atoms in more distant planes are darker. H atoms (not shown) take part in the hydrogen bridges (see text) indicated by sticks. **(c)** Natural ice, a hexagonal crystal with wurzite-like lattice.

The detailed mechanism of this catalytic action is not clear. It is conceivable that H or OH from dissociated water molecules assist in the oxidation reactions at the growing SiO_2/Si interface (Sections 5.5.3, 6.4.1 and 6.4.2), for example, by opening Si–Si bonds,[738] temporarily protecting broken bonds from reforming, or assisting strain relief and oxygen diffusion in the growing film by loosening up the SiO_2 network.

5.5.1. *Water, ice, and hydrogen bonds*

Water, the liquid of life, is indeed unique in its properties. For example, the high polarity of H_2O molecules makes water a nearly universal solvent. Interactions between water and solid surfaces has been a subject of intensive research; for an exhaustive review, see Ref. 858.

Each of the two O–H bonds in a water **molecule** can be rationalized as formed between a $2s2p^3$ hybrid of the oxygen and the $1s$ orbital of a hydrogen. This binds four of the eight valence electrons. The remaining four are localized on two sp^3 lone pairs (Fig. 5.5a). The opening angle of the H–O–H scissors is 104.5°, close to the tetrahedral angle of 109.45°.

This **localized orbital** (LO) sp^3 picture is intuitively compelling and helps to visualize chemical bonding to other molecules. However, the eigenstates of the H_2O hamiltonian differ from the LO orbitals. One reason is that the symmetry of a H_2O molecule is not tetrahedral (T_d), but C_{2v}. This symmetry lowering allows for mixing of LO orbitals. The eigenstates of the hamiltonian projected onto the LO subspace ("canonical" **molecular orbitals**, or MO) include a bonding and two weakly bonding orbitals, each of them with contributions from both H atoms. These three orbitals lie in the plane of the molecule and are fully occupied by six electrons. The remaining two electrons are localized on a nonbonding lone pair of O; this orbital has a p character and is perpendicular to the molecule plane.[858] These canonical orbitals are a good approximation of the real wavefunctions. As such, they provide the correct ionization potentials.

The slight deviation of the H–O–H angle from the tetrahedral value means that the localized (sp^3-like) lone pairs of oxygen have more s character than the bonding sp^3-like hybrides. A well-known interpretation of this deviation is that lone pair orbitals are "larger" than bonding orbitals, therefore they mutually repell.[873] In consequence, the angle between them changes from the 109.45° of sp^3 towards the 180° of sp. This corresponds to an increase of the s component in the lone paris and a decrease of the s component in the bonding pairs, that is, to closing of the H–O–H scissors (the angle between p orbitals is 90°).

The quantum-mechanical interpretation is that the H–O–H angle reflects a balance between the O–H bond strength, the electronegativity difference between H and O, and the energy difference between the $2s$ and $2p$ atomic orbitals of oxygen. The sp^3 hybridization of oxygen orbitals requires a promotion of a low-energy O($2s$) electron of oxygen to a high-energy O($2p$) orbital (cf. page 131). A wider angle between the lone pairs reduces the energetic expense due to the promotion, because it corresponds to a hybride with a smaller p content than an sp^3 hybride. At the same time, it also reduces the overlap between the orbitals participating in O–H bonds: the H–O–H angle closes and the bond changes from a strong O($2s2p^3$)–H($1s$) towards a weak O($2p$)–H($1s$). In addition, the hydrogen atoms are positively charged (oxygen is more electronegative than hydrogen), that is, a narrow $H^{\delta+}O^{\delta-}-H^{\delta+}$ angle leads to a strong electrostatic repulsion. As a consequence, the bond angle between sp^2 (120°) and pure p (90°) is chosen.

H_2O molecules in ice and in liquid water are interconnected by the so-called **hydrogen bonds**. In general, bonds of this type are formed between a bonding-antibonding pair on one side and a lone pair on the other side. Consider a regular bond X–H between H and atom X and described by an occupied bonding orbital σ_{XH} and its antibonding, unoccupied orbital σ^*_{XH}. The wave function ψ_H of a hydrogen bond X–H...Y formed between XH and a lone pair σ_Y of an atom Y combines these three orbitals:

$$\psi_H = \alpha\sigma_{XH} + \beta\sigma^*_{XH} + \gamma\sigma_Y. \tag{5.22}$$

Each OH bond and each lone pair in H_2O can thus be viewed as half of a **hydrogen bridge**, in which hydrogen is connected to one O atom with a molecular bond and to the other O atom with a hydrogen bond. As far as the bonding between molecules is concerned, the geometry of H_2O is nearly tetrahedral, with two O–H halves of hydrogen bridges pointing upwards and two lone-pair halves of hydrogen bridges pointing downwards (Fig. 5.5a). As a consequence, H_2O molecules in ice are arranged such that each O atom sits in a nearly tetrahedral configuration, being surrounded by four O atoms and connected with them by four hydrogen bonds. The crystallography of the cubic form of ice is analogous to that of bulk Si (Fig. 5.5b). Natural ice is similar, but the tetrahedrons are ordered as in wurzite, forming a hexagonal structure (Fig. 5.5c).

The length of a hydrogen bridge in ice is approximately 2.76 Å (measured as the distance between O atoms). In liquid water, which consist of long chains of H_2O molecules, this distance is longer, approximately 3 Å; the O–O distance in water dimers,[874] $(H_2O)_2$, is 2.95 Å. Each molecule in liquid water is, on average, connected to only two other molecules.[875] The strength of a hydrogen bond in water and ice is 0.2–0.3 eV.

Water molecules can adhere through hydrogen bonds also to OH moieties adsorbed on a surface. Such **hydrophilic** surfaces are, for example, Si surfaces covered with dissociated water (Sect. 5.5.2) or surfaces of SiO_2 layers produced by an RCA-like etch (Sect. 6.3.1). Short chains of H_2O (a few molecules) attached to adsorbed OH are strong enough to bind two flat Si wafers even if most of the surface is **hydrophobic**,[876] that is, passivated with H (Sections 5.2 and 5.3.1). Hydrogen passivation makes the surface hydrophobic because the charge of hydrogen adsorbed on Si is small[877] and negative[878] (H is more electronegative than Si), so that hydrogen bonds (Sect. 5.5.1) cannot be formed between them and the O atoms in H_2O.

Liquid water[879] contains, in addition to undissociated H_2O molecules, also protons (H^+) and hydroxyl radicals (OH^-). The protons are associated with water molecules in hydronium or hydroxonium ions (H_3O^+). These ions are, in fact, centers of more complex hydrogen-bonded and highly mobile formations.[879] From the mass action law it follows that the product of the concentrations C of H^+ and OH^- is a constant: $C(OH^-)C(H^+) = k =$

const. This product is equal to $k(H_2O) = 10^{-14}$ $g^2/litre^2$. Chemically neutral water has $C(OH^-) = C(H^+)$, alkaline (basic) water has $C(OH^-) > C(H^+)$, and acidic water has $C(H^+) > C(OH^-)$. Since the pH factor is defined as $pH = -\log C(H^+)$, the pH is 7 in neutral water.

5.5.2. *Adsorption and dissociation of water*

An analysis of adsorption and desorption of water will help us to view the origin of certain surface defects from a different perspective. Some water-related features are easily mistaken for intrinsic defects of a clean Si(001) surface. Light will be also shed on the atomistic details of interactions between molecules and surfaces. Some of this knowledge will help us to understand the process of wet oxidation.

Water adsorbs on Si(001) through a highly mobile precursor[538,880] state. This precursor promptly breaks up into H and OH. This is assisted by the dangling bonds of a surface dimer, both of which become then occupied by the decomposition products: 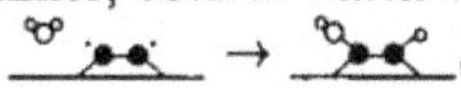, where the big circle indicates O and the small circles stand for H (Fig. 5.6e and 5.6f). The natural saturation coverage is thus 0.5 ML.[338] The dissociation process is facilitated if in the immediate neighborhood there are other dissociated H_2O molecules: adsorbed water tends to form islands on the surface.

The dissociative character of water adsorption is now agreed upon at least for temperatures above 90 K and for exposures which are sufficiently high to enssure that several water molecules have a good chance to come together. A recent *ab initio* DFT cluster study[870] has confirmed results of older semiempirical calculations[863] that water is unstable when adsorbed in molecular form. Electrostatic attraction orients the incoming water molecule with its p lone pair towards the partially empty "dangling" bond of the "down" atom of the dimer. The molecule can exist a while in this physisorbed state without forming any chemical bond to the surface. Ultimately, it attaches itself through its lone pair to the "down" atom (Fig. 5.6g). If one of the OH bonds of H_2O then becomes coplanar with the dimer bond, with the H atom above the dimer, this H atom is dragged towards the "up" dimer atom. At the same time electrons flow back from the "up" orbital towards the Si–O bond and the whole dimer becomes nearly symmetric (Fig. 5.6f). Once the Si–O bond between H_2O and Si is formed, the whole process proceeds with no barrier. Just as it was the case for H/Si(001), the need to break the dimer π bond upon adsorption reinforces the trend for both H and OH moieties of the broken molecule to occupy

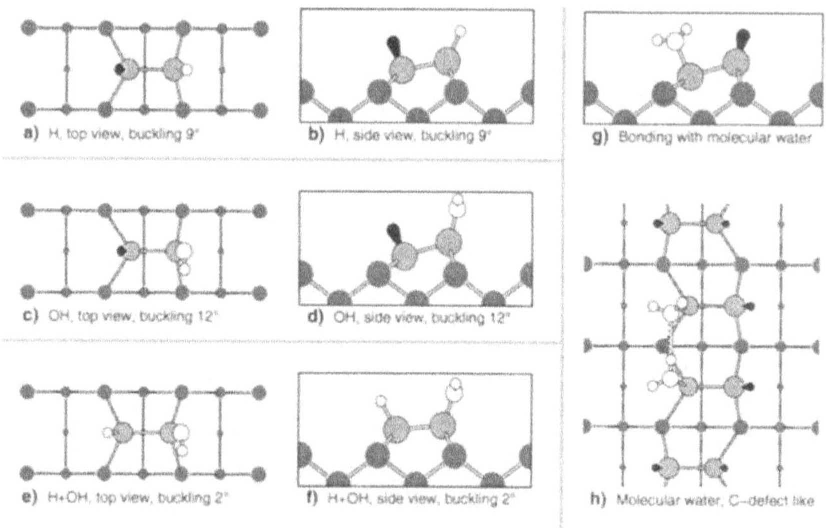

Fig. 5.6. Schematic sketch of water adsorbed on Si(001). Dangling bonds are black, bulklike Si atoms are dark gray, dimer Si atoms are gray, H and O atoms are white. Side views are displayed in frames. Panels (a) and (b) show isolated H. Panels (c) and (d) show isolated OH. Panels (e) and (f) show a dissociated H_2O decorating a Si dimer. The buckling angles are from *ab initio* DFT calculations.[712] Panels (g) and (h) show adsorbed molecular water. The hydrogen bond of the C-defect like arrangement of two H_2O molecules is drawn with broken line.

the same dimer:  + lost energy.

Once dissociated, H_2O does not migrate on the surface. Nevertheless, islands composed of apparently dissociated water can be observed.[538] This means that some form of interaction between the existing islands and the mobile precursor must exist. It is conceivable that this interaction helps the incoming molecule to align its O–H bond with the dimer bond. Naturally, the molecule may also be attracted to the island by electrostatic forces. According to results of *ab initio* DFT cluster calculations,[870] the energy of two dissociated H_2O molecules residing on adjacent dimers in the same dimer row is approximately 90 meV lower than the energy of these molecules residing on isolated dimers. The energy gain is due to the formation of a weak hydrogen bond between the OH moieties.

Step edges are relatively inert to water adsorption.[538] Water islands tend to grow away from them, but the edge dimers become reacted when the H_2O coverage approaches the saturation value of 0.5 ML.

Seen by STM, an island of dissociated water resembles STM micrographs of a 2×1 monohydride. In constrast to a monohydride dimer (), a hydrated dimer () seems to be slightly asymmetric. A single, hydrated dimer imaged among clean dimers is quite dark and can be easily mistaken for a missing dimer: . Particularly dark "missing dimers" may in fact be two H_2O molecules[537] chemisorbed on both ends of the dimer (). This is compatible with the *ab initio* derived dissociation path:[870] the molecule chemisorbed on one end of the dimer prevents the molecule adsorbed on the other end from being broken. But no ultimate proof of the existence of molecular water on Si(001) is known to us.

Circumstantial evidence for chemisorbed H_2O molecules exists, however. STM images of Si(001) samples exposed to small amounts of water show clearly that objects termed W-defects and closely resembling C-type defects (Fig. 3.15a, page 168) are created as the coverage increases.[537] A typical STM image of such a defect can be sketched as 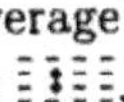, where the lines indicate the clean dimers (oblong spots in STM micrographs of occupied states) and the disks represent two neighboring, circular spots, sometimes poorly resolved. With time, some of these objects decay into a "dimer vacancy" and a clean dimer:[538] 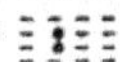 → . It has been proposed[537] that the W-defects are pairs of water molecules stabilized into a tiny "snowflake" by a hydrogen bond which connects them, maybe as in Fig. 5.6h. This model is not in conflict with the *ab initio* results, because the snowflake can be formed when two *physisorbed* molecules bump into one another. The resulting $(H_2O)_2$ dimer would dissociate when one of the O–H bonds ventures too close to the plane of the dimer bond.

An effect which may shed more light on the origin of W-defects was observed on nearly saturated surfaces.[538] When most of the surface is covered with water, some isolated areas of dimer size remain unreacted. Many of them look as bright, symmetric ovals, so they may be clean dimers or SiH_2 moieties[153] adsorbed on top of dimer rows (Sect. 5.5.3). But some of them are asymmetric in appearance and they may come in pairs, resembling then the W-defects. The bright spots of such pairs are *mobile*. They can switch their position in subsequent images: 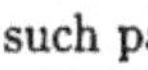 (*cis* arrangement) ↔ 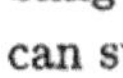 (*trans* arrangement) ↔ 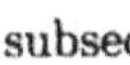 (*cis* arrangement); it is not clear whether these jumps are spontaneous or stimulated by the STM tip. An analogous behavior was recently observed[552] on a nominally clean surface for defects which were classified as C-defects but were probably due to contamination.

If the defects with mobile parts were W-defects, this would indicate that the model of Figure 5.6h is incorrect. When the hydrogen bond is broken (), both molecules should dissociate, particularly on a saturated surface. Since the defects survive, they can hardly contain molecular water!

Another option is that the jumping bright spots correspond to the unsaturated bonds of partially hydrated dimers (, where X stands for H or OH). *Ab initio* results indicate[712] that it would rather be OH. The energy barrier for such a motion of OH was computed to be 0.9 eV, while the corresponding H jump was found to be activated with a value of 1.1 eV.

Is it possible that W-defects, and maybe even all C-defects, are *cis* pairs of H or OH? In STM images of occupied and empty states, a single H atom adsorbed on a clean Si(001) 2×1 surface looks like half a C-defect.[549] Though the *cis* configuration of H atoms is unstable because the strength of the π bond of the clean dimer drives the H atoms from neighboring dimers to the same dimer (), *ab initio* data indicate that this instability is reduced at low coverages.[547] The same is probably true for two OH radicals; in addition, two *cis* OH may form a stabilizing hydrogen bond.

5.5.3. *Oxidation and etching by water*

H_2O molecules and OH^- ions are efficient oxidants. Hot steam is used as a fast oxidizing species (page 363) and liquid water can oxidize silicon even at room temperature (page 359). Boiling in water is a known method to improve the smoothness of polished and HF-etched Si(111) surfaces (page 268). However, an analysis of the interaction of water with silicon surfaces involves a number of different reaction paths and it is not immediately clear which of them dominates unter which conditions. *Ab initio* calculations would help to clarify the situation.

Once adsorbed dissociatively from the **gas phase** on the clean surface, H_2O molecules do not reassemble. Upon rising the temperature of the sample, H desorbs early as H_2, while O is evacuated at higher temperatures, in the form of SiO. Last traces of vibrations involving a Si–H bond disappear around 650°C,[872] that is, around the temperature at which SiO begins to desorb from weakly oxidized areas.[834] Thus, water oxidizes silicon.

The process of oxidation by adsorbed water has not yet been followed step by step in an STM study. Conclusions are drawn from analyses of surface vibrations and from the knowledge about interactions between water and clean Si surfaces at lower temperatures. However, these studies have brought a lot of information which is consistent enough that an atomic model of the water-mediated oxidation has been attempted.[872]

We thus return for a moment to the surface saturated at room temperature by dissociated water molecules. Some traces of oxide are detectable even without heating, which indicates that there is a certain probability that an OH moiety breaks spontaneously, liberating O and H. The O atom may then fall into a neighbouring Si–Si bond, while the H atom may saturate the dangling bond abandoned by the oxygen: 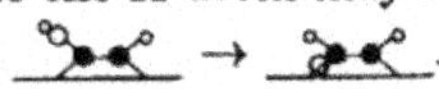.

An alternative path for low-temperature oxidation is suggested by the isomerisation which was deduced[881] to take place at somewhat higher temperatures. When the surface is heated to 150°C, vibrations characteristic of dihydride () appear. The end products of the isomerization may be the dihydride and a silanone complex: 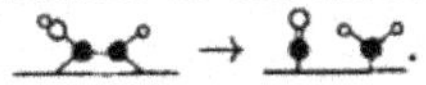. The silanone complex (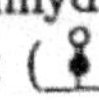), in which the O atom is double-bonded to a Si atom, is generally unstable and may decay into a Si monomer () and oxygen. The latter attacks a Si–Si bond in the neighborhood. The monomer, being highly mobile on a clean surface, is likely to diffuse away, unless its motion is hindered by the hydrogen saturating the neighboring surface dimers (Section 5.2.3). After H_2 desorbs from the dihydride (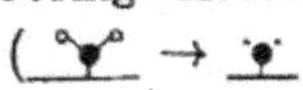 + H_2), another silicon monomer is produced and may diffuse away.[k]

Silicon monomers ejected in the course of room-temperature oxidation and adsorbed on top of surface dimers[774] have possibly been observed in STM images[537,538] as bright spots located centrally on top of dimer rows. According to *ab initio* data (Section 5.2.3), Si monomers can diffuse on a monohydride-covered surface. In the course of migration, they become covered by hydrogen. A monomer saturated by two H atoms and bonded to the dangling bonds of a dimer has a particularly low energy.[774]

Annealing a water-covered surface at various temperatures produces a whole spectrum of vibration lines which can be assigned to virtually all possible configurations of a hydrogen-saturated surface with oxygen atoms intervening in neighboring bonds.[872] The most stable arrangement attributable to hydrogen is $H\text{Si}O_3$, that is, three of the four Si valences are saturated by oxygen and one by hydrogen. This configuration is thus likely to correspond to a hydrogen atom terminating a surface dimer atom with oxygen atoms in both backbonds as well as in the dimer bond.

The message important for the understanding of the **wet oxidation**

[k] A dihydride monomer should be mobile as well, so the silanone and the dihydride may escape from the reaction site simultaneously, leaving a dimer vacancy behind them.

process is that oxidation by H_2O produces Si–H bonds which are relatively stable when the Si atom is surrounded by oxygen atoms, as in SiO_2. Other important information is that hydrogen can open Si–Si bonds (Sec. 5.2.1). The latter process prepares sites for adsorption of OH or O_2 from the oxidizing ambient, but it may also damage the SiO_2/Si interface during the operation of the device (Sect. 6.5.5d). At low temperatures, the desorption of hydrogen may be the rate-limiting step in wet oxidation.[872]

An important difference between oxidation in **liquid water** and oxidation by water molecules from H_2O vapor is that liquid water contains OH^- ions (hydroxyl radicals, Sect. 5.5.1). The radicals attack the surface, etching and oxidizing it. This process helps to produce atomically flat Si(111), but hinders the preparation of flat Si(001) surfaces (Sect. 1.4.2c).

Consider a Si(001) surface terminated by dihydrides and, ocasionally, by OH groups, as after an HF etch followed by rinse in water (page 279). The Si atoms terminated by OH are preferentially attacked by water.[745] This is because OH is a strongly electronegative species, more electronegative than H. Each Si atom bonded to an OH is positively charged and, in consequence, the Si–Si bonds in (Si_3)Si–OH are polarized. These bonds attract and pre-orient dipolar molecules, such as H_2O or HF (page 279), or ion pairs, such as OH^-+H^+. A H_2O molecule may then dissociate and open the polarized Si–Si bond:

$$\text{Si–Si–OH} + H_2O \rightarrow \text{Si–H HO–Si–OH} \qquad (5.23)$$

(compare reaction 5.21, page 280). But it is not clear if the reaction 5.23 is energetically favorable (Table 1.4, page 57). An alternative to it is an attack on the polarized bond by a $H^+ + OH^-$ pair:

$$\text{Si–Si–OH} + H^+ + OH^- \rightarrow \text{Si–H HO–Si–OH.} \qquad (5.24)$$

In other words, Si–Si bonds act here as recombination centers for H^+ and OH^- moieties. The energy gain is now apparent, but the process is hindered by the limited probability to form a close $H^+ + OH^-$ pair. Finally, a proton can be trapped in a less polarized backbond of a hydrogenated atom (Sect. 5.2.1) and this bond can be later attacked by an OH^- radical:

$$\begin{aligned} \text{Si–Si–H} + H^+ &\rightarrow \text{Si–}(H^+)\text{–Si–H} \\ \text{Si–}(H^+)\text{–Si–H} + OH^- &\rightarrow \text{Si–OH H–Si–H.} \end{aligned} \qquad (5.25)$$

In any case, repetitive reactions of these types will liberate soluble silicates and/or silanoles. Reactions resembling those of 5.23-5.25 are probably responsible for the removal of asperities on Si(111) in the process of boiling in water or during the etch in buffered HF (page 268).

There are also reaction paths which lead to oxidation instead of etching. An OH group may substitute the H atom in Si–H. As the next step, the two neighboring OH may react, producing H_2O and an oxidized bond:

$$\begin{aligned} \text{Si–H HO–Si–OH} + \text{OH}^- + \text{H}^+ &\rightarrow \text{Si–OH}\ldots\text{HO–Si–OH} + \text{H}_2, \\ \text{Si–OH}\ldots\text{HO–Si–OH} &\rightarrow \text{Si–O–Si–OH} + \text{H}_2\text{O}. \end{aligned} \qquad (5.26)$$

This is a plausible way in which water slowly oxidizes the surface. Another possibility seems to be that a H_2O molecule (or a H^++OH^- pair) attracted by a polarized Si–Si bond oxidizes the bond directly and liberates H_2:

$$\text{Si–Si–OH} + \text{H}_2\text{O} \rightarrow \text{Si–O–Si–OH} + \text{H}_2. \qquad (5.27)$$

Similarly, a Si–(H^+)–Si bond can be oxidized by OH^-:

$$\text{Si–(H}^+\text{)–Si} + \text{OH}^- \rightarrow \text{Si–O–Si} + \text{H}_2. \qquad (5.28)$$

Naturally, not only subsurface Si H bonds can be replaced by Si–OH. The same slow process (Si–H + H_2O → Si–OH + H_2) occurs on the surface. Within an hour or two,[745] neutral water at room temperature substitutes a significant fraction of Si–H bonds with OH groups. The surface is now prepared for further oxidation through the reactions of Eqs. 5.26–5.27. Ultimately, an ultrathin native oxide (Section 6.3.1) film is grown.

The character of the prevailing reaction (oxidation, etching), the specific reaction path, and the reaction rate depend on such factors as the pH (Sect. 1.4.2c) and temperature of water, its oxygen content, as well as the degree of polarization and distortion of the bond to be attacked (that is, surface orientation and roughness, density of steps, density of defects). These processes take place not only in neutral and chemically clean water, but also during etching of the surface in chemical solutions (Sect. 1.4.2c, page 268, page 279). Their variety enhance the possibility to control the quality of the processed surface by designing the composition of the etch.

5.6. Nitrogen: a Hope for a Better Gate Dielectric

The technological motivation to study interactions between Si surfaces and N is the formation of Si_3N_4 films, that is, nitridation. The most efficient nitridation schemes utilize NH_3 (ammonia) or NO (nitric oxide). NH_3 and NO on Si(001) are thus the main topics of this Section. Besides adsorption, we consider the mechanisms of nitridation. Apart from some remarks on deposition of Si_3N_4 layers and implantation of nitrogen atoms and molecules (Sect. 5.6.3), our focus is on thermal nitridation, that is, on reactions which involve silicon surfaces or interfaces during the whole growth process. Table 5.5 collects some references relevant to these subjects.

Si_3N_4 is a hard and stable ceramic used, for example, for chemical protection, as depth markers in planarization by polishing, or even as a dielectric. It would be beneficial if Si_3N_4 gate dielectrics for CMOS transistors could be grown.[882,18] Modern gate oxides are becoming so thin that maintaining their homogenenity is a challenge. Also electron tunneling and dopant diffusion through such thin oxides are troublesome. The oxide thickness is scaled down to increase the gate capacitance $C = \epsilon/t$, where ϵ is the dielectric constant and t is the thickness of the dielectric. This scaling is needed to keep the parameters of the miniaturized transistors within the optimal range and to reduce the power consumption. For instance, processors based on transistor with an effective gate thickness equivalent to ~2 nm (20 Å) of SiO_2 would operate at frequencies several times higher and would consume several times less power than contemporary processors. Since Si_3N_4 has a higher ϵ than SiO_2 (6–9 compared to ~4), nitride gate dielectrics would be thicker than an oxide with the same C. In addition, Si_3N_4 films are excellent diffusion barriers. Gate oxides can be nitridized to protect them from dopant penetration (Sections 6.3.4 and 6.5.5d).

An interesting *scientific* application of Si_3N_4 films grown on silicon is in studies of defect-mediated diffusion of dopants in Si bulk. The Si_3N_4/Si(001) interface is a source of bulk Si vacancies, maybe due to the strain generated during nitridation or to diffusion of Si into Si_3N_4 during the thermal nitridation process. These vacancies influence diffusion of dopants[883] in a way which depends on the atomistic details of the diffusion mechanism.[85] The diffusivity of species which migrate mostly by a vacancy-mediated mechanism is enhanced. Annihilation of vacancies with Si interstitial atoms retards the diffusion of species which migrate mostly

Ref year	Subject relevant to the N/Si system
884 1982	N/Si(001) adsorption, N_2 dissociation by electrons; AES, EELS
885 1984	Excimer laser nitridation, NH_3/Si(001); AES
886 1984	NO/Si(111) adsorption, geometries proposed; EELS, LEED, AES
882 1985	**Review**: Thermal NH_3 nitridation, kinetics, thickness, etc.
887 1985	NH_3/Si(001), nitridation kinetics, electron beam; AES
883 1985	Nitridation as a tool in studies of dopant diffusion in bulk
888 1986	N/Si(111) adsorption, N_2 dissociation by electrons; EELS
889 1987	NO, NH_3/Si(001), adsorption, nitridation; XPS, AES, ISS, UPS
447 1987	NH_3/Si(001), adsorption, dissociation; STM
890 1987	NH_3/Si(001) adsorption; EELS
891 1987	NH_3/Si(001) adsorption and dissociation; UPS, XPS
892 1988	NH_3/Si(001); temperature, pressure, exposure dependence; XPS
893 1989	NH_3 on Si(001) and Si(111), adsorption, decomposition; AES, ESD
894 1989	NH_3/Si(001), adsorption; STM, STS
895 1989	Fundamental information on Si_3N_4/Si(001)
896 1989	NO/Si(001) adsorption and nitridation; EELS
897 1991	NO and NH_3 on Si(001); adsorption, nitridation; SCLS
898 1991	NH_3 on Si(001) and Si(111); quantum chemistry, clusters
899 1992	NH_3 and N_2H_4 on Si(001), nitridation; AES
708 1994	**Book chapter**: Kinetics, structures, phase diagrams
900 1994	Bonding of N/Si(111) at steps; *ab initio* cluster
901 1994	NH_3 on Si(001) and Si(111), adsorption, nitridation; SCLS
902 1994	NO, NH_3/Si(001) adsorption and nitridation; Si(2*p*), N(1*s*), O(1*s*) SCLS
903 1995	N_2, N/Si(001) adsorption, nitridation; clean, Na-catalysed; AES
904 1995	Room-temperature deposition of Si_3N_4 by ECR CVD
905 1995	Excimer laser irradiation causes Si precipitation in Si_3N_4; XPS
906 1995	NH_3/Si(110) nitridation, temperature dependence; AES, LEED
907 1995	A model for thermal Si_3N_4/Si(001) growth
908 1996	NH_3 plasma nitridation; fixed charge, interface traps
909 1997	NH_3/Si(001) adsorption, atomic structure; photodiffraction
910 1997	NH_3/Si(001) adsorption; quantum chemistry, CI
911 1997	NH_3/Si(001) catalytic nitridation, W filament; XPS, electrical
912 1997	N_2O/Si(001) adsorption; EELS, TPD
913 1997	Adsorption of N_2H_4 on Si(001) and Si(111); STM
18 1998	**Review**: A realistic method to grow gate nitrides for CMOS

Table 5.5. Nitrogen on silicon; selected references

by an interstitial-mediated mechanism. Such investigations provide thus data useful in practical simulation of technological processes.[28]

A simple way to nitridize Si would be to let it interact with nitrogen gas. But Si is virtually immune to N_2 molecules even at very high temperatures.[911] A more practical method is to apply NH_3 or NO, which react with Si even at low temperatures, whereby Si–N bonds form.[889]

The formation of a thermal nitride *film* requires nevertheless high temperatures and is slow. This makes such an approach unsuited for production of MOSFET gate dielectrics. For example, to substitute gate oxides by gate nitrides in today's state-of-the-art commercial technology (0.25 μm minimal feature size, $t_{ox} \simeq 40$–50 Å), Si_3N_4 films of thickness $t \simeq 80$ Å are needed (page 355 and Fig. 1.4, page 11). Within less than a decade from now, t_{ox} may become 20 Å, which corresponds roughly to $t \sim 40$ Å of Si_3N_4. As a matter of fact, the first technologies which may need such thin dielectrics will appear as soon as around the year 2001. But to grow a 30 Å nitride film in NH_3 the wafer must be kept at 1100°C for about a minute.[882] The next 5 Å are added after half an hour, and further 5 Å would require four hours. A 40 Å SiO_2 film can be made within 20 min and at much lower temperature of 800°C. The need to keep the substrate at such high temperatures for a prolonged time (high "thermal budget") as needed for thermal nitridation interferes with the tendency to reduce thermal budgets. Exposure to elevated temperatures increases outdiffusion of dopant atoms from the implanted regions and facilitates contamination.

Alternative nitridation methods are therefore being studied. Examples include such deposition techniques as excimer-laser enhanced nitridation or plasma-enhanced CVD. The molecules may be decomposed by electron beam, by hot W filament, or by catalysers adsorbed on the surface.[911] Still, low-temperature nitrides made from NH_3 have a high hydrogen content, which deteriorates dielectric strength and reliability of the film. On the other hand, elastic and optical properties of the nitride depend on its morphology (size of grains, type of defects) and chemical composition (stoichiometry, H and O content). This opens the possibility to adapt the nitridation scheme to the specific application of the film.

In general, deposited nitrides are characterized by a high concentration of electrically active defects at the interface with the silicon substrate and in the film (fixed charges in the range of 10^{12} cm^{-2}, interface traps in the range of 10^{11}–10^{12} cm^{-2}, see Ref. 914 and references therein). This makes such a film unsuitable as a gate dielectric (Sect. 6.6). But the so-called Jet Vapor Deposition method[915] was recently used to produce ultrathin nitride films with quality (defects, homogeneity, reliability) comparable or exceeding that of standard thermal oxide films.[18] Technological implementations of this approach generally are considered as realistic.

An interesting aspect of NH_3 nitridation is that, in contrast to the Si(001) suface, the Ge(001) surface is totally inert to ammonia plasma.[908] A technique which utilizes this nitridation selectivity to grow self-assembled quantum wires on Ge/Si(001) has been proposed.[916]

5.6.1. *Nitrogen, silicon nitride, and related materials*

Nitrogen is the smallest of all group-V atoms. Its electronic configuration is $1s^2 2s^2 2p^3$. Gaseous N atoms combine into N_2 molecules. Such a molecule is characterized by a triple bond (:N:::N:), akin to the bonds in CO and SiO. This bond is strong (Table 1.4, page 57), rendering N_2 chemically inert.

The character of bonds between Si and N is interesting. Consider the stable compound of Si and N, silicon nitride (Fig. 5.7). It is straightforward to see that the composition Si_3N_4 satisfies all valences of both elements. Each Si atom in Si_3N_4 has four N neighbors and each N atom has three Si neighbors. The atoms are arranged such that Si atoms are in a nearly tetrahedral configuration (the hybridization of Si orbitals is nearly sp^3) and N atoms are in a nearly planar configuration (the hybridization of its orbitals is nearly sp^2). The latter may seem strange, because there is a strong electrostatic repulsion between the bonds and the two electrons in the lone pair p orbital. One might expect that N tends to leave the plane of its three Si neighbours, making a NSi_3 pyramid as in NH_3, so that the s component is added to the lone pair; this distortion would be limited by the bending of Si bonds. However, N *benefits* from the planar configuration.

This benefit comes from an additional bond made by the "lone-pair" $2p$ orbital of nitrogen. The nitrogen atom shares its lone-pair electrons with $3d$ orbitals of its silicon neighbors, forming $p\pi$–$d\pi$ bonds. This so-called **hyperconjugation**, an effect somewhat similar to the formation of hydrogen bonds (Sect. 5.5.1), keeps nitrogen in the planar sp^2 geometry.

The volume occupied by a Si atom in stoichiometric Si_3N_4 is 13.7 Å^3. This is less than the volume occupied by a Si atom in Si bulk (20.1 Å^3). A nitride film grown thermally on silicon has therefore a tendency to "steal" Si atoms from silicon bulk, that is, to eject Si vacancies into the substrate.

Silicon nitride, Si_3N_4, is so hard a ceramic that it is used in engines to protect the metal from abrasive wear-out. Its main role in the semiconductor industry is in masks for various purposes. For instance, LOCOS isolation process flow (page 377) uses Si_3N_4 layers to protect certain areas of pad oxide from further growth in the subsequent oxidation step. Another isolation technique, STI (page 378), uses Si_3N_4 to protect parts of SiO_2 and the substrate from etching and as a hard marker on which polishing stops. Si_3N_4 can be used in floating gate ("storage") dielectrics of some non-volatile memories and DRAMs[917] (Fig. 1.3). Si_3N_4 films deposited

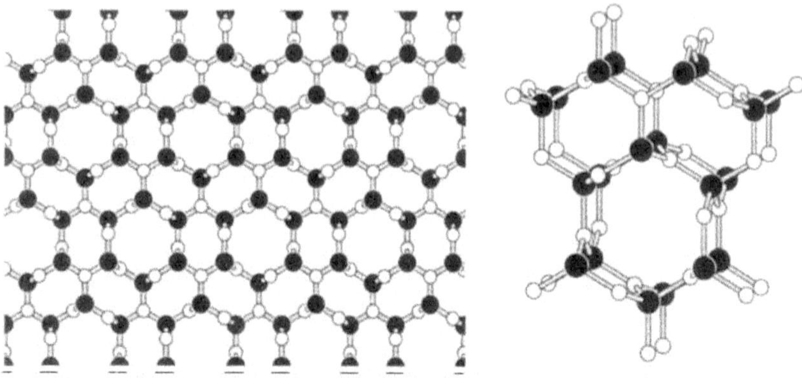

Fig. 5.7. The structure of β-Si_3N_4 (hexagonal: a=7.61 Å, c=2.91 Å). Si atoms are black, N atoms are white. The atoms are arranged as in phenacite (Be_2SiO_4), whereby Be and Si sites are substituted by Si atoms and O sites are substituted by N atoms. The length of Si–N bonds is 1.70–1.75 Å. **Left**: view along the c axis. **Right**: perspective view. The displayed structural element is larger than the unit cell. The other polymorph, α-Si_3N_4, is a derivative of the β form. The thermal expansion coefficient of Si_3N_4 is $\sim 3 \cdot 10^{-6}$ K^{-1}, relatively close to that of Si ($\sim 4.5 \cdot 10^{-6}$ K^{-1}). The band gap of Si_3N_4 is ~4.5 eV, approximately half as large as than that of SiO_2. Typical Si_3N_4 films are amorphous[918] or polycrystalline. Structural data from Ref. 919.

before ready chips are packaged in plastic provide ultimate chemical passivation and protection against penetration of moisture and alkali metals. Bibliography on applications of Si_3N_4 can be found, e.g., in Ref. 920.

Nitride films grown by Chemical Vapor Deposition (CVD) are often non-stochiometric and contain a significant fraction of hydrogen, oxygen, or water. The amount of Si in the films produced by various implementations of CVD may also differ. A non-stoichiometric silicon nitride is sometimes described as SiN, SiN_x, or SiN_x:H. When Si, N, and O are combined together, a stochiometric **oxynitride** (Fig. 5.8) can be formed. Oxynitride films (Sect. 6.3.4) of various composition can be deposited or produced by NO or N_2O nitridation. Nevertheless, the recently developed Jet Vapor Deposition (JVD) technique of $NiSi_2$ deposition (Sect. 5.6.3) may make Si_3N_4 a material which will challenge SiO_2 in its role as the gate dielectric. The important factor is here that the dielectric constant of Si_3N_4, which is between 6 and 9, is significantly higher than the dielectric constant of SiO_2 (~4). This helps to solve certain problems (page 355) associated with the miniaturization of MOSFET devices (Sect. 1.2.2a).

But up to now, Si_3N_4 films rarely touch the Si substrate or a metal

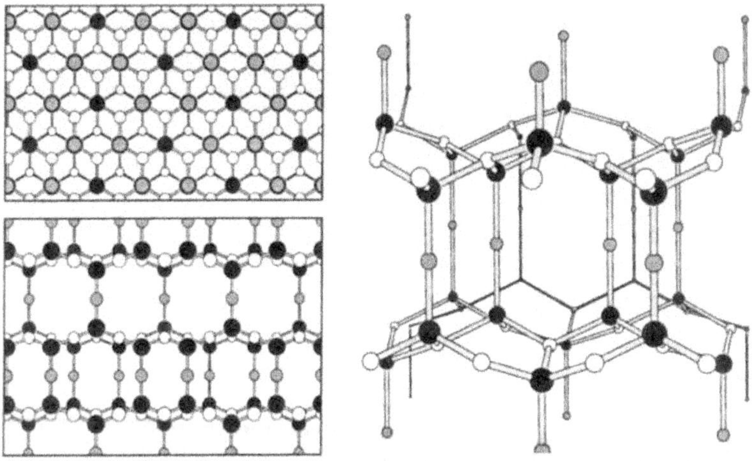

Fig. 5.8. A simplified structure of crystalline, stoichiometric oxynitride (Si_2ON_2, orthorombic: a=8.84 Å, b=5.48 Å, c=4.48 Å). Si atoms are black, N atoms are white, O atoms are gray. More distant atoms are shown smaller. SiN layers resembling (111) planes of zinc blende are interconnected by oxygen atoms intervening between Si atoms. As in Si_3N_4, Si–N–Si and N–Si–N bond angles are tetrahedral (109.45°) and planar (120°) within ~3°. Si–O–Si angles are close to 150°, as typical for SiO_2. For simplicity, they are rendered as 180° in the picture. Si–O bond lengths are ~1.65Å, Si–N bond lengths are 1.70–1.75Å. **Left**: view along the c axis (top) and a side view (bottom). **Right**: perspective view. The displayed structural element is larger than the unit cell. Structural data from Refs. 921 and 922.

line. Si_3N_4 is so hard that the stress generated during growth may produce misfit defects and even impede adhesion; the film peels and cracks. In order to accommodate the stress, a SiO_2 buffer layer is commonly placed below the nitride. An interesting exception is when a polysilicon buffer layer separates the nitride and the pad oxide, as in the so-called polybuffered LOCOS scheme.[923] But again the purpose of the silicon is to reduce stress produced during the nitridation, in order to assure that no extended defects are generated in the substrate below the Si_3N_4/polysilicon/SiO_2 structure. In this case, the details about the interface between Si and Si_3N_4 are of no importance; it suffices that the deposited nitride sticks to the polySi film.

Another exception, though far beyond the scope of our discussion, is the production of free-standing membranes. When micromachining techniques such as anisotropic etching of Si are applied, a nitride layer deposited directly on silicon substrate can be converted into such a membrane.[924]

A Si_3N_4 layer may be deposited on silicon or on a few atomic layers thick "native" SiO_2 film on silicon when it is deposited on the polysilicon gate. It acts then as a diffusion barrier between the gate material and a BPSG glass contact-level dielectric.[925] The nitride does a good job: only 15 nm of Si_3N_4 protects the gate as efficiently as 50 nm of a SiO_2 film (TEOS). But again, the interface between Si_3N_4 and silicon plays no important role in this case. What matters is that the thin Si_3N_4 film does not let B and P atoms from the BPSG through. Similarily, N segregates to the interface with Si when SiO_2/Si(001) films are nitridized, so that a thin nitride/Si(001) interfacial layer is formed. The purpose of such nitridation is again to make the gate oxide resistant to boron penetration.

5.6.2. *Adsorption of nitrogen-containing molecules*

Early studies[889] indicated that N atoms are freed from impingent NH_3 or NO molecules even below at 90 K by the catalytic action of surface dangling bonds and become incorporated in subsurface sites. Today, the consensus seems to be[909] is that though **ammonia** does adsorb dissociatively at least at room temperature, the NH_2 and H moieties attach themselves to surface dimers (). NH_3 is in this context similar to water (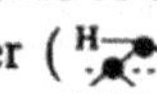, Section 5.5.2). The 2×1 LEED pattern remains unchanged after NH_3 exposure, though traces of dyhidride (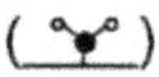) can be detected below ~320°C,[891] that is, when dihydrides are stable (Section 5.2).

The low-temperature initial sticking coefficient of NH_3 is close to 1 and decreases with increasing temperature. Room temperature adsorption covers the surface with half a monolayer[902] of ammonia or less.[901] 0.5 ML coverage, that is, all "dangling" bonds of the surface silicon dimers saturated by NH_2 and hydrogen, can be achieved[901] by room-temperature annealing of the sample reacted at 90 K. NH_3 deposited at 90 K condenses easily on Si(001) and Si(111), forming a multilayer solid film[901] which contains many species, including *atomic* nitrogen. Some of these nitrogen atoms make it to the substrate and seem to react with silicon atoms. This reaction path becomes severed when the solid film evaporates.

When ammonia is deposited around room temperature, the only source of nitrogen atoms are the NH_2 moieties on the surface. Dissociative adsorption of NH_3 is catalysed by the "dangling" bonds of the Si(001) dimers.[889] The nitrogen uptake (the amount of adsorbed nitrogen) reaches a minimum when NH_3 is deposited at 300°C. The adsorption mechanism is then changed and formation of nitride (Section 5.6.3) begins to dominate over adsorption in the form of NH_2 and H fragments.[892] Still, some NH_2

was found[926] on Si(001) even around 330°.

Quantum chemistry calculations confirm that NH_3 which arrives at a clean site of Si(001) 2×1 breaks into NH_2 and H. Ammonia is chemisorbed as NH_3 without any barrier[910] and then decomposes with a small activation energy.[898,910] The dissociative adsorption of an ammonia molecule was computed[910] to be highly exothermic, releasing more than 3 eV. The calculated[898] angle α between the Si–N bond and the surface normal, $\alpha = 25°$, compares favorably with the experimental value of $\alpha = 21° \pm 4°$ determined recently by photoelectron diffraction.[909] Note, however, that since the angle between the dangling bond of a symmetric Si dimer and the surface normal is approximately 20°, both the computed and measured values are close to what should be expected. The measured bond length, 1.73±0.08 Å, is somewhat smaller than the computed one of 2.04 Å.

Adsorption of **hydrazine**, N_2H_4, on Si(001) has recently invoked some interest. The N–N bond in hydrazine is relatively weak, making this molecule quite reactive. It is hoped that this increased reactivity will allow to conduct nitridation at lower temperatures.[899]

According to a HREELS study,[927] hydrazine adsorbs at very low temperatures as N_2H_4 and occupies a bridging site between dimer rows (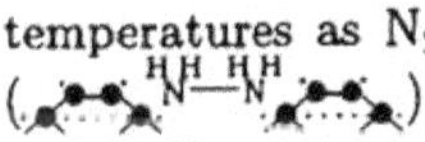

). But even at 100 K some partial dissociation into H and N_2H_3 moieties is noticeable. After this dissociation, the geometry is probably: 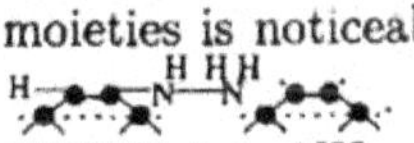

. The N_2H_3 moiety decays between at 220 K and 450 K (180°C) into NH_2 and NH, whereby NH may join the neighboring dimer rows: 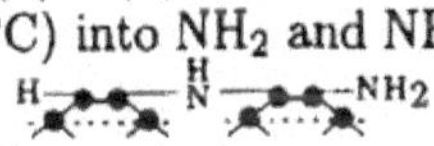

. H–N_2H_3 and H–NH–NH_2 coexist up to 600 K (330°C), at which temperature all N–N bonds dissociate.

This is in principle consistent with room-temperature STM images acquired after submonolayer adsorption at room-temperature.[913] Saturation coverage (0.6 ML) preserves the 2×1 periodicity of the clean surface. The occupied states of the adsorbed molecules have a node between dimer rows, which agrees with the character of the highest occupied orbital of N_2H_4 (theoretical data for N_2H_3 moieties are not available).

However, the geometry of NH as deduced from HREELS must be corrected. After a sample covered with a small amount of hydrazine (0.05 L) had been annealed at 330°C, bright STM spots *on top* of dimer rows were imaged.[913] They are incompatible with the structures invoked in the original HREELS analysis, but may correspond to NH intervening in dimer bonds[913] and/or in dimer backbonds.[301] Such adsorption sites would then resemble adsorption sites of oxygen (Section 5.4.3). The analogy between electronic configurations of H and O makes this assignment plausible: Si–(NH)–Si ~ Si–O–Si (compare, for example,  and 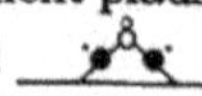).

Besides the bright spots, the annealed sample showed bright, needle-like clusters of unresolved structure. Building upon the analogy between NH and O, we speculate that these clusters are composed of NH_2 (to account for HREELS data), NH, and ejected silicon (Section 5.4.3a). Si may be ejected from the substrate when N_2H_3 break and form new bonds, just like when O_2 molecules break. Indeed, the STM images taken after annealing show an increased concentration of dimer vacancy complexes, apparently etch pits. The bright clusters grow across dimer rows, as the clusters formed from the reaction products of oxygen etching.

Reports on the character of **nitric oxide** (NO) adsorption on Si(001) are rather ambiguous. Some of them favor a dissociative and some a molecular adsorption.[l] There is no consensus on even such fundamental observables as the periodicity of surface reconstruction (2×1 or disordered) and the coverage acquired at comparable conditions. It is well possible that some details of surface preparation, the technique of measurement, and the quality of the applied data processing scheme may have a decisive influence on the conclusions.[928,902] Though this is definitely an exciting subject, we prefer to restrain ourselves from taking sides in this discussion.

Let us only mention that interesting arguments have been raised in a recent photoemission study which used synchrotron radiation and was analysed particularly carefully.[902] This work supported the opinion that NO dissociates at room temperature and that a reaction beyond the topmost monolayer takes place. Results of MNDO cluster calculations[929] may be interpreted as indicating that electrostatic interactions between the molecules and, possibly, interactions between the molecules and surface defects, may strongly affect the microscopic path of the adsorption process. Note also that low-temperature (90 K) adsorption of NO on Si(111) 7×7 is partially molecular[930] and that NO species were detected[931] in photodesorption from NO/Si(001). It is imaginable that molecular NO is stabilized when it adsorbs at a metallic site, for example, at Si(111) 7×7 or on a C-type defect (Section 3.3.1b). Electrostatic interactions between such molecules, being only weakly screened on the semiconducting Si(001) 2×1 surface, may impede further adsorption if the metallic sites are very numerous.

[l] See references in Ref. 902, and the debate in Refs. 928 and 932.

In any case, it seems to be accepted that when the substrate temperature is sufficiently high (such as 230°C), NO molecules dissociate when they hit the clean surface.[929] An ultrathin oxynitride (Fig. 5.8, page 313) layer, albeit nonstoichiometric and disordered, appears to be formed.[902] EELS can detect vibrations attributable to Si–O–Si and Si_3–N even after room temperature adsorption.[896] Annealing at ~1000°C desorbs oxygen from this film, but N is still adsorbed.[896,929] The annealed sample recovers a 2×1 LEED pattern,[896] appears to be terminated by silicon,[889] and reacts with NO similarly to a fresh sample.[889,902] This reactivation is most probably related with a gross reorganization of atoms in the topmost layers. Islands of Si_3N_4 are nucleated and separated by patches of the clean Si(001) surface.[897,902] Since oxygen desorbs from silicon in the form of SiO (Section 5.4.4), such reactivation must be accompanied with a substantial etching and roughening of the surface.

Nitrous oxide (N_2O, known as laughing gas because it causes mild hysteria when used as anesthetic) interacts with Si(001) in a completely different way. N_2O oxidizes the surface instead of nitridizing it. This is not surprising, because N_2O is one of few gases other than oxygen that support combustion. N_2O molecules are easily dissociated into N_2 and O.

Below 60 K, N_2O *physisorbs* on the surface.[933] Multilayer coverage is possible.[912] Two molecularly *chemisorbed* species of N_2O, denoted α-N_2O and β-N_2O, form disparate bonds to the substrate and are characterized by distinct vibrations and stability.[912] The α-N_2O molecule, probably attached via its O atom, decomposes at 100 K into O and volatile N_2. The β-N_2O molecule, possibly attached via one of its N atoms, decomposes into O and N_2 at 150 K. The sticking coefficient of N_2O decays with temperature and there is little oxidation at room temperature:[934,912] the initial reaction probability at 300 K is as low as 2×10^{-5}. Nitridation by N_2O becomes possible at higher temperatures (page 370).

Due to the low reactivity of N_2, studies of adsorption of **atomic N** from N_2 are rare. The molecules must be forced to dissociate, for example, by high substrate temperatures,[935] by electron beams[884], or by catalizers such as adsorbed alkali metal atoms.[903] It is natural to expect that the interaction of atomic N with silicon is similar to that of atomic oxygen: unless temporarily trapped at surface dangling bonds, the atoms attack Si–Si bonds, as O atoms do. Indeed, adsorption of atomic nitrogen progressively destroys the 2×1 LEED pattern of the clean surface.[884]

5.6.3. *Nitridation of silicon*

Though surface Si–N bonds are formed even at room temperature, the nitridation starts when the N atoms diffuse into the substrate.[907] Investigations of N uptake, vibrational spectra, and atomic core levels indicate that this begins to happen around 300-400°C (see, for example, Refs. 892, 897, and 936). In this temperature range some N atoms acquire three Si neighbors[891] and some Si atoms acquire more than one N neighbor. The NSi_3 configuration prevails[892] at 800°C, but a complete nitridation with all but interfacial Si atoms fourfold nitridized is possible only at temperatures[902,901] above 920°C; it may well be that 1200–1300°C is needed.

Thermal nitridation is, naturally, limited also by the need to dissociate the impinging molecules. Since this dissociation is done by the "dangling" bonds of the Si dimers, the nitridation stops when all these bonds become saturated,[889] be it with H from NH_3 or with O from NO. Heating at 430–530°C reactivates the surface by H_2 desorption when NH_3 was used; in the case of NO nitridation, the reactivation takes place at ~920°C, when SiO desorption cleans the surface from oxygen.[889,902] A way to get around this limitation may be to dissociate the incoming molecules by some external agent, such as by an electron beam,[887] laser light,[885] ion bombardment, adsorbed alkali metals,[903] or a hot W filament.[911]

Thermal nitridation is self-limited. Though the initial growth of nitride is apparently determined by the reaction rate,[882] further advancement of the Si_3N_4/Si interface is hampered by the migration of nitrogen through the Si_3N_4 film.[907] Since Si_3N_4 is a very efficient diffusion barrier also for N atoms, the reaction virtually stops after a few monolayers have grown. For example, Ref. 902 estimates that the nitride thickness is about 16Å and more or less independent of the nitridating species (NH_3, NO) and of the details of the nitridation procedure (direct nitridation at 920°C or cycles of room-temperature deposition and 930°C reactivation). Similar conclusions (~3 ML of Si_3N_4 grown thermally from NH_3 at 930°C) were drawn from an AES study.[887] Measurements at a slightly higher temperature of 950°C yielded ~25 Å of nitride.[882,907]

Like thermal oxidation, thermal nitridation is a first-order phase transition. The surface reactivation separates two phases:[892,902] the nitride precipitates in flat islands surrounded by clean surface. The nitride islands are thinner and broader when the temperature is lower; the thinner islands are also less stoichiometric. When all islands coalesce, the nitridation stops.

This mechanism is virtually the same in NH_3 and in NO processing. In general, the main difference between high-temperature NH_3 and NO nitridation is that the latter proceeds faster.

Sufficiently high NH_3 pressures or sufficiently low temperatures make a homogeneous growth of the nitride kinetically possible.[892]. On the other hand, if the temperature of the substrate is too high and the pressure of NH_3 too low, ammonia may become an etching agent. It was proposed[892] that it is then easier for N atoms to bind two Si atoms in a volatile Si_2N molecule than to find one another and form a stable seed of nitride (compare Sect. 5.4.4). The key experimental observation invoked to support this thesis was that the nitrogen uptake is reduced under such conditions and that it decreases further if the exposure is kept constant but the temperature is increased. However, no analysis of the desorbed material was performed, so volatile species other than to Si_2N may be involved. Indeed, the Si_2N desorption threshold[884] was reported to be 1120°C, while the N uptake reduction was observed already at temperatures as low as 550°C. On the other hand, the desorption threshold is likely to depend on the N coverage, similarly to that of SiO.

Poor stoichiometry of low-temperature nitrides may hamper some of the efforts to develop a low-temperature nitridation scheme. However, note that certain approaches, such as electron-beam induced growth, excite N atoms; this may sufficiently increase their reactivity and diffusivity to make the problem less dramatic. Also the presence of oxygen in the ambient may possibly assist the low-temperature growth,[911] either by assisting the migration of nitrogen through the film by providing nitride-free diffusion paths in oxynitride, or by thinning the Si_3N_4 film by oxidation.

A method to reduce exposure of the substrate to high temperatures is to apply Rapid Thermal Nitridation (RTN, page 365). For example, RTN in NH_3 for 60 seconds at 850–950°C produces a film which can be then thickened by **deposition** from NH_3 with SiH_4 or SiH_2Cl_2. The dielectric quality of such dual films is superior to standard deposited nitrides.[917] To keep the nitridation time short, the surface should be free from oxygen. The oxide must be removed before NH_3 nitridation can start.[937] The standard low-pressure NH_3 CVD is done at ~800°C. Good low-temperature deposition schemes would be in any case beneficial.[938,911,18]

In general, Si_3N_4 films deposited directly on the silicon substrate are not suited for gate dielectrics because such films have too many defects in the bulk and at the interface with silicon.[914,939] Probably the only known method to produce gate nitrides in an industrially feasible technological process is the Jet Vapor Deposition[915] (JVD). This technique applies a gaseous mixture of N_2, N (generated from N_2 in a plasma cavity), SiH_4, and He expanding with a supersonic speed towards the substrate so that premature formation of nitrides in the gas phase is suppressed. Since the impinging molecules and atoms carry high kinetic energy (of the order of 1 eV per particle), the deposition can be done nominally at room temperature. This approach leads to a Si_3N_4 film of electrical quality comparable to that of a good thermal oxide.[18] In addition, the film has a high dielectric strength, a low stress, and is impenetrable for B dopant atoms from the transistor gate. The JVD procedure can be integrated with a CMOS process, whereby a reduction of gate leakage current and boron penetration with respect to thermal gate oxidation is achieved.[19]

The physical mechanisms which assure the high quality of JVD gate nitrides are still unknown. If they are not related with the slow growth of the nitride in the experimental setup, this technique may one day become popular in the industrial production of integrated circuts.[19]

An alternative to JVD is to grow an extremely thin oxide buffer and deposit Si_3N_4 on top. Good quality MOSFETs have been recently prepared in this way.[939,940] The oxide buffer (6–7 Å) was made by oxynitridation in NO. The oxidation and Si_3N_4 deposition (from NH_3 and SiH_4) were done at 800°C. From the point of view of CMOS processing, this is a very reasonable temperature. The equivalent oxide thickness of the film was 19 Å (1.9 nm; see page 355 and Fig. 1.4 on page 11).

A potential problem with nitridation schemes which use hydrogen-containing species such as NH_3 or SiH_4 is the hydrogen contamination of the film.[941,942,943] This is particularly of concern in low-temperature deposition. N–H and Si–H bonds are relatively weak and their presence may deteriorate the dielectric strength of the nitride. Carriers tunnelling through a thin film may break these bonds and the electrically active defects created in that way will assist in the tunnelling of more carriers. The hydrogen content decreases with increasing substrate temperature. In processes such as electron-beam assisted CVD hydrogen begins to be released from some Si–H and N–H bonds[920] at substrate temperatures as low as 100°C and the number of H-saturated bonds decreases from ~60% at 100°C to ~20% at 450°C. The nitridating species also play an important role. For example, when in the latter process NH_3 was substituted by N_2, the hydrogen content was reduced tenfold. Cyclotron resonance CVD[904] with 400 kHz frequency eliminates Si–H bonds even at room temperature. On the other hand, JVD nitrides do contain a measurable fraction of Si–H and N–H bonds. However, these bonds seem to be very stable and, apparently, do not lead to reliability problems.[18] The reason for this stability remains unknown.

It is not always necessary to get rid of hydrogen. For instance, low-temperature plasma-enhanced deposition of SiN_x:H is used in back end processing.[944] The variable chemistry and morphology of deposited nitrides allow the property of the film to be tailored to the actual needs. The deposition of Si_3N_4 can be modeled by commercial process simulation tools,[945] and the fundamental information on the chemical reactions can be obtained from *ab initio* quantum chemistry calculations.[946,947]

An interesting alternative method of nitridation with a possible industrial future is **implantation** of nitrogen (see, for example, Ref. 948, 949, 950, and 951). Either N^+ or N_2^+ ions can be implanted.[952,948,953,954] Stoichiometry close to Si_3N_4 can be achieved at implantation angles between 0° and 30° to the surface normal.[948]

An important advantage of nitrogen implantation is that if the substrate is afterwards oxidized, the **oxidation** rate of the depends on the nitrogen implant dose. In this way, one can obtain different gate oxide thicknesses on the same substrate.[955] This gives one unique opportunities to optimize the circuit design. For example, one can put in the same circuit transistors which need a higher supply voltage, work faster, but dissipate more power, and transistors which need a lower supply voltage, work slower, but dissipate less power. As a consequence, one can save power consumption without sacrifying much of the operational speed.[956]

Oxidation of the implanted silicon results in a gate oxide with nitrogen piled up at the interface. When compared to N_2O thermal oxides, ultrathin (below 30 Å) oxide films grown thermally on silicon implanted with N_2 doses in excess of $5{\cdot}10^{14}$ cm^{-2} are characterized[950] by smaller oxide thickness variations, reduction of the tunneling current, and improved resistance against dielectric breakdown (including a decrease in the number of defects which cause an early failure of the dielectric).

Ref year	Selected literature on interaction bewteen TM and Si
957 1982	Kinetics of silicidation, phase transformation; theory
958 1984	Mechanism of Schottky barrier formation; theory
959 1990	Ordered TM overlayers, Si(111) and Si(001); ARPES
960 1995	**Review**: Silicides in CMOS technology
961 1996	**Review**: Silicidation, phases, practical problems
267 1996	**Review**: Silicidation, phases, TM/TM and TM/Si interfaces
962 1996	Formation of polymorphic phases; phenomenological model
963 1998	**Review**: Ta_2O_5 thin films for high-K (high-ϵ) dielectrics
103 1998	Interaction of Ti with Si(001) 2×1; *ab initio* DFT-LDA
964 1998	Optimization of silicide-source/drain structure; process simulation

Table 5.6. Transition metals (TM) and silicon; selection of references.

5.7. Transition Metals: Contacts, Dielectrics, Impurities

With the exception of gold and silver, all transition metals (TM) react with Si, building a host of silicide compounds. These silicides are usually highly conducting, which makes them an excellent material for contacts to crystalline or polycrystalline Si in microelectronic devices. On the other hand, transition metals diffuse easily in silicon and special precautions must be taken to keep the active areas of transistors free from them. For example, a single Fe atom segregated to the gate oxide of a 0.1 μm MOSFET would kill the transistor. Laboratory samples can be easily contaminated with such metals as Cu or Ni. A touch with a stainless steel tool can inject enough Ni into the sample to affect the arrangement of surface atoms. The famous 7×7 reconstruction of Si(111) was long thought to be due to nickel, and many of Si(001) missing dimer defects are likely to be induced by Ni.

Nickel interacting with Si(001) is the main subject of this Section. Though Ti/Si(001), W/Si(001), Pt/Si(001), and Co/Si(001) are definitely the silicide systems of the major technological interest, they are somewhat off the main thread of this monograph, which leads from the clean surface through adsorption to MOSFET gate dielectrics. We will therefore devote only a few remarks to them. We will focus on nickel for two reasons. First, stabilization of dimer vacancies due to nickel promotes the $2\times n$ reconstruction of the Si(001) surface (Section 3.4.1). Second, the structures of nickel silicides and Ni/Si interfaces resemble those of cobalt, the metal which is likely to substitute titanium in future technologies. Table 5.6 collects some general bibliography on TM atoms and silicides on Si(001) and Si(111).

The characteristic feature of a TM atom from n-th row of the periodic

Name	Symbol	Row	Electrons	Main valences	Melting
titanium	Ti	4	$3d^2 4s^2$	+2, +3, +4	1675°C
vanadium	V	4	$3d^3 4s^2$	+2 to +5	1890°C
chromium	Cr	4	$3d^5 4s^1$	+2, +3, +6	1890°C
manganese	Mn	4	$3d^5 4s^2$	+2, +6, +7	1244°C
iron	Fe	4	$3d^6 4s^2$	+2, +3	1535°C
cobalt	Co	4	$3d^7 4s^2$	+2, +3	1495°C
nickel	Ni	4	$3d^8 4s^2$	+2	1453°C
copper	Cu	4	$3d^{10} 4s^1$	+1, +2	1080°C
zirconium	Zr	5	$4d^2 5s^2$	+4	1852°C
niobium	Nb	5	$4d^4 5s^1$	+5	2468°C
molybdenum	Mo	5	$4d^5 5s^1$	+2 to +6	2610°C
technetium	Tc	5	$4d^6 5s^1$	+4, +6, +7	2172°C
ruthenium	Ru	5	$4d^7 5s^1$	+3, +4, +6, +8	2250°C
rhodium	Rh	5	$4d^8 5s^1$	+1, +3	1966°C
palladium	Pd	5	$4d^{10} 5s^0$	+2, +4	1552°C
silver	Ag	5	$4d^{10} 5s^1$	+1, +2	961°C
hafnium	Hf	6	$5d^2 6s^2$	+4	2150°C
tantalum	Ta	6	$5d^3 6s^2$	+5	2996°C
tungsten	W	6	$5d^4 6s^2$	+2 to +6	3410°C
rhenium	Re	6	$5d^5 6s^2$	+3, +4, +5, +7	3180°C
osmium	Os	6	$5d^6 6s^2$	+2 to +6, +8	3000°C
iridium	Ir	6	$5d^9 6s^0$	+1, +3, +4	2410°C
platinum	Pt	6	$5d^9 6s^1$	+2, +4	1769°C
gold	Au	6	$5d^{10} 6s^1$	+1, +3	1063°C

Table 5.7. Transition metals, except rare earths.

table of elements (Fig. 5.7) is an interplay between its $(n{-}1)d$ and ns valence electrons. The $(n{-}1)d$ shells of TM atoms are energetically close to the ns shells and usually incompletely occupied (with the exception of Ir, Pd and the noble metals Cu, Ag, and Au). The d^1 elements (group IIIb: Sc, Y, La, Ac), known as rare earths and sometimes also included in the category of transition metals, are not consider here.

The term "transition" has no particular chemical significance. It is used to designate the elements in which d orbitals dominate the properties of the atom: compare the second most reactive metal, the alkali metal rubidium ($[Kr]5s^1$) and the noble, inert silver ($[Kr]4d^{10}5s^1$). Transition metals are moderately or weakly reactive and most of them are hard, refractive (difficult to melt), and lustrous in appearance. TM atoms with partially filled *d* orbitals may show magnetic properties. Bulk Fe, Cr, and Ni are ferromagnetic at room temperature. All transition elements are metals, which means that they are quite electropositive. They easily give electrons away, for example to silicon or to metallic bonds.

We will now make a brief introduction of the chemistry, atomic structures, and morphology of silicide films, focusing on $TiSi_2$, WSi_2, PtSi, and particularly on $NiSi_2$ and $CoSi_2$ (Sect. 5.7.1). Afterwards, we will describe the behavior of nickel (Sect. 5.7.2) and titanium (Sect. 5.7.3) on silicon surfaces. Finally, we will address the search for alternative materials for CMOS gate dielectrics (Sect. 5.7.4). This subject is closely related to the next Chapter which is devoted to oxidation of silicon.

Unfortunately, we had to leave out such interesting systems as Pd/Si and Fe/Si. The former is probably the "scientific standard" silicide interface. Since the Schottky barrier in contacts between Pd_2Si and *p*-type Si is low (0.37 eV), it was suggested[965] that silicide/source and drain/drain terminals in deep-submicron nMOSFETs might be substituted by **palladium** silicide source and drains made solely of Pd_2Si. In this case, the standard two *pn* junctions joined by the gate-controlled channel (Fig. 1.2, page 8) are substituted by two Schottky contacts separated by a tunneling distance, whereby the tunneling current is controlled by the gate voltage. The electric parameters of such a device would be comparable to those of a raised source-drain MOSFET, but the choice of the silicide must indeed be such that the Schottky barrier height is minimized.[966] To our knowledge, working devices of this type have not yet been demonstrated. **Iron** silicides are studied partially because of hope that the *direct* band gap of $FeSi_2$ will afford integration of optoelectronic devices into the Si technology.

5.7.1. *Transition metal silicides on silicon*

The wide use of transition metal silicides in CMOS technology started when it turned out that the classical Si–Al contacts are unreliable in submicron devices. The problem was that Si escapes from the substrate into aluminum, deteriorating the electrical integrity of the electrodes, and Al diffuses fast in the silicon substrate, troubling the dopant distribution. Ti was then used to separate the two materials. The most important silicides are $TiSi_2$ and WSi_2 in CMOS and PtSi in bipolar designs. Table 5.8 lists some references.

Silicide contacts have favorable chemical properties and low resistivity. They are also a convenient element in the manufacturing process and in the

Ref year	Selected literature on some TM silicides
967 1982	**Review**: Epitaxial $NiSi_2$, $CoSi_2$, $PdSi_2$, PtSi
968 1982	Ni_2Si, NiSi, $NiSi_2$ on Si(001), (111), crystallography; TEM, LEED
969 1983	Flat $NiSi_2$/Si(001) and $NiSi_2$/Si(111); template technique
970 1983	Atomic mechanism of silicidation, $NiSi_2$/Si(111); SEXAFS
971 1984	Schottky barriers and interface structure; $NiSi_2$/Si(111)
958 1984	Mechanism of Schottky barrier formation; theory
972 1984	**Review**: Electronic structures, spectroscopic methods
973 1985	$NiSi_2$ fluorite and adamantane structures; *ab initio*
974 1987	$CoSi_2$, $NiSi_2$, fluorite and adamantane structures; *ab initio*
975 1988	**Book article**: Silicidation, fundamental concepts
976 1989	Si dimers at $CoSi_2$/Si(001) interfaces; TEM
977 1991	$CoSi_2$ in the adamantane structure; EXAFS
520 1991	$CoSi_2$/Si(111) growth mode; demonstration of LEEM abilities
978 1992	**Review**: Epitaxy of $CoSi_2$ and $NiSi_2$ on Si(111)
979 1993	$CoSi_2$/Si/$CoSi_2$ heterostructure for tunable detectors
980 1994	Ni/Si(001), $NiSi_2$/Si(001); STM
981 1994	$TiSi_2$ C49 to C54 phase transformation on polySi; XRD, TEM
982 1994	Epitaxial $CoSi_2$/Si(001) with Ti membrane, phase sequence
707 1995	NiSi technology, ring oscillator fabricated
983 1995	$TiSi_2$ C49 to C54 phase transformation, linewidth dependence
984 1995	CoSi/Si(111) can exist in the CsCl structure; TEM
985 1996	$TiSi_2$/Si(001), epitaxial C49 grains hamper C49 $\rightarrow$ C54 transition
986 1996	Scalability of $TiSi_2$ and $CoSi_2$ salicidation; resistance
987 1996	Ti and growth of $CoSi_2$/Si(001); XRD, SIMS, RBS, SEM, resistance
988 1996	Si amorphization faciliates C49-C54 $TiSi_2$ transition; implantation
961 1996	**Review**: Silicides in CMOS technology
989 1998	Interface energy favors C49-$TiSi_2$/Si(001) over C54/Si(001); *ab initio* DFT
710 1999	**Review**: Perspectives for $TiSi_2$, $CoSi_2$ in CMOS

Table 5.8. TM silicides on silicon; selected references

design. Simple silicidation methods exist (Sect. 1.2.2b). Certain silicides, such as $TiSi_2$ and WSi_2, are quite resistive to reactive ion etching, therefore they act as natural etch stops. Rectifying properties of PtSi/Si contacts are used to speed up bipolar transistors by clamping down slow carriers.

Naturally, silicidation has its disadvantages. When overheated, the silicide agglomerates in islands separated by re-grown silicon, and the contact deteriorates. This effect restricts process windows (thermal budget). Too shallow junctions may be punched through by the rough silicide. This limits on the source and drain (S/D) depths. Finally, the silicide may grow through the gate and reach the gate oxide, petrurbing the threshold voltate and sometimes attacking the SiO_2. This puts limits on gate thicknesses.

Besides the specific resistance of the film, an important electrical parameter of a silicide is the height of the Schottky barrier Φ_B at the contact.

In CMOS, this barrier should ideally be the same in contacts to n-type and p-type S/D. This means that the interface states which pin the Fermi level should be as close to the middle of the Si forbidden gap as possible. $TiSi_2$ and $CoSi_2$ are fine in this respect. In spite of the high Φ_B (the gap of Si is ~1.1 eV), the contact resistivity is low due to high carrier concentration n_d in the S/D electrodes. Indeed, since the barrier width (that is, the width of the depletion region) is proportional to $1/\sqrt{n_d}$, sufficiently heavy doping enables tunnelling of carriers under the barrier. The contact then becomes ohmic, with the resistance proportional to $\exp(\Phi_B)$.

The tendency of silicides to agglomerate cause an interesting problem: when the contacts become narrow and long, surface tension easily destabilizes the film. Any thickness fluctuation in a cylindrical rod of a liquid decreases its surface energy; the same fluctuation in a blanket film would lead to increased energy. The interface energy and stability against agglomeration may therefore become an important issue in future technologies. Already nowadays, degradation of the contacts by roughening due to agglomeration limits the process windows. A roughened S/D contact would be leaky, and a roughened gate contact can damage the gate oxide.

Another general problem related to silicidation is the interaction between the silicide and the dopants in the substrate.[267] A lot of dopants can be drawn into the silicide during the contact formation process. Particularly troublesome are here boron and $TiSi_2$, which react to form a stable titanium diboride, TiB_2, but other dopants mix with the silicides as well. The interdiffusion of dopants, occuring mostly along grain boundaries, affects the film growth. Acceptors diffusing through the blanket metal film deposited on the wafer in the salicidation process (Sect. 1.2.2b) may lead to counter-doping of n-type contacts. Even worse, the substrate close to the contact becomes depleted from the dopant ("dose loss"), increasing the contact resistivity. This may be a problem in 0.1 μm and smaller devices.[28]

Microscopic processes of silicidation are complex. Silicide films on Si(001) are polycrystalline. A typical effect is that grains of silicide phases with varying compositions and/or crystallographic structures are formed at different stages of the process. Phases with low silicon content appear first and then they transform into phases with higher silicon contents. The mechanism of this transformation, including the major diffusing species, may change in the course of the process. Consider the case of cobalt as an example. The growth of the first phase, Co_2Si, takes place at about 400°C and is diffusion limited, with cobalt diffusing into silicon. The growth of the second phase, CoSi, is also diffusion limited, but this time silicon diffuses. The growth of the third phase, $CoSi_2$, which requires 600–700°C, is governed by nucleation and reaction, and the diffusing species is again cobalt. The mechanism of growth may also depend on the substrate: the thickness of $TiSi_2$ grown on a single crystal increases linearly with time t, indicating a reaction-limited process, while on polySi it increases as $\sqrt{t}$, indicating a diffusion-limited growth (Eq. 6.32, page 382). A realistic theoretical modeling of silicidation is a difficult task: such a simulation must describe the transport of material between and along grains with changing volumes and moving boundaries, and between phases which coexist only temporarily.

Atomistic calculations could possibly contribute to solutions of problems related to the dopant loss and to the interfacial roughness. A successful application of atomistic theories requires that the physical effects involved are decoupled from the complex processes of silicide formation.

a. *Titanium, tungsten, and platinum silicides*

Silicide contacts are either formed in a self-aligned salicide process (Section 1.2.2b) in the course of reactions between a blanket TM film deposited on a Si surface patterned with SiO_2, or they are directly grown in a chemical vapor co-deposition from silicon- and TM-containing precursors. A typical CMOS process (Section 1.2.2b) produces contacts to MOSFET sources and drains (Fig. 1.2, page 8) by $TiSi_2$ salicidation.[990] Gate contacts may be made concurrently, but they are often manufactured earlier, by co-deposition of WSi_2 (polycide process). The process flow, though more complicated, becomes then easier to control. Namely, sources and drains (S/D) are parts of a single-crystal silicon, while gates are made of polysilicon. Silicidation of crystalline and polycrystalline materials is optimal under different conditions. Since the CMOS process window is further narrowed by the need to make equally good contacts to both n-type and p-type silicon, the separation of gate and S/D silicidation steps is often beneficial.

Titanium disilicide, $TiSi_2$, has one of the lowest resistivities among silicides (Table 5.9). Unfortunately, this is only true for the orthorombic phase of $TiSi_2$. This so-called C54 phase[m] has the specific resistivity of $\rho \simeq$13–20 $\mu\Omega$ cm, which is less than that of pure titanium ($\rho \simeq 55$ $\mu\Omega$ cm).

[m] C54 is the name of this type of structure as catalogued in Landoldt-Bornstein.

Property \ Silicide	$TiSi_2$	$CoSi_2$	$NiSi_2$	PtSi	NiSi
Resistivity, $\mu\Omega\,$cm	[a]13–20	14–20	35–50	28–35	14–20
Si consumption, nm	2.3	3.6	3.6	1.3	1.8
Formation temperature, °C	750–950	550-900	750–850	300–400	350–750
Schottky barrier, eV	0.61	0.64	0.67	0.87	0.67

Table 5.9. Selected silicides: resistivity ([a] C54), Si consumption per nm of TM, process window on crystalline Si, Schottky barrier on n-type Si. See Table 5.8 for references.

For comparison, the specific resistivity of highly doped Si is $\rho \simeq 500\ \mu\Omega\,$cm, while aluminum has $\rho \simeq 2.7\ \mu\Omega\,$cm, copper has $\rho \simeq 1.7\ \mu\Omega\,$cm, and silver, the best conductor, has $\rho \simeq 1.5\ \mu\Omega\,$cm. But $TiSi_2$ has also another allotrope, the hexagonal and usually highly defected C49. When silicon reacts with a titanium film, C49 forms first, around 600°C. Since the specific resistivity of C49 $TiSi_2$ is $\rho \simeq$60–300 $\mu\Omega\,$cm, it should be transformed into C54. This transformation,[981] which can be done by annealing between ~650–850°C (usually above 800°C), causes problems, some of them serious.

First, Si is driven into the Ti film while TiSi (presumably) and then the C49 $TiSi_2$ phase form. Since the salicidation process covers the wafer with a blanket Ti film, the neighboring contacts may be shortcut due to silicon diffusion in the Ti film and coalescence of silicide stripes. This problem is partially solved by nitridation of Ti done simultaneously with silicidation: the growth of TiN from the top competes with the growth of $TiSi_2$ from the bottom. But since a successful conduction of this procedure requires that the temperature is not too high, the process of $TiSi_2$ salicidation must be split into a low-temperature part, during which the C49 allotrope is formed, and a high-temperature part, during which the C54 phase is made. This technique minimizes also the number of dopant atoms transported between n-type and p-type contacts through the Ti layer.

More importantly, the C49 → C54 phase transformation is nucleation-limited. This means that the time needed to complete the transformation increases when the dimensions of the contact decrease. Though the conversion rate increases with temperature, the upper limit for heating of a $TiSi_2$ contact to a polysilicon gate is 850°C, when $TiSi_2$ becomes unstable toward agglomeration. In addition, some of the C49 grains can grow epitaxially. Such grains are particularly difficult to transform into C54. It is because of this poor scaling of the C49 → C54 phase transformation that alternative silicides, notably $CoSi_2$, are being investigated. Cobalt salicidation has been succesfully integrated into deep submicron CMOS technologies; see, for example, Ref. 991 and references therein.

Pre-amorphization of the silicon substrate[n] facilitates the C49 → C54 phase transformation.[988] This observation indicates that crystallinity of silicon favors silicidation reaction paths which lead to the C49 phase. The $TiSi_2$/Si(001) interface seems to play an important role in this effect. According to results of recent *ab initio* calculations,[989] the C54 phase is destabilized at the interface. In other words, the $TiSi_2$/Si(001) interface energy is responsible for the relative stability of the C49 phase and opposes the C49 → C54 transition, making the transition difficult in thin films.

Difficulties notwithstanding, Ti silicidation has a number of advantages. We already noted that $TiSi_2$ films have low resistivity and that the interface with silicon pins the Fermi level close to the mid gap. Moreover, $TiSi_2$ is relatively stable thermally. It is more stable than, for example, PtSi and NiSi, other low-resistivity silicides. It is also beneficial that titanium easily reduces native oxides on the silicon surface. This makes silicidation more reliable. Also the amount of the substrate used up in the reaction is moderate: a Ti film of 1 nm thickness would consume only 2.3 nm of Si.

For practical reasons, S/D junctions are formed in standard CMOS processing before silicidation.[o] Silicidation then becomes an important factor limiting the junction depth. This is due to Si consumption in Ti silicidation and in subsequent etching of the unreacted Ti, and to roughness of the silicide surface. The S/D junction must be at least 100 nm deep in order to assure low junction leakage after silicidation. Continued use of silicidation, particularly in conjunction with Co silicidation, will soon (maybe around the year 2006[22]) require elevated S/D structures.[992,993,22] The substrate will have to be shaped by selective epitaxy or etching of silicon in such a way, that the silicon surface before salicidation is higher in the S/D region than in the gate region (compare the standard structure of Fig. 1.2 on page 8, sketched here as S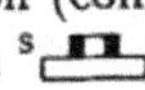D, and the raised S/D structure, S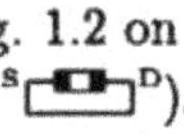D). The growth of $CoSi_2$ consumes even more silicon than the growth of $TiSi_2$: a nanometer of Co/Si(001) film uses 3.6 nm of Si upon conversion to $CoSi_2$. It is worth noting that standard TCAD software lacks the database for cobalt processes. A well-designed fundamental research project in this field could thus be attractive to industry.

Tungsten dislicide, WSi_2, is the most important material for poly-

---

[n] Si can be locally amorphized by implantation of approximately a monolayer of ions. However, the price for that is increased Transient Enhanced Diffusion[710] (page 421).

[o] But silicides have been also applied successfully as diffusion sources.[992]

cide gate contacts. Though its specific resistivity, $\rho \simeq$60–70 $\mu\Omega$ cm, is higher than that of $TiSi_2$, tungsten disilicide is much more stable on polysilicon than $TiSi_2$. This is due to the high melting temperature of WSi_2, 2160°C ($TiSi_2$ melts already at 1500°C). Indeed, it has been established[994,995] that the temperature at which the silicide becomes unstable on polysilicon is roughly proportional to the temperature of melting point of the silicide.

WSi_2 is commonly deposited from silane (SiH_4) and tungsten hexafluoride (WF_6). When annealed above 600°C, such films acquire a tetragonal structure and become very stable (at least to 900°C) and smooth.[996] However, F atoms from WF_6 (and, possibly, also H atoms from SiH_4) migrate through the gate into the gate oxide. This facilitates penetration of boron from the highly doped gate through the gate oxide into the substrate. Boron penetration degrades electrical parameters and reliability of the transistor (Sect. 6.5.5c). A solution is to nitridize ultrathin gate oxides (Sect. 6.5.5d) and/or use SiH_2Cl_2 instead of SiH_4 (see Ref. 961 for bibliography).

Platinum silicide, PtSi, is widely used in bipolar transistors. PtSi contacts to n-type silicon are used as Schottky diodes. In bipolar devices, the carriers are electrons and holes. When the potential which drives the current is switched off, the carriers equilibrate by recombination, which is a slow process. Being a one-carrier device, a Schottky diode installed parallel with the base and the collector increases the working speed of the transistor. Schottky barrier height of a PtSi contact to n-type silicon, Φ_B=0.87 eV, is well suited for this purpose. A major problem with PtSi is its low thermal stability: it degrades already at 400°C.

b. CaF_2 *structures: Ni and Co disilicides and interfaces*

We now consider one specific class of silicides in more detail. Two silicides, $NiSi_2$ and $CoSi_2$, crystallize in a simple fluorite structure, closely related to that of diamond. $CoSi_2$ is of a potential technological interest. It is nearly as stable thermally as $TiSi_2$. In contrast to $TiSi_2$, its formation does not slow down with decreasing dimensions of the contact. Its electric properties are similar to that of $TiSi_2$ (Table 5.9, page 328). Disadvantages of $CoSi_2$ include lower chemical reactivity and larger Si consumption during salicidation. The latter means that making shallow junctions is more difficult with $CoSi_2$ than with $TiSi_2$, while the former means that the initial surface must be more carefully cleaned. Cleaning (Sect. 1.4.2c) is always a somewhat problematic, unreliable procedure. $NiSi_2$ has no technological significance. It is less stable than $TiSi_2$ and has a higher resistivity (Table 5.9). But since the lattice misfit between Si and $NiSi_2$ is small, $NiSi_2$/Si(111) and $NiSi_2$/Si(001) are prototypes of the simplest silicide/silicon interfaces.

$NiSi_2$ and $CoSi_2$ films have been grown on Si(001) and Si(111) by a number of different techniques. Apart from thermal growth from a blanket metal film deposited on the (patterned) silicon substrate (salicidation) and deposition of the silicide on (usually) polysilicon material (polycide process), techniques such as mesotaxy[997] (annealing after ion implantation), endotaxy[998] (Co diffuses from a deposited Co layer through a deposited Si layer to buried $CoSi_2$ seeds), and Solid-State Amorphization Reaction (SSAR, described below) have been successfully applied.

Can one make an **epitaxial** $NiSi_2$ or $CoSi_2$? In principle, yes. Fluorite and diamond structures are closely related and the lattice dimensions of these silicides fit those of bulk Si quite well.[p] But the salicidation reaction path is very complex. It proceeds through two other phases, first Ni_2Si (Co_2Si) and then NiSi (CoSi), and the final transformation to $NiSi_2$ ($CoSi_2$) starts at many nucleation centres. In consequence, the ultimate silicide is in general polycrystalline, though some degree of epitaxy of $NiSi_2$ and of the preceeding phases exists,[968] particularly on Si(111). The degree of epitaxy is higher when silicide is deposited.[999,1000]

Very thin epitaxial $NiSi_2$ films (a few monolayers, whereby a monolayer of $NiSi_2$/Si(111) is ~3.1 Å) grow even from a deposited Ni layer.[969] The phase sequence typical for thick films (Ni_2Si stable at 200–350°C, NiSi stable at 300–750°C, and $NiSi_2$ stable above 750°C) can be then probably suppressed due to smaller vertical distances. The interface advances smoothly into the substrate and planar growth becomes possible. Again, Si(111) is a better substrate for $NiSi_2$ than Si(001). Flat $NiSi_2$/Si(001) was reported[969] only if the Ni film was in the range between 10 and 30 Å. Once such an epitaxial template is formed, it would act as a nucleation center for a new silicide. It can be therefore covered by a thicker Ni film which then reacts with the substrate to form an epitaxial $NiSi_2$/Si(001).

An important role of the template is that it separates the substrate from the metal source, controlling the supply of the metal atoms. A somewhat similar effect is known to produce epitaxial $CoSi_2$/Si(001) films. A semi-permeable solid layer[1001] ("membrane") which slowly passes Co atoms enables the formation of uniform silicide with a smooth interface to Si(001).

[p] $NiSi_2$ ($CoSi_2$) lattice would have to expand by 0.4% (1.2%) to match than that of Si; $a/2$<001> dislocations spaced by ~0.09 μm would relieve the misfit in $NiSi_2$.[1002]

This SSAR (Solid-State Amorphization Reaction) growth of $CoSi_2$ was observed for the first time for thin (100–300 Å) titanium membranes.[1003] Other transition metals, such as zirconium, can also be used, maybe even more efficiently.[1004] Though the details, including the sequence of phases below the membrane, may depend on the substrate temperature, the main allotrope which serves as a template for the epitaxial $CoSi_2$ is epitaxial CoSi.[1005]

The SSAR technique is compatible with the standard CMOS process flow.[1006] It has at least three advantages over simple Co salicidation: the thermal stability of the film is better, the specific resistivity is lower, and the Ti film binds oxides and impurities from the Si surface, making the requirements for the cleaning less stringent. On the other hand, the process becomes more complicated and higher temperatures may be needed.

A similar trick[1007] can force NiSi to grow epitaxially on Si(111), but not on Si(001). On the other hand, a simple NiSi *salicidation* was proposed as a solution to the anticipated technological problem with Si consumption (Section 5.7.1a). While $CoSi_2$ is here more troublesome than $TiSi_2$, NiSi uses up less silicon than $TiSi_2$ (Table 5.9). The main practical problem with NiSi is conversion of the monosilicide into $NiSi_2$ above ~750°C. Deep-submicrometer technologies may be, however, able to avoid such high temperatures after salicidation. Indeed, a working test circuit (ring oscillator) with NiSi contacts has been demonstrated.[707] Naturally, industry will abandon $TiSi_2$ and switch to $CoSi_2$, NiSi, or to another silicide only when it is convinced that such a change is economic.

Apart from their higher thermal stability and generally lower resistivity, epitaxial silicides offer an opportunity to study their interfaces with silicon, for example, by cross-sectional TEM. We will now make a brief excursion into atomic structures of $NiSi_2$ and its interfaces with Si.

The **electronic configuration** of Ni is $[Ar]3d^84s^24p^0$ (Table 5.7, page 323). Nickel in bulk $NiSi_2$ may seem to behave like an atom with the valency of eight: it is surrounded by eight Si atoms, as if it tried to complete its atomic shell to that of krypton, $[Ar]3d^{10}4s^24p^6$. But eight electrons is a significant amount and, since Ni is a metal, the bonding in $NiSi_2$ is mostly metallic, with admixture of covalent and ionic characters. Some electrons are promoted from $4s$ to $4p$. The Fermi level lies above the maximum of the density of states (DOS), close to the top of a Ni d band; the DOS is still rather high there.[974] The same is true for Co, which has one electron less

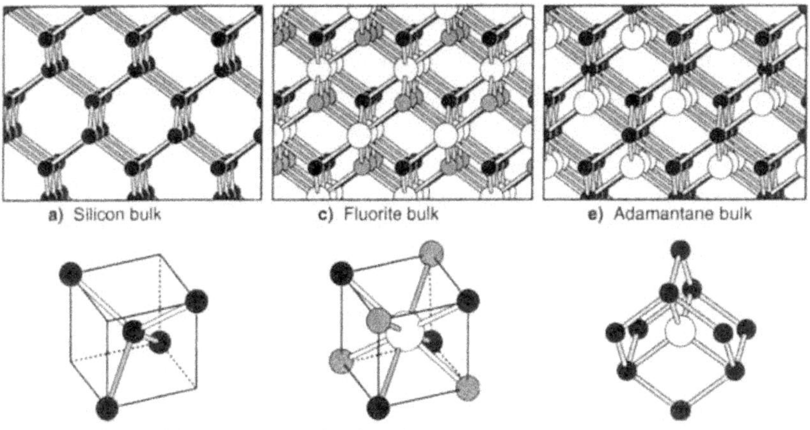

Fig. 5.9. Fluorite (CaF_2) and adamantane structures of $NiSi_2$ and $CoSi_2$ compared to that of silicon. Si atoms are black in the diamond lattice sites and gray in the interstitial sites, Ni (Co) atoms are white. The metal-Si bonds are less directional than the ball-and-stick models suggests. **(a)** Silicon bulk and **(b)** a $SiSi_4$ tetraheder of Si bulk, circumscribed by a cube. The empty corners of the cube are the tetrahedral interstitial sites. **(c)** Fluorite (CaF_2) structure and **(d)** a $NiSi_8$ cube of $NiSi_2$ bulk. Ni substitutes Si and is surrounded by diamond-lattice Si atoms and Si interstitials. **(e)** A hypothetical adamantane-type silicide and **(f)** "adamantane cage": interstitial Ni surrounded by Si lattice atoms (the formula of adamantane is $C_{10}H_{16}$). The Si lattice constant is larger than that of fluorite silicides and smaller than that of adamantane silicides.

than Ni ([Ar]$3d^7 4s^2 4p^0$); in fact, the remarks below apply to Co as well.

To illustrate the characteristic **atomic structures** of $NiSi_2$ and its interfaces, let us sketch a [110] cross-section through the Si crystal as 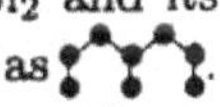. The disks are Si atoms and the lines are the bonds, shown for simplicity only within the area of interest. The [001] axis points upwards (like ↑). Only three of the four neighbors of the central Si atom are visible from this perspective (): one of the two tetraheder atoms aligned on the same [110] axis (the atoms with gray bonds in Fig. 5.9b) and the remaining two tetraheder atoms (the atoms with white bonds in Fig. 5.9b). All vertical bonds (such as 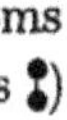) sketched further down represent thus in fact two bonds: one to an atom above, and one to an atom below the plane of the sketch.

Two structures have been proposed for thin $NiSi_2$ and $CoSi_2$ films: the bulk-like fluorite (Fig. 5.9c) and adamantane structures(Fig. 5.9e). A **fluorite**-type $NiSi_2$ crystal is built on the basis of the diamond lattice with (001) planes occupied alternatively by Si and Ni. The atoms replaced by

Ni move to tetrahedral interstitial sites surrounding the nickel (Fig. 5.9d).

Indeed, the diamond lattice contains a lot of empty space. The tetraheder of the diamond structure (Fig. 5.9b) can be viewed as a centered cube with only half of the corners occupied by atoms. The empty corners are the four tetrahedral interstitial sites. Each lattice atom in bulk silicon is thus surrounded by four Si neighbors and four interstitial sites. In the silicide, the metal atoms which substitute Si atoms in every second (001) plane of the Si crystal (the hollow circles in 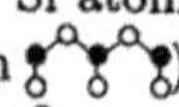) are surrounded by four lattice and four interstitial Si atoms: 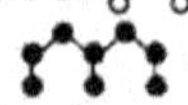 → 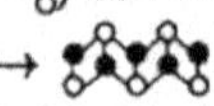 (Figs. 5.9a and 5.9d). Since every second Si atom of the diamond lattice is substituted by Ni, only half of the interstitial sites are occupied.

In a hypothetical **"adamantane"** structure,[q] Ni occupies the interstitial sites, while Si remains in the diamond sites:  → 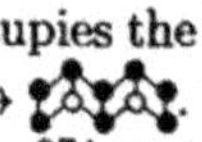. Though the bulk adamantane structure is energetically unfavorable,[974] there is a lot of evidence that a structure different to fluorite appears in $NiSi_2$/Si(001) and $CoSi_2$/Si(001) films (see Ref. 1008 for references). $NiSi_2$ or $CoSi_2$ with a non-fluorite structure may be stabilized by the presence of the surface, or may represent a metastable configuration which facilitates the transport of metal atoms during phase transformations or agglomeration.

It is not clear whether the actual film is adamantane-like or not. Experiments did not determine the atomic positions conclusively. The calculations may not necessarily be reliable, if only because the bonding in both structures is very different. Since in such cases the energy difference may be significantly affected by correlation effects, careful studies with state-of-the-art techniques are need. The experimental data may be dominated by surface contributions; the calculations were done for bulk phases.

In the following, we assume the fluorite structure. It appears to be established at least for epitaxial $NiSi_2$/Si(111) films.[969] Since the lattice relaxation in the interface region is constrained by the presence of bulk silicon and bulklike silicide, and since the calculations suggest a high instability of bulk adamante structures, we will also use the fluorite model to construct the $NiSi_2$/Si(001) interface.

In spite of the good geometric match, the interface energies of $NiSi_2$/Si(001) and $NiSi_2$/Si(111) are rather high. The reason is a perturbed bonding at the interface. Let us consider the (001) interface first.

[q] The molecule of adamantane ($C_{10}H_{16}$, or tricyclo[3.3.1.1^{3,7}]decane) forms a carbon "cage" closely resembling that in which the interstitial Ni is placed (Fig. 5.9f).

In $NiSi_2$ bulk, all Si valence electrons are used in bonds. The situation is different at the $NiSi_2$/Si(001) interface. One can sketch this interface simply by changing the lowest plane of Ni atoms into silicon ($NiSi_2$/Si(001) ≡ 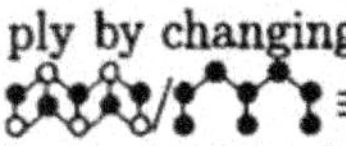 ≡ ). Since two nearest neighbors of each "interstitial" Si at the interface are changed from Ni to Si, two valences of such an atom are now unused. This resembles the unreconstructed Si(001) 1×1 surface. The interface Si atoms should thus join in pairs:  → 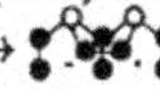), forming dimers similar to those on Si(001). The main difference is that the dimers belong to the $NiSi_2$ side of the interface and are backbonded to Ni atoms.

A $NiSi_2$/Si(111) interface has one *nickel* dangling bond per Si(111) 1×1 cell. This is because each Ni atom is surrounded by four Si intersitials, all in <111> directions. One of these neighbors disappears at the interface, so that interfacial Ni have only seven neighbors: $NiSi_2$/Si(111) ≡ 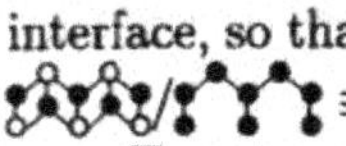 ≡ ; the dots mark the (111) boundary plane.

Two types of $NiSi_2$/Si(111) exist. The A-type interface is created when the diamond lattice sites in the silicon and in the silicide are in registry, while the B-type interface corresponds to a stacking fault at the boundary. The type of the interface can be deduced from LEED, RBS, and ion channelling data and can be controlled by the thickness of the initial Ni layer deposited on the surface.[969] What interface type will be formed is most probably determined by an interplay between interface energy and the reaction path. Pioneering *ab initio* calculations predicted[973] that A-type and B-type interfaces have nearly identical energies.

It is remarkable that even such a small difference in the interface structure as that between $NiSi_2$/Si(111)-A and $NiSi_2$/Si(111)-B results in a significant difference[1009] in the Schottky barrier height Φ_B: the value for type A is 0.65 eV, and for type B is 0.79 eV. Since a usual interface contains areas of both types, this difference explains why there is a spread in the measured Φ_B reported in the literature for $NiSi_2$/Si(111).

5.7.2. *Nickel on silicon surfaces*

Already small amounts of nickel, equivalent to tiny fractions of a monolayer, affect the reconstruction of Si surfaces. Since laboratory conditions make Ni contamination easy, the Ni–Si system has been extensively studied. Table 5.10 lists some bibliography. Ni coverages higher than a few per cent are unstable: first islands and then complete silicide films grow (Section 5.7.1b).

Nickel contaminates Si so easily[343] that touching the sample with a stainless steel tool brings enough impurities to invoke a 2×n superstructure

Ref year	Subject relevant to nickel on silicon
602 1985	Trace amount of Ni on Si(001) leads to 2×8 reconstruction; LEED
595 1988	Ni/Si(001) 2×8, dimer vacancy, split-off dimer; STM, LEED
1010 1989	Si(111), Si(001), Si(110), Ni diffusion, structures; LEED, AES
1011 1990	Ni/Si110) 5×1, 4×5, formation after annealing; STM, RBS
708 1994	**Book chapter**: Reconstructions, segregation, phase transformations
1012 1995	Nickel diffuses faster in the bulk Si than on the surface; AES
541 1996	Ni/Si(001) 2×8, split-off dimer, dimer vacancy, roughening; STM

Table 5.10. Nickel on silicon; selection of references.

on Si(001) or the 1×5 reconstruction of Si(110). Ni is more troublesome than other constituents of steel (Fe, Cr) because Fe and Cr form silicides which are vaporized during annealing while Ni survives even above 1200°C. Nickel can also arrive from a sample holder or can be backsputtered from walls of the UHV chamber. Thermocouple alloys can give off Ni vapor. Methods to keep surfaces free from nickel are described in Ref. 541.

Defying the common wisdom that surface diffusion is easier than diffusion in bulk material, Ni spreads over the surface migrating mostly through the bulk, apparently in the interstitial form.[1010,1012] The diffusivity and solubility of Ni atoms in silicon is very high. Ni particles dissolve in the crystal at temperatures above ~300°C; many Ni atoms then become immobilized at various traps in the volume.[1013] Flashes or high temperature anneals liberate them and subsequent cooling makes them segregate to the surface, due to a lower Ni solubility at lower temperatures. Distributed evenly, the surface nickel is now stabilized in ordered or quasi-ordered structures.

When more atoms are segregated than it is needed to saturate the surface, the excess impurities agglomerate in $NiSi_2$ islands. Cooling with a rate above ~50 K/s, that is, fast enough that surface diffusion is rather inefficient, leaves then a mixture of two phases: small $NiSi_2$ islands ~1 μm in diameter and the silicon reconstructed in a Ni-induced pattern. Slower cooling rates allow for the larger islands to grow at the expense of those smaller (Oswald ripening). Also the nickel from the surface structures becomes drawn into the islands and the phase picture changes: large islands (above 10 μm in diameter) are separated by the clean surface.[r]

Annealing/cooling cycles can thus lead to a dramatic increase of the Ni surface concentration Θ_{Ni}. The ultimate Θ_{Ni} depends on the number

[r] Since the 3D islands cover a tiny fraction of the surface, the Ni AES signal drops. The misleading impression is that the nickel atoms diffused into the bulk.

and the maximum temperature of anneals and on the cooling rate.[541]

The Ni atoms diluted in the Si bulk occupy predominantly interstitial sites[1013] (Fig. 5.9f, page 333). Also the Ni atoms *deposited* on silicon may possibly acquire such "adamantane" places in the subsurface layer.[1014] This interstitial nickel, while electrically inactive, is a source of compressive stress; indeed, the lattice constant of the hypothetical adamantane $NiSi_2$ would be larger than that of the bulk silicon by ~1%.[974]

However, neither the surface structures of Ni nor the atomistic mechanisms of their formation are really understood. Maybe the most clear case is the **Si(111)** $\sqrt{19}\times\sqrt{19}$-$R23.4°$–Ni, for which suggestive STM images and a compelling intertpretation exist.[1015,1016] This structure is most probably complete with three subsurface Ni atoms per unit cell; lower Θ_{Ni} coverages produce a disordered structure, possibly a mixture[1017] of the $\sqrt{19}\times\sqrt{19}$–Ni and the clean Si(111) 7×7 surfaces. The role played by Ni on Si(001) remains rather mysterious, while the models for Si(110) are either controversial or hardly cross-checked in independent studies.

The behavior of Ni on **Si(001)** may seem rather simple. Nickel introduces defects which appear as dimer vacancy clusters with a split-off dimer, (1+2)DV: a dimer divacancy and a single dimer vacancy are separated by a single Si dimer (Section 3.3.1a and Figure 3.29 on page 196). If the top view of Si(001) is sketched as , an STM image of the split-off dimer defect looks like . The abundance of these defects is directly proportional to the nickel coverage,[541] whereby each (1+2)DV corresponds to one or maybe two Ni atoms. At high concentrations, these defects form $2\times n$ lines closely resembling the $2\times n$ lines of the clean surface, only more regular and stable (Section 3.4.1; compare Figures 3.29b and 3.29c). The value of n decreases with increasing Θ_{Ni}, staying typically between 6 and 10.

The problem begins when it comes to atomic structures and to the role of nickel in their formation. Does Ni belong chemically to (1+2)DV, or does it merely lower the formation energy of dimer vacancies by exerting compressive stress? Are the (1+2)DV structures exclusively related to Ni, or are they simply thermodynamically stable complexes of usual missing dimers? An early survey[595] found that the double vacancy part of the complex, 2DV, appears "deeper" to the STM tip than the single-vacancy part. More so, the tip descended lower while crossing over the 2DV fragments of $2\times n$ rows than when traversing a monatomic step. The conclusion was that the 2DV trenches are *two* atomic layers deep. The bottom of such a trench should be dimerized, just as the lower terrace of a simple biatomic step, D_B-s (Fig. 3.18, page 175). The only difference is that the bottom of the trench would be a single dimer wide. Recent STM studies[544,607] confirmed that the trenches look deeper than a monolayer.

While biatomic-high walls and dimerized bottoms of the divacancy rows would explain at the same time the apperent depth of the trenches and the tendency of the (1+2)DV defects to line up, we believe that a shallow, monatomic-depth configuration (Fig. 3.13d, page 164) is more plausible. First, split-off dimers are a typical element of simple monatomic steps, S_B-s (Section 3.3.2a and Fig. 3.20b, page 179), while at biatomic steps they are not observed. Second, biatomic steps have a higher formation energy than monatomic steps. In addition, the simple D_B-s step edge, which would form the deep trench walls, is the less favorable of the two types of D_B; it is the other, rebonded version (D_B-r) which is formed on stable vicinal surfaces. Third, the bottom of an isolated, deep (1+2)DV defect would consist of a single Si atom between two highly energetic D_A steps. This would make this geometry even more unstable energetically. We are not familiar with a mechanism by which one or two Ni atoms could stabilize this geometry. Fourth, the shallow (1+2)DV was computed to be the lowest-energy configuration of a dimer trivacancy as well as of a dimer divacancy–vacancy pair (page 163).

On the other hand, the presence of the nearby nickel may influence the electronic states and/or the tunnelling barrier of the trenches, making the 2DV fragments *appear* deeper. After all, STM does not see atoms, it only detects the current which tunnels between the electronic states of the tip and of the sample. Finally, even if the evidence for the shallow (1+2)DV defect is only circumstantial, so is the evidence for the deep configuration; given this situation, we prefer to apply the Ockham's economy principle and stick – by default – to the simpler, "shallow" model.

The problem of the "depth" of the Ni-induced vacancies remains, however, open. The same concerns the fundamental question: are these vacancies "clean" or directly bonded to nickel? Interestingly, *titanium* atoms adsorbed on Si(001) seem to create dimer vacancies by forming direct bonds with the Si atoms at the vacancy bottom (Sect. 5.7.3). One cannot *a priori* exclude the possibility that nickel atoms stabilize the dimer vacancies (or at least some of them) in a similar way.

Reconstructions of other silicon surfaces are also dramatically affected by nickel. Approximately 0.01 ML of Ni changes the 2×16 periodicity of the step up – step down arrays of the clean **Si(110)** into 1×5, whereby the direction of the steps changes but the same structural elements as on the clean surface (Section 1.5.5) are probably retained.[1011] Higher nickel concentrations transform the 1×5 into such structures as 4×5, 5×4, 1×9, 1×7, or 1×2.[1018,342] The actual periodicity depends on the amount of Ni and on the substrate temperature. When Ni is added to or removed from the surface region, reversible transitions among these phases take place.

Similarly, the 3×2 reconstruction of the clean **Si(113)** surface (Section 1.5.4) is changed by nickel contamination[1019] to a mixture of 3×2 and 3×1 at $\Theta_{Ni} \simeq 0.004$ ML (0.4%), to a mixture of 3×1 and 2×2 around $\Theta_{Ni} = 0.016$ ML, and to a pure 2×2 around $\Theta_{Ni} = 0.02$ ML. The atomic structures of these Ni-induced reconstructions are not known, but we speculate that the 2×2 geometry corresponds to consecutive rows of Si(113) tetramers and adatoms (Section 1.5.4). This structure has a low surface energy already when the surface is clean and is likely to become favorable when the compressive stress produced by the interstitial nickel destabilizes the Si interstitials which make up the 3×2 geometry (Figure 1.8c, page 75). Since the results of our *ab initio* calculations indicate that the 2×2 structure has lower energy than the 3×1, and since the only difference between the 3×1 and 3×2 reconstructions is the presence of Si interstitials in the latter, we suggest that the 3×1 phase is metastable. It would appear due to the kinetic limitations imposed by the need to transport many Si atoms across the surface. A similar effect stabilizes, for example, the 5×5 reconstruction of clean Si(111) on surfaces with low miscut (Section 1.5.2).

5.7.3. *Titanium on Si(001)*

An interesting system which one may want to recall when discussing interactions between nickel or other transition metals and Si surfaces is an isolated titanium atom adsorbed on Si(001). STM images[1020] show that, at temperatures below 170°C, Ti adsorbs between dimers on dimer rows, that is, on the pedestal site H: 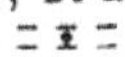. The adsorbate is rather strongly bonded at H. Indeed, the barrier for surface diffusion of titanium atoms adsorbed on top of Si(001) 2×1 is close to 2.0 eV (numerical *ab initio* result[103]).

When the temperature is raised above 170°C, the adsorbed atom changes its geometry: objects resembling dimer vacancies (DV, ) appear. We denote such a defect as **Ti–DV**. The center of each Ti-related dimer vacancy is *brighter* than the center of a clean DV defect.[1020]

What is the atomic structure of Ti–DV? Let us sketch the structure of an unreconstructed dimer vacancy DV*, seen from top, as:  (the dots mark the undimerized atoms at the vacancy bottom, cf. Fig. 3.12a, page 159). A plausible geometry of the Ti–DV defect is: 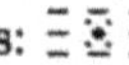. The Ti atom is bonded to the four Si atom at the unreconstructed vacancy bottom. We will call this adsorption site the DV* site.

Ab initio calculations show that the energy of Ti(DV*) is by ~0.1 eV lower than the energy of Ti adsorbed at the H site.[103] The energy barrier separating Ti(DV*) and Ti(H) in the reaction Ti(H) → Ti(DV*) + 2Si(M) → Ti(DV*) + 2Si(μ_{Si}) was computed to be ~0.3 eV lower than the surface diffusion barrier. Si(M) stands here for a Si monomer (Sect. 4.2.1a) and Si(μ_{Si}) is a Si atom incorporated at a surface step edge. This energy difference is larger than the uncertainty expected in this type of calculations.

If the Ti(DV*) model is correct, it means that that a titanium atom ejects two silicon atoms when it reaches the silicon surface. Such ejected silicon atoms have been observed in STM images.[1020] They form ad-dimers and ad-dimer islands. As expected, there are approximately twice as many ejected silicon atoms as Ti–DV defects.

Ti adsorption at the DV* site is energetically favored also over placement of the Ti atom in the third subsurface layer at the tetrahedral intersitial site (cf. Fig. 5.9f, page 333). The corresponding *ab initio* energy difference[103] is ~0.15 eV. Let us note that it is *not* a big number: kT is ~0.1 eV at 1000°C. With the binding energy as low as that, the titanium atoms would dissolve in the silicon bulk. But the binding of titanium at the surface may be somewhat stronger than the *ab initio* result suggests. Namely, the Ti atom occupying the tetrahedral interstitial site in the third subsurface layer may have a lower energy than the same intestitial atom in the bulk silicon. This is because the presence of a crystal surface helps to relax the mechanical stress around the impurity atom and because the overlap of strain fields caused by the surface reconstruction and by the presence of the impurity can produce an attractive force. As a consequence, the impurity becomes bonded to the surface by elastic interactions.[63]

On the other hand, one cannot exclude the possibility that the geometry of the Ti–DV centers and/or their relation to titanium atoms are not as simple as pictured above. A puzzling feature of the Ti–DV defect is that it diffuses along dimer rows with the diffusion barier of 1.8±0.1 eV, which is close to the barrier for migration of a clean dimer vacancy, 

(Sect. 4.2.1c). It is not clear why the incluence of the titanium atom adsorbed at the DV* site ($\Xi\,\Xi\,\Xi$) on the diffusion of the vacancy part of the complex should be so small.

5.7.4. *Transition metal oxides: new gate dielectrics?*

The semiconductor industry shows a growing interest in transition metal oxides.[1021] This stems from the expectations that such oxides can replace SiO_2 gate dielectric in deep-submicron CMOS technologies (see also page 355).

At first, the new dielectric will be used mostly to reduce the MOSFET gate leakage current. Within a couple of years from now, MOSFETs will be so miniaturized and, as a consequence, the gate dielectric so thin that electrons would tunnel through it if it was made of SiO_2. By very modest gate voltages of 1 V, the density of the current tunneling through a 1.2 nm thick SiO_2 (the thinnest gate oxide in a properly working CMOS transistor known to us) reaches a few ampere per square cm! Even when the SiO_2 film is twice thicker, 2.4 nm, the tunneling current can be a good fraction of mA/cm^2. Such high leakages mean high power losses and, possibly, reliability concerns (Sect. 6.6.2a). Power losses can be to some extent compensated by designing the circut in such a way that the transistor is powered only when it is processing data, but it is clear that the problem becomes obviated when the leakage current is reduced.

Later, within little more than a decade from now,[22] the gate "SiO_2" will have to be less than three (!) atomic layers thick (Fig. 1.4, page 11). This means that SiO_2 dielectrics will cease to be viable. A material with the dielectric constant of $\epsilon = 20$–30 and with the quality comparable to that of thermal SiO_2 (Sect. 6.6) will be needed. This material must be technologically compatible with the CMOS manufacture process (Sect. 1.2.2b). Several candidates are being considered in research, for example, zirconium-tin-titanium oxide[1022] ($\epsilon \simeq 60$), titanium dioxide[1023,1024] (TiO_2, with $\epsilon \simeq 25$), or tantalum pentoxide[1025,1026] (Ta_2O_5, with $\epsilon \simeq 25$; see Refs. 963 and 1027 for recent reviews). Gate dielectrics made of a material with such a high ϵ would be more than five times thicker than SiO_2 gate oxides and approximately three times thicker than Si_3N_4 gate dielectrics.

In most cases, transition metal oxides (TMO) are deposited by CVD. For instance, tantalum pentoxide films[963] can be made from tantalum ethylate precursor, $Ta(C_2H_5O)_5$ (see Refs. 1025, 1028, 1029, 1030, 1031, 1032, or 1033). Such CVD precursors as $Ta(CH_3O)_5$, $TaCl_5$, TaF_5, or

$TaC_12H_30O_5N$ have been adapted as well (see Ref. 963 for references and discussion). Oxygen is usually delivered in form of O_2 or N_2O, sometimes diluted in H_2 or an inert gas such as N_2. Deposition of Ta_2O_5 by sputtering[1034,1035,1036] and TiO_2 by the JVD process[18,1024] (Sect. 5.6.3) is also a considered as a possibility.

One of the concerns related to carbon-containing precursors in CVD is that the carbon content in the film should be kept low. Otherwise, the leakage currents will be much higher than expected.[1032] But a well-prepared TMO can reduce the leakage current by three to five orders of magnitude compared to the SiO_2 film with the same capacitance.[1033,1026,1024]

When the oxides are deposited directly on silicon, the density of interfacial traps is often high.[1023,1036] This would be reflected in poor performance of the transistors (Sect. 6.6). Interestingly, when Ta_2O_5 is deposited from TaF_5 precursor,[1037] the interface state and oxide charge densities can be kept quite low, at $7{\cdot}10^{10}$ $eV^{-1}cm^{-2}$ and $5{\cdot}10^{10}$ cm^{-2}. Other problems are associated with interfacial reactions between silicon and the oxide.[1038] For instance, although silicon does not reduce tantalum pentoxide,[1039] the substrate may be attacked by the oxygen from the CVD gas mixture and oxidized in a difficult to control way. The solution is to grow the TMO on a buffer film which assures a good interface to silicon and protects it from oxygen. One possibility is to use JVD nitride[1024] (Sect. 5.6.3). Maybe simpler options are to grow a nitridated oxide[1026] (Sect. 6.3.4), oxidized nitride,[1033] or oxynitride (Sect. 5.6.3). These processes are not too distant from the industrial practice. This means that TMO dielectrics may enter the CMOS process as extensions rather than independent substitutions of SiO_2. Therefore, the knowledge on the oxidation of silicon (next Chapter) is likely to remain technologically relevant even when SiO_2 gate dielectrics will eventually disapear from CMOS production lines.

As an example of a problem related to the presence of SiO_2 buffers between the dielectric and the substrate consider the fact that the capacitance of the buffer adversely affects the capacitance of the whole film. Indeed, the $TMO/SiO_2/Si$ stack corresponds to two capacitors connected in series. The effective dielectric constant of the film becomes

$$\epsilon = \frac{(t_{ox} + t_{tmo})\epsilon_{ox}\epsilon_{tmo}}{t_{ox}\epsilon_{tmo} + t_{tmo}\epsilon_{ox}}, \tag{5.29}$$

where t_{tmo} and ϵ_{tmo} are the thickness and the dielectric constant of the

TMO layer. The equivalent oxide thickness t^*_{ox} is now:

$$t^*_{ox} = t_{ox} + t_{tmo}\frac{\epsilon_{ox}}{\epsilon_{tmo}}. \qquad (5.30)$$

In practice, this means that the capacitance of the stack made of 1 nm SiO_2 and 20 nm Ta_2O_5 film corresponds to the capacitance of a 4 nm SiO_2 film. When the Ta_2O_5 thickness is reduced to 7 nm, the equivalent SiO_2 thickness becomes 2 nm. Scaling to equivalent oxide thicnkesses around 1 nm and below requires that the buffer thickness is further reduced. However, a 1 nm thick silica has only about three atomic layers of SiO_2...

Reactions at the interface between SiO_2 and Ta_2O_5 are also a potential problem, particularly when high-temperature (above ~800°C) anneals are involved.[1040] In this case, the interfacial region may become partially decomposed. Silicon and oxygen atoms diffuse from SiO_2 into Ta_2O_5, some oxygen is lost (possibly as H_2O and SiO), and tantalum atoms move away towards the bulk of the Ta_2O_5 film (at least, when the Ta_2O_5 film is oxygen-rich). Migration of Si into Ta_2O_5 is probably associated with substitution of Ta atoms by Si atoms, which leads to the creation of an acceptor level.[1041] Such high-temperature anneals are also likely to induce crystallization of Ta_2O_5. This effect is undesirable, because it leads to enhanced inhomogeneity and roughness. A Si_3N_4 buffer stabilizes the interface between Ta_2O_5 and silicon.[1040]

It follows that deposition of Ta_2O_5 directly on silicon or SiO_2 may be not practical for applications which require high-temperature anneals after the dielectric film has been prepared. Stability problems of this kind are typical in deposition of many TM oxides. A systematic analysis of thermodynamic stability of binary oxides, conducted using tabulated thermodynamic data (free energies of formation of a given elementary TM, its binary oxide, silicides, ternary phases $TMSi_xO_y$, SiO_2, and silicon) revealed that most binary transition metal oxides are unstable in contact with silicon at 700°C. Notable exceptions include Zr and Hf (ZrO_2, HfO_2), aluminum (Al_2O_3), most of alkaline earth, column IIIB, and rare earth oxides.[1042]

Those who admire the massive, rigid bone structures of dinosaurs should remember that jellyfish still enjoy their very secure ecological niche.

– Beau Sheil, "Power Tools for Programmers"

CHAPTER 6

On the Road to Devices: $SiO_2/Si(001)$

J. Dąbrowski and H.-J. Müssig

Physics of SiO_2/Si interfaces viewed from the perspective of basic science and semiconductor technology. Methods and mechanisms of oxide growth. Interface roughness, atomic and electronic structures, and interaction with foreign atoms. Influence of interfaces on MOSFET electrical parameters. Reliability issues.

6.1. Introduction

The end of the 20th century marks the beginning of the Silicon Age. It is the inexpensive silicon CMOS technology (Sect. 1.2.2) that initiated and maintains the information revolution. However, the particular role played by silicon in our civilization is not due to its "superior" properties as a semiconductor. Consider the mobility of charge carriers as an example. High mobility of carriers is important for a good high-frequency performance of microelectronic devices. The mobilities are higher in germanium than in silicon. So why do we have silicon rather than germanium memory and microprocessor chips in our personal computers and mobile phones?

Historically, germanium was indeed the favorite of the semiconductor industry. The spectacular success of silicon, the material which has eventually outsted germanium, is due to the superior properties of *silicon dioxide*, SiO_2, and to the quality of interfaces between SiO_2 and silicon. An oxide layer is a natural electrical insulator and a natural chemical protec-

Ref year	Selected books and reviews on oxidation of silicon
821 1967	**Book**: Physics and technology of semiconductor devices
1043 1978	**Book**: Physics of SiO_2 and its interfaces
1044 1982	**Book**: MOS physics and technology
1045 1986	**Book**: Low temperature oxidation
1046 1988	**Conference series**: Physics and chemistry of SiO_2 and SiO_2/Si
1047 1988	**Book**: Physics and chemistry of the SiO_2/Si system
1048 1988	**Review**: Physics and chemistry of the SiO_2/Si system
265 1988	**Book chapter**: O/Si(001) and O/Si(111), atomistic aspects
831 1991	**Fundamentals**: O/Si(001) and O/Si(111), atomistic aspects
1049 1993	**Book article**: SiO_2 films: composition, structure, morphology
242 1995	**Review**: Oxidation of silicon for modern microelectronics
268 1995	**Book**: MOS device physics, practical issues related with oxide layers
1050 1996	**Conference proceedings**:Physics and chemistry of SiO_2 and SiO_2/Si

Table 6.1. Oxidation: general references, books, reviews

tor. Germanium has failed chiefly because its oxide proved to be too active chemically to assure the required reproducibility and stability of the final device.[3] Silicon dioxide, an excellent electrical insulator, is not only easy to produce, but it is stable and chemically very inert.

Silicon dioxide plays an essential role in the semiconductor business. It fullfils such general technological requirements as:

- Chemical compatibility with the silicon substrate;
- Processing compatibility with the metals forming contacts to silicon (Ti, W, Pt) and with the materials wiring the chip (Al, W, Cu);
- Compatibility with lithographic processes;
- Chemical, physical, and structural stability during chip manufacture;
- Chemical homogeneity and low concentration of defects;
- Easy, efficient growth and deposition processes.

Many books and reviews have been devoted to the oxidation of silicon (see examples in Table 6.1). We do not attempt to rewrite this material. Instead, we focus on some interesting aspects of the oxidation process and on its technological role. Section 6.2 gives an overview of the properties of SiO_2 and of the importance of SiO_2/Si(001) films in a standard CMOS process (see Sect. 1.2.2 for an introduction to CMOS technology). Section 6.3 considers subjects related to practical oxidation. It describes oxidation methods, isolation techniques, and morphology of oxide films. Section 6.5 presents the atomistic view point of the formation of the SiO_2/Si interface and the growth of a thermal oxide. Finally, Section 6.6 addresses the quality of SiO_2/Si(001) from the perspective of CMOS devices.

Our knowledge of the systems discussed in this Chapter is often more practical than fundamental. This is natural: they are so complicated that the computational and experimental techniques able to access them on the atomic scale are still in the devolpment stage. But the technological weight of oxidation processes is significant. Relatively crude but nevertheless efficient methods to treat them evolved therefore in a natural way.

In fact, it seems that the best chances of survival in the rapidly changing environment of our technocratic world have such research methods which, as the jellyfish from the motto, are simple-minded enough to get food after an environmental disaster. Precise but expensive tools may one day share the fate of the dinosaurus, unless they develop a symbiotic relation to a jellyfish (Sections 7.3.3 and 7.3.4). *The ability to deliver high-quality knowledge is not a survival guarantee.* Knowledge becomes food only when sold, and it can be sold in sufficient quantities only when it is cheap enough and delivered fast enough to the right customer. Oxidation of silicon in particular and dielectric formation for CMOS applications in general is one of those areas of research where such sophisticated but at the same time high-cost approaches as *ab initio* calculations cannot stand alone. However, when combined with simpler tools, they may become extremely useful.[28]

6.2. Silicon Dioxide in Nature and Technology

As we said, an oxide layer is a natural electrical insulator and a natural chemical protector. SiO_2 films can be grown on silicon substrates relatively easily, are sufficiently inert chemically, and their dielectric properties are good enough to assist the development of silicon CMOS technology into a giant holding nowadays 95% or more of the microelectronic market value.

SiO_2 is so inert that silicon wafers which leave vendor sites are protected simply by very thin (10–15 Å, that is, a few atomic layers) oxide films. These protective layers are obtained by dipping the wafers in uncomplicated chemicals (Sect. 6.3.1). Later in the laboratory or factory, this oxide can be removed, for example, by a dip in HF solution.

The silicon industry uses silicon dioxide for such diverse purposes as:

- Primary passivation layers (0.5 to 1.5 μm thick) for chemical and electrical stabilization of the surface;
- Secondary passivation layers (0.5 to 1.5 μm) deposited for corrosion pro-

tection and as diffusion barriers on the primary layers, on metallization layers (phosphosilicate glass, $SiO_2 \cdot P_2O_5$), and on polycrystalline Si;
- Stress-relaxing pad oxides for nitride layers (0.003 μm);
- Cleaning (in sacrificial oxidation);
- High-quality gate insulators (around 0.004 μm) in MOSFETS;
- Insulation layers (0.5 to about 2 μm) between electrical interconnects and between active devices;
- Auxiliary masks (0.1 μm) for various applications;
- Planarization layers (0.05 μm, often of phosphosilicate glass);
- Diffusion sources for doping the subtrate.

SiO_2 and oxynitrides may be used to make optical waveguides on silicon chips,[1051] which may facilitate integrtation of optoelectronics with silicon microelectronics. Last but not least, SiO_2 precipitates in the substrate act as internal gettering centers for impurities, notably iron. This so-called relaxation gettering is the standard method used to keep impurities far away from the active regions of microelectronic devices.[a]

Diffusion sources was an important application of various silicate glasses. For example, P atoms from a $SiO_2 \cdot P_2O_5$ layer deposited on Si were allowed to diffuse into the substrate, where they acted as shallow donors. But this method of doping is difficult to control when applied to small devices and is often relatively dirty. It is now used mostly in the production of power devices, where junctions are too deep to be implanted, and in the cases when dopants must be distributed near to surfaces of deep trenches etched in the substrate. Recently, in-diffusion of dopants into Si from a silicate glass has been succesfully used to produce *ultrashallow* junctions; see, for example, Refs. 991 and 1052, and references therein. Formation of such junctions (less than $\sim$ 100 nm below the surface) by ion implantation is difficult[28] because the damage made to the crystalline lattice by the implanted ions caused a significant broadening of the dopant profile (page 421). However, recent advances in implantation techniques[33,34] may overcome this problem without reverting to diffusion from solid sources.

A notable exception in the chemical inactivity of silica is that SiO_2 does react with **aluminum**. Aluminum is used in "bitlines" which connect the devices into an integrated circuit. If allowed to interact with SiO_2, aluminum steals oxygen from it[919] and the bitline is partially converted

[a] As devices become smaller, the acceptable impurity levels are reduced. As a consequence, this classical relaxation gettering must be augmented by new techniques.[28]

into an insulating Al_2O_3. The silicon atoms liberated in the reduction of SiO_2 dissolve in aluminum, leaving behind them voids that deteriorate the SiO_2 layer. Fortunately, the oxidation of aluminum is self-limited and these effects are negligible when the bitline is thick enough.

As a matter of fact, the original metallization scheme was to deposit Al directly on silicon (for contacts) and directly on SiO_2 (for bitlines). Along with the progressive miniaturization, TiN or TiW liners separating Al from SiO_2 and Si were introduced. Their original purpose was to improve adhesion of Al, assure that the material beyond a certain point is not removed during etching, and adsorb light during lithography. The important additional role of these liners is now to prevent diffusion of Si atoms from SiO_2 films and from the silicon substrate into the aluminum layers.

A weakness of SiO_2 is that it is nearly transparent to **alkali metals**. The alkalis floating in the gate oxide can dramatically change its electrical parameters, or at least make the transistor change its parameters in time (page 358). Phosphorus deposited on top of a sealing SiO_2 layer produces a phosphorosilicate glass, $SiO_2{\cdot}P_2O_5$, which efficiently binds the alkalis.[821]

A somewhat similar problem is that certain dopant atoms (notably boron) may diffuse through ultrathin SiO_2 films, such as those serving as modern gate oxides (Sect. 6.5.5c). The solution is **nitridation**. The presence of nitrogen during oxidation or annealing of the oxide in a nitridizing atmosphere leads to the formation of oxynitride layers (Sect. 6.3.4). Such a layer makes the oxide practically impenetrable to boron atoms (Sect. 6.5.5c). At the same time, it improves the atomic-scale smoothness of the interface and allows one to control the oxide thickness.

6.2.1. *Silicon dioxide, SiO_2*

Silicon dioxide (silica) is the main component of more than 95% of the known rocks and constitutes 59% of the mass of Earth's crust. A number of atomic structures of SiO_2 are known,[1053] such as fused and amorphous silica and the fundamental crystalline polymorphs: quartz, tridymite, and cristobalite. Quartz is by far the most abundant polymorph. Our discussion will concentrate on the forms most closely related to thin SiO_2 films on silicon, that is, on vitreous silica, tridymite, and cristobalite.

Other silicas include coesite (a characteristic mineral of meteoritic shock), stishovite (the most dense form of SiO_2), keatite (metastable and artificial), melanophlogite (cubic, with many voids filled with compounds of carbon, sulfur, and hydrogen), and fibrous silica (lussatite and chalcedony).

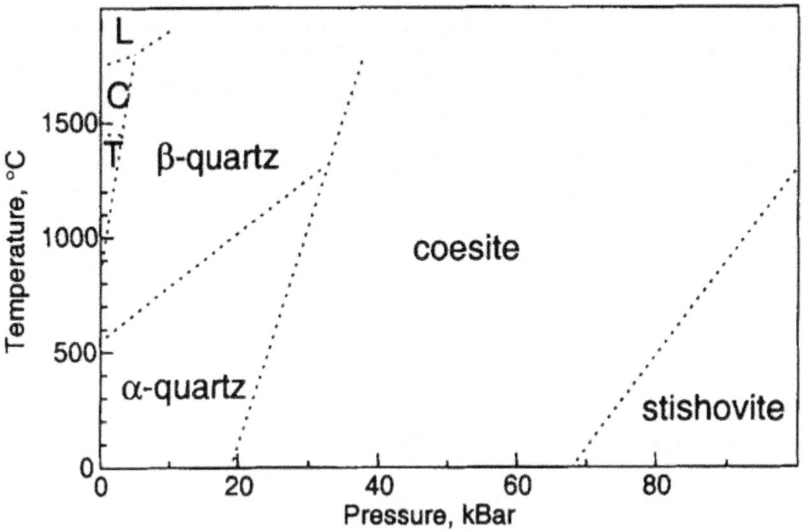

Fig. 6.1. Phase diagram of SiO_2. T stands for tridymite, C for cristobalite, and L for liquid silica. Adapted from Boyd and England.[1054] 1 atmosphere $\simeq$ 1 kBar.

Lussatite is a very fine-grained form of cristobalite; chalcedony, known for its usage by gem engravers, is a very fine grained form of quartz. The play of colours in opal comes from orderly arranged, spherical microcrystallites of hydrous cristoballite or hydrous tridymite (there is approximately 4–9% H_2O in opal), tens to hundreds of Ångstrøms in diameter. Quartz, trydimite, coesite, cristobalite, and stishovite are represented in the phase diagram (Fig. 6.1) together with liquid SiO_2.

Quartz, cristobalite, and tridymite have two allotropic forms each, labelled α (the low-temperature forms) and β (the high-temperature forms). If the pressure is held at one atmosphere, α-quartz is the stable polymorph at temperatures up to 573°C, β-quartz in the range 573–870°C, β-tridymite in the range 870–1470°C, and β-cristobalite above 1470°C until it melts at 1720°C. The low cristobalite and tridymite can be supercooled, whereby β-cristobalite transforms into α-cristobalite at 250°C (this metastable form can exist down to 200°C), while β-tridymite transforms into α–tridymite at 117°C. Mineralogists distinguish two types of β-tridymite: β_1-tridymite (which exists metastably in the range 117–163°C) and β_2-tridymite (the usual β form). Their atomic structures differ by details which are not completely established. We will use the generic names tridymite and cristobalite and we will focus on the major structural properties of these polymorphs.

The characteristic **atomic feature** of SiO_2 structures is that each Si atom is surrounded by four O atoms, forming a SiO_4 tetrahedron. Each O atom bridges two Si atoms: ⋗Si–O–Si⋖. This kind of bonding is known as "4:2". Minerals with the SiO_4 tetrahedron as the dominant structural element are known as **silicates** (for example, borosilicate glass, $SiO_2 \cdot B_2O_3$). The spatial arrangement of the tetrahedrons in SiO_2 determines the overall symmetry, which may be hexagonal (β-quartz, β-tridymite, coesite), tetragonal (α-cristobalite, keatite), cubic (β-cristobalite, melanophlogite), monoclinic (α-tridymite, coesite), orthorombic (fibrous silica), or none at all. Only stishovite is unique in that each of its silicon atoms is surrounded by six oxygen neighbors (rutile structure). This is the most dense, high-pressure, low-temperature polymorph (Fig. 6.1).

The degree of order in noncrystalline (vitrous or glassy) SiO_2 may differ in its various forms. Fused quartz (fused silica) exibits a short-range ordering of the SiO_4 tetrahedrons. In amorphous silica, which is a supercooled liquid, even short-range order is absent. The distinction between fused and amorphous silica is somewhat arbitrary. In the context of silicon technology, a reference to fused silica usually means a properly grown oxide, particularly thermal oxide (Sect. 6.3.2), while a poorly grown deposited oxide (Sect. 6.3.3) is likely to be described as amorphous. But we will use the term "amorphous" in the meaning "non-crystalline".

A typical Si–O bond in SiO_2 is 1.61 ± 0.02 Å long. This bond length is rigid, but the Si–O–Si angles are quite flexible.[1055] This flexibility promotes the variability of silica forms and probably is responsible for the good structural quality of SiO_2/Si interfaces.[b] The Si–O–Si angles vary between $\sim 150°$ in quartz, cristobalite and tridymite to 120° in coesite.

Since oxygen is one of the most electronegative elements in the periodic table, Si–O bonds are partially ionic. Each Si atom in SiO_2 is symbolically described as being in the +4 **oxidation state** (Si^{+4}) and each oxygen atom as in the O^{-2} state. The oxidation state (or number) should not be confused with the charge state of the atom. None of group VIa atoms, and certainly not oxygen, forms purely ionic bonds with nonmetals. Valence electrons of Si and O are shared in bonds, and electrons in a bond are formally assigned to the more electronegative atom, that is, to oxygen. The actual charge z on oxygen, computed as the excess charge on oxygen

[b] It is noteworthy that germania (GeO_2) has much fewer forms than silica.

orbitals by Mulliken analysis, depends on the Si–O–Si bond angle and may be much smaller than −2. Simple calculations[1056] indicate $z \simeq -1.3$ for ideal cristobalite (see below), $z \simeq -1$ for quartz, and $z \simeq -0.8$ for coesite.

For our purpose, the most interesting structures of silica are those of tridymite, cristobalite, and of amorphous SiO_2. The structures of tridymite and cristobalite are closely related to the structure of silicon. The Si atoms in the so-called ideal cristobalite occupy diamond lattice sites, and the oxygen atoms intervene in Si–Si bonds. Each resulting Si–O–Si bridge is a straight line, with an Si–O–Si angle of 180°. Similarily, ideal tridymite is built on the basis of the wurzite (ZnSe) lattice, with Zn and Se atoms substituted by Si and oxygen atoms located between Si atoms as in ideal cristobalite. The wurzite lattice is a close relative of the diamond lattice. In diamond, subsequent (111) planes are stacked as ABCABC, while in wurzite they are stacked as ABAB. In reality, straight Si–O–Si bonds are unstable and the Si–O–Si angles in real tridymite and cristobalite structures are close to 145°. The unit cells of their low-temperature forms are larger (that is, encompass more atoms) than the unit cells of the ideal crystals.

Like diamond and wurzite, crystals of tridymite and cristobalite can be viewed as being built of six-fold rings. By a "fold" we count an interatomic bond in diamond and wurzite and a Si–O–Si unit in cristobalite and tridymite. Amorphous silica has a more open structure. It contains rings built of three to nine Si–O–Si units.[1057]

The typical intrinsic structural **defects** in SiO_2 are the E′ center and the peroxy radical (PR) defect. The E′ center is an oxygen vacancy with a hole trapped on one of the two affected silicon atoms: $O_3Si\cdot\ {}^{+}SiO_3$. The precursor of this defect is a silicon bridge, that is, a rebonded oxygen vacancy: $O_3Si\text{–}SiO_3$. The PR defect is a hole trapped by a negatively charged O_2 peroxy molecule: $O_3Si\text{–}O\text{–}O{:}^{+}\cdot SiO_3$. Its precursor is the Peroxy Linkage Defect (PLD): $O_3Si\text{–}O\text{–}O\text{–}SiO_3$. A similar defect, related to the presence of hydrogen or water, is the Nonbridging Oxygen Hole Center (NBOHC): $O_3Si\text{–}O{:}^{+}\ H\text{–}O\text{–}SiO_3$. For the literature on defects in SiO_2, see Refs. 835, 1058, and 1059.

Typical for SiO_2 defects is a strong electron-phonon coupling. The defects change their atomic structure in response to capture of electric charge. The reader may be familiar with a similar behavior of certain defects in semiconductors. The established examples are the Si vacancy in silicon[554,1060] and the DX center in GaAs and GaAlAs alloys.[556,1061] This interesting property of SiO_2 defects plays a role in the response of the oxide film and the SiO_2/Si interface to the current leaking through the gate oxide of a transistor (Sect. 6.5.5) and, possibly, in the process of formation of a space charge in the oxide during thermal oxidation of silicon (Sect. 6.5.4).

Densities of the different varieties of silica vary in a wide range,[1055] from 2.2 g/cm^3 (vitreous silica) and 2.3 g/cm^3 (α-cristobalite), through 2.6 g/cm^3 (α-quartz) and 2.9 g/cm^3 (coesite), to 4.3 g/cm^3 (stishovite). Since silicon contributes about 50% of the mass of SiO_2, and the density of silicon bulk (2.33 g/cm^3) is comparable to a typical SiO_2 density, it follows that the partial volume of Si in SiO_2 is about twice as large as in bulk silicon. It is interesting to note that the lightest form of SiO_2 is a commercial form of silica gel, a highly porous, amorphous material used mainly as a desiccant. The porosity of silica gels gives them a bulk density of about 0.7 g/cm^3.

The density of SiO_2 is related to the **refractive index** n. A higher density results in a higher n. The density of fused silica[4], 2.20 g/cm^3, corresponds to $n \simeq 1.46$ for light with $\lambda \simeq 0.5$ μm and to 1.55 for $\lambda \simeq 0.2$ μm. The **dielectric constant** $\epsilon = n^2$ of fused silica[4] is 3.75 at 100 MHz and 4.1 at 1 MHz. A low-density SiO_2 with $\epsilon \simeq 1$ (xerogel) has been recently used as low-ϵ interlayer dielectric[1062] (cf. Sections 1.2.2b and 6.3.7).

The **resistivity** of SiO_2 grown on silicon[1063] exceeds $10^{16}\Omega$cm. Optimized annealing techniques (Sect. 6.3.6) can keep the density of fixed oxide charge and bulk neutral traps in the 10^{10}/cm^2 range (for the terminology, see page 355). The density of interfacial charge traps is in the 10^9–10^{10}/cm^2 range, as estimated from the density of states at the mid gap of bulk Si.

Similarly to silicon surfaces terminated by hydrogen atoms, clean SiO_2 **surfaces** are not wetted by water, that is, they are **hydrophobic**. However, this is only true for *dehydrated* SiO_2. In contrast, a typical SiO_2 surface contains approximately five OH bonds per nm^2. This is enough to make the surface **hydrophillic**. When a piece of silica with a hydrophillic surface is immersed in liquid water, it becomes covered by an ice-like, three monolayer thick film of hydrogen-bonded H_2O molecules.

6.2.2. *SiO_2/Si(001): MOS technology views and limits*

Oxide layers and Si/SiO_2 interfaces critically influence parameters and performance of MOS transistors.[1064] The most obvious examples refer, naturally, to the quality of gate oxides and to interactions between dopants in the substrate and in polysilicon gates with the pad and gate oxides.

First, irregularities of the electric field due to the roughness and to electrically active defects facilitate dielectric breakdown of the gate oxide (Sect. 6.6.2a). These defects can be native to the system (broken bonds, misplaced atoms), caused by contamination (metal atoms), or by dopants (segregation of boron from the gate to SiO_2, Sect. 6.5.5c). The electric charge captured "forever" on such defects, the so-called oxide charge, induces unfavorable changes in the electrical parameters of the device[21,1065] (Sect. 6.6). Electrically active defects, which are closer to the interface than the sources of the oxide charge, can trap and release carriers, producing a low-frequency noise[1066] (Sect. 6.6.2c).

Second, the interaction between SiO_2 and the dopant atoms affects the spatial distribution of dopants below and above the gate oxide.[1067] This influences electrical parameters of the transistor. For instance, "shoveling" of dopants by the oxidizing interface (dopant pileup) and segregation of dopants from the channel to the interface increases the threshold gate voltage V_T by which the device switches, contributing to the so-called reverse short-channel effect (Sect. 6.5.5c). The redistribution of dopants is reinforced by enhanced diffusion of dopants induced by point defects (Si self-interstitials) emitted from the oxidizing interface[c] (Oxidation Enhanced Diffusion, OED) and produced during implantation (Transient Enhanced Diffusion, TED).[85] Even as much as 50% of the implant dose can be lost to the oxide! Moreover, stress fields induced in the substrate by oxide strips[1068] may affect dopant diffusion in regions adjacent to the oxides. In addition to the huge volume mismatch between SiO_2 and silicon, there is a five-fold difference in thermal expansion coefficients. This easily produces compressive as well as tensile hydrostatic stress in silicon. This stress is of the order of 10^9 dyne/cm^2 (100 MPa).[1069] Stresses of this magnitude can change diffusion properties of a dopant.[1070]

Finally, scattering by atomic-scale roughness of the SiO_2/Si interface and by electric field irregularities caused by charged defects in the oxide and at the interface can significantly degrade the mobility of carriers moving immediately below the gate oxide.[1071,842] This leads to a decreased maximal current ("drive current", I_{Dsat}) which can flow between the source and the drain when the transistor is in its ON state. As a consequence, interface roughness increases the time needed to re-load the capacitors, decreasing the operational speed (Sect. 6.6.2b).

Interfacial defects are generated in the production line during gate

---

[c] They are emitted to accommodate the increase in molar volume[1072] (Section 6.5.5b).

oxide growth and during subsequent technological processes. Examples are lithography, which can produce irradiation damage, and electrodeposition of metals and plasma processing (for instance, plasma eching or plasma-assisted deposition), which can overbias the MOS capacitors[37] and reduce the final product yield by 1-7%, or by as much as 50% if a very damaging process is used.[38] The well-assessed reason for such damage is local charging caused by plasma inhomogeneities or shadowing effects. Ultraviolet light produced during plasma processes may also lead to oxide degradation.[1073] The bias induced by an accidental electrostatic discharge can lead to a serious damage or weakening of the oxide.[1074] The interface and the oxide are changed by processes conducted at temperatures high enough to allow for diffusion and segregation of dopants to the interface (Sect. 6.5.5). In particular, diffusion of boron atoms through SiO_2 is facilitated by the presence of fluorine or hydrogen atoms in the film. Fluorine can originate, for example, from HF etch (Sect. 5.3.1), from WF_6 molecules used in the polycide process (Sect. 5.7.1a) or from BF_2 implants (Sect. 1.2.2b).

Defects are also generated during the operation of the device (Sect. 6.6.2a). For instance, carriers from the current that leaks through the gate oxide may cause damage deteriorating the parameters of the transistor or limiting its life time. Some carriers may acquire enough energy in the MOSFET channel to be injected into the oxide[1075] or to damage the interface. This can happen even at low applied voltages.[1076]

Interactions between the oxide and the silicon substrate are also important in regions other than the critical gate oxide layer. An integrated circuit contains an array of devices which are separated by "field oxide" (FOX) isolation gaps (Fig. 1.2). Naturally, this isolation is not ideal and some leakage current can flow between two neighboring transistors. This current depends not only on the spatial separation between the devices, but also on the dopant profile below the FOX.[1077] As in the case of gate oxides, this dopant profile is influenced by dopant segregation and misfit-induced or thermal strain. In addition, a high strain field may be partially relieved by formation of extended defects, which is a very undesirable effect.

Modern technologies can handle SiO_2/Si films and interfaces very efficiently. This is a result of many decades of intensive research and practical experience. Let us only mention that, in the dawn of semiconductor microelectronics, the density of the oxide charge was close to 10^{13} per cm^2 (one charged defect per hundred interface atoms), making n-channel MOSFETs not practical: such a MOSFET would be ON even with no gate voltage applied. Today, values as low as a few 10^9 per cm^2 (a few charged defects per million interface atoms) can be achieved[d] This is as if a city of the size of New York had only about a score of criminals.

The charge traps in the gate oxide are of a particular practical interest to us (Sections 6.6 and 6.3.6). Formally, they are divided into:[1078]

1. **Interface trapped charge**, localized at the interface on states within the energy gap of bulk Si. Dry oxidation produces 10^{11} to 10^{12} states per cm^2eV in the mid gap. This density can be reduced by conducting the oxidation at a higher temperature, but *not by low-temperature annealing in inert ambient.* The density of trap states at SiO_2/Si(001) is usually about three times smaller than at SiO_2/Si(111). N_2 pre-annealing may bring it down to the 10^9 range.[1079] Post-oxidation annealing in H_2 (Sect. 6.3.6) can reduce it even to a few times 10^9 $cm^{-2}eV^{-1}$.
2. **Fixed oxide charge**, usually positive, localized within 30–100 Å from the interface. It does not vary with the surface potential, that is, with the applied voltage. This charge is apparently associated with non-stoichiometric SiO_2 and may be due to ionized excess Si. It can be reduced by annealing in weakly oxidizing ambients (Fig. 6.2, page 376). Very high oxidation temperatures (close to 1200°C) keep it at a minimum, which is $2{\cdot}10^{11}/cm^2$ for Si(111) and $1.1{\cdot}10^{11}/cm^2$ for Si(001).
3. **Oxide trapped charge**, bulk charge is associated with defects in SiO_2. It can be positive or negative. In contrast to the fixed charge, it changes with the surface potential. The defects may arise from ionizing radiation, hot carrier injection, high currents through the oxide, or other reactions that break Si–O bonds or lead to carrier trapping on the existing traps. It can be annealed out at low temperatures (450–500°C).
4. **Mobile ionic charge**, positive, mostly from Na and K. It can and should be kept below $10^9/cm^2$. Precautions include: clean processing, gettering, neutralization of Na by oxidizing in chlorine-containing ambients, using Si_3N_4 or oxynitride passivation films which act as diffusion barriers for Na, routine tests for Na contamination.

Atomic smoothness of the interface correlates with the atomic smoothness of the surface before oxidation and with the thickness variations of the grown film. Thickness variations of ultrathin gate oxides are becoming

[d] The number of all defects is much larger, but their great majority is electrically inactive. Such defects do not affect the performance the device directly.

one of the **main problems** in CMOS oxidation.[22,1080] As CMOS devices are smaller and smaller (Sect. 1.2.2a and Fig. 1.4 on page 11), the oxide thickness, t_{ox}, is reduced as well, approaching the regime in which direct tunneling of electrons is possible (Fig. 6.12, page 435). Since the current of carriers tunneling under a potential barrier depends exponentially on the barrier width, that is, on t_{ox}, a local thinning of such an oxide can strongly enhance the leakage current.[991,1081] Moreover, this current will flow through a small area of the oxide film, increasing the probability that dielectric breakdown will occur (Sect. 6.6.2a). Other key issues of Si oxidation for CMOS are those related to isolation techniques (Sect. 6.3.7) and the penetration of boron through ultrathin gate oxides (Sect. 6.5.5c).

When will the CMOS technology arrive at the oxide tunneling regime? Very soon (Fig. 1.4, page 11). Direct tunneling begins[270] around t_{ox} ~3 nm (about ten atomic layers of SiO_2). Contemporary state-of-the-art technologies (180 nm minimum feature size) use t_{ox} between 3 and 4 nm. Integrated circuits made with a 180 nm technology need t_{ox} between 2 and 3 nm; such chips should appear on the market in the year 2002. In the year 2005 the minumum feature size will be 100 nm, with t_{ox} between 1.5 and 2 nm.

Scaling the oxide thickness has its natural limit:[1082] three atomic layers of SiO_2. Thinner SiO_2 makes just two adjacent SiO_2/Si interfaces. But before this limit will be reached, the gate leakage currents will become unacceptably high (Sect. 6.6.2a) and SiO_2 will have to be substituted by another dielectric which can be made thicker. This alternative dielectric will be characterized by a high dielectric constant ϵ. A material with a high dielectric constant is needed because the reason for the reduction of t_{ox} is that the capacitance C_{ox} of the MOS capacitor must increase concurrently with the miniaturization. Indeed, the power supply voltage must be reduced in order to prevent hazardous increase of electric fields in a smaller device; the electric current drivability would then decrease if C_{ox} did not increase. Since $C = \epsilon/t$, where t is the thicknes of the dielectric, the dielectric film made of the material with $\epsilon = n\epsilon_{ox}$ will have to be n times thicker than the corresponding SiO_2 film ($\epsilon_{ox} \simeq 4$, Sect. 6.2.1). A plausible candidate for the immediate follower of SiO_2 is JVD Si_3N_4 (Sect. 5.6.3); also Al_2O_3 is a promising material.[1083] Gate dielectrics made of Si_3N_4 or Al_2O_3 will be nearly twice as thick as their SiO_2 counterparts. But this is only a temporary solution. A pursuit for a suitable dielectric with $\epsilon \simeq 20$ has already started (page 341). In any case, since many of these new materials are oxides and since /silica/Si(001) interface has good electrical properties, it is likely that an ultrathin silica film (maybe only one monolayer thick) will act as a natural buffer between the dielectric and the silicon substrate at least in the first implementations of these new technologies.[e]

6.3. Preparation of Oxide Films

Oxide films on silicon can be produced either by deposition (the deposited SiO_2 layer contains no Si from the substrate) or by oxidation (the upper layers of the substrate are converted into SiO_2). In the latter case, one can use reactions with aqueous solutions of chemicals containing oxygen. Alternatively, the silicon surface can be exposed to a reactive atmosphere containing oxygen. Methods which do not involve water are known as "dry" oxidations, in contrast to "wet" methods. In MOS manufacturing, oxidation in aqueous solutions takes place during cleaning. Steam or dry oxidation at high temperatures (above ~800°C) is used to grow oxides which remain in contact with silicon. Finally, deposition is used for isolation, masking, planarization, and filling. We briefly describe all these methods, focusing on dry processes. A selection of references is listed in Table 6.2.

Of main interest to us are gate oxides. As a consequence of CMOS miniaturization (Sections 1.2.2a and 6.2), these oxides will be within the next few years (Fig. 1.4, page 11) scaled down to thicknesses of only a few atomic layers. Therefore, studies of the initial oxidation are expected to yield important information on the critical processes which determine the morphology and microscopic defects in the film. When the SiO_2 layer grows, oxygen must penetrate through the oxide to the oxidizing interface. The uniformity of the interface is, thus, dependent on the uniformity of the oxygen supply to the moving SiO_2/Si boundary. The latter depends on the smoothness of the surface and on the uniformity of the oxide formed so far.

Indeed, if **atomic-size smoothness** of the interface is required, care must be taken to control the quality of the initial surface of the wafer. The native oxide (Section 6.3.1) is a source of roughness. It should be either carefully annealed before oxidation[1079] or stripped competely.[844] The best surfaces are prepared by UHV flashing and annealing (Section 1.4.2c). Though this procedure is useful only under laboratory conditions, it can

[e] JVD Si_3N_4 was also proposed as a buffer film,[1024] in TiO_2/Si_3N_4/Si(001).

Ref year	Subject relevant to preparation of oxide films
1084 1965	Growth kinetics of thick oxides: linear-parabolic (Deal-Grove) law
1043 1978	**Book**: The Physics of SiO_2 and Its Interfaces
1085 1981	Oxidation Enhanced Diffusion (OED)
1044 1982	**Book**: MOS Physics and Technology
1086 1985	Growth rate enhancement in thin oxides
1087 1985	Pinholes in SiO_2 films
1088 1987	Initial oxidation rate for five different orientations
1089 1987	Excellent SiO_2/Si(001) interfaces can be grown; TEM
1090 1988	**Review**: Rapid Thermal Oxidation (RTO)
1091 1991	**Book**: Silicon-on-Insulator (SOI) technology
834 1992	Thermal decomposition of ultrathin SiO_2/Si(001); TPD, isotopes
1079 1992	Annealing before oxidation smoothens the surface, greatly improves film
1049 1993	**Book article**: SiO_2 films (composition, structure, morphology)
1092 1993	**Book**: Practical simulation of the oxidation process
1093 1993	**Book**: Physics and technology of MOS transistors
1094 1994	Control of native oxides; XPS
844 1994	SiO_2/Si(001) interface for different oxidation processes; TEM
1095 1995	Quality of SiO_2/Si(001) films on implanted material; TEM
268 1995	**Book**: Growth and reliability of gate oxides
955 1995	N implantation for oxide thickness control
842 1996	Initial surface quality and interface roughness; TEM, STM, AFM
1096 1997	Roughness and thickness control of ultrathin (below 3 nm) oxides
1097 1998	Wet Rapid Thermal Oxidation for gate dielectrics
1098 1998	**Review**: MOS devices with gate oxides in direct tunneling regime
1099 1998	**Review**: Reliability problems with extremely thin gate oxides

Table 6.2. Preparation of oxide films: selection of references

access the regimes of initial roughness which may be reached in production only in many years from now.

In addition to roughness, a key issue is **cleanness** (Sect. 1.4.2c). Impurities may have a dramatic influence on the performance of the gate oxide. Organic contaminants facilitate dielectric breakdown of the oxide.[239] This effect is most probably due to the formation of stable silicon carbide precipitates on the surface of the wafer and depends on the hydrocarbon species involved.[253,239] Alkali and alkaline-earth metals[268] such as Na and Ca dissolved in SiO_2 were identified as the source of mobile charges which plagued the dielectric in early MOS. One of the most spectacular and most meaningful successes which made the modern CMOS technology feasible was the containment of this contamination. Implementation of research results[1100,1101,1102,821] allowed a reduction in the areal density of oxide charges[f] from nearly $10^{13}/cm^2$ (by which MOS technology is hardly

[f] See Sect. 6.6 for a discussion of the practical importance of these and other defects.

possible) to barely above $10^9/cm^2$ (which is almost negligible). The major contribution to this success came from cleaning procedures.

Oxygen content in the subsurface region of the wafer is also an important quantity. Oxygen is needed in the wafer, because SiO_2 precipitates act as internal gettering centers. They keep the active regions of the devices clean from metal impurities.[28] However, when these precipitates are too close to the growing gate oxide, its quality will be diminished. This is the reason why epi-wafers (wafers covered with an epitaxial Si layer) are better than wafers simply cut from the crystal. In contrast to the bulk material, epitaxial layers can be considered free from oxygen.

The simplest one-step method best suited for fabrication of ultrathin (below 10 nm) films with high-quality interfaces as needed for gate oxides is **dry oxidation**. Relatively thick (~2 μm) oxides for LOCOS isolation (page 377) are made more efficiently by **steam** oxidation, which is faster than dry oxidation, typically several times. Steam or dry oxygen can be used to grow thin (below 100 nm) protective pad oxides. A steam oxidation procedure for ultrathin gate oxides has been recently proposed (page 364). Thermal oxides in shallow trench isolation (page 378) can be grown in this manner, as well. The thermal oxide covers the trench walls, while the trench volume is filled by **deposited** material. The growth of extremally thin (~1 nm) oxides is efficiently controlled by **nitrogen** content (Sect. 6.3.4).

Below, we introduce the major oxidation methods, including remarks on the formation of native oxides, deposited SiO_2, and isolation techniques. We also mention some issues related to the morphology of oxide films.

6.3.1. *Native oxides and oxidation by chemical etch*

Oxide is intentionally or unintentionally grown between different process steps on surfaces exposed to air, water, chemicals, process gases, or even to vacuum.[1094] This is the so-called **native oxide.**

Native oxides protect the sensitive Si surface from contamination between technological steps. However, they must be carefully removed before epitaxial Si or SiGe layers are grown. A native oxide can be a source of contamination from metallic impurities. Moreover, since the native oxide growth is non-uniform, the surface is likely to be much rougher after the removal of the native oxide than it was before. Thus, advanced technological processing of Si wafers must include some control of native oxidation.

Up to 2 nm (20 A) of not necessarily stoichiometric oxide grow during a day on an initially clean wafer stored in air under normal conditions.

In a dry atmosphere this process is even slower, around 0.1 nm (1 Å, or less than a monolayer) per day. The charge trap density at the interface between silicon and the native oxide grown in humified air at room temperature after HF etch is $\sim 2\cdot 10^{13}$ $cm^{-2}eV^{-1}$ at the mid gap.[1103,1104] This is approximately ten times more than after thermal oxidation process of industrial quality. The growth of native oxides is practically self-limited and longer storage does not lead to significantly thicker films: the growth continues on a logarithmic time scale.[1046] In very dry air, room temperature oxidation reactions are limited to the surface atoms.[867,1105]

Water is a strong oxidant, apparently due to the chemical activity of OH and hydrogen (Sections 5.5.3 and 6.5.5d). Silicon oxide grows, albeit slowly, even when the sample is immersed in room-temperature ultrapure water with the dissolved oxygen concentration in equilibrium with air.[1105]

In practice, water is used in two entirely different oxidation methods. It is an efficient active oxidant when applied as **steam** to a silicon surface at red heat (page 363). Also numerous **wet chemical** methods exist in which water acts mostly as a solvent for the oxidizing chemical. Silicon wafers are habitually protected by native oxide layers produced in such a solution. These native layers are removed and re-grown as many times as needed in the technological process.

A protective chemical oxidation is made as soon as the wafers are cut from the Si crystal and polished. The corresponding chemical procedure has two purposes: clean the wafer surface of contaminants (such as microparticles and hydrocarbon molecules; see Sect. 1.4.2c) and passivate it against the influence of the environment. The details of these procedures are proprietary, but the surface is likely to be treated in a way resembling the RCA etch described below.

An **RCA etch**[1106] (named after the patent holder, Radio Corporation of America, and known also as Kern cleaning) consists of two dips. The wafer is first immersed in an alkaline, ammonia-peroxide solution (hot water solution of H_2O_2 and NH_4OH), which removes particles from the surface. This step is also known as SC-1 or Ammonia-Peroxide Mixture (APM). Next, it is treated by an acidic peroxide solution (hot water solution of H_2O_2 and HCl), which removes metal precipitates from the surface (Sect. 1.4.2c) and completes the oxidization. This is known as SC-2 etch. The passivating oxide layer can be later stripped, for instance, by etching in HF:H_2O or buffered HF (that is, HF:NH_4F:H_2O). The HF solutions dissolve the oxide and temporarily protect the surface with hydrogen atoms (Sect. 5.3.1). Alternatively, an RCA-like etch can be applied again to remove the old oxide and produce a fresh, ultrathin SiO_2 layer. But it is expected[235] that in CMOS manufacture, an HF-based sequence which uses a diluted HF:H_2O or HF:HCl:H_2O etch and H_2O/O_3 rinse/oxidation steps (IMEC clean[244] or dilute dynamic clean,[249] Section 1.4.2c) will substitute RCA in the procedure of surface cleaning prior to gate oxidation.

Other chemicals are often used, as well, for example, sulfuric acid (usually in form of Piranha etch: a hot 4:1 mixture of H_2SO_4:H_2O_2) or HNO_3. We cannot discuss all these chemical methods of oxidation in detail. Let us only mention that a properly designed chemical treatment can produce a very smooth surface. Surface roughness of oxidized commercial wafers is routinely[1049] only about 2 Å rms. Chemicals which etch silicon (such as the basic SC-1) contribute to roughness more than substances which oxidize but do not etch (such as the acidic SC-2).

Chemical oxide films grow particularly well[1107,1108] on Si(111). Can one use such films as gate oxides? Very smooth ultrathin oxide layers can be obtained by chemical oxidation in H_2SO_4:H_2O_2, HCl:H_2O_2, or HNO_3 solutions. The interface trap densities (see Sect. 6.6 for the importance of interface traps) are[1108] then in the range of 10^{12} $cm^{-2}eV^{-1}$ for Si(111); for Si(001) oxides the trap densities are somewhat higher. Such values are comparable to those obtained in standard thermal oxidation (cf. Fig. 6.2 on page 376). However, trap densities in this range are too high for gate oxides and must be reduced by a post-oxidation treatment (Sect. 6.3.6). The trap density at SiO_2/Si(001) decreases to $\sim 10^{10}$ $cm^{-2}eV^{-1}$ in the course of low-temperature annealing in H_2 (Sect. 6.3.6). This seems to be the thermodynamical equilibrium value. Since the thermodynamical equilibrium density of interface traps is higher for Si(111) oxides than for Si(001) oxides, it is at the moment not clear that chemical oxidation of Si(111) may become a useful procedure to produce gate oxides.

Moreover, some Si(111) chemical oxides are likely to suffer from realiability problems due to the presence of hydrogen in the film (Sect. 6.6.2a). Though XPS measurements found that H_2SO_4:H_2O_2 oxides are practically free from hydrogen, (see Ref. 1107 and references therein), HNO_3 and HCl:H_2O_2 oxides do contain hydrogen.[1107] In addition, the density of chemical oxides is apparently lower than thermal oxides,[1108] which also makes the former inferior as gate dielectrics.

The mechanism of roughening by wet etching is possibly similar to the step pinning mechanism discussed on page 295 in the context of the initial stages of dry oxidation. Roughening is, therefore, due not to the etching itself, but rather to local inhomogenities in the etching rate. Certain impurities in the solution, like Ca or Fe, can make the etching process very non-uniform and, thus, enhance the surface roughness considerably.[1049]

6.3.2. *Thermal oxidation*

In contrast to the chemical methods, which require temperatures of around 100°C and produce an ultrathin native oxide, the temperature needed to make a thicker oxide within a reasonably short time is much higher. Such oxidation is conducted in a reactive atmosphere consisting of O_2, often with an admixture of HCl and sometimes with a sizeable H_2O content.

Oxides grown above 400°C are called thermal (TOX). Temperatures during industrial oxidation usually range from 800°C to 1100°C. At 900-1000°C, a 100 Å (10 nm) thick film may take several minutes to grow.

Oxygen molecules interact chemically with clean silicon surfaces even at the temperature of liquid nitrogen, but no oxidation takes place at such low temperatures. The O_2 molecules are chemisorbed, but do not dissociate (Sect. 5.4.2). The oxidation rate is slow at room temperature; a chemisorbed molecule needs many hours to dissociate. The oxidation rate can be enhanced by tricks like pre-bombardment of the surface with Ar^+ ions,[1109] but this does not solve another, more severe problem. Namely, since the full oxidation reaction path includes a number of thermally activated processes, kinetic bottlenecks lead to non-uniformity of low-temperature films, rendering them useless as gate oxides. A better solution, with possible industrial applications, is low-temperature remote plasma oxidation.[1110]

High temperatures facilitate the breaking of Si-Si bonds. Local, atomic-scale smoothness of the interface is also improved. The transition layer of non-stoichiometric oxide at the interface to silicon (Sect. 6.5.2a) becomes narrowed to a nearly monatomic thickness. Such parameters as the mesoscopic interface roughness and the fixed charge density (page 355) are improved as well. A complete growth cycle ends with annealing steps (Sect. 6.3.6), first at high temperature and low oxygen pressure in order to reduce the amount of the fixed charge, then at low temperature in H_2 in order to passivate the interfacial defects.

Exposure to high temperatures has unwanted side effects. In particular, significant out-diffusion of dopants from previously prepared structures may occur, a problem which gains importance as the dimensions of devices decrease.[28] Broadening of the dopant profiles due to the so-called transient enhanced diffusion (page 421) may be of the order of 0.1 μm. This is comparable to the size of critical features (channel length, junction depth) in contemporary devices. Lower oxidation temperatures slow down the growth of oxidation-induced stacking faults (Sect. 6.5.5b) and, perhaps more importantly, help to contain the diffusion dopants and of impurities. Modern technologies tend to reduce the process temperatures and duration of high-temperature steps ("thermal budget").

The presence of **water** in the oxidizing ambient facilitates the breaking of dangling bonds and speeds up the oxidation (Sect. 5.5.3). For a fast and versatile oxidation, silicon can be exposed to hot (700-1000°C) pyrogenic steam[1111] made by burning hydrogen in oxygen ambient. Alternatively, oxidizing H_2O can be obtained by bubbling oxygen gas through hot water. Water, and not oxygen, is then the oxidizing species. The industry uses steam chiefly for field oxidation.

The diffusivity of water is low in SiO_2, ten to five times lower than the diffusivity of oxygen.[1084] Nevertheless, the solubility of water in SiO_2 is three decades higher than the solubility of O_2 and the result is that steam can oxidize much faster than dry oxygen under comparable conditions.

This high oxidation speed is advantageous when thick oxides are grown, but makes the process difficult to control when gate oxides (a few nanometers thick) are needed. Interestingly, recent developments in oxidation techniques[1112] allow for controlled growth of wet oxides down to 4 nm. But "wet" interfaces are generally believed to be of a worse quality than "dry" interfaces. Though hydrogen improves the interface by saturating troublesome dangling bonds (Sect. 6.6), this passivation must be done as late as possible in the production cycle (Sect. 6.3.6). Otherwise, the passivated bonds are likely to be damaged. Steam oxidation is thus typically used to fabricate field isolation (Fig. 1.2, page 8). Gate oxides, which are only a few nanometers thick, are made without water.

Dry/wet/dry sequences were often used in production lines: dry to get a good interface, wet to save time by fast oxide growth. When thinner oxides came into play, the process flow was reduced to dry oxidation.

Interestingly, a dual-layer dry/wet technique[1113] can lead to a *better* oxide than dry oxidation alone. A sample procedure is: grow dry oxide

(3.7 nm, 8 min at 820°C), then grow a wet oxide (4.3 nm at 820°C), and finally anneal the resulting film at 900°C. A film prepared in this way had a lower density of interfacial states (5.5 versus $8 \cdot 10^{10}$ $cm^{-2}eV^{-1}$), lower leakage current (by about 50%), longer time to dielectric breakdown (by about 30%), and lower interfacial roughness (3.7 versus 7.2 Å rms) than a control 8.2 nm dry oxide grown and annealed at similar conditions (850 and 900°C).

How can this be reconciled with the poor quality of wet gate oxides? The answer is simple: H degrades the dielectric strength of the oxide if it is located near to the anode terminal (Fig. 6.12, page 435) of the nMOS capacitor, that is, near to the SiO_2/polysilicon gate interface. (Sect. 6.6.2a). The dry/wet sequence places the hydrogen near to the cathode, keeping the anode hydrogen-free. At the same time the number of interfacial states is reduced by H and OH passivation (Sect. 6.3.6). The interface is smoothed, possibly due to a better stress relaxation (Sect. 6.5.4c) in hydrated silica.

More puzzling may seem a recent finding[1097] that advanced single-step steam oxidation can produce a several times more reliable 4 nm gate oxide than a control dry procedure[9] We think that a smoother interface, typical for wet oxides, was responsible for this effect. The reported reliability improvement illustrates that thicknes variations due to interface roughness may dominate the breakdown characteristics of dry oxides.

As its name suggests, **dry** oxidation is done in a water-free atmosphere. The substrate is placed in a furnace filled with dry O_2. The silicon surface is at first protected by a native oxide layer (Sect. 6.3.1), as after an RCA etch, or by H atoms, as after an HF etch (Sect. 5.3.1). Both types of protections have their drawbacks. On one hand, native oxides tend to be inhomogeneous. On the other hand, the oxide produced on an initially hydrogenated surface may contain some hydrogen atoms. The incorporation of hydrogen takes place at low temperatures, when the cool wafer is slowly loaded into the oxygen-filled, hot furnace (ramping up). Naturally, this incorporation is avoided when the gas which fills the furnace during the ramping up does not contain oxidizing species, so that the hydrogen escapes before the oxidation takes place. But residual oxygen in this inert atmosphere will etch and roughen the surface (Sect. 5.4.4). Such "ultradry" oxides are, therefore, inferior to conventional dry oxides in which no attempt was made to eliminate all hydrogen from the film.[819]

---

[9] The oxides were formed by Rapid Thermal Oxidation, described later in this Section.

Thin dry oxides (50 nm or less) seem to grow differently than thick oxides (Sect. 6.4.2). In general, it is more difficult to grow a thin oxide than a thick oxide and the process is less understood. An additional problem appears when the oxide thickness, t_{ox}, is intended to be only a few nanometers, as in modern gate dielectrics.[h] One of the reasons is the presence of the native oxide on the silicon surface (Sect. 6.3.1). Native oxides are not only inhomogeneous, but their thickness (about 1.5 nm, or 15 Å) can often be only estimated. 1.5 nm is a significant fraction of the total t_{ox}, even in contemporary sub-micrometer technologies.

Well-defined and uniform thickness of the film can be best maintained when the native oxide is removed and the exposed surface is covered by a smooth, oxygen-free, epitaxial layer of silicon. However, epitaxy makes the process flow more complicated, more expensive, and more sensitive to unavoidable random variations in conditions during production. Today, there is a clear tendency to avoid epitaxy in production lines. This may change in the future if an overall cost reduction is achieved in this way.

One of the problems encountered during oxidation in a furnace is that wafers must be slowly inserted into the hot furnace and slowly removed from it. Too fast "ramping" would create defects due to thermal strains. Thus, the price for acceptably low thermal strains is an increased thermal budget. A promising technique which obviates this problem is **Rapid Thermal Oxidation** (RTO), one of many applications of Rapid Thermal Processing[1090] (RTP). In RTP, the substrate is briefly heated to a high temperature by infrared lamps. The total exposure to high temperatures is reduced nearly to the minumum needed to conduct the reaction. RTP is a popular method for annealing. Though it may not be the standard oxidation scheme yet, it will become one in the nearest future. It is suited not only for precise annealing, dopant activation,[33] and dry[1114] and steam[1112] oxidation, but also for nitridation[1115] and oxynitridation (Sect. 6.3.4).

Oxidizing ambients used in CMOS gate oxidation have a small admixture (about 1-4%) of a **chlorine**-containing agent, typically HCl. Chlorine stabilizes the threshold voltage and improves uniformity and dielectric strength of the oxide[1116] by removing alkali metals and other impurities and reducing the interfacial roughness (page 416). The prob-

[h] An industrial state-of-the-art t_{ox} is today 4–5 nm. One of the thinnest oxides in a demonstrably working transistor[30] (late 1997) is 1.3 nm. The channel length of this transistor is 60 nm. See Ref. 1098 for other examples.

lems caused by oxidation-enhanced diffusion and oxidation-induced stacking faults (Sect. 6.5.5b) are also diminished in this way.

Chlorine accelerates dry[1116] and slows down wet oxidation.[1117] The later effect is apparently due to dilution of the oxidizing species by HCl: both Deal-Grove parameters B and B/A (Sect. 6.4.1) scale linearly with the concentration Θ of HCl as $1-\Theta$. The enhancement of dry oxidation, usually by a factor of 1.3–1.5, may be caused by H_2O produced in the reaction of HCl with O_2. It is also plausible that HCl catalyzes the oxidation reactions (page 416) and that it assists the diffusion of oxygen in SiO_2.

Since no widely accepted physical model for the rate enhancement by HCl during oxidation exists, professional process simulation tools[1092] deal with chlorine ambients in the simplest way, by multiplicative factors in the formulas for B and B/A. Their dependence on temperature and the HCl percentage is stored in a look-up table which contains experimental data.

6.3.3. *Growth by deposition*

Another class of methods by which oxide films are made is deposition. Numerous techniques have been developed, such as sputtering, high-density plasma (HDP) CVD, subatmospheric and atmospheric pressure CVD with TEOS/O_3 chemistry[i] low-pressure CVD with TEOS/O_2 chemistry, or electron beam deposition. The advantage of deposited oxides is that the process temperatures can be as much as by a few hundred degrees C lower than in the thermal oxidation of silicon, the film grows fast, and the substrate silicon is not consumed. Moreover, by adjusting the deposition method, oxide films of various densities, chemical compositions, morphologies, or residual stress magnitudes can be obtained (for literature see, e.g., Ref. 1118 and Ref. 1119). SiO_2 films can be deposited not only on silicon and SiO_2, but in principle on any relevant material, such as Si_3N_4 or TiN.

SiO_2 deposition temperatures may be indeed low. An interesting example of a catalyzed CVD growth of SiO_2 at **room temperature** was recently reported.[1120] Plasma enhanced CVD of SiO_2 can be done[1119] at 360°C. It should be noted, however, that an unoptimized plasma-assisted deposition may cause gate oxide damage and degrade the product yield by several percentage points.[38]

Deposited oxides are used as isolating, masking, or protecting layers. Covering metal surfaces with ultrathin SiO_2 layers may be an in-

---

[i] TEOS stands for tetraethyl orthosilicate, $(C_2H_5O)_4Si$.

teresting technological option which would allow to unify anticorrosion protection.[1121] Deposited SiO_2 can be used as charge storage films in non-volatile memory cells[23] (Fig. 1.3, page 9). Deposition of a few nanometers of oxide on top of a thermal gate oxide may significantly improve the dielectric strength of the film by disrupting micropores which thread through the oxide (Sect. 6.3.5). Since a good quality of the interface between the thermal and the deposited oxide is needed, such dual-layer oxides should be reoxidized in oxygen atmosphere. This reduces the number of defects at the SiO_2/SiO_2 boundary.[1122]

Industrial production of silicon microchips does not involve deposition of silica or silicate glass films on bare crystalline silicon. This will probably not change. Although a careful preparation of the initial surface can keep the density of defects low,[j] hydrogen atoms present in the precursors are incorporated into the oxide and at the interface. These hydrogen atoms are likely to cause reliability problems (Sect. 6.6.2a).

In certain process designs (as in the shallow trench isolation, page 378) the deposited SiO_2 may be separated from the silicon substrate only by a thin layer of a thermal oxide. This requires certain care, because the deposited film can be a source of a significant stress which can influence the silicon and lead to the formation of defects and deterioration of the gate oxide. These difficulties have been overcome only recently.

6.3.4. *Nitridation and oxynitridation*

A major problem which emerges as gate oxides are scaled down (Sect. 1.2.2) is boron penetration (Sect. 6.5.5c). Boron can diffuse from the polySi gate through the thin oxide into the MOSFET channel. This happens during high-temperature annealing, chiefly during boron activation in implanted gates (Sect. 1.2.2b). The dopant profile is thereby perturbed (Sect. 6.5.5c) and the dielectric is deteriorated (Sect. 6.6). A small amount of N hardens the oxide against B penetration (Sect. 6.5.5c). In comparison to standard oxides, nitridated oxides have also an increased dielectric permittivity ϵ and strength,[k] are better barriers against diffusion of impurities, are less reactive with gate electrodes, and are more resistant to ionizing radiation.[941] The latter is important in space applications, for example, for telecommunication satellites. Last but not least, oxynitridation can be used as a means to

---

[j] This is efficient on Si(111). Plasma-assisted CVD at 300°C can then produce a SiO_2 layer with the interface trap density comparable to that in thermal oxides.[1123]

[k] See Section 6.6 for the pracitcal quality issues of SiO_2/Si films.

control the growth of extremely thin (a few monolayers) oxide buffer films between silicon and high-K (high ϵ) gate dielectrics (Sect. 5.7.4).

There are three major ways to introduce N into thin oxide layers. The first is nitridation of thermally grown SiO_2 films on silicon. The second is a direct growth of oxynitride films. The third is implantation of N into the substrate before oxidation; this useful approach was adressed on page 321.

Succesful **nitridation of oxide films** was demonstrated[941] in 1984. SiO_2 layers (9–100 nm grown dry at 900–1000°C) were annealed in pure ammonia at 1150°C for up to 6 h. The distribution of nitrogen in films thinner than ~30 nm (300 Å) was nearly uniform (~27 at.%). In thicker films the nitrogen content decreased with the distance from the outer surface. A small nitrogen enrichment was observed at the interface.

The problem with this procedure is similar to the problem encountered in NH_3 nitridation of silicon (Sect. 5.6.3). Namely, the presence of hydrogen induces high densities of electrically active defects.[941,942] To reduce their number, an extra reoxidation step can be performed.[1124,1125]

Such problems are obviated when NO or N_2O is used instead of NH_3 (Sect. 5.6.1). Moreover, with NO and N_2O it is possibile to combine oxidation and nitridation into one step, oxynitridation. SiO_2 films grown from N_2O were reported[1126] as early as in 1969, but only in 1990 it was discovered[1127] that some N from N_2O is incorporated into the oxide.

Properly applied N_2O or NO can significantly improve the quality of the interface.[1128,1096,1129,939,940] A good approach is to begin the oxidation in O_2 ambient and complete it by annealing in a nitridating ambient at 800–900°C.[1129,1130] In comparison with nitridated oxides, oxynitrides grown from NO or N_2O are characterized by a more pronounced nitrogen pileup at the oxide/substrate and a smaller nitrogen content near to the outer surface of the oxide. For the sake of carrier mobility in the channel, excess amounts of nitrogen in the interfacial layer should be avoided.[1131] In general, modern oxynitrides are less prone to wearout (interface-state generation, trap creation, low-field leakage current increase) and boron penetration than usual oxide films. As the gate oxide thickness is reduced towards 1 nm, these properties are gaining practical importance.[1132]

Nitrogen-containing oxides can be grown efficiently by Rapid Thermal Processing (page 365) in the range 800–1200°C. Oxynitride is produced directly by Rapid Thermal Oxidation[1127] (RTO). Nitridation of a previously grown oxide is done by Rapid Thermal Nitridation.[1133]

The **mechanism** of nitrogen incorporation is unclear. N atoms may be stabilized at the interface by the interfacial stress (Sect. 6.5.4c). N and O bond differently to Si: N forms planar NSi_3 plates (Sect. 5.6.1) while O forms linear OSi_2 "boomerangs" (Sect. 6.2.1). The availability of both structures at the strained interface may help to accomodate the stress. However, if this is true, it is not clear why the presence of N in the first interfacial layer degrades the carrier mobility in the MOSFET channel.[1131]

Formation of interfacial N–O bonds has been suggested as well.[1128] Such bonds might help to relax the stress.[1134] On the other hand, they are much weaker than Si–O and Si–N bonds.[4] A stoichiometric oxynitride (Si_2ON_2, Fig. 5.8 on page 313) contains only Si–O and Si–N bonds, and NO molecules dissociate at Si(001) surfaces (Sect. 5.6.2). Measurements of N(1*s*) core-level binding energies at the interface[889,1135,1136] indicate that the local bonding configuration of N is similar to that in Si_3N_4. This conclusion is corroborated by recent *ab initio* calculations of nitrogen core-level shifts for a number of interfacial N geometries.[1137] We note, however, that a peak at a binding energy comparable to the energy computed *ab initio* for the Si_2NO configuration was found in at least one experiment.[1138]

Si_3N_4 and Si_2ON_2 can be oxidized to SiO_2. This means that N incorporated in the bulk of the film is attacked by O during growth. Studies of reoxidized oxynitride films may therefore provide information on the growth mechanism.[1139] Reduced by oxygen, N diffuses in the film, drifts to the sample surface, and escapes. In order to explain the behavior of N during reoxidation, the existence of a concentration gradient of defects favoring the migration of N atoms towards the surface was proposed.[1139]

Nitrous oxide (N_2O) is a popular hydrogen-free nitridating agent. Similarly to oxidation (Fig. 5.3, page 294), N_2O **oxynitridation** has three regimes:[1134] active (surface etching), passive (smooth film growth), and intermediate (rough film growth). But the temperature dependence of the critical pressure which limits the active and passive regimes is more complex than for O_2 oxidation. In practice, exposure to low-pressure N_2O at low temperatures (such as 600–800°C and 10 mbar) leads to etching by SiO desorption, because the reactivity of N_2O is then low and so is the oxidation rate. At higher temperatures, (800–1100°C) etching dominates as well if the N_2O pressure is below ~0.01 mbar. A good film grows at temperatures above ~800°C and pressures above ~0.1 mbar. Nitrogen content in the film depends on the thickness t of the reacted layer.[1134] Stoichiometry of the film shifts from oxygen-rich ($t \lesssim 1$ monolayer) towards nitrogen-rich ($1 \lesssim t \lesssim 10$ ML) and, finally, oxygen rich material ($t \gtrsim 10$ ML).

The kinetics of N_2O oxynitridation is very different from that of O_2 oxidation: the former is seemingly self-limited, while the latter is not. The growth rate decreases as N content increases. This is due either to blocking of the reactive sites at the oxidizing interface by N or to inhibition of oxidant transport through the reactive layer (Section 6.5.4), again by nitrogen.[1140] When the N content rises to the equivalent of a surface monolayer, the reaction appears to stop. At temperatures around 900°C, it takes around 100-200 seconds to reach this stage and only a few nanometers of the film are produced.[1141] However, when the sample is kept in the furnace for a longer time (many minutes or hours), it becomes clear that the film still grows, albeit slowly.[1142] Both kinetic regimes can be unified into one phenomenological model; however, the assumption of steady-state flow of the oxidant, typical for oxidation models (Sect. 6.4), must be released.[1142]

Also **nitridation** in N_2O has a self-limiting tendency. Nuclear reaction analysis shows that nitrogen in RTN films is mostly incorporated during an early stage of growth. The N-rich region appears to retard nitridation and oxidation.[1143] The total amount of built-in nitrogen[1128] depends on oxide the thickness and, for 10 nm oxides, is around a monolayer. The N content increases linearly with the oxidation temperature between 800°C and 1200°C by roughly one third of a monolayer per 100°C. One monolayer is incorporated at ~1000°C if the oxide is 10 nm thick. N is localized mostly near to the interface in an interlayer a few nanometers thick. Its distribution is almost independent of oxide thickness (up to 20 nm) and basically unchanged[1144] by H or water, at least at 1150°C. Furnace processing distributes N more evenly throughout the film. In contrast to O_2 or wet oxidation rates, RTN rates are *independent* of the substrate orientation.[942]

Highly reliable inter-polysilicon oxide (polyoxide) for nonvolatile memories (Fig. 1.3, page 9) can be made using Electron Cyclotron Resonance (ECR) N_2O plasma. It has a lower leakage current, higher breakdown field, less electron trapping, and much larger charge-to-breakdown (up to Q_{BD}=10 C/cm^2) than thermal polyoxide.[1145] The lower trapping rate and larger Q_{BD} are attributed to the N-rich layer and to a smoother interface.

The other hydrogen-free nitridation ambient, **nitric oxide** (NO), has also been demonstrated[1146,1147,1129] to produce high-quality gate dielectrics. NO has a significantly higher N incorporation efficiency compared to N_2O. This higher concentration of N is probably due to the complete dissociation of NO and the reaction of N and O directly with Si. The reaction with N atoms from N_2O is indirect and less probable,[1148] chiefly because N_2O dissociates into atomic O and inert N_2 (Section 5.6.2).

Ultrathin (below 2 nm) oxynitrides for gate dielectrics can be also grown in a low energy electron-beam enhanced oxidation at moderate temperatures (below 600°C) from low pressure (less than $8 \cdot 10^{-6}$ mbar) NO gas.[1149] The growth at room temperature is characterized by the formation of an oxygen-rich ultrathin film. The nitrogen content increases as the temperature is raised up to 600°C. This result confirms other reports that synchrotron radiation at room temperature efficiently promotes oxidation from low-pressure NO gas, while thermal activation above 1000°C promotes nitridation by NO.[1150,1151]

NO and N_2O oxynitrides differ not only in their nitrogen content, but also in the interface roughness and the growth rate.[1138] NO-made SiO_2/Si(001) interfaces are generally smoother than O_2-made interfaces. MOS capacitors with NO-nitrided gate dielectrics show highly suppressed trapping of charges, high immunity to interface state generation, and excellent breakdown characteristics compared to RTP oxides grown from O_2.

Nitride and oxynitride films can be **deposited** by a soft vacuum electron beam plasma assisted deposition in SiH_4, NH_3, and N_2 mixture.[920] The oxynitride deposition is governed by two parallel and competitive surface reactions, one leading to the formation of Si–N bonds and the other to the formation of Si-O bonds. When the substrate temperature is below 100°C, the oxygen species reaction rates are dominant, leading to the formation of SiO_2. When the substrate temperature increases beyond the critical temperature, the reaction rate of the major nitrogen species, such as NH_3 and NH radicals, rapidly becomes dominant. Residual oxygen then leads to an amorphous, single-phase mixture of Si, N, H, and O, the so-called $SiN_xO_yH_z$. A very promising deposition technique is the so-called **Jet Vapor Deposition** (JVD). It is probably the only known method to deposit gate-quality Si_3N_4 dielectric films on silicon substrates (Sect. 5.6.3).

Molecular nitrogen (N_2) is very inert. It is generally not suited for oxynitridation or nitridation, though successful attempts have been made (Sect. 5.6.2) to activate it by breaking N_2 into atomic nitrogen. On the other hand, **implantation** of N^+ or N_2^+ ions has a great potential, chiefly as a method to control locally the thickness of gate oxides (page 321).

6.3.5. *Silicon dioxide film morphology*

SiO_2 morphology and the related practical quality of the film is influenced by the silicon substrate and the conditions of oxidation. An important example are "**pinhole**" defects: micropores in the oxide which extend through the whole film. They are caused chiefly by microparticles contaminating the surface before oxidation. Microscopic defects due to impurities (as nitrogen) and flaws in the substrate and in the film may be responsible as well. The stress induced in the film and in the substrate during thermal oxidation promotes the formation of various **atomic-size** defects, including fixed oxide charges and interfacial traps (Sect. 6.5.4c), is responsible for the ejection of Si interstitials (Sect. 6.5.5b), and possibly also for the segregation of dopant atoms to the interface (Sect. 6.5.5c). If the oxide contains **inhomogeneities** (Sect. 6.3.5c), it may be roughened during further processing. This may eventually cause failure of the transistor.

a. *Pinholes*

Pinholes lead to early electrical breakdown of the device and are real "yield killers". A pinhole can short-circuit the MOS capacitor at electric fields below a typical operational value (Section 6.6.2a).

Pinholes nucleate at microparticles adsorbed on the surface. Moreover, their formation is promoted by mechanical stresses generated during oxidation, when the volume expands as Si is converted into SiO_2. Also cooling is hazardous, because bulk Si and SiO_2 have very different thermal expansion coefficients (Sect. 6.3.5b). The risk of pinhole-induced breakdown can be reduced by preparing the oxide in two steps, by two oxidation methods, so that there are no pinholes which punch through the whole film. The dielectric strength of the oxide is then significantly improved.[1122] Though it is difficult to use such dual oxidation to grow ultrathin gate oxides (a few nanometers thick), a successful fabrication of 8.2 nm dry/wet oxide film for nonvolatile MOS memories has recently been reported.[1113]

Pinholes can be also caused by cooling of the oxidized substrate from the oxidation temperature in ambient with too low partial pressure of oxygen. Some oxide can then escape from the hot material.[1087,834,836] This process may begin in an area which accidentally contains excess silicon and is thus more prone to emission of volatile SiO.

The oxide can be tested for the presence of pinholes by propane infiltration.[1152] The molecules attack the substrate through the micropores. This is a destructive test: after exposure to propane, the oxide is stripped and SiC islands are observed with an optical microscope.

b. *Stress field in SiO_2*

The importance of stress-related problems for silicon technology has been recognized.[1153] In particular, oxidation is a source of stresses which may he hazardous. The huge difference in molar volumes (roughly the factor of two) and in thermal expansion coefficients of Si and SiO_2 cause problems. The linear expansion coefficient of silicon at room temperature (2.4×10^{-6} K^{-1}) is five times larger than that of fused silica. The oxide layer is compressed during growth and subsequent cooling. At the same time the substrate is subjected to tensile stress. The effect is so strong that the wafer may bow. For example, 500 nm of flat SiO_2 grown at 1000°C can bend a 0.3 mm thick Si(001) wafer of 200 mm diameter by 0.25 mm. This corresponds to the curvature radius of 20 m. The hydrostatic stress is then $1.8 \cdot 10^7$ dyne/cm^2 (1.8 MPa) in Si and $2.7 \cdot 10^9$ dyne/cm^2 (270 MPa) in SiO_2.

The distribution of stress in a thermal SiO_2/Si system depends on thickness and lateral dimensions of the SiO_2 film, thickness of the substrate, growth conditions, cooling rate, annealing, orientation of the substrate, and temperature during measurement. For example, stress σ_{ox} in the oxide increases with decreasing film thickness t_{ox} as $\sigma_{ox} \sim \exp(-\sqrt{t_{ox}})$. Similarly, stress in the substrate decreases with decreasing film thickness as $\sigma_{Si} \sim \exp(\sqrt{t_{ox}})$. When the temperature exceeds ~950 °C, the bulk of the oxide film is virtually free of stress, because around this temperature viscous flow of SiO_2 becomes efficient. Note that this does not imply that atomic-scale, local stresses at the interface are relaxed as well.

The stress in the atomic-scale distance from the interface is caused by the mismatch between the partial density of silicon in Si and SiO_2 and by topological incompatibility. It is difficult to arrange the interfacial atoms in such a way that their valences, preferred bond angles, and preferred bond lengths are all satisfied. The interfacial stress has an important, maybe decisive, contribution to the formation of the SiO_x transition layer, interface traps, and the fixed charge (Sections 6.5.2a and 6.5.4c), it seems to adversely influence the oxide reliability[1154] (Sect. 6.6.2a), and may be responsible for segregation of dopants to SiO_2/Si interfaces (Sect. 6.5.5c). Residual stresses in SiO_2 further from the interface may be affected by the electrostatic repulsion of fixed oxide charges[1155] (Sect. 6.4.2b and Fig. 6.4, page 390). Process simulation software can model the continuous distribution of stress in the oxide as well as in the substrate,[1092] but the assesment of interfacial stresses requires computationally expensive, atomistic calculations.[1156]

It appears that stresses in the oxide promote the pinhole formation and stresses in the substrate may produce dislocations. The stress field in the oxide film influences the oxidation process (page 389). This may lead to problems when tiny mesoscopic features like trench edges in shallow trench isolation must be oxidized (pages 378 and 389). Accounting for stress fields allows for a reliable simulation of such processes.[1157] Stress in the substrate may also affect diffusion of dopants in an up-to-now uncontrollable way. For example, residual stress of the order of 10^9 dyne/cm^2 (100 MPa) remains after LOCOS isolation (Sect. 6.3.7) in the oxide as well as in the substrate and may influence the response of the system to further technological processes. It is also of possible relevance that the stress field frozen into the substrate changes the electronic structure in the channel, directly bearing on electric properties of the transistor.[1158] Though the stress exerted on the substrate by ultrathin gate oxides is small and mostly produced by the oxide edges, one should keep in mind that sides of the MOSFET channel are isolated from the rest of the circuit by a thick oxide.

c. *Microcrystallites and homogeneity*

Another important morphological issue is the degree of order in the film. Oxides thicker than a few atomic layers are amorphous, but thiner films may show a remarkable degree of crystallinity (Sect. 6.5.2). Their local structure may resemble tridymite and cristobalite embedded in an amorphous matrix. Films grown by various techniques, such as thermal oxidation (particularly low-temperature), oxidation in air, or electron-beam evaporation, may contain numerous interfacial SiO_2 microcrystallites[1159] with diameters around 100 Å. X-ray scattering patterns from these microcrystallites resemble those of α-cristobalite. The mutual orientation of the microcrystallites is correlated, reflecting the influence of the substrate lattice.

Microcrystallites can be detected even in layers many hundreds Ångstrom thick.[1159] Virtually all SiO_2 polymorphs are then represented. These intrusions nucleate around impurities such as alkali metals and hydroxyl radicals (OH). No relation between the density of dislocations in Si and the crystallization tendency has been observed. The nucleation process is activated with an energy barrier of approximately 3 eV, and is initiated during high-temperature processes such as annealing steps.

Such inhomogeneities are troublesome, particularly in oxides which are later etched or polished. A film with crystalline islands covering as little as 1% of the surface is useless for microelectronic purposes. Such a film will certainly roughen during subsequent etching steps: the microcrystallites have various chemical reactivities, different from the amorphous "matrix" of the film. Insulators are often exposed to multiple etch processes which may selectively dissolve certain parts of an inhomogeneous oxide, leading to macroscopic roughness. If the film contains voids, as can happen, for example, during low-pressure TEOS CVD for shallow trench isolation,[1160] subsequent etching may open them, further degrading the layer.

In addition, there is an interplay between microcrystallites and stress. Not only their specific densities differ from that of the amorphous SiO_2 but their thermal expansion coefficients do not match that of the matrix. For example, at room temperature the spatial thermal expansion coefficient is about twenty times larger for α-quartz ($34.4 \cdot 10^{-6}$ K^{-1}) than for vitreous silica ($1.3 \cdot 10^{-6}$ to $1.5 \cdot 10^{-6}$ K^{-1})

6.3.6. *Post-oxidation treatments*

The objective of post-oxidation treatments is to improve the gate oxide by reducing the number of interface traps, fixed oxide charge, and microroughness, which adversely affect the MOSFET parameters (Section 6.6). Low-temperature annealing (below 500°C) reduces the density of interface states, but the fixed charge anneals only above 500°C and in a different ambient. Separate high- and low-temperature annealings are therefore needed.

The amount of fixed charge Q_f which survives a **high-temperature** annealing (20–5 min at 900–1100°C in oxygen-containing ambient) is a function of temperature, time, and oxygen partial pressure. It does not depend on the oxidation method (wet or dry) but it is different for various surface orientations. The minimum achievable Q_f is $\sim 2 \cdot 10^{11}/cm^2$ for Si(111) and $\sim 1.1 \cdot 10^{11}/cm^2$ for Si(001); Si(110) performs between these extremes.

The fixed charges are created during oxidation (Sect. 6.4.2b). They are most probably due to holes trapped permanently on incompletely oxidized silicon atoms embeded in the SiO_2 matrix in the *fresh* oxide, that is, close to the interface (Sect. 6.5.4). It is therefore natural that the annealing temperature should be high enough to allow for reorganization of the interatomic bonds. The annealing ambient must be oxidizing to improve the stoichiometry, but the oxygen partial pressure must be sufficiently low

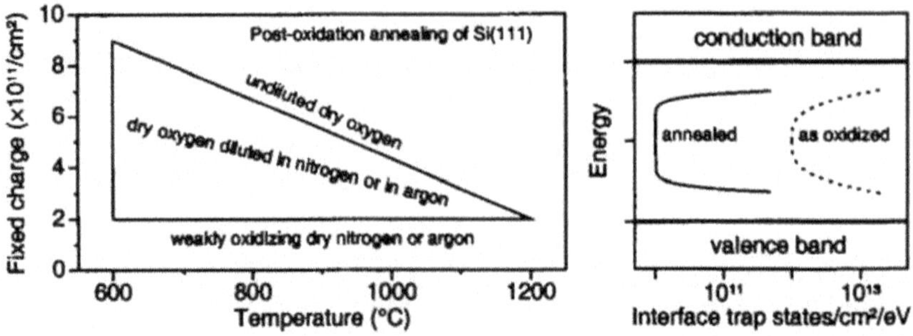

Fig. 6.2. Post-oxidation treatments of SiO_2/Si(001). **Left**: Schematic Deal triangle,[1161] which specifies the dependence of fixed oxide charge on the annealing conditions. **Right**: Schematic distribution of interface trap states within the silicon band gap after dry thermal oxidation and after low-temperature anneal (adapted from Ref. 268).

to prevent the advancement of the SiO_2/Si interface[l] (Fig. 6.2, left).

Annealing in a completely inert ambient (without any oxygen) may even increase Q_f, particularly if the temperature is too high (above ~1000°C) and the initial Q_f was low. There is a **critical partial pressure** of oxygen below which damage is done.[1162] This is analogous to the existence of the active and passive oxidation regimes of Si surfaces (Sect. 5.4.4). In the passive regime, the oxide grows (or defects are annealed). In the active regime, SiO is desorbed (or defects are created).

The **low-temperature annealing** is different, though some parallels to the high-temperature step exist. Let us take the well-studied SiO_2/Si(111) interface as an example. Consider a typical density of interface trap states in the region near the mid-band, where this density tends to be the lowest for thermal oxides (Fig. 6.2, right). Directly after oxidation, this density is between 10^{11} $cm^{-2}eV^{-1}$ (oxidation in steam) and 10^{12} $cm^{-2}eV^{-1}$ (oxidation in dry oxygen). Since the band gap of silicon is ~1 eV, this corresponds to at least 10^{11} and 10^{12} electrically active defect centers, assuming that each center introduces a single state. Post-oxidation annealing in H_2 can reduce it to 10^{10} $cm^{-2}eV^{-1}$. The latter value is reasonably independent on the previous history of the oxide, and therefore it may represent a thermodynamic equilibrium. It depends only on the orientation of the surface, whereby the ratio between Si(111), Si(110), and Si(001) densities of interface states is approximately 3:2:1.

[l] If the interface advances, new defects are created.

The key components of the ambient are now: hydrogen and, if water is present, OH radicals. H atoms and/or OH moieties saturate dangling Si bonds. These bonds are localized mostly at the interface (Sect. 6.5.4). After passivation, they turn into electrically inactive weak spots of the oxide which can transform into charged defects when the device is biased[1163] (Sections 6.5.5d and 6.6.2a). Water from technological steps may be responsible for this type of damage.[1164] For that reason it is important that no excessive biasing of the circuit follows the passivation step.

In practice, hydrogen passivation is done during other processes, such as sintering after metal etch or formation of protective oxides. A typical temperature and ambient are: 400–450°C, 10% H_2 and 90% N_2. A 450°C exposure takes about an hour. D_2 ambients provide a better passivation than H_2, (Sect. 6.5.5d), but the exposure should not be too long.[1165]

The **roughness** of as-grown SiO_2/Si(001) oxides used to be high (10-15 Å rms at 900°C oxidation). It can be reduced by post-oxidation annealing. Annealing in an inert gas at 900°C decreases[1166] the interfacial roughness to 2.2 Å rms at spatial frequencies above 0.2 nm^{-1} (page 438).

6.3.7. *Isolation techniques*

The elements constituting a processor or a memory chip must be electrically insulated from each other and then interconnected by metal wires. The wires cannot short-circuit, the propagating signals should not cross talk, and parasitic capacitances should not slow down the signals by capturing the carriers. In contrast to the gate dielectric, the interlayer dielectric should have a dielectric constant as low as possible, so that the capacitances are minimized. The insulating material in contemporary designs is usually a silicate glass. In the future, dielectrics with lower permittivity k (low-k materials) will have to be adapted for this purpose.[22,1021]

Naturally, we cannot discuss interconnects here. Our main concern is the formation of SiO_2/Si interfaces. A special case of interest is for us the Shallow Trench Isolation (STI) technique, the new scheme which substitutes the classical LOCalized Oxidation of Si (LOCOS) isolation in technologies in the quarter-micron linewidth regime and below.

The classical **LOCOS** process flow[1167] starts on a wafer covered with a thin (about 50 Å, or 5 nm) pad oxide. The areas allocated for devices are protected by Si_3N_4 masks. Steam oxidation is performed at ~1000°C to thicken the oxide in the unprotected areas to ~2 μm. The nitride is then removed. In this way, field oxide such as in Fig. 1.2 (page 8) is produced.

Stress from LOCOS can deteriorate the gate oxide.[923] But a major problem is that the oxide grows not only vertically into the silicon, but also expands laterally below the nitride mask, forming "bird's beaks" (the triangular terminations of FOX on both sides of Fig. 1.2, page 8). Large dimensions of these features limit the usefulness of LOCOS in sub-0.25 μm designs. A lot of research has been devoted to this problem. For example, it was observed that oxidation by plasma anodization keeps the oxidation front from defocusing and no beak is formed.[1155]. A more classical approach is to use thicker nitrides. A thicker, stiffer Si_3N_4 film squeezes the beak, so that it does not open so wide. In such a case a polysilicon buffer is often deposited between the nitride and the pad oxide to adsorb the extra stress.

STI[1168,1160] avoids bird's beaks in a different way. The device areas are masked by Si_3N_4, as sketched by the thick lines in  (side view). Next, the pad oxide and some of the underlying silicon are etched away from the unmasked areas (). Each device will be thus fabricated on an island surrounded by a shallow (~0.5 μm or ~5000 Å) trench.

Before these devices are constructed, the trenches must be filled with a stable material. This is done in three major steps. First, bottom and walls of the trenches are thermally oxidized to 70–150 Å. Second, the trenches are filled with deposited oxide or polysilicon (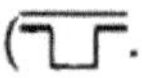. Third, excess filling is removed by Chemical Mechanical Polish (CMP) planarization,[1169] which stops at the hard nitride layer (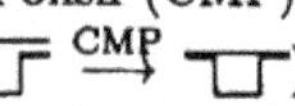). Finally, this nitride is stripped and the isolation is complete (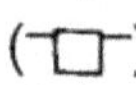) with no bird's beaks.

However, STI is more complicated than LOCOS. Therefore, the real advent of STI began only when LOCOS became impractical, that is, when the minimal feature size of MOSFETs was scaled down to 0.25 μm.

From the point of view of interface formation, STI poses other difficulties than those typical for LOCOS. By consuming a large volume of silicon, LOCOS produces many Si interstitials and a long-range strain field. The interstitials enhance diffusion of dopants, intensifying the effects of interface-dopant interactions (Sect. 6.5.5). They can also accumulate into stacking faults, which may lead to collector-emitter electrical shorts in bipolar transistors. Thermal oxidation in STI is much shorter and the amount of ejected interstitials is correspondingly lower. But the stress induced in the substrate[1170] by the oxidation may adversely affect the homogeneity of the gate oxide grown subsequently between the isolation fields.

Moreover, STI devices suffer from reliability and performance degradation caused by trench corners (). The oxide grows slower in corners than on a flat surface (page 389). This can result in sharp corners and/or locally thinned liner and gate oxides. Such effects are undesirable. Too thin an oxide at the corner can even cause a failure of the dielectric. Indeed, the isolating trench encompasses the whole transistor, that is, it is adjacent to the gate oxide. The gate polysilicon strip extends above the corner, as indicated in this top view of a MOSFET isolated by STI: 

(the thin line is the edge of the STI trench surrounding the source S and drain D of the transistor, the thick line marks the gate strip). This means that the corner oxide is exposed to the electric field originating from the gate voltage. Moreover, sharp corners disturb the electric field in the active device and the thinning of the gate oxide reduces the threshold voltage along the trench edge. This creates parasitic transistors (the broken lines in 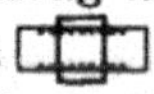), which turn on earlier than the parent transistor, so that "humps" appear in the subthreshold I–V characteristics[1171] ("kink effect").

In order to optimize an STI structure, one should properly design the thermal oxidation step.[1172,1173] This often requires that a simulation is done in which a model capturing salient features of curvature-dependent oxidation is applied. A standard Deal-Grove model (Sect. 6.4.1) extended to include strain-dependent rate parameters can describe the oxidation process properly.[1157] It can be easily implemented in professional TCAD software. An alternative approach is based on space-charge effects[1174] (page 387).

A simple and efficient way to circumvent many problems of STI is to change the sequence of the manufacture steps and begin with the wafer covered by *gate oxide and polysilicon gate.* This is the idea behind the Extended Trench Isolation Gate Technology[1175] (**EXTIGATE**). The trenches are then etched through the gate poly and the gate oxide, and the trench walls are oxidized. This oxidation causes an effect similar to the bird's beak formation in LOCOS: the thin gate oxide becomes locally thickened at the trench corner because the oxidant diffuses into the gate oxide film. The EXTIGATE bird's beak is, however, much smaller than the LOCOS bird's beak, and it's role is positive: it rounds the corners and thickens the oxide around the trench edge. The kink effects are gone.

An additional advantage of EXTIGATE is that the gate oxide is prepared on a very smooth surface of a fresh wafer. Traditional technologies include a number of cleaning and oxidation sequences before the actual gate oxide is grown. Each oxidation and each cleaning roughens the surface to some extent. EXTIGATE has therefore the potential to fully utilize the fact that the practical quality of the gate oxide (Section 6.6) and the microscopic quality (smoothness and cleaness) of the initial surface are correlated.

Other trench-based advanced oxidation approaches, such as **recessed thermal oxidation** (trenches are etched prior to LOCOS-type field thermal oxidation, so that a nearly planar isolation is achieved) and **offset trench isolation** (the shallow trench is offset from the nitride by aligning it to an oxide sidewall spacer around the Si_3N_4) have been proposed as well; see Ref. 1176 and references therein. Also **selective epitaxial growth** can be utilized for isolation purposes. Isolation areas are masked on an oxidized wafer, then holes are etched and silicon is grown epitaxially in the holes. Apart from process complexity, one of the difficulties in this case is that the cleaning which must be done between the etch and the epitaxy steps attacks the SiO_2/Si interface. This produces "undercuts" which deteriorate I–V characteristics of the device. The problem can be eliminated if the isolation oxide is TEOS annealed in N_2O; this works probably because it minimizes the interfacial stress.[1177]

Finally, we must mention the isolation technique known as **SIMOX**, or Separation by IMplanted OXygen. As the name suggests, the SiO_2/Si interface produced in by SIMOX is buried under the surface of the wafer. The implanted oxygen atoms stop at the depth defined by the energy of the ion beam. Subsequent annealing heals the implantation damage in the silicon film above the oxygen-rich layer and agglomerates the oxygen atoms into a buried SiO_2 layer.[1178,1179] This technique is an efficient way to produce Silicon-On-Insulator CMOS (**SOI CMOS**) devices[1091,1180] and may become a mainstream technology. SOI CMOS offers 20–35% gain over a conventional, "bulk" CMOS implemented with the same linewidth.[1181] As a consequence, SOI circuits can operate at lower voltage and use much less power (by the factor between two and three).

6.4. Models of Thermal Oxidation

Various views of the thermal oxidation process have been proposed. Since their verification is difficult, the detailed physical mechanisms of oxidation remain controversial. This was not an urgent technological problem, because a very simple phenomenological model is used successfully to simu-

Ref year	Subject relevant to models of thermal oxidation
1182 1946	Fundamentals of space-charge effects in oxide growth
1183 1948	Fundamentals of space-charge effects in oxide growth
1184 1961	Diffusion of O_2 in SiO_2
1084 1965	Growth kinetics of thick oxides: linear-parabolic (Deal-Grove) law
1185 1978	Atomic oxygen and interfacial reactions
1186 1980	Space-charge limited model of growth rate
1187 1982	Space-charge limited model of growth rate
1188 1984	Space-charge limited model of growth rate
1086 1985	Growth rate enhancement in thin oxides
1088 1987	Initial oxidation rate for five different orientations
1189 1987	Role of electrons in interfacial oxidation reactions
1190 1988	Space-charge limited model of growth rate
1191 1989	Space-charge limited model of growth rate
1174 1992	Space-charge limited model: power-parabolic law
1092 1993	Practical simulation of the oxidation process
1192 1995	Space-charge effects in rapid thermal oxidation
1193 1995	Fractal model of silicon oxidation
907 1995	Defect-mediated mechanism of oxidation
242 1996	Oxidation model unifying defects, stress, diffusion, and reaction rates
1157 1997	Stress dependent growth of thin oxides; dual Deal-Grove model
1110 1998	Excellent oxides by remote plasma oxidation; TEM, AFM, C–V, I–V
1194 1999	Diffusion-reaction approach to silicon oxidation

Table 6.3. Models of thermal oxidation: selection of references

late the oxidation rates (Sect. 6.4.2c). An issue of growing importance is a controlled formation of very thin, homogeneous oxides for gate dielectrics. This is probably the reason why the research in this field, after an apparent stagnation, is once more gaining impetus. Table 6.3 lists some references.

Ultrathin oxides is now an attractive, evolving field of research. "Ultrathin" means here a thickness of a few nanometers or less. Studies are motivated by the fact that the quality of such an oxide (morphology, concentration and type of defects, thickness variation) and of the quality of the interface (roughness, stoichiometry, electron traps) are more critical in a small device than the quality of a thicker oxide and its interface in a bigger device. In addition, it is both interesting and of practical relevance that, as the oxide thickness decreases, the essential parameters of the oxide and of the interface are increasingly dependent on the atomic-scale state of the surface before oxidation. Finally, extremely thin oxides (down to one monolayer) will be possibly used as buffer layers between the silicon substrate and a gate dielectric which will substitute SiO_2 gate dielectrics in CMOS transistors around the year 2012 or earlier (page 355).

Our discussion is restricted to dry oxidation from O_2 ambients, with a few remarks on steam oxidation. Thermal oxidation in dry O_2 is the established method used to produce gate oxides for CMOS devices. Let us only note that excellent dry thermal oxides grown in the flux of electrically neutral **oxygen atoms** (remote plasma oxidation) have recently been reported.[1110] Although these so-called "O^* oxides" grow many times faster than conventional O_2 oxides, thin films (6.6 nm) with interfacial roughness and dielectric breakdown characteristics remarkably better than the control O_2 oxides have been prepared. Maybe more importantly, O^* oxides are electrically reliable down to oxidation temperature of 700°C. This method or its derivative may find an application in CMOS processing.

Information on other oxidation methods can be found: for **water-** and steam-containing ambients, in Sect. 5.5.3, Sect. 6.3.1 (chemical oxidation), and Sect. 6.5.5d; for **nitrogen**-containing ambients, in Sect. 5.6.3 (including oxidation of nitrogen-implanted silicon), Sect. 6.3.4, and Sect. 6.5.5d; for **chlorine**-containing ambients, in Sect. 6.5.5b; for oxide **deposition**, in Sect. 6.3.3; for oxygen **implantation**, in Sect. 6.3.7.

6.4.1. *Deal-Grove model of growth rate*

From a phenomenological point of view, the growth rate of the oxide film during a reactive oxidation is determined by two basic processes: diffusion of the oxidant species through the existing SiO_2 layer and reaction of the oxidant with the substrate at the interface. When the oxide layer is already relatively thick (a few hundred Ångstrøms, or several times the thickness of contemporary gate oxides), the rate of SiO_2 growth obeys the well-known linear-parabolic Deal-Grove law.[1084] This law states that the growth rate dX_{ox}/dt falls off hyperbolically with the oxide thickness $X_{ox}(t)$:

$$\frac{dX_{ox}}{dt} = \frac{B}{2X_{ox} + A}, \text{ or} \tag{6.31}$$

$$t + \tau = X_{ox}^2 B^{-1} + X_{ox}\left(\frac{B}{A}\right)^{-1}. \tag{6.32}$$

Here, t is the oxidation time, τ is a correction for the presence of an initial oxide layer, B is known as the parabolic rate constant, and B/A as the linear rate constant. The Deal-Groove equation is derived by solving a simplified case of oxidant diffusion in SiO_2. The parameter B describes the diffusion of the oxidant and the parameter B/A enters as a consequence of the boundary conditions at the moving SiO_2/Si interface. The Deal-Grove

Parameter or process	Dry oxidation	Steam oxidation	
		below 950°C	above 950°C
$E_{B/A}$	2.0 eV	2.05 eV	1.60 eV
E_B	1.23 eV	0.78 eV	1.17 eV
O_2 diffusion	1.23 eV	-	
H_2O diffusion	-	0.78 eV	

Table 6.4. Approximate standard SUPREM values of activation energies in the Deal-Grove model[1142] and activation anergies of H_2O and oxygen diffusion in SiO_2.

expression has been used also for description of oxidation of other systems, for example, for oxidation of silicon nitride.[1195]

This model employs assumptions which are sometimes not fulfilled (Sect. 6.4.2). In particular, the time offset τ is in fact a simple device used to account for the presence of the native oxide (about 15 Å) and for the growth of the first several hundred Å of oxide. The Deal-Grove model does not work properly in this "initial" regime. Some of these assumptions can be relaxed in alternative approaches described in Section 6.4.2.

The value of B is proportional to the oxidant diffusibility in the SiO_2 film and to the partial pressure p of the oxidant. The value of B/A describes interfacial reactions and is proportional to the rate constant of interfacial reactions. For steam oxidation, $B/A \sim p$, but this dependence is sublinear for O_2 oxidation: $B/A \sim p^{0.75}$. This sublinear dependence may reflect the relative contributions of atomic and molecular reactions.[1196] It breaks the assumptions needed to derive the equations 6.31 and 6.32.

Both B and B/A are thermally activated. They depend on temperature with activation energies E_B and $E_{B/A}$, respectively (Table 6.4). While the dry process is characterized by the same activation energy in the whole range of technologically relevant temperatures,[1117] the presence of steam in the ambient causes a change[1111] in the activation energies at ~950°C. The low-temperature value of E_B coincides with the activation energy for diffusion of water in SiO_2; similarly, E_B in dry oxidation is remarkably close to the activation energy of oxygen diffusion in SiO_2.

The value of $E_{B/A}$ is often interpreted as breaking of Si–Si bonds at the interface, accompanied by stress generation. The change of activation energies at 950°C in steam oxidation is sometimes explained by viscous flow of SiO_2 above this temperature. Viscous flow[1197] releases a lot of the stress, apparently facilitating diffusion of the oxidant and rearangement of bonds at the interface. Water softens the SiO_2 network, assisting the flow.

The physics that controls the ratio B/A is of particular interest to us because it reflects interface processes. The reactions may seem to be uncomplicated, but atomic-scale details are not easy to resolve experimentally. Moreover, theoretical calculations are difficult and still rare. As a consequence, our knowledge remains more phenomenological than physical.

Its limitations and failures notwithstanding, the oxidation model formulated by Deal and Grove is an example of how a simple theory which is based on sufficiently physical assumptions can improve predictability of a technological process simulation, making at the same time the models easier to use in practice. Before Deal and Grove published their analysis of the oxidation process, several mutually conflicting, ad-hoc models were used. They did work within certain experimental data windows, but their formal parameters (as activation energies) scattered widely, giving an impression that oxidation is full of anomalies. A solid physical analysis cleaned the scene. Though the Deal-Grove model has severe limitations (Sect. 6.4.2), a well-defined meaning of its parameters helps to pinpoint the sources of the observed discrepancies and provides the conceptual assistance needed to extend the practical knowledge to new areas.

6.4.2. *Beyond the Deal-Grove model*

The key assumptions built into the Deal-Grove model are:

- the diffusing species do not dissociate/associate at the oxide surface;
- the oxidation reactions take place at the interface;
- the reaction rate depends linearly on the oxidant partial pressure;
- the influence of electric fields on diffusion is negligible;
- the influence of stress fields on diffusion is negligible;
- the flow of oxidants is steady.

Processing details determine the extent to which these conditions are met.

The issue which has been fascinating researchers for a long time is the behavior of oxidation rate during the initial stage of growth. When oxidation is done in a *dry* O_2 ambient, the interface moves faster than predicted by the standard Deal-Grove model until the film becomes 20-50 nm thick (which is at least several times more than a contemporary gate oxide thickness). For instance,[1188] at 950°C and under atmospheric pressure, the oxidation front of a 100 Å (10 nm) thick oxide moves nearly twice as fast as calculated from the Deal-Grove expression. At 500 Å (50 nm), the enhancement is still about 10%. On the other hand, *wet* processes obey the

Deal-Grove rule even for oxides as thin as 30 Å (3 nm).

Different models have been proposed to describe this fast growth phase. They explore mechanisms invoking (1) a reactive interfacial layer of a finite thickness, (2) the electric field generated during the oxidation, (3) the dependence of oxidation parameters on local stress, (4) enhanced solubility of O_2 in thin films, (5) the role of electron supply in the interface reactions, (6) diffusion of silicon to the oxide outer surface. The published models can be conceptually as divergent as they are numerically compatible. It is difficult to judge which of them gives the most correct physical picture of the actual process. It is plausible that the oxidation is indeed a complicated process with a rich spectrum of physical mechanisms which may manifest themselves in a similar way.

A simple experiment[1198] shows that some kind of a removable obstacle accumulates during oxidation: when the growth is interrupted and the sample annealed, the oxidation rate takes off with a speed about ten times greater than it was when the growth was stopped! This observation favors the first three mechanism (reactive layer, electric field, and stress) as capturing the major physical effects which control the rate enhancement. We will therefore analyse them in some more detail.

a. *Diffusing and reacting species*

Let us first identify the reactive species ("oxidants"). The candidates are: silicon, molecular[1184] or atomic oxygen, or water. Here we focus on oxygen and silicon. The role of H_2O was addressed in Section 5.5.3; moreover, it is dry oxidation that leads to the "anomalous" growth rate of thin SiO_2 films.

How could diffusing silicon come into play? Oxidation at the interface releases a certain amount of interstitial, that is, unreacted, silicon atoms (Sect. 6.5.5a). It is sometimes assumed[1157] that a fraction of these migrates through the SiO_2 layer. When the Si atoms reach the outer surface of the oxide, they can react with oxygen and contribute to the growth of the film.

In principle, isotope tagging can verify if new oxide is formed on the SiO_2 surface. When the usual ^{16}O ambient[m] is changed to ^{18}O-rich gas, the rare isotope builds itself into the chemically active region. When a source of ^{18}O is turned on in the last phase of oxidation, most of the ^{18}O atoms are found incorporated close to SiO_2/Si, as expected. But a weak ^{18}O peak can be detected close to the outer surface of the oxide.[1199] This can be

[m] The major isotope of O is ^{16}O, with almost 99.8% abundance. Oxygen has two other stable isotopes: ^{17}O (about 0.04% in natural substances) and ^{18}O (0.2%).

interpreted as a manifestation of a growth on the SiO_2 surface, that is, Si diffusion through SiO_2.

Can silicon diffuse through SiO_2 fast enough to reach the SiO_2 surface before it is oxidized and incorporated into the oxide network? Diffusivity measurements indicate[1200,1201] that lattice silicon migrates through SiO_2 extremely slowly: with its diffusivity $D \simeq 4{\cdot}10^{-4}$ Å^2/s at 1100°C, a Si atom needs nearly ten hours to be displaced by a the distance of one monolayer. Though the diffusivity of interstitial Si in SiO_2 seems to be orders of magnitude faster (Ref. 1202 and references therein), ^{18}O tracer studies[1203] indicate that the interstitial Si travels in SiO_2 less than 30 Å before oxidizing under typical conditions. Therefore, the diffusion of Si through the growing oxide may contribute to the enhancement of the oxidation rate in extremely thin oxides, but it can hardly lead to oxidation on SiO_2 surfaces of films with thicknesses in the ten nanometer regime or thicker.

For similar reasons, the surface peak is not caused by the diffusion of ^{18}O through the SiO_2 network.[1204] A recent analysis[1205] suggests that the peak is due to minute traces of water. This peak is hardly observable when care is taken to eliminate all water from the ambient.[1206] The residual water, apparently also enriched with ^{18}O, exchanges O with the SiO_2 network.[1205] This produces a surface ^{18}O-tagged region, which slowly extends into the film (water diffuses in SiO_2 slower than oxygen; Sect. 6.3.2).

Thus, the **major diffusing species** in dry oxidation is oxygen. During an important fraction of its life time, this oxygen is negatively charged, which manifests itself in a dependence of the oxidation rate on an external electric field.[1207] It is, however, unclear in which form does the oxygen diffuse: as O_2^- (superoxide ion[1188]), O_2^{2-} (peroxide ion[1207]), or as O^-. Some authors favor the oxide ion O^{2-}, though its electron affinity (Table 6.5) is negative, about −8 eV, that is, energy is lost when an electron is taken from *vacuum* and localized on the atomic orbitals of a free atom. However, atoms dissolved in a *solid* interact with the host and it is imaginable that SiO_2 stabilizes double negatively charged interstitial oxygen atoms. A series of good *ab initio* calculations could clarify this dillema.

The actual oxidation takes place when the oxidant comes close to the SiO_2/Si boundary. The **chemical reactions** at the oxidizing interface seem to involve atomic oxygen.[1185] This may be the reason for the sublinear dependence of the oxidation rate on the partial pressure p_{O_2} of dry O_2. *This non-linear dependence contradicts one of the fundamental assumptions*

electron capture			η [eV]
$O + e^-$	$\rightarrow$	$O^- + \eta$	1.45
$O^- + e^-$	$\rightarrow$	$O^{2-} + \eta$	-8
$O_2 + e^-$	$\rightarrow$	$O_2^- + \eta$	0.45
$O_2^- + e^-$	$\rightarrow$	$O_2^{2-} + \eta$	1.5

Table 6.5. Electron affinities of oxygen after Ref. 873. Electron affinity η is defined by the reaction $X^m + e^- \leftarrow X^{m-1} + \eta$, where X is a chemical species and m is its charge state. This charge state should not be confused with the *oxidation number* used in redox reactions, where oxygen is assigned the value -2 and is written as O^{-2}.

needed to derive the linear-parabolic Deal-Grove model: that the oxidation rate is proportional to p_{O_2}. Therefore, the parameters in the Deal-Grove expression may have a properly defined physical meaning only for H_2O oxidation and in the parabolic regime of dry oxidation. In the former case, the assumption about linear reactions holds. In the latter case, oxygen reacts faster than it can be supplied, so that interfacial reactions do not influence the oxidation rate: the process is diffusion limited.

b. *Space-charges, stress, and defects*

Interestingly, the growth-rate variations can be consistently explained solely in terms of ionic transport and chemical potential differences, with no explicit incorporation of interfacial reactions. The corresponding model is based on **space-charge limited** growth and employs an analogy with electrochemical cells.[1191,1174] It leads to a power-parabolic growth law:

$$t = X_{ox}^2 B^{-1} + X_{ox}^{2-\alpha}\left(\frac{B}{A}\right)^{-1}. \tag{6.33}$$

The exponent $0 < \alpha < 1$ is proportional to the amount of space charge close to the interface during the ionic transport. For $\alpha = 1$, the power-parabolic law is identical with the Deal-Grove expression. The condition $\alpha = 1$ is fulfilled in wet oxidation. The condition $\alpha \simeq 0$ corresponds to predominantly parabolic growth, which takes place at very high temperatures or at low O_2 pressures. Intermediate values of α are common for dry oxidation.

That the diffusing oxygen molecules or atoms are charged was revealed in an experiment reported[1207] already in 1962. When an external electric field is applied to the oxide during dry oxidation, the growth is either enhanced or slowed down, depending on the sign of the voltage. The retarding field, if sufficiently high, can completely stop the oxidation.

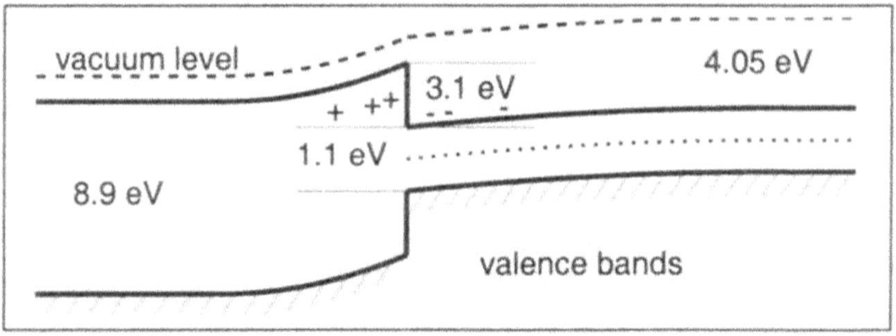

Fig. 6.3. Band diagram of a typical SiO_2/Si(001) oxidizing interface. The fixed oxide charges are indicated by "+", the compensating negative charges in Si are indicated by "-". To generate the compensating charge, the crystal bends its electronic bands such that additional electrons flow towards the interface or holes move away from it. This charge can be partially localized on interface states and on surface states of the oxide (not shown). The energies correspond to approximate band offsets and band gaps for thick to ~40 Å thin oxides.[1208,1209]

Given that oxygen is the most electronegative species after fluorine, one would expect that oxygen in SiO_2 is negatively charged. Indeed, the field which slows down the growth would oppose migration of negative charges from the surface towards the interface.

According to the space-charge model, the oxidation rate is limited by the positive charge which builds up in the interfacial layer (Fig. 6.3) during oxidation. The localization of this positive charge close to the interface is formally a result of an interplay between the continuous mass transport of the charged oxidant through SiO_2 and nonlinear dependence of the diffusivity on the electric field.[n] The electric field becomes unstable and disproportionates in such a way that most of the voltage drop is confined to a narrow SiO_2 layer immediately behind the interface. Physically, the field disproportionation comes from a positive space charge. This charge marks the defect sites from which the oxygen effectively takes its excess electrons. The resulting space charge is the origin of the fixed oxide charge in the vicinity of the interface.

Formal descriptions notwithstanding, it is natural that the width of the space charge region is limited. The charge brought to the interface by a charged oxidant must be taken back to the place of origin. Otherwise, a growing electric field will retard the oxidation. The process of charge

[n] The nonlinearity is a natural consequence of the energy barrier for diffusion.

neutralization is a possible candidate for a bottleneck which may limit the oxidation rate (for example, at low temperatures). On the other hand, if the oxygen diffuses through SiO_2 as a molecule, the charge collected by O_2 may help in the interfacial reaction. Superoxide molecules (O_2^-) are less stable than neutral O_2 and they dissociate easier. The bond strength of O_2^- is only 1.5 eV, to be compared to 5.1 eV of the neutral O_2 molecule.[873]

The space-charge model gives a unified description of oxidation rates of curved or stepped surfaces as in shallow trench isolation (page 378), formation of "bird's beaks" (page 377 and Fig. 1.2, page 8), and influence of chemical pretreatment on the growth rate:[1155]

- **Chemical cleaning**. The surface is charged and mobile ions (such as NH_4^+ for NH_4OH solution) are inserted into the oxide. This influences the electric field in the oxide, affecting the oxidation rate.
- **Curved or stepped surfaces**. Near the corner of a step in the oxide the electric field is inhomogeneous. Formally, this corresponds to an extra space charge. The field grows stronger at corners, making them oxidize slower because the oxidants are deflected towards flatter regions.
- **Bird's beaks** in LOCOS (Sect. 6.3.7). The high field caused by the space charge at the oxidizing interface reflects the oxidant particles. They stray under the nitride layer and a "bird's beak"grows.

A space-charge model for rapid thermal oxidation, that is, under illumination by very strong infrared light, is described in Ref. 1192.

Step coverage and bird's beaks can be modeled also by the local influence of **stress** on the growth rate. The stress-dependent scheme which could reproduce oxidation of trench walls and corners assumes[1157] that besides oxygen diffustion to the interface, also silicon diffuses through the oxide to the surface (dual Deal-Grove model). Note that this contribution to the oxidation rate, which may be non-vanishing for oxides several nanometer thick (Sect. 6.4.2a), is missing in the charge-controlled model.

Stress-controlled models have their advocates[1210] and adversaries.[1155] As an argument against the stress and in favor of the space charge one can quote the case of "bird's beaks" in LOCOS. This troublesome effect is usually explained in terms of stress buildup. But it is possible to prepare strongly strained LOCOS structures which show virtually no trace of any bird's beaks.[1155] This is done by plasma anodization, when the electric field helps to localize the oxidation below the openings in the nitride.

The two apparently different parameters (stress and space charge)

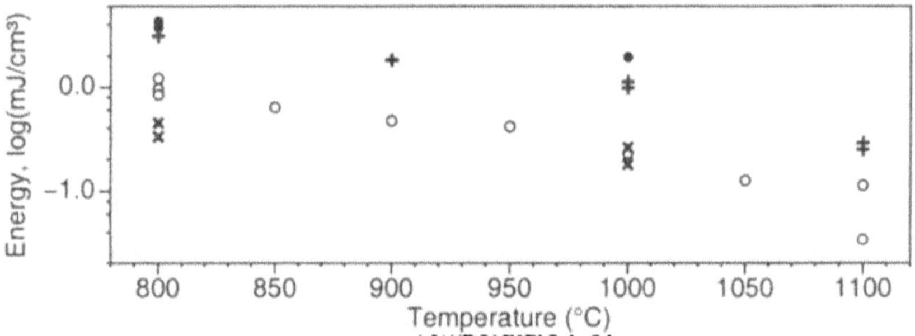

Fig. 6.4. Energy densities of electrostatic repulsion due to the fixed charge in SiO_2/Si(001) oxides (open circles) and of mechanical stress (other symbols). The mechanical stress corresponds to oxides of various thicknessess t_{ox}. Full circles: $t_{ox} \simeq 36$ nm, X-crosses: $t_{ox} \simeq 24$ nm, crosses: $t_{ox} \simeq 9$ nm. Data from Refs. 1155 and 1211.

that control the growth in these two models may be in fact interrelated to a considerable degree. There is a remarkable agreement[1155] between the magnitudes of the energy accumulated in the electrostatic repulsion of the fixed charge and in the mechanical deformation of SiO_2 (Fig. 6.4). On the other hand, consider the excessive elastic stress imposed on the SiO_2 by, for example, a stiff Si_3N_4 layer (deposited during LOCOS to suppress bird's beaks) or by a macroscopic step in the geometry of the substrate (such as the trench edges in STI, cf. page 378). This stress changes the chemical potential of reacted interfacial oxygen, the chemical potential of oxygen in SiO_2, and the viscoelastic flow of the oxide. This means that the energetic and dynamic aspects of the oxidation are affected. We believe that a full descritption of the oxidation process should include the mechanical stress and the electric fields in the SiO_2 as partially coupled variables.

Stress-dependent Deal-Grove models appear to be favored by Technology Computer Aided Design (TCAD) users and developers. The motivation appears to be simple: "don't fix it if it ain't broke". As long as a model works in a practical environment, it is not likely to be substituted by a new, even more realistic one. New models are destined to live in research labs until their predecessors become too troublesome to be useful.

Regardless of whether the space charge controls the oxidation rate, the formation of the fixed charge layer is a real effect.[o] It is clear that the fixed charge is localized on some defects in the oxide. The observation that

[o] But only a part of the charge predicted by the space charge model is retained in the form of fixed charge when the substrate is cooled down to room temperature.

a cloud of defects follows the oxidizing interface is at the root of the model of defect-mediated growth.[907] The defected interfacial layer is treated as the region in which the oxidation takes place. The relevant processes include charge transfer, bond breaking, oxidation, and relaxation of stress. A related computational model based on the diffusion-reaction formalism has been recently developed.[1194] We discuss the interfacial reactive layer in Section 6.5.4c, after analysing the atomic interfacial structures.

c. *The practical approach*

Given this complex situation, professional TCAD software like SUPREM-IV takes a pragmatic approach, adding a correction $K \exp(-X_{ox}/L)$ to the Deal-Grove rate equation.[1092] The parameter $L \simeq 70$ Å is an empirical decay length, approximately independent of temperature. The rate constant K is thermally activated: $K = K_0 \exp(E_A/kT)$, with $K_0 \simeq 6 \cdot 10^{10}$ Å/min being roughly independent of the orientation and E_A ranging from about 2.3 eV for $SiO_2/Si(001)$ and $SiO_2/Si(111)$ to 1.80 eV for $SiO_2/Si(110)$.

A natural drawback is that the physical meaning of the model parameters is obscure, if it exists. Naturally, a properly calibrated model is very useful, since it allows one to compute the evolution of the device geometry. This works fine for existing technological processes, yet it does not provide a reliable assistance in the development of new processes for new designs. For instance, such information as the dependence of local thickness variations on process conditions must be extracted from new experiments. A designer looking for innovative solutions still cannot count on computer programs but must rely on his or her experience, sixth sense, and a bit of luck.

6.5. Formation and Structures of SiO_2/Si Interfaces

The microelectronic circuits which revolutionize our world are built exclusively on wafers with the (001) orientation. Why this devotion to Si(001)? The answer is hidden at SiO_2/silicon interfaces. Table 6.6 lists some references pertinent to the microscopic structure of this boundary.

As the superiority of silicon dioxide over germanium dioxide won the microelectronics market for silicon, the superior structural quality of $SiO_2/Si(001)$ interfaces over $SiO_2/Si(111)$ interfaces[1064] won this market

Ref	year	Subject relevant to SiO_2/Si interfaces
1212	1955	Interface scattering and carrier mobility; theory
1161	1967	Pressure, temperature dependence of fixed charge: Deal triangle
1213	1977	SiO_2/Si interface is very sharp; TEM
1043	1978	**Conference series**: The Physics of SiO_2 and Its Interfaces
1214	1979	Fourfold Si–O–Si rings near the interface, sixfold rings further away
1215	1983	Structure and stoichiometry of $SiO_2/Si(111)$; ion channeling
1064	1984	Quality of SiO_2/Si interface is crucial for MOSFET
1216	1984	Midgap states at SiO_2/Si are due to dangling bonds
1217	1986	Interface traps investigated by STM
1089	1987	Atomically smooth $SiO_2/Si(001)$ interfaces can be grown; TEM
1218	1987	Atomic models of $SiO_2/Si(001)$; Keating potential
1219	1987	P, B, As segregation at $SiO_2/Si(001)$; SIMS
1159	1988	α-cristobalite–like microcrystallites at $SiO_2/Si(001)$; X-rays
867	1989	Layer by layer growth of $SiO_2/Si(001)$; XPS
1220	1990	Models of tridymite/Si(001) interfaces; semi-empirical theory
1221	1990	Dangling bonds at β-cristobalite/Si(001) interface; *ab initio* theory
1071	1991	Smoothness of SiO_2/Si interface is crucial for MOSFET
1049	1992	**Review**: SiO_2 films: composition, structure, morphology
1174	1992	Analytical expression to estimate the fixed oxide charge
1222	1993	**Review**: Fixed charge, interface traps, terminology, structure, noise
835	1993	**Review**: Point defects and charge traps in SiO_2/Si
1094	1994	Control of native oxides; XPS
844	1994	$SiO_2/Si(001)$ interface for different oxidation processes; TEM
841	1995	Ordering of thermally grown oxide films; X-ray diffraction
1067	1995	Dopant segregation to SiO_2/Si; overview, SIMS analysis for P
842	1996	Importance of initial surface quality; TEM, STM, AFM
1223	1996	Core-level shifts at $SiO_2/Si(001)$ interfaces; *ab initio* theory
1224	1996	Microvoids found at $SiO_2/Si(001)$; positron annihilation
1050	1996	**Review**: SiO_2 defects, formation mechanism, role in diffusion
1225	1997	Si=O at β-cristobalite/Si(001) interface; *ab initio* theory
1166	1997	Influence of annealing on $SiO_2/Si(001)$ roughness; TEM
1226	1997	Tridymite/Si structures; *ab initio* theory
1156	1998	SiO_x, Si bridges and interstitials; *ab initio* molecular dynamics

Table 6.6. SiO_2/Si interfaces: selection of references

for (001)-oriented wafers. A good gate oxide with a good interface[p] should be nearly atomically smooth, have as few electrically active defects as possible, and a minimal concentration of carrier traps which can be charged either permanently (fixed charges) or temporarily (fast states). Flaws at the interface and in the oxide destabilize the technological process, impede the reliability of the devices, and shorten their life time. The fast traps, located at or near the interface, cause noise.[1227,1086] The fixed oxide charges, located somewhat deeper in the oxide, deteriorate the electrical character-

[p] See Section 6.6 for practical problems.

istics of the transistor. Since their influence on the electric field in the MOSFET channel is the same as that of a gate voltage, they affect the threshold voltage V_T. In a particularily bad case they may even lock the transistor in a single logical state. The charge bonded temporarily by the fast states also contributes to this deterioration of V_T. In addition, both types of traps reduce the mobility of carriers in the channel (by Coulomb scattering). Microroughness of the interface[1071] worsens the mobility as well. At comparable preparation conditions, the density of interfacial defects is the lowest for (001) interfaces and the highest for (111) interfaces.

The SiO_2/Si(001) interface is buried below the oxide film and, thus, not accessible to direct microscopic measurements. The most straightforward data on its atomic structure comes from X-rays.[1228] The measured core-level shifts are usually interpreted as reflecting the oxidation states of Si atoms, that is, the number of oxygen neighbors. This is consistent with *ab initio* analysis of binding energies of 2*p* electrons on Si atoms in the interface region.[1229] One might therefore suppose that the abundance ratios of oxidation states provides information on the atomistic details in the interface region. This is, however, not necessarily the case because the presence of various oxidation states may be related to intrinsic, atomic-scale roughness of the SiO_2/Si boundary.[1166]

Several important structural models of the clean interface have beeen investigated by *ab initio* methods (for example, Refs. 1221, 1223, and 1156). In addition, results of numerous model calculations are available (for instance, Refs. 1220, 1230, and 1231). For extremely thin oxides (around 2 nm, or 20 Å) also STM measurements have been reported.[1217,1232]

In spite of the huge lattice mismatch, the thermal oxidation of Si(001) may proceed in an orderly manner.[838] Layers of astonishingly good quality and nearly perfectly joined with the substrate can be produced,[1213,1231] with no more than one dangling bond for several hundred interface atoms. Most of these dangling bonds can be later saturated with hydrogen atoms; eventually, only several interface atoms in a hundred thousand have unsaturated, broken bonds. Partially *ordered* layers may grow up to about 5 Å (0.5 nm, or roughly two monolayers) thick.[1089] Such quasi-epitaxial interfacial oxides have also been observed in wet oxides and in room-temperature dry oxides.[1233,1228,1234] However, conductance measurements indicate that oxidation leads to a substantial atomic-scale roughening.[1235]

It seems clear that the SiO_2/Si interface energy tends to keep the interface smooth. The roughness is most probably caused by stochastic fluctuations which are not effectively flattened out by the interfacial energy because the growth conditions are far from thermal equilibrium. In addition, the electric fields due to the space charge created during oxidation (Sect. 6.4.2b) may enhance the tendency of the interface to roughen.

In general, there exists an empirical rule: *faster oxidation rate means more roughness and more defects.* This applies to surfaces with the same crystallographic orientation, oxidized at different oxidant pressures, as well as to surfaces with different crystallographic orientations oxidized at the same conditions. Since the atomic mechanism of the interface formation is not fully understood, physical mechanisms beyond this rule remain largely elusive. In general, it seems to be related to deviations from thermal equilibrium: the higher is the oxidation rate, the less equilibrated is the growth.

Would a complete explanation of the relation between oxidation rate, roughness, and defect density be of practical importance? Though it is not particularly likely that research aimed to understand such differences by comparison between various orientations of Si will have a *direct* impact on the development of semiconductor technologies, it may still affect it indirectly. This may be done through a contribution of such studies to the physical knowledge about SiO_2/Si(001) processes. Physical insight gained in this way may bring a major reward when transferred to the right people at the right time. Unfortunately, finding "the right people at the right time" is often the most difficult part. But the researcher who surmounts this barrier may discover behind it a straight path leading to an impressive success (see the impact made by the Deal-Grove model).

6.5.1. *Interface formation: a simple analysis*

Silicon planes of different orientation vary in atomic densities (Table 1.3, page 49), in abilities to accommodate strain, and they are interconnected by bonds arranged in different ways. Thus, truncation of the bulk at these planes leads to different systems of "dangling" orbitals (Fig. 1.5, page 48). It is therefore not astonishing that the oxidation process depends on the orientation of the crystalline-silicon side of the interface.

When the oxidized layer is not too thin, the speed with which interfacial reactions consume the substrate is the highest in the SiO_2/Si(111) system and the lowest in the SiO_2/Si(001) system, with other orientations falling between these limits. A (111) plane has more densely spaced atoms (~0.16 atoms/Å^2, cf. Table 1.3 on page 49) than a (001) plane (~0.07

atoms/Å^2). It follows that an oxidant attacking along (111) has nearly 2.3 times more targets than an oxidant attacking along (001). An often invoked argument says: an oxidant on a denser plane oxidizes faster, hence the oxidation rate of Si(111) is higher than the oxidation rate of Si(001).

How does it work? Assume that each oxidant particle approaches many Si atoms before it finally chooses one as an object of attack. If the areal density of atoms, σ, increases, the average time between oxidation attempts decreases, and so does the life time of the oxidant at the interface. The functional character of the life-time dependence on the atomic density is related to the diffusion path at the interface. If the particle moves along bonds, then the attempt frequency is independent of the atomic density, because the distance covered by the particle between individual attempts is always the same. If the particle moves at random in the interface plane, the life time τ varies as the reciprocal average lateral distance between atoms: $\tau \sim 1/\sqrt{\sigma}$. In a general case, $\tau \sim \sigma^{-\alpha}$, where $0 \leq \alpha \leq \frac{1}{2}$. On the other hand, the number of successful attacks N needed to oxidize a plane is proportional to the atomic density, $N \sim \sigma$. When a single plane of targets is destroyed, the interface moves forward by a distance d equal to the monatomic step on this surface. Monatomic step height d is proportional to σ (because the ratio of the areal density σ and the step height yields the bulk density of the material, which is a constant). This means that, depending on the details of the interaction between the oxidant and the interface atoms, the areal density σ translates into the oxidation speed $d/(\tau N) \sim \sigma^{\alpha}$, that is, between 1 and $\sqrt{\sigma}$. For (111) and (001) planes this corresponds to enhancement by the factor between 1 and 1.5.

The upper bound of 1.5 is somehow below the experimental result of 1.7. This indicates is that the preexponential factor of the interface reaction rate is affected by diffusion paths or by interface reconstructions. It follows that a quantitative description of oxidation requires atomistic knowledge about interface structures at different stoichiometries and for different interface orientations (which is not very astonishing). Nevertheless, already the geometry of a "truncated-bulk" interface gives us some insight (which is nice): if for some reasons the next plane can be attacked *only* when the previous plane is consumed, then the difference in the oxidation rate is roughly explained by differences in inter-planar distances. Such layer-by-layer advance of the interface, consistent with the fact that SiO_2/Si interfaces are very sharp, has indeed been observed.[867] Experiments show that the interface has a tendency to self-smoothing: the interface roughness of oxides thicker than 200 Å (20 nm) is independent of the oxide thickness.[1236]

Can one easily envision atomistic mechanisms which produce such a "stepwise" motion of the interface? Assume that the reaction probability is higher when more bonds of the involved Si atoms have been already oxidized[q] A collective motion of the interface through step flow or is thereby promoted, because protrusions become unfavorable. In addition, model calculations[1237] indicate that completely oxidized planes have lower elastic energy. In fact, any model for which a flat interface has a lower energy than a stepped interface will reproduce the tendency to keep the interface smooth. Reconstructions of interface atoms as well as reaction pathways may stabilize layer-by-layer growth of the oxide, just like surface reconstructions on a clean or surfactant-covered surface can stabilize layer-by-layer epitaxy. The destabilizing factors are entropy and deviations from thermodynamical equilibrium.

An oxidizing interface is thus in many aspects similar to a surface on which new material is deposited (Sect. 4.4). By analogy, we would therefore expect that, depending on the interfacial reaction rate and availability of the oxidant, either the step flow may dominate the growth, leading to a smooth interface, or nucleation and growth of "islands" will prevail, leading to a rough interface. Annealing should make the interface smoother. This is indeed the case:[1166] $SiO_2/Si(001)$ interfaces produced in dry O_2 at 900°C are rough (10–15 Å rms), but annealing at the same temperature in an intert ambient (N_2) reduces this roughness to 2.2 Å.

Oxidation rates, interface roughness, interface defects, and fixed charges appear to be correlated through orientation of the surface: a surface orientation with a higher oxidation rate yields a rougher interface with more defects and more fixed charges in the oxide. The correlation between orientation and roughness is the easiest to rationalize. It suffices to note that the unit of roughness is the surface-step height! Assuming that the physical mechanisms which govern the oxidation of Si(001) and Si(111) are the same, the roughness of Si(111) should exceed the roughness of Si(001) by a factor of 2.3. Indeed, a direct comparison[1064] yields a factor of 2–3.

Approximately the same enhancement ratio has been observed[1064] for the density N_{ST} of electrically active interface defects (apparently silicon dangling bonds) and for the density Q_f of fixed charge. The latter is due

[q]For example, because stress and/or charge re-destribution weaken the remaining bonds.

to defects located in the oxide within 30–100 Å from the interface. To understand this correlation we may invoke some simple thermodynamic aspects of the interface propagation. An equilibrium layer-by-layer growth is possible when the oxidant supply is sufficiently low and the temperature is sufficiently high to allow for re-distribution of interfacial oxygen to configurations which minimize the interface enthalpy. When the supply of oxygen is too high or the temperature is too low, local fluctuations cannot be smoothed out and the interface roughens.

This explains behavior of surfaces with the *same* orientation oxidized at different oxidant pressures and at different temperatures. In agreement with experiment, one would predict that too fast oxidation means not only rough interfaces but also high concentration of defects. The interesting question is: can one use this to explain the correlation between oxidation rate and defect densities for *different* surface orientations?

If additional assumptions are not made, then the answer is "No". Namely, the quantity which can measure the extent of thermodynamic equilibrium is the ratio between the areal oxidation rate $r_o \sim 1/(N\tau)$ and the lateral diffusion rate $r_d \sim 1/\tau$ of oxygen. A lower r_o/r_d means more efficient equilibration and less defects. Since $r_o/r_d \sim 1/N \sim 1/\sigma$, an orientation-independent lateral diffusion rate would lead to a more equilibrated oxidation of high-density planes. In other words, $SiO_2/Si(111)$ interfaces would be expected to contain *less* defects than $SiO_2/Si(001)$ defects prepared at the same conditions. Apparently, this prediction is wrong.

A possible solution is that the diffusion of oxygen proceeds faster along $SiO_2/Si(001)$ than along $SiO_2/Si(111)$. This diffusion should depend on the details of the atomic structure of the interface. Even these structures are still controversial. Predictions about the diffusion rate simply do not exist.

Another explanation is that a completely oxidized, denser plane accumulates more stress than a completely oxidized, less dense plane. Put differently, denser planes are expected to have a higher stress contribution to their interfacial energy than less dense planes. Higher stress energy can promote the formation of less stoichiometric oxide, which is likely to be accompanied by a higher density of interfacial traps and fixed charges. Since the equilibrium values of defect concentrations are also orientation-dependent, the structural, atomistic aspects of are certainly important. But again, even qualitative predictions require reliable information about interfacial atomic structures.

The conclusion is that some very fundamental aspects of interface reactions, like the enhancement of defect density at SiO_2/Si(111) compared to SiO_2/Si(001), still escape understanding. This makes hand-waving explantions of orientation-dependent growth rate and roughness questionable. Either the knowledge needed to understand what is happening at SiO_2/Si interfaces is not there, or a simple model description which could capture the salient qualitative consequences of interfacial reactions is missing.

6.5.2. *Atomic structures of SiO_2/Si(001)*

The atomic structure of the SiO_2/Si(001) interface is known only approximately. The arrangement of atoms at the interface depends on the oxidation conditions (for example, see Refs. 829 and 836 and references therein), that is, on the oxidizing ambient, temperature, and mechanical stress. The structure of the SiO_2/Si(001) boundary may resemble locally an interface with tridymite[1089] (Sect. 6.5.2c) or cristobalite[1159] (Sect. 6.5.2d), particularly at lower oxidation temperatures; even as much as 10% of the interfacial area may be ordered. The remaining area may represent a rather straightforward extension of bulk silicon, obtained by insertion of oxygen atoms into Si–Si bonds and rearangement of more distant SiO_2 atoms into a random network[1237] (Sect. 6.5.2e). Oxidation at higher temperatures, as during gate oxidation for commercial CMOS devices, or post-oxidation annealing tend to relax the interface by introducing more disorder. Upper, bulk-like layers of thicker thermal SiO_2 films are fully amorphous.[1238]

Below we collect and then put together different pieces of data. Our attempt to reconstruct a part of the jigsaw-puzzle should by no means be taken as the complete solution. Yet, since SiO_2/Si interfaces are subject to intensive studies, there are good chances that the nearest future will present us either with the missing pieces or even with a sketch of the picture which the puzzle should display when finally solved. One may hope that when SiO_2/Si(001) structures and the physical mechanisms governing their formation are understood, an efficient scheme to produce better interfaces will emerge. An optimal interface would minimize the density of electrically active defects and precursors of such defects in order to increase reliability of the gate oxide (Sect. 6.6), and would have a low stress in order to reduce segregation of dopants (Sect. 6.5.5c). The spectacular improvement of the oxide and interface quality which was achieved over the last few decades provides some rational footing to these hopes.

a. *Non-stoichiometric layer:* SiO_x

To start with, we consider the stoichiometry of the interface.[836] The transition from SiO_2 to silicon takes place within the so-called SiO_x layer. This layer is up to 7Å thick.[829] The index x indicates that silicon atoms in this region have varying numbers of oxygen neighbors. The thickness of the SiO_x region depends on the temperature of growth or post-growth annealing. With increasing temperature, it decreases and the interface becomes chemically sharper. The presence of the SiO_x layer and the compressive stress in it have an adverse effect on the reliability of gate oxides:[1154] weaker bonds are more easily damaged (cf. also Section 6.6.2a).

Experimental data on the stoichiometry of SiO_x comes from core-level spectroscopy. The $2p$ electrons of Si atoms in the transition layer have different energies than their counterparts on the Si atoms in SiO_2, that is, than atoms in the 4+ oxidation state. The observed binding energies are peaked around 0.9, 1.8, and 2.5 eV above the $2p$ binding energy on the Si^0 atoms of the silicon bulk.[1239] They are traditionally interpreted as coming from atoms with one, two, and three oxygen neighbors. This view has been confirmed by *ab initio* calculations[1223] as valid for at least two classes of the interface (with tridymite and with cristobalite). The assumption that there exists a thin transition region between stoichiometric Si and stoichiometric SiO_2 is thus consistent with the current status of *ab initio* theories. However, this does not exclude the possibility that this region may be intrinsically related to atomic-scale roughness of the interface rather than to a specific atomic arrangement in "perfect" areas of a smooth interface.

The interpretation of the $2p$ core-level energy spectra as representing many different ionization states of silicon has been questioned.[1240] The proposed alternative is that differences in the binding energy reflect the various atomic configuration of atomic next-nearest neighbor shells. This opens a possibility that the transition region is in fact much sharper than usually supposed, that is, that the existence of a "SiO_x" layer is caused solely by crystallographic constraints. If this is true, then the simplest of the structural models discussed below could be correct and attempts to extend them to reproduce the measured oxidation spectra would be mistaken. However, this interesting analysis is based on indirect experimental evidence and it is at variance with the available *ab initio* results.[1223] On the other hand, one must keep in mind that SiO_2 belongs to computationally demanding materials. The theoretical results, though convincing in our private view, must be treated with care. *Vide* the didactive story of the atomic structure of the clean Si(001) surface (Chapter 2)...

b. *Three classes of SiO_2/Si(001) models*

A number of atomic models have been proposed for the SiO_x transition region and for the SiO_2/Si(001) interface. We will focus on a few structures, chosen because they are relatively simple and because we believe that they represent the configurations which are the most relevant to this system.

TEM images can be interpreted in terms of a tridymite (hexagonal) structure, rotated and strained to fit the Si substrate.[1089] In contrast, X-ray studies[1159] reveal interfacial microcrystallites resembling cristobalite. This is not necessarily a contradiction, since the structure which is actually realized must depend to certain extent on the oxidation conditions. An ordered interface is a metastable configuration imposed during oxidation by kinetic constraints. Only a fraction (as 10%) of the whole interface area is ordered, and this fraction is reduced by annealing. The rest of the interface is most probably amorphous on the SiO_2 side. Such a "random network" can utilize the various structural constructs typical for interfaces of Si(001) with the crystalline polymorphs of SiO_2.

As in previous Chapters, we will illustrate some structures by sketches showing [110] cross sections of the interface. Thus, the abruptly truncated Si(001) is sketched as:  (Fig. 3.1, page 130). The solid line at the bottom indicates the Si substrate, and the surface Si atoms are symbolized by full circles. Each of them is connected to the substrate with two backbonds; viewed from this perspective, all backbonds lie in the plane perpendicular to the plane of section, so only one backbond per atom is visible. The two electrons from the two broken bonds per atom are marked by small dots; the broken bonds lie in the plane of section, hence both of them are visible. Bond lengths and angles displayed in our sketches are chosen for convenience and are not related to their real or optimal values.

c. *Class 1: tridymite-based models*

Tridymite-based structures[1089] provide a natural local description of at least the first interfacial SiO_2 layer. The relative orientation between the tridymite and the substrate is such that the areal density of Si in each tridymite layer parallel to the interface is the same as in an {001} plane of bulk silicon. The silicon atoms in the first tridymite layer are separated from the substrate by oxygen atoms intervening in the Si–Si bonds.

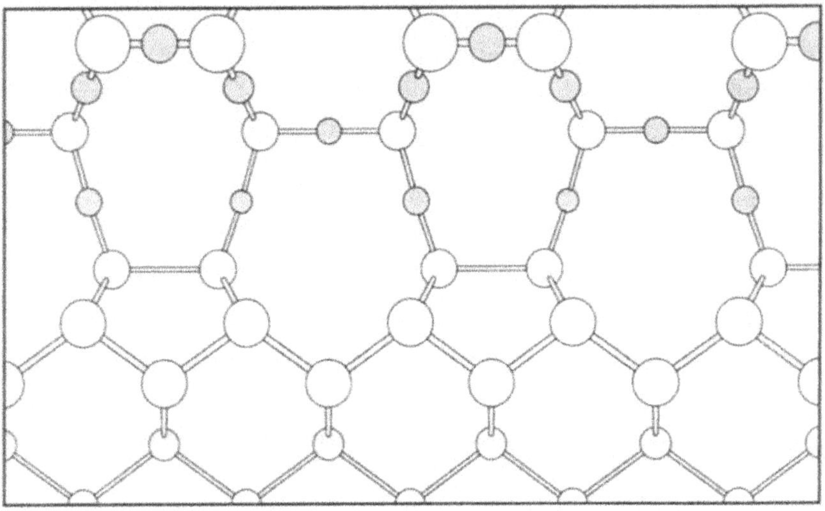

Fig. 6.5. Ball-and-stick model of the interface between ideal tridymite and Si(001). Silicon atoms are white, oxygen atoms are gray, atoms further from the observer are smaller. The formation of interface dimers, resembling those of Si(001), places the interfacial silicon atoms in positions reasonably close to the silicon sublattice in tridymite. The Si–O–Si angles, shown as 180°, vary in reality between 140°and 150°.

Interfaces with tridymite involve dimerization of the boundary Si atoms, as on the Si(001) 2×1 surface (). This helps to relax strain by pointing the bonds of the substrate Si in more suitable directions (Fig. 6.5). The dimers are ordered in rows as on Si(001) 2×1. The top view of the interface plane is: 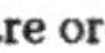, where dashes symbolize the dimers.

The tridymite film is compressed. The *c*-axis of the hexagonal tridymite is parallel to the dimer bonds. With the lattice dimensions $c = 8.23$ Å of tridymite and $2a_s$=7.68 Å of Si(001), the mismatch between Si(001) and tridymite is about 7% along the dimer bonds. In the direction perpendicular to the dimer bonds, the mismatch is about 13%.

In the simplest model of a tridymite/Si(001) interface,[1089] oxygen atoms of the tridymite side are connected with the "dangling" bonds of the dimers: 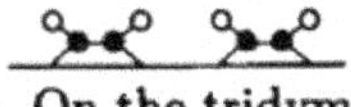 (Fig. 6.5). The oxygen atoms are indicated by open circles. On the tridymite side, each boundary oxygen atom is bonded to a single Si atom of SiO_2; these bonds are not shown in the sketch.

This interface introduces only single oxidation states; these are due to the Si^+ atoms of the dimers (). It is straightforward to extend this model to contain also double oxidation states: when an oxygen atom intervenes in a dimer bond (), both Si atoms of the attacked dimer become doubly oxidized, that is, Si^{2+}. Alternatively, the intervening atom may be located in one of the dimer backbonds (); in this case the affected dimer atom becomes Si^{2+} and its former neighbor in the substrate becomes Si^+. Both configurations can be now combined to produce a triple oxidation state, Si^{3+} ().

The tridymite cannot be easily continued when the interface moves into the crystal. Assume that the bonding of the SiO_2 film is frozen in the tridymite-like structure in Fig. 6.5 while the oxidation goes on below that film. Three in four newly oxidized silicon layers must then be completely reorganized in order to grow a crystalline tridymite. Only every fourth layer of the substrate silicon can match this extended frozen tridymite by dimerization isomorphic to that in Fig. 6.5. It seems much more probable that the film above the interface becomes amorphised during growth. Local tridymite-like regions may appear and disappear in this process.

d. *Class 2: cristobalite-based models*

The crystallographic structure of crystobalite is even more closely related to that of silicon than the structure of tridymite. Ideal crystobalite crystals are cubic, with silicon atoms located in the diamond lattice and joined by straight Si–O–Si joints.[r] But the dimensions of the crystobalite unit cell fits badly to the Si lattice. To see how a crystobalite microcrystallite and the Si substrate are arranged in SiO_2/Si(001), imagine that crystobalite is made on (001)-oriented silicon by insertion of O atoms into Si–Si bonds. Such films do not exist in nature; they would be too strongly compressed. Imagine now that this hypothetical film is cut off from the substrate. The free film relaxes, expanding substantially. Let us then rotate the film by 45° around the surface normal and re-attach it (Fig. 6.6). The SiO_2 stretches somewhat and attaches itself through O bridges to every second atom of the Si(001) surface. The characteristic feature of cristobalite-based models[1221] is thus that the density of Si in the first SiO_2 layer is twice *lower* than in an {001} plane of bulk Si. This rotated and stretched cristobalite fits much better the substrate than the hypothetical, compressed film.

Perfect interfaces with cristobalite have 2×2 periodicity. This 2×2 or-

[r]The Si–O–Si angles in real cristobalite are 140–150°. This does not affect our discussion.

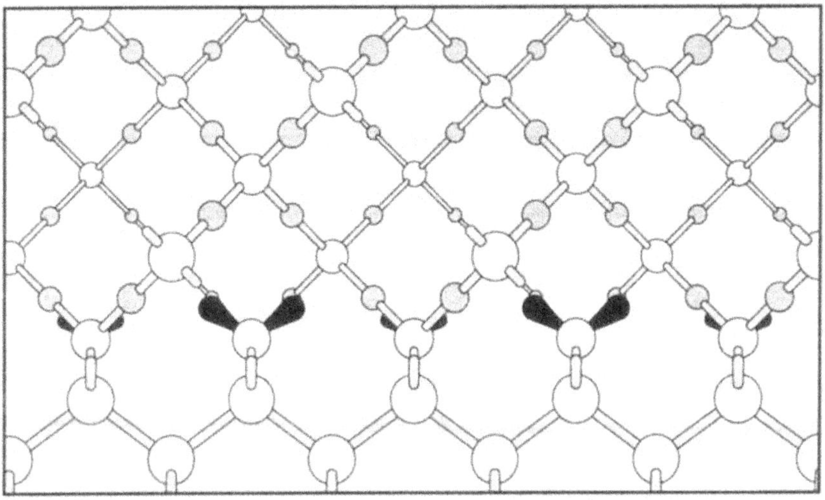

Fig. 6.6. Ball-and-stick model of a simple interface between ideal cristobalite and Si(001). Silicon atoms are white, oxygen atoms are gray, "rabbit ears" dangling bonds are black, atoms further from the observer are smaller. The Si–O–Si angles are 180° in ideal cristoballite; in reality, they vary between 140° and 150°.

dering is intrinsic to the arrangement of atoms in the boundary plane. The SiO_2 film is *expanded* by 7% with respect to a free cristobalite crystal.[1225] This seems to indicate that cristobalite is not the dominating structure: experimental reports mention only compressive stress in flat $SiO_2/Si(001)$ films. However, note that the electrostatic repulsion between fixed charges may be a significant source of compressive stress (Fig. 6.4, page 390).

In the simplest model of a cristobalite/Si(001) interface, the boundary atoms of the silicon side are not dimerized. The interface can be sketched as: ; the bonds connecting the interfacial oxygen atoms (open circles) with Si atoms in cristobalite are not shown. The 2×2 periodicity is caused by the $c(2\times2)$ ordering of the interface Si atoms connected through the oxygen atoms to the cristobalite: the top view of the interface plane is like: 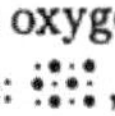, where the thick dots indicate the silicon-SiO_2 connection points, and the small dots indicate the Si atoms that make no bonds to O. The symmetry of this centered surface cell is then reduced to the primitive 2×2 by the cristobalite half-infinite crystal.

It is possible to construct cristobalite/Si(001) interfaces which have no broken bonds. Two of these constructs will be mentioned below.

First, the broken bonds can be saturated by H atoms or OH radicals (when water is available), or by oxygen atoms:[1225] . Each saturating oxygen forms a double bond with the substrate, creating a silanone complex (, compare Section 5.4.3a). The silanone oxygen atoms are *not* connected to the SiO_2 side of the interface. Though all interface Si atoms in this model are double oxidized, one can easily add other oxidation states, for example, by inserting oxygen atoms into Si–Si backbonds.

The other model of the interface, the oxidized dimer structure proposed by Ohdomari,[1218] is more complicated, but it illustrates that a simple rearrangement of bonds at the interface can reduce the number of unsaturated electrons. To visualize this geometry it is convenient to begin with the simple interface (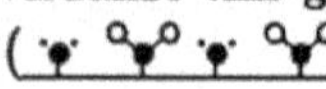, Fig. 6.6). Each of the silicon atoms at the Si–SiO_2 connection points is bonded to SiO_2 through two oxygen atoms. Each of these oxygen atoms makes a bond to a Si atom in SiO_2. These Si atoms of SiO_2, not shown in the sketch but easy to find in Fig. 6.6, are the nearest silicon neighbors, Si^{NN}, of the interface on the cristobalite side of the boundary.

We will now switch the bonds between the atoms such that all dangling bonds disappear. The reader may want to repeat the steps described below on a model made of real balls and sticks.

We begin by severing the bonds between the Si^{NN} atoms and a half of the interfacial oxygen atoms. The Si^{NN} atoms now have a single broken bond each. Next, we join the broken bonds of the oxygen atoms with half of the dangling bonds at the silicon side of the interface: . Finally, we join the broken bonds of Si^{NN} atoms with the other half of the dangling bonds of Si atoms on the substrate side. The final structure may be represented by: 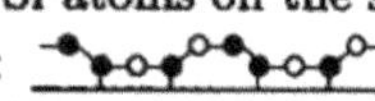, where the Si^{NN} atoms are indicated as full circles in the upper part of the sketch.

The characteristic feature of the Ohdomari model are Si–Si bonds between triply oxidized Si^{NN} atoms of cristobalite and singly oxidized interfacial atoms of the substrate. This model has the same number of Si^{3+}, Si^{2+}, and Si^{+} atoms. The model employs an interfacial dimer with an in-bonded O atom (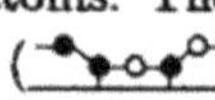). This is similar to the oxidized interfacial dimer in tridymite/Si interfaces, which we have symbolized as: .

e. *Class 3: random network interface*

Crystalline SiO_2 is built of six-fold rings (counting a Si–O–Si segment as a "fold"), like crystalline silicon (counting a Si–S bond as a "fold"). If one allows for formation of rings containing other numbers of Si–O–Si segments (from four to eight), as in amorphous silica, it is possible to construct a random-network SiO_2/Si interface which has no dangling bonds and contains relatively little strain.[1237] Such models may resemble tridymite/Si(001) models in the sense that they may conserve the silicon bulk-like density in the first SiO_2 layer. The main difference is that the upper SiO_2 layers form atomic rings of variable length. Dimerization of Si atoms is possible. Recent *ab initio* molecular dynamics calculations[1156] have shown that in addition to dimers, also interstitial-like atoms may be incorporated at the interface in order to minimize the interfacial stress and to provide the best topological match between the two materials.

The additional freedom which the amorphization gives to the SiO_2 side of the boundary leads to a conceptually simple interface. If we neglect variations in bond lengths and angles, occasional complications such as dimers, bridges, and especially the presence of Si interstitials, then we can view the silica-side Si atoms in the nearest neighborhood of the interface as occupying their bulk Si sites. The main difference between the interface and the Si bulk is then that oxygen atoms intervene in Si–Si bonds.

Thus, if the oxygen atoms are ignored, the interface may be schematically represented by: , where the upper thick dots are the Si atoms in the amorphous SiO_2 and the lower thick dots are the Si atoms in the silicon substrate. A sketch of the perfect Si bulk would look exactly the same. By inserting oxygen atoms (empty circles) into the interfacial Si–Si bonds, we obtain a structure like: 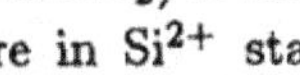. The resulting stress is largely relieved by relaxation in SiO_2. This is facilitated by the freedom to make randomly distributed rings with a variable number of "folds".

The most characteristic feature of the random network interface[1237] is that it is atomically sharp. The deviation from the ideal oxidation states of zero (as in bulk Si) and four (as in bulk SiO_2) is limited to the interfacial atoms on the substrate side, which are in Si^{2+} state. If this model is fundamentally correct, the SiO_x transition region is due to microroughness of the interface caused by the stochastic process of oxidation. But recent *ab initio* results[1156] show that Si–Si bonds appear on the SiO_2 side as a result of atomic dynamics in the interfacial region of the oxide. More such studies are needed to give us a full picture of the interface formation process.

Interfaces with amorphous SiO_2 can adopt the reconstructions typical for cristobalite and tridymite interfaces. For example, it is straightforward to see how the Ohdomari construction can be used. If every second oxygen atom is removed from the interface, a strained structure like  is obtained. Next, one must remove every second Si atom from the silica-side neighborhood of the interface: . The neighboring broken bonds of Si and O atoms rebond and, after relaxation, the Ohdomari interface is formed: . Naturally, Si–O–Si rings in the vicinity of the interface must be rearranged to avoid dangling bonds in SiO_2.

The Ohdomari model was originally formulated for the interface with amorphous silica. Estimations of elastic energy have shown that optimal structures were obtained when most (but not all) of the amorphous interface area was transformed as described above. The percentage of Si^{3+}, Si^{2+}, and Si^{+} atoms was then 29%, 43%, and 29%, respectively.

6.5.3. *Electronic states in the gap*

There are two effects responsible for the presence of electrically active states at the interface between SiO_2 and Si. First, the bonds between silicon atoms in the interfacial region are strained. This creates interfacial states which tail from the conduction and valence bands of silicon into the band gap. A recent study[1241] provides an illustration (probably exaggerated) of this effect: metallic densities of states were computed for highly strained interfacial atomic structures. The other natural source of interfacial electronic states in the energy range of the forbidden gap of silicon are broken bonds. Gate-oxide quality oxidation of Si(001) produces at most a few times 10^{12} cm^{-2} of broken bonds, that is, a broken Si–Si bond per roughly a hundred interfacial Si atoms. An isolated dangling bond of silicon is expected to have a half-occupied electronic state in the middle of the Si forbidden gap (cf. Fig. 3.5, page 144). After thermal oxidation, such states pin the Fermi level at the mid-gap position. Annealing in hydrogen-containing ambients (Sect. 6.3.6) passivates most of these states and unpins the Fremi level.[829]

The energy position of a dangling-bond state ψ depends on the interaction between ψ and the neighboring defects and on the local atomic structure of SiO_2 in the interface region. First, an increased defect-defect interaction tends to split the interfacial states into two groups moving away from the mid gap, one towards the conduction band and the other towards the valence band (see Refs. 1108 and 1103, and references therein). Second, the atomic structure of the interfacial SiO_2 affects the local electric field, which influences the energies of dangling bond states.[1225]

Silicon dangling bonds appear to be the most important objects responsible for SiO_2/Si defects which can trap charges (Sect. 6.6). These unsaturated orbitals give rise to an Electron Paramagnetic Resonance (EPR) signal.[835] A dangling bond of a silicon atom with three O neighbors is known as an E′ center. When the Si atom with the dangling bond has three Si neighbors, the EPR-active defect is known as a **P_b center**. The concentration of P_b is directly proportional to the concentration of defect states in the middle of the forbidden gap of silicon.

The anisotropy of the P_b signal shows that the dangling bonds are perpendicular to $SiO_2/Si(111)$ interfaces. Also for $SiO_2/Si(001)$ and $SiO_2/Si(110)$, the directions of the unsaturated orbitals coincide with directions of Si-Si bonds in the bulk. This is consistent with the interpretation of P_b as dangling bonds of Si with three Si neighbors.

P_b centers are localized predominantly at the interface. A smaller amout of these states can also exist in the SiO_x region. From measurements of interfacial density of states it follows that the density of broken bonds is much lower (by a factor of 2-3) at $SiO_2/Si(001)$ interfaces than at $SiO_2/Si(111)$ interfaces. Microvoids possibly related to P_b defects have been recently discovered at $SiO_2/Si(001)$ by positron annihilation.[1224]

Interestingly, EPR does not detect any interfacial signal coming from defects typical for bulk SiO_2: oxygen vacancies and E′ centers. But it does not necessarily follow that these defects are absent in the interface region. Namely, when interpreting EPR data, one should not forget that sensitivity of EPR is too low to observe defects in oxides produced by an industrial-quality technique. The density of interface defects must be enhanced by special preparation methods, like quenching from high temperatures, irradiation with X-rays, or irradiation with electrons.

Damage similar to that induced by X-rays can be generated in a real CMOS process during back-end steps (such as plasma processing or Cu electroplating). Electrons leaking through the oxide or speeding through the substrate may create such interfacial defects also later, when the device works. The accumulation of damage during the life of the device may eventually lead to the failure of the element (Section 6.6.2a).

Defects at and close to the interface are potentially able to exchange charge with the silicon substrate. Response times ranging from less than 1 μs to many years have been reported. From this point of view, electronic states of charge traps can be separated into two groups: "interface states", which can exchange electrons within a finite time, and "fixed charges", which keep their charge infinitely long. The meaning of "finite" and "infinite" is pragmatic and depends on the experimental method applied.

One should keep in mind that this classification[s] does not necessarily reflect the actual difference in the physical, atomic structure of the defects. The characteristic time needed for an oxide trap to exchange an electron or a hole with the silicon decreases exponentially with the activation energy of the trap as well as with the distance to the interface; also the electric field across the oxide affects the response time. Two defects with the same atomic structure but located at different distances from the interface can thus exert very different electrical influences. Electronic states of the defect which is close to the boundary may act as interface states, while electronic states of the more distant defect may act as traps for fixed oxide charge. To avoid confusion, it is therefore helpful to reserve the word "defect" to indicate a physical imperfection, and use the word "state" when its electronic response is meant.

One should also remember that certain defects can appear in multiple charge states. Some defects, like those with silicon dangling bonds, are of amphoteric character, that is, they can donate or accept electrons. Different types of states can sometimes be generated by the same defects located at various distances from the interface.

6.5.4. *Oxide defects and growth of the oxide*

We will now try to sketch a picture of the processes which lead to the formation of SiO_2/Si(001) interfaces.[242,907] We will begin with the growth at low temperatures. Next, we will observe changes indroduced by higher temperatures. Finally, we will discuss the formation of defects.

a. *Growth at low temperatures*

Interfaces grown "thermally" at low temperatures may be of no industrial relevance. But this does not prevent them from delivering important information, which may shed light on the physical mechanism behind such key technological steps as growth, annealing, and passivation of the oxide.

X-ray scattering shows that a fraction (about 10%) of the interface formed by dry oxidation at room temperature has 2×1 periodicity (page 393). When the initial surface is carefully prepared[t] one can infer

[s] See page 355 for detailed terminology of traps in gate oxides.

[t] Clean, with clear 2×1 LEED patterns.

from experimental data that the 2×1 interfacial superstructure is limited to a narrow layer (~3 Å, that is, roughly a monatomic layer of SiO_2). On the mesoscopic scale, the interface resembles an array of large terraces separated by monatomic steps.[1234] This resembles the morphology of the Si(001) surface from which the oxidation started. It follows that even a low-temperature growth, when properly initiated, can produce a smooth interface. But the most exciting scientific result is the periodicity.

Just like the sign of the stress in the film (Sect. 6.5.2), the observed 2×1 periodicity favors models with dimerized tridymite/Si interfaces. If cristobalite/Si is represented at real interfaces in a significant amount, it must be much more disordered than the tridymite/Si areas.

Tridymite-based models can be viewed as the first approximation of amorphous structures with ordered interfaces (Sect. 6.5.2c). Upper layers of thicker SiO_2 films undergo a transformation to amorphous silica.[1238] The amorphization allows for an efficient relief of the stress produced at the interface. This is fortunate not only for the MOS technology, but also for calculations of the interfacial properties. A Si(001) interface with a properly grown SiO_2 film can be, to a first approximation, simulated by an interface with an ultrathin SiO_2 layer terminated by a free, fully relaxed surface. This reduces the size of the system which must be considered in the calculation.

The presence of hydrogen (delivered, for example, by prolonged exposure to humid air) eliminates the 2×1 structure.[1228] This may be interpreted as an indication that the bonds responsible for the double periodicity are relatively weak and can be preferentially broken and subsequently passivated by hydrogen atoms and OH radicals, so that the periodicity disappears.

However, if these bonds were so weak, then why should they survive the regular oxidation? Should not the radicals attack rather these bonds which are usually attacked by oxygen? But water interacts quite actively with SiO_2, so the reason of the water-induced disorder may be quite banal: randomization of the tridymite lattice in the very neighborhood of the dimers, where stress is the largest. In other words: water may leave the interfacial dimers unbroken, but it may intervene in the bond between an oxygen atom and the dimer in a reaction like: Si–O–**Si**–**Si**–O–Si + H_2O → Si–OH + HO–**Si**–**Si**–O–Si, where –**Si**–**Si**– symbolizes an interfacial dimer.

Let us now assume that low-temperature SiO_2 films are tridymite-like, at least close to the interface. The next thing we have to consider is how to get rid of the excess stress. A part of the lateral compression might be compensated by cristobalite interfacial intrusions. The vertical component of the stress is best relieved by amorphization of upper layers.

b. *Effect of high temperatures*

Amorphization is a process which requires the transport of material across the solid and is accompanied by the rearrangement of bonds. The rigid, six-fold rings of crystalline SiO_2 inherited from the diamond structure of silicon must be reorganized into a more flexible system of shorter and longer loops. The structural flexibility given to SiO_2 by the Si–O–Si "boomerangs" makes this task feasible. High temperatures help to rotate and shift SiO_4 tetramers in a way which resembles viscous flow. When the temperature is high enough (above 950–975°C), the amorphous SiO_2 film can relax most of its stress by viscous flow.[268]

Stress relaxation destroys the order in SiO_2/Si(001) interfaces. The 2×1 superstructure is best observed in films grown at low temperatures. It follows that it represents a metastable, intermediate state induced by kinetic bottlenecks which enforce structural constraints on the growing film.

The temperature at which the 2×1 superstructure begins to disappear, 550°C,[1228] is much lower than the temperature in which viscous flow begins to be noticeable in amorphous SiO_2 (~950°C). This suggests that weak structural elements are more numerous in the vicinity of the interface than deeper in the film. It is natural to associate these weak points with such defects as broken bonds and Si–Si bonds. Both are expected to abound in the transition region,[1156] where the diamond lattice of the substrate has just been converted into a lattice of crystalline or quasi-crystalline SiO_2.

c. *Defects, regrowth, and stress*

The most important defects at chemically clean SiO2/Si(001) interfaces seem to be Si dangling bonds, Si–Si bridges and vacancies,[u] and, maybe, Si interstitials.[1156] The non-stoichiometric region is traditionally explained by Si–Si bridges. The total amount of all these defects depends on the oxidation method and constitutes percents of a monolayer. Their energetics and dynamics have been addressed by recent *ab initio* calculations.[848,1156]

This SiO_x region is limited to a narrow strip, at most a few atomic layers thick (Sect. 6.5.2a). As the oxidation front moves into the silicon, new non-stoichiometric defects are created in the freshly formed SiO_2 while

[u] See Refs. 1226, 1156, and 848 for *ab initio* models with Si–Si bridges and Si vacancies.

the older defects become buried in the growing oxide. There they do a favor to the film, facilitating rearrangements of the SiO_2 network. The film pays back by killing them: it was the stress that protected them from oxidation, and once this stress is gone, the defects become exposed to the oxidant and gradually disappear. The observed thickness of the SiO_x decreases with increasing oxidation temperature, even to a nearly monatomic layer. Indeed, higher temperatures mean more efficient relaxation of the stress.

Excess silicon can be annihilated at the interface, for example, at steps or in interstitial-like sites.[1156] Though the process of adsorption of Si interstitials at interfacial steps, known as **regrowth**, is to a certain extent hindered by the presence of the oxide layer, it may be easier to acommodate many such atoms at a step than in a flat region. However, compelling evidence exists (see Ref. 1242 for discussion and references) that the volume of the amorphous oxide film constitutes a more efficient sink for the excess Si than the interface alone. Si atoms can penetrate approximately 30 Å into the growing oxide until they become oxidized and incorporated into the SiO_2 network. Oxonium atoms (triple-coordinated oxygen, cf. Section 6.5.5d) may play a role in this diffusion and reorganization.[1156]

These mechanisms of silicon incorporation open the possibility to relax stress by formation of cristobalite-like microcrystallites at the interface (Sect. 6.3.5c). The density of Si atoms in a cristobalite-like interface is lower than in a tridymite-like interface, and the excess silicon must be eliminated.

And what happens deeper in the film? From the fact that a few atomic layers away from the interface the concentration of non-stoichiometric defects drops below the detection limit of core level spectroscopy it does not follow that all such defects are destroyed. Some of them definitely survive and take a revenge on the oxide, degrading its dielectric quality.

How does it work? A silicon dangling bond is electrically active. It introduces two electronic levels, (+/0) and (0/−), in the forbidden gap of silicon. But electrons of a Si–Si bridge are expected to behave like electrons in the σ bond of a Si(001) dimer. This means that the electronic states of the bridge are located in the energy range of the valence and conduction band of silicon. An interfacial Si–Si bridge should thus remain electrically neutral. However, a Si bridge in SiO_2 is *electrically active*. This is because the forbidden gap of SiO_2, nearly 9 eV, is much larger than the forbidden gap of bulk silicon (1.1 eV). Electrons on the σ orbitals and holes on the antibonding σ^* orbitals are localized within the large gap of the insulator.

Such electrically active states deteriorate the dielectric strength of the insulator. Strong electric fields in the gate oxide[v] assist tunneling, and states in the gap act as intermediate stops for electrons, further reducing the tunneling barrier.[1243,1244] A halted electron can dissipate its energy at the defect site, facilitating local rearangements of bonds. This may damage the dielectric. Another consequence is that defects can capture electric charge permanently, because the electron-phonon coupling is strong in SiO_2. In other words, the SiO_2 network around a defect relaxes strongly on capture or emission of an electron by the defect. This charge trapping causes many problems. They are described in more detail in Section 6.6.

Besides the tunneling of carriers from the gate or from the channel, there are other sources of this charge (Sect. 6.6). First, the oxidation itself is an electrochemical process which involves ions, electrons, and/or holes. Second, certain processes in MOS technology temporarily charge the gates, which can also lead to injection of carriers into the oxide. Third, electrons can be injected during electron beam lithography, a technique which may be used in future MOS technologies. Finally, when the device is operating, some of the electrons which travel from source to the drain of a MOSFET can heat up so much that they can overcome the energy barrier between the conduction band of silicon and SiO_2 (about 3.1 eV) or tunnel below the barrier when the gate is biased.[1245] Such an electron can be captured by a defect or it can ionize an electrically neutral defect.

6.5.5. *Reactions at the interface and in the oxide film*

Several fascinating and at the same time technologically important effects are related to chemical reactions which take place at the SiO_2/Si interface and in the SiO_2 film during and after oxidation (Table 6.7). These reactions may lead to significant changes in the dopant profiles below and above the gate oxide. A typical consequence is that the MOSFET threshold voltage V_T is affected, sometimes by as much as 50% of its ideal value. Also the leakage currents flowing through the gate oxide and below the field oxides which separate the neighboring transistors are sensitive to them.

SiO_2/Si films and interfaces influence the distribution of dopant atoms and are in turn influenced by dopants and other foreign atoms. First, boron atoms diffuse from the gate through the oxide into the channel; this is en-

---

[v] Gate voltage is usually equal to the power supply voltate, which is of the order of a volt (2.5 V in contemporary circuits). Given that the oxides are a few times 10^{-7} cm thick, the electric field in the gate oxide of an operating transistor is several MV/cm!

Ref year	Subject relevant to chemical reactions at SiO_2/Si
1246 1972	B diffusion through thin SiO_2 films
1247 1974	Emission of Si interstitials from SiO_2/Si into silicon
1248 1975	N_2 annealing and oxide charges
1249 1975	Nitrogen can degrade the SiO_2 film
1250 1978	**Book article**: $SiO_2/Si(001)$ films and interfaces, and P, B, As
1251 1979	Oxidizing species prevent nitrogen from degrading the SiO_2
1252 1981	**Book article**: Cl, interstitials, stacking faults
886 1984	N incorporation at $SiO_2/Si(001)$; SCLS
1253 1985	Sb, As segregation to $SiO_2/Si(001)$ interfaces; RBS, SIMS, TCAD
1219 1987	P, B, As segregation to $SiO_2/Si(001)$ interfaces; SIMS
1254 1989	P, As segregation at $SiO_2/Si(001)$; TCAD
1242 1992	Si interstitials in SiO_2 and silicon; diffusion/reaction model
1067 1995	Dopant segregation, dose loss; overview, SIMS analysis for P
1142 1996	Oxynitridation reactions
1255 1996	N, H, and excess O and Si at $SiO_2/Si(001)$; *ab initio*-based analysis
1256 1997	Oxynitridation reactions
1096 1997	Very high quality ultrathin $SiO_2/Si(001)$ by N_2O oxidation
1257 1997	Effect of interface traps and N on electron mobility in MOSFETs
1179 1997	N_2 annealing not good for SIMOX
1258 1997	Rapid Thermal Annealing and fixed charge
1077 1997	P, As segregation at $SiO_2/Si(001)$; process simulation
801 1997	Fluorine at $SiO_2/Si(001)$
1059 1997	Boron penetration, effect of N, H, F; TCAD
1259 1997	Boron penetration and device reliability issues
1260 1997	Boron penetration through nitridated oxides; modeling
1261 1997	Boron penetration prevention through N implantation into Si
1262 1997	Instabilities related to boron-doped polysilicon MOS gates
1263 1997	Cyanide passivation of $SiO_2/Si(001)$ interfaces for solar cells; XPS
1264 1999	Dopant pairing model of segregation and dose loss; *ab initio*, AES, SIMS

Table 6.7. Chemical reactions at SiO_2/Si: selection of references. TCAD stands for Technology Computer Aided Design (process simulation).

hanced by the presence of fluorine which may contaminate the SiO_2 film, for example, after BF_2 gate implantation (Sect. 1.2.2b) or silicide contact deposition from WF_6 (Sect. 5.7.1a). Nitrogen atoms incorporated at the interface control the oxidation rate (Sect. 6.3.4 and page 321), tend to reduce the interface roughness, and act as a diffusion barrier for boron atoms. Second, an oxidizing interface emits Si interstitials, Si_i. This enhances diffusion of certain dopants in Si and may also cause electrical shorts through *pn* junctions. Chlorine atoms in the oxidizing ambient can minimize this problem, apparently by etching silicon and producing "anti-interstitials", that is, Si vacancies (V_{Si}). Third, some dopants preferentially segregate to SiO_2 or to the SiO_2/Si interface, while other dopants are expelled from

SiO_2. The presence of certain impurities, like alkali atoms, in the oxide can invoke serious instabilities of the device parameters. Fourth, hydrogen or deuterium atoms can deactivate dangling bonds at the interface and in the oxide, improving the reliability of the device.

The importance of the chemical reactions at $SiO_2/Si(001)$ is not limited to the CMOS technology. For example, consider solar cells: a cyanide treatment of the surface before the oxidation was recently reported[1263] to improve the cell performance, apparently due to the passivation of the interface states by cyanide ions (CN^-).

a. *Main reactions during oxidation*

There are three main groups of oxidation reactions at the interface. First of all, oxygen may simply attack Si–Si bonds, converting silicon into SiO_2:

$$Si + O_2 \rightarrow SiO_2.$$

This produces a lot of misfit stress and should be characterized by a relatively high activation energy. In extreme cases, part of the stress is relaxed via formation of extended defects, such as dislocations. Second, oxidation injects Si self-interstitials into silicon bulk when the oxide layer is thin (short oxidation times). Third, when the oxide is thick, Si vacancies are injected.[1265] This happens after long oxidation at high temperatures, such as[1266] $\sim$ 18 h of oxidation at 1160°C. Vacancy ejection is typical for thick $SiO_2/Si(111)$ films. Below we will consider solely the case of short oxidation times, comparable to those used in CMOS processing of Si(001) wafers.

Silicon vacancies (V_{Si}) help to accomodate the mismatch in the partial density of Si on both sides of the interface. In reactions:

$$Si + V_{Si} + O_2 \rightarrow Si\text{–}O + O$$

$$Si\text{–}O + V_{Si} + O \rightarrow SiO_2$$

vacancies are adsorbed to neutralize the volume expansion on the SiO_2 side. These reactions are thought to be responsible for enhanced oxidation of heavily doped (such as $n = 10^{20}cm^{-3}$) Si, where oxidation can proceed about ten times higher than the oxidation of an intrinsic material.[1267] This is a Fermi level effect. Besides being electrically neutral, V_{Si} can be positively or negatively charged, therefore V_{Si} formation energy depends on the Fermi energy E_F. To be more specific: when a charged defect is produced, it must donate or accept an electron from a "reservoir" of electrons, that is, from E_F. The difference between the initial (E_F) and final energy (on defect) of this electron contributes to the energy balance (Fig. 3.15b). As a consequence, the concentration of V_{Si} is enhanced in doped material.[w]

Alternatively, the volume mismatch can be reduced by emission of Si self-interstitials (Si_i) in reactions:[1247]

$$Si\text{–}Si + O_2 \rightarrow Si\text{–}O + O + Si_i$$

$$Si\text{–}O\text{–}Si + O \rightarrow SiO_2 + Si_i.$$

These reactions can be viewed as an incomplete oxidation, which leaves two silicon atoms unreacted (not incorporated into the new SiO_2 network).

The fate of Si_i produced at the interface is not completely known. A significant fraction may diffuse into SiO_2[1242,1268] (Sect. 6.4.2a). These atoms become oxidized by the oxidant species from the ambient and incorporated into the SiO_2 network. Some of them may escape oxidation to be trapped in the oxide close to the interface, where they are supposed to act as a source of the fixed charge. One can interpret Si–Si bridges as trapped Si_i. The consensus is that only approximately one Si atom in a thousand is ejected into the substrate.

b. *Chlorine and fluorine: dielectric strength, dopant diffusion*

Once ejected from the interface into the Si bulk, Si_i atoms migrate through the silicon crystal relatively easily.[1269,1270] This alone would be of little interest to us, but in addition, these interstitials interact with dopant atoms.[85,1271,1272] A dopant can be kicked out from its substitutional site and becomes an impurity interstitial, while the self-interstitial comes to rest in the lattice. Substitutional dopants can hardly migrate in the crystal, but as interstitials they can cover considerable distances until they find an unoccupied lattice site (V_{Si}) or until they eventually manage to replace a silicon atom from the lattice ("kick-in"). The overall result is an **Oxidation Enhanced Diffusion** (OED) of dopants.[1247,1273,1274,1242,1268]

Less than 1% of Si atoms in each oxidized Si(001) plane are converted to interstitials during oxidation. But even one Si atom in a thousand corresponds to a volume concentration of $5 \cdot 10^{19}$, comparable to the concentration of dopant atoms. The amount of ejected Si interstitials is, thus, high enough to enhance diffusion of the dopant species (boron and phosphorus) which migrate predominantly by the interstitial-mediated mechanism.[85] At temperatures around 900°C, the enhancement can be as large as a factor of ten. It becomes even more pronounced at lower temperatures, at which the intrinsic concentration of interstitials drops to very small values.

---

[w] Dopants on both sides of the interface may help to relax stress (Sect. 6.5.4c). Indeed, SiO_2 deposited for isolation purposes is often doped to make the glass reflow easier.

Excess interstitials emitted by the oxidizing interface precipitate in **stacking faults**. Stacking faults, as typical for extended defects, tend to collect around them an "atmosphere" of dopants. This atmosphere is in fact a thin platelet of material with enhanced conductivity. It is thus clear that an oxidation induced stacking fault can short-circuit a *pn* junction if it happens to cross one. If a defect like that extends across the base of a bipolar transistor, the transistor cannot work properly.

Oxidation-induced stacking faults grow with an activation energy E of about 2.3 eV: $L \sim \exp(-E/kT)$, where L is the fault length. It is interesting that this simple law breaks down at very high temperatures (around 1200°C). The length L reaches a peak at a certain critical temperature T_R, and then begins to fall down sharply[1275] (**retrogrowth**). For Si(001) wafers, $T_R = 1240°C$ in dry oxygen and 1300°C in steam. Vicinal Si(001) has a slightly lower T_R. For example, a 5° misorientation reduces T_R to 1175°. Si(111) surfaces have $T_R = 1220°C$.

Such a behavior indicates that there exists another process which competes with the stacking faults in collecting Si_i. This process is characterized by a high activation energy. Diffusion of V_{Si} has been originally suggested to be the relevant factor.[1276] Indeed, the excess interstitials deplete the near-interface region of the silicon crystal from vacancies. The influence of HCl ambient (see below) and of high oxidation temperatures would then be conceptually similar: both conditions help to deliver V_{Si} which annihilate with Si_i and shrink the stacking faults. Another possible candidate for the high-temperature retrogrowth of oxidation-induced stacking faults is thermal emission of Si_i from the faults.[1252]

A remedy to problems with the interstitials created during oxidation is **chlorine** (cf. page 365). When added to the oxidizing ambient (usually as HCl), chlorine reacts with silicon atoms at the substrate side of the interface, producing volatile $SiCl_n$ molecules and silicon vacancies. This slows down the ejection of interstitials and can even lead to partial evaporation of stacking faults if such defects were formed before HCl was added. Four per cent of HCl in the O_2 ambient at 1100°C reduces the measured length L of stacking faults by a factor of ten.[1252]

Another role of the vacancies created by chlorine is that they accommodate the compressive stress induced by the oxidation. This prevents stress-induced roughening of the film and the interface and, as a consequence, enhances the dielectric strength of the oxide (page 438). A somewhat similar effect may be caused by **fluorine**. The fluorine atoms are introduced through HF etching prior to oxidation (page 279). They tend to remain in the vicinity of the silicon surface during the oxidation;[819] in other words, they segregate to the SiO_2/Si interface. Since each fluorine atom is attached to a single silicon atom, the presence of fluorine losens the network of interactomic bonds. In this way, the oxidation stress is more efficiently relaxed and the roughness of the oxide is reduced. This resembles the action of H and OH (Sect. 6.5.5d). The resulting fluorinated gate dielectric is somewhat more resistant to hot-electron damage (Sect. 6.6.2a) than gate oxides prepared by a conventional dry oxidation.[819]

c. *Dopants: Fermi level, penetration, segregation*

Interactions between dopants and SiO_2 films have many consequences. Dopant atoms in the substrate affect the oxidation rate, possibly by increasing the equilibrium concentration of Si vacancies (page 414), but maybe also by assisting the relaxation of the interface. Boron from the gate can escape through the gate oxide into the channel of the MOSFET. Oxidation injects Si interstitials into the channel, enhancing the mobility of the channel dopants. Phosphorus and arsenic are expelled from the growing oxide and accumulate in front of the interface, changing the dopant profile in the channel. During annealing steps, dopant atoms are brought to the interface by interstitials or vacancies and become electrically inactive (deactivated). This leads to dose loss: after the implantation and activation (Sect. 1.2.2b), the segregated atoms are removed together with the pad oxide by the chemical etching, or, if they segregate to the interface with the gate oxide, they are useless as dopants. The majority of the segregated P and As is trapped under the interface on the Si side,[1277,1264] but some atoms (under typical conditions, ~10% in the case of P) diffuse into the oxide.[1278] Too much P in the substrate can even deteriorate the dielectric.[1279] Table 6.7 on page 413 collects some references pertinent to these subjects.

SiO_2/Si interfaces of MOSFETs are located next to doped regions of the substrate (Fig. 1.2, page 8). They are an integral part of the MOSFET's "command deck", the gate capacitor. This capacitor controls the electric current between the source and the drain electrodes. Since the current flows very close to the oxide, the spatial distribution and the electronic structure of the dopant atoms below the SiO_2 film affect electrical parameters of the transistor and must be carefully designed and managed.

The interaction between dopants and the oxide can change the distri-

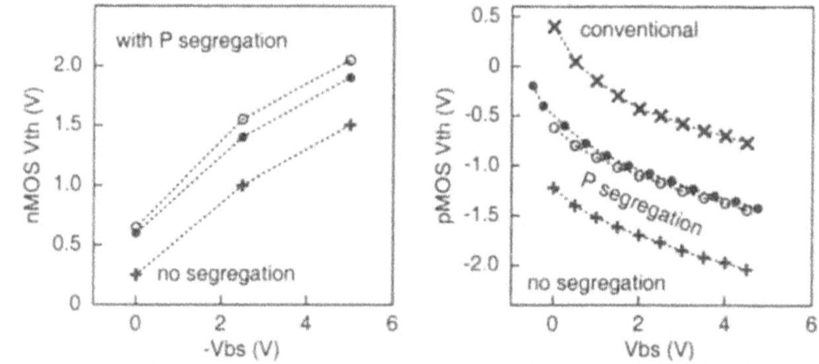

Fig. 6.7. Influence of P segregation to SiO_2/Si(001) on MOSFET threshold voltage. Measured (full circles) and simulated (other symbols) threshold voltages V_T are plotted against the substrate voltage (the source electrode is grounded). **Left**: nMOSFET (adapted from Ref. 1077). Simulations with P pile-up and dose loss (open circles) and only with P pile-up (crosses). **Right**: burried-channel pMOSFET (adapted from Ref. 1280). Simulations with P pile-up and dose loss (open circles), only with P pile-up (+-crosses), and with neither pile-up nor dose loss (×-crosses).

bution of dopants in the substrate so drastically that the threshold voltage (the gate voltage by which the transistor switches) computed by a standard simulation software deviates from the measured value by more than 50% (Fig. 6.7). Models to compute dopant profiles close to SiO_2/Si(001) exist and have been succesfully applied in present CMOS designs (e.g., Ref. 1254, 1280, 1077, and 1281). However, these models assume that the interface has a fixed number (~1 ML) of sites, which can trap dopants. This is apparently inconsistent with the measured dependence of the dose loss on the implanted dose (Fig. 6.8a). This inconsistency indicates that such models would fail when used to simulate the dopant concentration which changes strongly along the interface (as under gate oxide sidewalls, Fig. 1.2 on page 8), even though they work over a limited concentration range.

Some unsolved problems of practical importance may be related directly to the electrical activity of atoms at SiO_2/Si(001). A high concentration of *ionized* dopants at the interface affects the threshold voltage and the mobility of carriers below the gate oxide. Quasi-random exchange of electrons between deeper but still electrically available interface states and the conducting channel below the oxide causes a characteristic low-frequency noise ("flicker" or $1/f$ noise, see Section 6.6.2c). Such states may be in

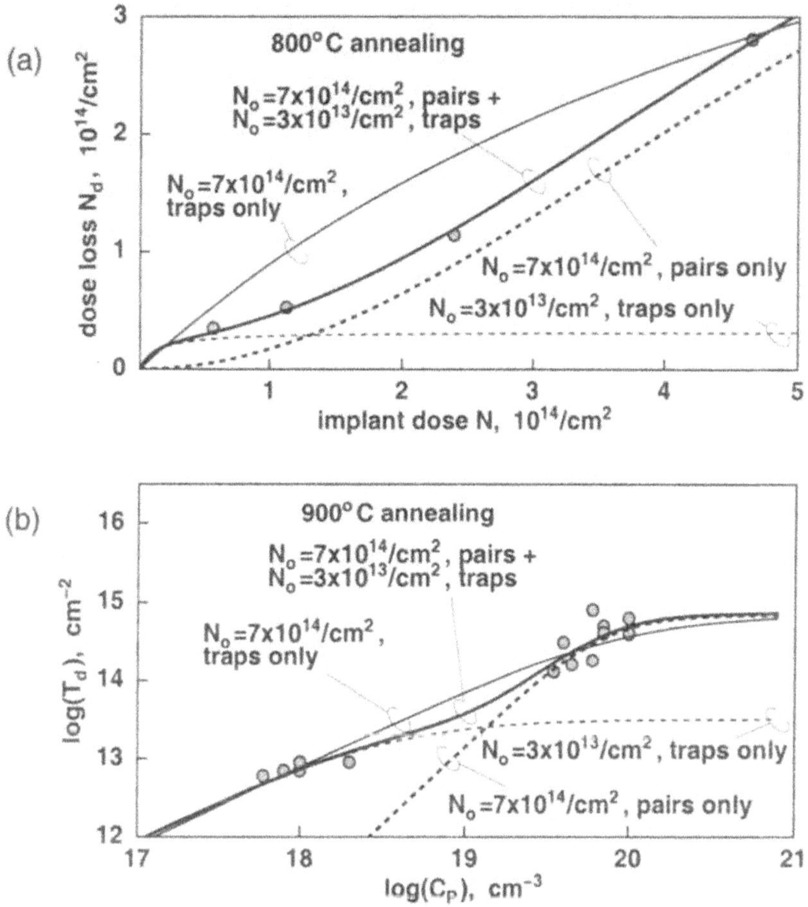

Fig. 6.8. Dopant pairing and dose loss. **(a)** Dependence of P dose loss N_d on P implant dose N, SIMS data after Ref. 1067. If traps only are assumed, the functional dependence is qualitatively wrong. Pairing and traps together give an excellent fit. **(b)** Dependence of P dose loss N_d on P concentration C_P close to the interface, SIMS data after Refs. 1278 and 1281. The trap-dominated (low C_P) and pairing-dominated (high N_d) regimes are clearly visible. (Adapted from Ref. 1264.)

principle induced by the segregated donors.

Below we consider in some detail the most characteristic effects of these interactions: boron penetration, dopant segregation, and dose loss. We focus on physical effects; for practical issues see Section 6.6.

Boron penetration of gate oxides is a troublesome phenomenon, which is at work at high temperatures. For example, B atoms in SiO_2 diffuse approximately 2-3 Å (0.2-0.3 nm) during 20 min annealing at 900°C.[1262] Boron atoms diffusing from a heavily *p*-doped polysilicon gate through the gate oxide into the silicon substrate change the dopant profile in the MOSFET channel[991] and deplete the gate from dopants.[1063] This affects the threshold voltage of pMOSFETs[1282,1283] (Sect. 6.6.2d; see page 10 for MOSFET naming convention). Boron atoms trapped in the oxide deteriorate the dielectric[1284] by promoting electrical shortcuts (Sect. 6.6.2a).

Boron has a lower energy in SiO_2 than in silicon, so it accumulates in the oxide, where it seems to occupy predominantly interstitial sites.[1285] Boron atoms become mobile in SiO_2 around 900°C and diffuse through it into the channel. Boron migration appears to be enhanced in oxides of thicknesses comparable to those of contemporary gate oxides.[1286] Also the presence of hydrogen[1287] and, particularly, of fluorine[1288,1282] enhances boron diffusivity. The penetration of B into gate oxides is retarded by nitrogen, but this effect becomes small when the fluorine content is large.[1059]

A recent phenomenological model[1059] assumes that boron diffuses via Peroxy Linkage Defects (PLD, Sect. 6.2.1). PLD would act as low-energy doorways passing boron from one interstitial site to the neighboring one: B_I + O_3Si–O–O–SiO_3 → O_3Si–O–B–O–SiO_3 → O_3Si–O–O–SiO_3 + B_I. The enhancement of B diffusivity in thin oxides may be then explained by a higher concentration of PLDs in the interfacial region. These defects can be generated by volatile SiO molecules produced at the oxidizing interface and then diffusing into the oxide.[1289,1286] SiO is metastable in SiO_2 and breaks into a PLD and a Si interstitial. Fluorine may increase the abundance of PLD, while hydrogen is responsible for the NBOHC defects (Sect. 6.2.1) which may act similarly to the PLDs. On the other hand, nitrogen should block the PLD doorways, hindering the diffusion of boron. This model can explain the dependence of B diffusivity on the concentration of F, N, and H.[1059] However, no *ab initio* studies which verified its assumptions are known to us. Another model, which involves no interstitial B and is based on *ab initio* results for small clusters, has been very recently suggested.[1290]

Contrary to boron, **phosphorus** and **arsenic** atoms are expelled from the growing oxide.[1250] They are "shovelled" by the oxidizing interface and **pile up** in front of the SiO_2/Si boundary. If the silicon is boron-doped *p*-type with an appreciable degree of compensation of boron by phosphorus, it is clear that an *n*-type (phosphorus-rich) inversion layer forms quite readily by oxidation.

Some of the P and As atoms become incorporated in the oxide.[1278] Very high (above 10^{20} cm^{-3}) concentrations of phosphorus into the substrate can deteriorate the oxide dielectric strength.[1279] Smaller concentrations improve the quality of oxides by shortening the reaction-controlled region, by stress relaxation, and by enhancement of SiO_2 viscous flow.[1279]

Further thermal processing, for example, annealing of the gate oxide, results in a redistribution of dopants in the channel. The dopant atoms **segregate** to the interface. This is enabled by the Transient Enhanced Diffusion (TED) effect:[1291] Si self-interstitial atoms, Si_i, produced during implantation and agglomerated in the so-called {113}-planar defects, are gradually released and enhance the mobility of dopants, notably P and B. Recent results[1292] show that the influence of TED on MOSFET parameters can be reduced when a few times $10^{14}/cm^2$ carbon atoms are implanted into the subsurface region (200-300 nm, implant energy ~100 keV). The reduction of TED takes place because the carbon-implanted layer acts as a sink for the defects created during all implantation steps.

When the interstitials arrive at the interface, they are likely to "carry" with them dopant atoms (Sect. 6.5.5b). Since the interstitials either recombine at the interface or, when oxygen is available from the annealing ambient, they enter the SiO_2 film to become oxidized (Sect. 6.5.4c), there is a Si_i flux towards SiO_2/Si, accompanied by a dopant flux. The interfacial region thus becomes enriched in dopant species.[1293] This effect is somewhat similar to the formation of the "atmosphere" of dopants around stacking faults. Also dislocations are known to be surrounded by dopants. The dopants are probably held there by strain and remain electrically active, but they must have diffused somehow to the dislocation. Just like the SiO_2/Si interface, a dislocation is an excellent sink for interstitials or vacancies.

The atoms segregated to the SiO_2/Si(001) interface do not behave as shallow dopants. The consequence is the **dose loss** effect: Even 50% of the implanted dose may "disappear".[1067] The interfacial area can accumulate more than $2{\cdot}10^{14}$ dopant atoms per cm^2, nearly as many as there are interfacial lattice sites available (1 ML $\simeq 7{\cdot}10^{14}$ cm^{-2}). Typical implantation doses are around 10^{13}–10^{15} cm^{-2}, that is, the total amount of dopant atoms may be comparable to a monolayer (Fig. 6.8).

The dose loss is more pronounced when the annealing ambient contains oxygen (needed to heal defects in SiO_2). This may be related to the enhanced incorporation probability of interstitial silicon into the SiO_2 lattice in the presence of oxygen (Sect. 6.5.5a). An inert ambient (N_2) may be critical at very high temperatures (above 1000°C), such as during activation of dopants.[28] Similarly, large amounts of dopants may be lost to the pad oxide after implantation.[1067] The resulting increase of series resistance becomes a problem for channel lengths below 0.25 μm. One might try to avoid this by implanting more dopants, but the atoms migrate in *all* directions, so that a higher dose means more diffuse dopant profiles. Special capping layers and annealing schemes should thus be studied.

We will now summarize our recent findings on segregation of dopants. They are based on *ab initio* calculations.[301,1264] These results suggest a new microscopic mechanism of segregation and deactivation of phosphorus. Experimentally, we have observed for the first time the segregated phosphorus atoms in Auger Electron Spectroscopy through an ultrathin oxide layer. The spectra show that P segregation does not produce P-O bonds. This is consistent with the suggestions from the literature and our calculations.

The segregation process begins when a P atom becomes weakly bonded (~0.2 eV) in a "precursor" state directly below the interface, while retaining its substitutional character. The bonding is caused by small changes in the structure of the substrate in the vicinity of the interface. As explained below, we have reason to believe that the precursor can diffuse along the interface faster than in the bulk, and that this diffusion needs no point defects. This enhancement facilitates the completion of trapping.

A phosphorus atom becomes permanently trapped when it exchanges site with an interfacial silicon atom, which has a dangling bond and is not bonded to oxygen. This is analogous to the trapping and deactivation of P at Si vacancies or at a free surface. This effect can account for the permanent segregation of a small amount of phosphorus. However, the number of such interfacial trap sites is below 10^{13} cm^{-2} (Sect. 6.5.3), while experiments show that the amount of segregated P can exceed 10^{14} cm^{-2}.

There exist at least two atomic configurations of phosphorus, which allow coverage of a perfect interface with nearly a full monolayer of impurities ($\sim 5 \cdot 10^{14}$/cm^2). These structures involve *no pre-existing defects*, neither in silicon nor in the oxide. One of them contains two threefold-coordinated, electrically neutral P atoms (Fig. 6.9a). This complex employs a local rearrangement ({113}-like rebonding[215]) of Si lattice bonds, resembling the atomic configuration acquired temporarily near the barrier along

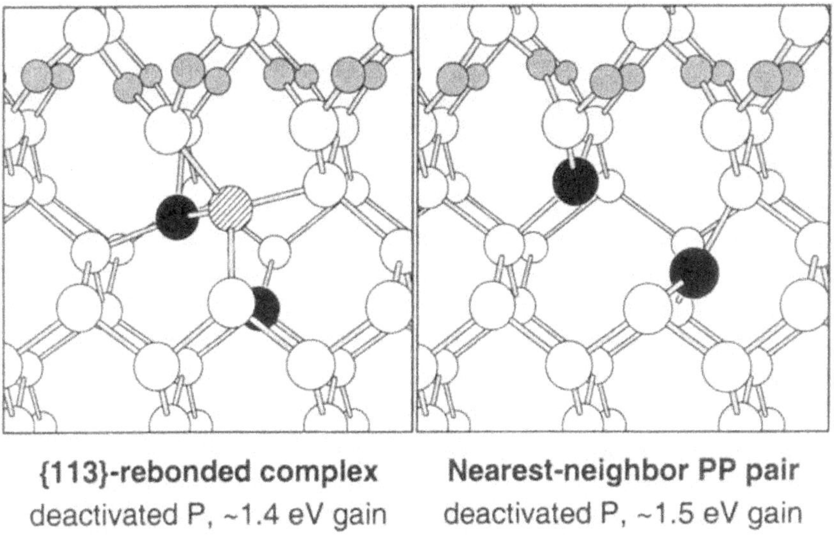

Fig. 6.9. Atomic structures of phosphorus (black) trapped and deactivated under a model $SiO_2/Si(001)$ interface. These complexes trap two electrons from the Fermi level, so that their P atoms become electrically neutral. When the Fermi level is at the conduction band edge, the binding energy of such a complex is ~1.5 eV. (**a**) The dashed Si atom and its P neighbor form a pair of {113}-rebonded lattice atoms mentioned in the text. (**b**) A pair of threefold-coordinated P atoms (a distorted nearest-neighbor pair).

the concerted-exchange path of Si self-diffusion.[1294] The other structure is simpler. It is a distorted nearest-neighbor PP pair (Fig. 6.9b).

These complexes are unstable in the bulk. But they are stabilized near the interface. The effective value of the segregation energy due to the pairing, estimated by us as $E_s \simeq 0.5$ eV per defect in intristic and $E_s \simeq 1$ eV in n-type material, depends on the oxide and interface structure. The uncertainty in the computed E_s should not exceed ~0.3 eV per defect, including an estimate of the latter dependence. The stabilization takes place because the oxide helps to accommodate the angular stress caused by the deformation of the bonds around the defects. We believe that the same effect enables an efficient concerted-exchange diffusion of substitutional P below the interface. Numerical verification is in progress.[301]

We used a simple thermodynamical model to estimate the dependence of the segregated dose N_d on the concentration C_P of active P under the interface. We assumed that: dopant atoms can be deactivated by pairing or trapping, the corresponding reaction constants are thermally activated, and the active and inactive dopants under the interface are in equilibrium (activation rates = deactivation rates). We ignored the small amount[1278] of P that enters SiO_2. When N_o deactivation sites exist under the interface, this leads to $N_d = N_o C_P^2/(C_P^2 + B^2 \exp(-2E_s/kT))$ when segregation is dominated by pairing and to the usual $N_d = N_o C_P/(C_P + B\exp(-E_s/kT))$ when trapping on pre-existing defects dominates.

Our predictions compare favorably with the published experimental data (Fig. 6.8). The segregation energy due to the pairing is ~0.4 eV/defect from Fig. 6.8 (800°C and 900°C) and from the data in Ref. 1295 (1000°C, not shown), the segregation energy due to trapping is ~1.3 eV. These values are close to our *ab initio* estimates for pairing and trapping of phosphorus at broken bond sites, respectively. The $N_d(C_P)$ dependence is recovered for the whole concentration range under the gate and sidewall oxides.

In the dual (pairing and trapping) model the density N_o of "deactivating sites" for pairing corresponds to a monolayer. The value of N_o for traping is $\sim 3{\cdot}10^{13}/\mathrm{cm}^2$, which is several times more than the typical number of electrically active interfacial defects. This may viewed as indication that such defects as nonstoichiometric sites (Si–Si bridges) act as dopant traps. Maybe more realistic interpretation of the vigh value of N_o for trapping, based also on *ab initio* calculations, is that Si vacancies created during implantation segregate to the interface and act there as donor traps.[1296]

These findings[1264,1296] implicate that dopants will segregate to the interface with any dielectric (not only SiO_2) that introduces interfacial stress. Since dose loss is promoted by this stress, the losses should be minimal if the surface passivation film is stress-relaxed. Moreover, the loss can possibly be reduced by a modified etch that does not remove surface P atoms. This is because the P atoms are separated from the oxide by a Si layer; during an HF etch, this Si layer is removed together with the oxide (Sect. 5.3.1). Finally, the influence of donor segregation on flicker noise in MOSFETs should be negligible. This is because the electrons are localized on deactivated phosphorus near to the valence band of silicon.

These results are, however, subject to further verification. It is, therefore, still correct to say that the knowledge on details of interactions between the dopants and the interface is fragmentary. It remains unclear how many of the interfacial or nearly-interfacial dopants are active electrically, what are their atomic and electronic structures, what are their energies, and which microscopic processes and reaction paths are responsible for the segregation and pile-up effects. Answers to such questions would be welcome by those who make practical simulation of technological processes.[28]

d. *Hydrogen, nitrogen, oxygen: control of stress and traps*

An important role of nitrogen and hydrogen in gate oxides is in the control of charged defects.[x] Hydrogen can either induce or deactivate such defects. Nitrogen hardens the oxide against trapping of charge, unless hydrogen is present in the nitridizing atmosphere. Annealing of oxides first in weakly oxidizing ambients and then in hydrogen is the standard way to reduce the number of charged defects (Sect. 6.3.6). In addition, hydrogen opens Si–Si bonds (Sections 5.2.1 and 5.5.3), assisting the process of stress relaxation during wet oxidation of silicon (Sections 6.3.1 and page 363). Last but not least, nitrogen slows down the oxidation of silicon. This has interesting applications in designs of circuit architecture[956] (page 321).

There are many ways by which **hydrogen** can be introduced into the gate oxide, intentionally or unintentionally. Probably the most natural source of H is steam in wet oxidation (Sect. 6.3.2). But H can diffuse to the gate oxide in the course of many technological processes, such as CVD from hydrogen-containing precursors, or spin-on glasses.[1164] Process-induced stress influences the diffusion of H from passivation layers to the interface, accelerating the device wearout when this stress is compressive.[1297]

Contemporary gate oxides are made in dry O_2 (Sect. 6.3.2), precisely to exclude hydrogen because it is believed that H must have a deteriorating effect in this stage of processing. But hydrogen acts in many ways and different oxidation procedures may emphasize different aspects of hydrogenation. Recently, a reliable RTP steam gate oxidation was reported[1097] (page 364). In this case, it seems that the positive effect of interfacial stress relaxation during oxidation exceeded the detrimental effect of hydrogenation and hydration. It remains to be seen if this Wet Rapid Thermal Oxidation approach becomes verified and establishes itself in the industry.

An important source of hazardous hydrogen is SiH_4 used in polysilicon deposition (Sect 1.2.2b). Also, further deposition steps may release hydrogen from reactive gases, as in the case of the tungsten polycide process (page 329). Intentional exposure to hydrogen is done later in the manufacturing,[1165] when there is no danger of damaging the hydrogenated gate oxide by process-induced electrical or mechanical stress. Hydrogen

[x] We describe this effect mostly on the basis of the recent model reported in Ref. 1255.

then becomes incorporated mostly within 20–40 Å from the interface.[1298] Annealing at 400–450°C in an ambient containing H_2 or, preferably,[1165] D_2, is sufficient to passivate most of the interfacial traps (Sect. 6.3.6).

In order to understand this ambivalent role of hydrogen, we consider the atomic configurations which hydrogen atoms may form with Si and O atoms in SiO_2 and at the SiO_2/Si interface. Atomic H may passivate a Si dangling bond: ⋗Si· + ·H → ⋗Si–H, where ⋗Si· indicates a triply bonded Si atom with one electron on its fourth, unused orbital. Another possibility is that hydrogen becomes inserted into a Si–Si bond and loses an electron: ⋗Si–Si⋖ + H + h^+ → ⋗Si–(H^+)–Si⋖, where h^+ is a hole. Similarly, a proton may be cracked by a hot electron (Sect. 6.6.2a) from a Si-H bond and attack a Si–Si bond: ⋗Si–Si⋖ + H^+ → ⋗Si–(H^+)–Si⋖.

These reactions have their analogues in the reactions known from studies of hydrogen with clean silicon surfaces (Sect. 5.2.1). The passivation is equivalent to the formation of a monohydride on the clean surface of silicon, while the attacks on Si–Si bonds is equivalent to the first step that leads to H etching of silicon. However, the strength of the Si–H bond is likely to be smaller at the interface than on the surface; the estimated values are: 3.1–3.4 eV for H adsorbed on Si(001) dimers (Sect. 5.2.1) and 2.6 eV for H adsorbed at SiO_2/Si(111) defects.[1299]

The passivation and the insertion are the simplest examples of the helpful and the troublesome roles of H. The broken bonds are responsible for interface traps (Sect. 6.5.3) and they disappear after passivation. This is exactly the intention behind hydrogen anneals. On the other hand, a proton trapped in a Si–Si bond produces a particularly dangerous defect. First, it is charged. Second, when it captures an electron, there is a good chance that the H atom will not be ejected but that a Si–H bond breaks, producing a Si active broken bond: e^- + ⋗Si–(H^+)–Si⋖ → ⋗Si–H ·Si⋖.

It is not possible to passivate all dangling bonds with hydrogen. Since the energy of a H–H bond is higher than the energy of a Si–H bond (4.5 eV compared to ~3.2 eV), the reaction H + ⋗Si–H → ⋗Si· + H_2 is energetically favorable. Though the H_2 molecule will later dissociate in SiO_2, the thermodynamic equilibrium state will always contain a fraction of unsaturated dangling bonds. This is well known from surface studies (Sect. 5.2.1).

Such surface studies have resulted in an idea of significant practical importance. STM investigations[1300,1301] revealed that **deuterium** atoms are much more difficult to be removed from Si surfaces by the STM tip than hydrogen atoms. This is a dynamic effect. The desorption of hydrogen (or deuterium) induced by STM proceeds via vibrational excitations by tunneling electrons.[1302] The relevant vibrations of deuterium (bending of Si–D bonds [1303]) are short-lived, in contrast to those of hydrogen.[1304,1305] Though the dissociation barrier is the same for Si–H and Si–D, deuterium has a lower probabilty to surmount it because it has less time to try.

The same process is responsible for hot-electron induced desorption (Sect. 6.6.2a) from SiO_2/Si interfaces.[1303,1306] This makes D-passivated bonds more stable against hot-electron damage in comparison to H-passivated bonds.[1300] Recent experiments have shown that deuterium passivation has another beneficial effect: the ratio of unsaturated to saturated dangling bonds becomes lower.[1307] In brief, deuterium processing can increase device lifetimes by up to 50 times in comparison to H processing!

An important contribution to hot-electron damage in modern CMOS circuits possibly comes from H^+ ions (page 437). The mechanism sketched above allows for the generation of protons from vibrationally excited, passivated Si dangling bonds. A wagging Si–H induces energy levels near to the valence and conduction band, that is, impact ionization of the defect requires ~1 eV, much less than 13.6 eV needed to ionize an isolated H atom. Impact ionization can be caused by hot electrons leaking through the gate oxide, or by electrons colliding with the oxide above the channel in the drain region (Sect. 6.6.2a). Ionization may proceed as well by the capture of holes (page 436). The ionized complex can dissociate with a significantly lowered energy barrier into a dangling bond and a proton in a neighboring Si–Si bond.[1305] Such in-bond protons easily diffuse through the silicon lattice.[1305] It is therefore plausible that, if H^+ is produced at the gate/oxide interface, it diffuses in the same way through the SiO_2 film to the SiO_2/channel interface, where it is trapped and inflicts further damage.

Since hydrogen- or deuterium-passivated bonds can be destroyed by hot carriers, the concentration of H and D should be kept low. Fortunately, many of the electrically active defects at the interface and in the oxide are caused by non-eqilibrium growth. Post-oxidation annealing in low-pressure **oxygen** can re-oxidize some of the weak spots (Section 6.3.6). At the same time, a part of the mechanical stress is relaxed, at least in the bulk of the film. The interface does not move now (that is, the oxide film does not grow thicker), but the oxygen heals slowly the poorly oxidized parts of the film, so that the amount of the fixed charge is reduced. It was also reported that rapid thermal annealing (page 365) improves the oxide, allegedly because the interstitial oxygen from the oxide bulk segregates to the interface.[1258]

Interactions between hydrogen and oxygen can lead to charged defects. For example, when a hole and a hydrogen atom arrive at about the same time near to an oxygen atom in a ⋗Si–O–Si⋖ bridge, both can be trapped: ⋗Si–O–Si⋖ + H + h^+ → ⋗Si–(O^+H)–Si⋖. This happens because a singly ionized O atom (**oxonium**) behaves approximately as a group-V atom: it readily uses three of its five electrons in bonds with other atoms and keeps the remaining two electrons on a lone pair. A Si dangling bond terminated by an OH group can be attacked in a similar way: ⋗Si–OH + H + h^+ → ⋗Si–O^+H_2. Both reactions can occur with only little perturbation of the bonds with the silicon atoms, that is, they are relatively easy.[1255] This is confirmed by *ab initio* calculations done for small clusters.[1308]

Nitrogen hardens the interface against charge trapping. The stabilizing factor here is the $p\pi$–$d\pi$ bonding between Si and N (Sect. 5.6.1). This bonding places the nitrogen atom in a planar configuration with its Si neighbors (Fig. 5.7, page 312). Though a triple-bonded N can in principle undergo similar reactions with H as a double-bonded oxygen, the reaction $(⋗Si)_3N$ + H + h^+ → $(⋗Si)_3(N^+H)$ would require that the N atom moves away from the planar geometry into a tethrahedral geometry. Such a distortion requires a significant straining of the structure and hence it is most probably characterized by a significant activation energy.[1308,1255]

This stabilization of the interface against charge trapping does not work if NH_3 is used during the nitridation. When N–H bonds already exist in the film or at the interface, the nitrogen atom can easily change to N^+ (**nitronium**) when it has an opportunity to react with H: the atomic relaxation required to complete the reaction ⋗Si–(NH)–Si⋖ + H + h^+ → ⋗Si–(N^+H_2)-Si⋖ is small and does not induce any excessive stress.[1308,1255]

A careful nitrogen processing is thus a promising technique to improve gate oxides.[1129] It makes the oxide more resistant to boron penetration (Sec. 6.5.5c), to formation of charge traps, and it even appears to smoothen the interface. However, it was reported that mechanical stress invoked during the technological process increases the interface trap density in nitridated gate oxides (see references in Ref. 1309). No such stress effect has been found in SiO_2/Si gate oxides.[1309]

6.6. Practical Quality Issues of SiO_2/Si

Apart from the effects caused by interactions between dopants and the interface (Section 6.5.5), three types of atomic-scale interfacial imperfections are of practical importance: roughness, interfacial charge traps, and fixed oxide charge.[y] They seem to be interrelated[1064] (page 394). Models unifying defects, stress, diffusion, and reaction rates have been recently discussed.[242,1194] But the knowledge on the physics behind this correlation is still somewhat speculative (Sect. 6.5). In the following, we focus on the influence of these defects on the fundamental parameters of MOS transistors: threshold voltage, drive current, noise, and time to breakdown.

Troublesome, trapped charges can be created by ion implantation (for example, during polysilicon doping), electron beam deposition of metal films, X-ray lithography, and, maybe the most importantly, plasma processing[37,38,1284,1310] and electrostatic discharge[1074] which send currents through the circuit. The ionizing radiation encountered in space flights is also of concern, mostly in telecommunication (if we ignore military satellites). Degradation of oxide films due to high Fowler-Nordheim (page 67) tunneling currents through ultrathin gate oxides is an emerging problem.

6.6.1. *Experimental access*

Defects in the interfacial region influence electrical parameters of MOS devices (Sect. 1.2.2) and can be investigated by analysis of electrical (I–V and C–V) characteristics of capacitors, diodes, and transistors. An interested reader may want to consult Ref. 1311 where five methods to measure **interface trap** density (high-low frequency capacitance, conductance, subthreshold current, charge pumping, and low-temperature capacitance) are evaluated. XPS can provide valuable information on interface states at interfaces with ultrathin oxides (see Ref. 1108 and references therein). Flicker noise (Sect. 6.6.2c) offers another way to examine interface traps and oxide damage.[1312] STM measurements of individual trapping sites have been conducted for tunneling oxides.[1217,1232] The so-called Ballistic Electron Emission Spectroscopy (BEEM), which is a version of STM, has been used to investigate oxide defects, including their distribution in the film.[1313] **Fixed charge** density can be inferred from C–V characteristics.[1314]

Information about the **atomic structure** of some defects and about

[y] See page 355 for the terminology of charge traps.

the character of the involved electron orbitals can be obtained from electron paramagnetic resonance (Section 6.5.3). Irradiation by gamma rays ($\sim$ 1 MeV), which generates defects, can be used to enhance the signal. Charge trapping can be conveniently addressed in this way as well.

Interface roughness is measured by electron microscopies and by SPA-LEED.[1064] AFM or STM can be used after the oxide is stripped (Sect. 1.4.2c). But a key experimental problem is how to monitor the **thickness** and **thickness variation** of ultrathin ($\sim$2 nm) oxides.[22,28,1315] A combination of AFM and STM allows one to measure the local oxide thickness and interface roughness simultaneously: the AFM detects the topology of the oxide surface and the tunneling current flowing through the AFM tip provides the thickness data.[501] The classical way to obtain oxide thickness is by optical methods, chiefly ellipsometry (Sect. 1.3.2j). As an alternative, analyses of Fowler-Nordheim tunneling currents were proposed.[1316,270]

6.6.2. *Influence on device parameters*

Charges in the oxide cause three main problems: they facilitate dielectric breakdown of the MOS capacitor, degrade carrier mobility in the channel by Coulomb scattering, and affect the threshold voltage. The influence of interfacial traps is similar, but in addition they contribute to noise in the current. Interface microroughness can decrease mobilities in the MOSFET channel and facilitate dielectric breakdown in thin (below 10 nm) gate oxides. Particularly hazardous are local thickness variations in tunneling oxides (page 355), because anomalously large currents flow through the weak points were the oxide is thinner.[1081,991] Metals dissolved in the oxide can be a culprit, because they not only trap charge but also move slowly in SiO_2 under the electric field and gradually change the parameters of the device during its exploitation.[1100,1101,1102,821] Dissolved boron does not move at the temperatures at which devices operate, but it can trap charge.[1262]

These problems have captured the attention of industry.[1049] A showcase story about how research helped the development of technology is about the mobile charge: after its relation to alkali metals had been discovered, clean processes which lead to stable, reproducible threshold voltages could be designed (page 358).

a. *Oxide failure and leakage current*

Though SiO_2 is an insulator, some current flows through it when the device is biased. This current may be used to load or unload a nonvolatile memory

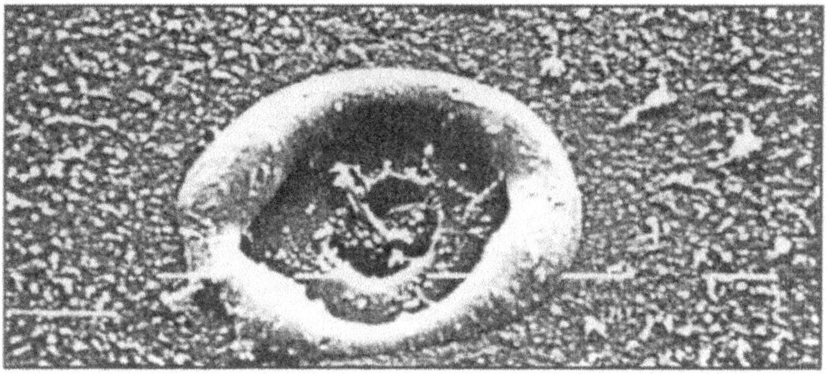

Fig. 6.10. SEM image of a MOS breakdown spot (from Ref. 1317). The size of the dash is 1 μm. When the energy stored in the capacitor was released, the polysilicon electrode melted away, spreading the material over the surface. Interestingly, some capacitors may be still working after such a hard breakdown. The polysilicon may evaporate from the damaged region, so that there is no short-circuit[1318] ("self-healing" effect).

cell (Fig. 1.3, page 9), but in general it is just a troublesome leakage. In practice, the gate leakage current density should not exceed 0.1 μA/cm^2. One obvious problem with high leakages is that they lead to power losses. It can be partly solved by proper process and circuit design,[956] Another problem is that the leakage current damages the dielectric. In particular, charge is trapped in the oxide. When a critical value Q_{bd} of the charge passed through the oxide is reached, the capacitor fails (Fig. 6.10).

As a matter of fact, in order to be considered as useless, the gate oxide need not to be destroyed by a hard punch-through such as the impressive one depictured in Fig. 6.10. Long before such an event happens (if it happens at all[1232]), the electric stress causes the gate leakage current to increase. This is the so-called **Stress-Induced Leakage Current**[1319] (SILC). Such a behavior is typical for ultrathin oxides (below 10 nm).

SILC is a complex phenomenon which evolves in a few steps. First, the gate leakage current increases only a few times. This is the classical SILC effect. Next, pre-soft breakdown events appear:[1320] the current rises several times for a brief period, and then decreases again, but to a higher value than before. The device still functions, though its operation may be erroneous. As a consequence, some bits stored in the memory may be lost occasionally. After several such pre-breakdown events, the leakage current rises strongly (hundred or thousand times) and does not decrease

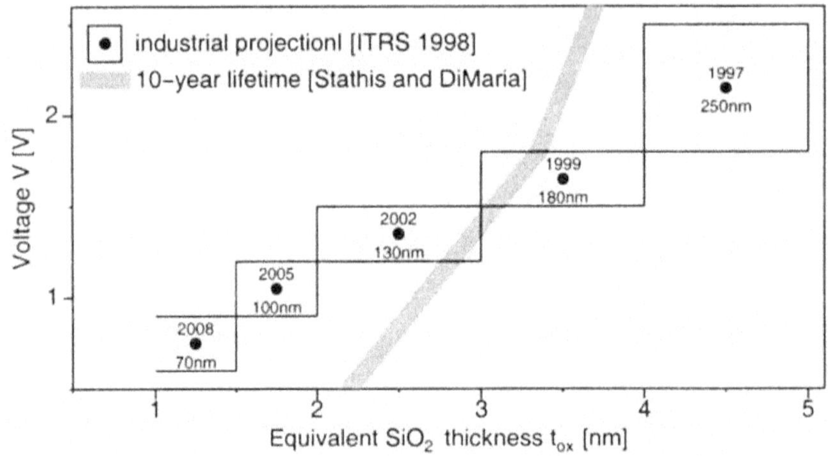

Fig. 6.11. Reliability of ultrathin oxides. The boxes mark the range of gate oxide thicknesses and power supply voltages for CMOS technologies, as projected by semiconductor industry.[22] The gray line[1099] estimates the gate oxide reliability. It corresponds to 100 ppm failures of nMOSFETs in the population with the total gate oxide area of 0.1 cm^2 after 10 years of operation under constant gate voltage V.

any more.[1321,1322] This is the so-called **soft breakdown**. Finally, the hard breakdown may occur: the oxide ceases to be an insulator, the gate current is extremaly high even for very small gate voltages.

Some researchers believe that transistors with oxides even as thin as $t_{ox} = 1.3$–1.5 nm may be practical.[30,991,1098] Although such an oxide is in the direct tunneling regime (page 355), the high leakage current flowing through it under normal operating conditions is hoped to be relatively mild on the dielectric. However, the electron trapping probability scales only as the second or third power of oxide thickness,[1063] while the leakage current density increases exponentially. A recent study[1099] estimated that, in order to work reliably, circuits made of transistors with $t_{ox} \simeq 2.5$ nm must operate with gate voltages not exceeding one volt (Fig. 6.11). This means that the power supply voltage that the industry projects to use in 0.13 μm CMOS technologies may be too high to assure a reliable 10-year operation of the circuit. This result is alarming if only because 0.13 μm chips are expected[22] to appear on the market as soon as in the year 2002. Another recently signalled difficulty is that Q_{bd} and the breakdown field are worsened when the stress is applied at elevated temperatures.[1323] A significant effect was seen even at 180°C. On the other hand, ultrathin oxides ($t_{ox} \lesssim 3$ nm) are more resistant to brief electrical stress than thicker oxides: little or no process-induced plasma damage was observed when the oxide was thin enough.[1324,1310] But the damage effects *increase*[1324,1325] with decreasing t_{ox} until $t_{ox} \simeq 3.5$ nm.

One reason why it is difficult to reach a consensus is that it is hardly possible to wait for a transistor to fail under usual working conditions if the expected failure rate is 100 parts per million in ten years! Therefore, methods to speed up failures have been designed. Test devices are electrically overstressed until they are damaged. The problem is that this can be done in many ways, and different schemes can give qualitatively different dependence of the breakdown behavior on t_{ox}. For example, the Constant Current Stress (CCS) yields significantly *improved* breakdown characteristics for thinner oxides, while Constant Voltage Stress (CVS) results in significant reliability *losses*.[1322] An extrapolation from thicker to thiner oxides must often be done, which assumes that breakdown of operating, ultrathin oxides is the same as breakdown of highly stressed, thicker oxides. We feel that the issue is still open, but prefer not to take sides in the discussion. Let us only note that, according to a recent study in which transistors were stressed by constant fields of 7–11 MV/cm during three years,[1326] it is the electric field rather than current that causes the degradation.

Electrons with energies high enough to inflict oxide damage can originate also from the source-drain.[1075,1245] A proper distribution of the electric field in the channel by a careful design of the dopant profile (Lightly Doped Drain extensions, LDD) can minimize this hot-electron effect.[1327,268] A reduced power supply voltage in modern submicrometer devices makes reliability degradation by channel hot electrons less problematic.[1328]

Electric fields below 8–10 MV/cm should not damage a modern oxide. Good oxides can withstand fields up to ~12 MV/cm or even higher. With supply voltages around 3 V and oxide thicknesses of ~10 nm, as typical for 0.35 μm circuits, the operating fields are 2-3 MV/cm; one volt aplied across a two nanometer thick oxide corresponds to 10 MV/cm. If the oxide contains a pinhole (Sect. 6.3.5), it may fail even at voltages below 1 MV/cm. **Testing** of the circuit is likely to detect such defective devices.

A typical time distribution of **failure rates** has three regions: early failure, steady state, and wearout. The time dependence of cumulative failures in the steady-state and wearout phases can be usually fitted by a Weibull distribution obtained from the Weibull function for the instantaneous failure rate,[z] $\lambda(t) = \frac{\beta}{\alpha}t^{\beta-1}$, where α and β are parameters.[268]

Early failures ("infant mortality") happen after the testing. The killer defect is likely to be created during back-end fabrication steps (Sect. 1.2.2b) which include the application of a voltage (like in plasma processing[37,38,1284] and during an electroscatic discharge event[1074]) or ionizing radiation (during lithography). Carriers may then be injected into the oxide or/and charge traps may be created or activated there. As we will see, such injected carriers and oxide defects are the main factors which weaken the dielectric. Also, contamination originating from front-end processes may lead to hazardous charges. For example, Na^+ ions (which may arrive at the wafer surface from W filaments of the furnace[268]) can reduce the dielectric strength of the oxide. It is worth noting that the hot carrier degradation induced by technological processing through charging of the gates can be minimized by insulating the back side of the wafer.[38] Since the gates *and* the substrate are charged during processing, insulating the back side reduces the voltage drop across the gate oxide. This minimizes the current flow across the oxide, so that less damage occurs. However, lot of damage can still be caused by plasma nonuniformities across the wafer surface. They lead to charging currents between two transistors rather than between a transistor and the substrate.[1329]

The possibility of inflicting damage increases with increasing complexity of the interconnects, that is, with increasing number of back-end process steps.[1330,1329,1284] Interconnects become more complicated as the devices themselves become smaller. A contemporary circuit may have three to five levels of metal bitlines, running one above another like lanes in a crossing of many freeways. Within the first decade of the new millenium, the number of metal layers may increase to seven.

The gate oxides which survive the early failure phase are highly defect-free and likely to endure until they **wear out**. The accumulated damage then reaches the critical limit and the device breaks down.[1320,1322] Various failure modes exist, from degradation of the I–V characteristics, through significant increase of the leakage current, up to a complete punch-through ("hard breakdown"). When the damage done to the oxide by the breakdown current is not too extensive, a hard breakdown event opens an atomic-size conduction channel, which can be considered as a point contact.[1331]

[z] The instantaneous failure rate is defined as the rate during an interval of time divided by the population remaining at that time.

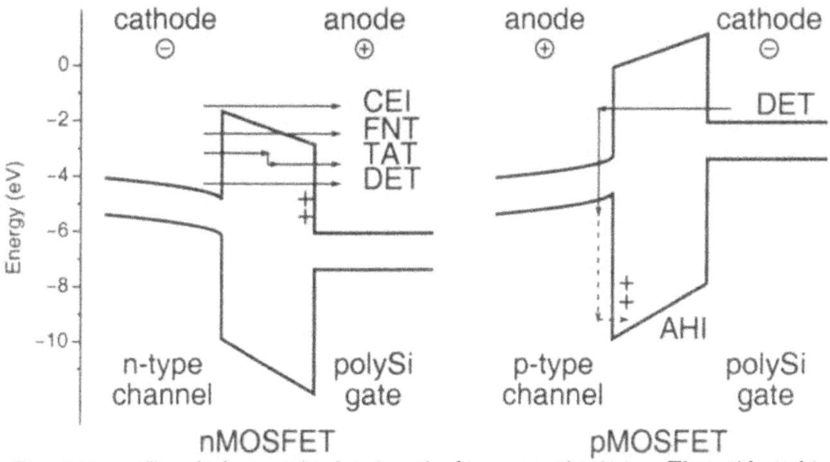

Fig. 6.12. Gate leakage and related oxide damage mechanisms. The oxide is biased with 2 V. Assuming that the gate oxide is 2.5 nm thick, this corresponds to the electric field of 8 MV/cm. Band bending in the gate is ignored. The charged defects (hot carrier damage) are indicated as ++. **Right**: Paths of leakage currents shown in nMOSFET band diagram. Classical Electron Injection (CEI), Fowler-Nordheim Tunneling (FNT), Trap-Assisted Tunneling (TAT), and Direct Electron Tunneling (DET) mechanisms are indicated. Electrons emerging from SiO_2 can ionize defects at the anode interface. **Right**: Anode Hole Injection (AHI) damage shown in pMOSFET band diagram. The emerging electon recombines with a hole, the energy is used to excite another hole which is afterwards trapped in the oxide.

We will now examine the **physical mechanisms** of dielectric breakdown. These mechanisms are not fully understood. However, some basic principles are known or at least widely accepted. We focus on the oxides which are good enough to survive at least until the steady-state phase. Most probably, the processes which make them fail are the same as those which cause "infant mortality", the main difference being in the initial condition.

The key concept (Fig. 6.12) is that electrons are injected into the oxide from the cathode (the negatively charged terminal), are accelerated by the electric field in the oxide, and produce charged defects. Moreover, hot holes are injected from the anode and can be trapped in the oxide. The resulting electric fields and local changes in the electronic structure promote higher injection rates. Eventually, this positive feedback leads to a short-cut.

The current flowing through the dielectric (Fig. 6.12) is termed the **leakage current**. The carriers may be injected either over the barrier (classical injection) or through the barrier (tunneling). At the point of injection, they can be either **hot** (have energy much higher than the lattice temperature T) or **cold** (have energy comparable to T). Since the band offset between Si and SiO_2 is much larger for the valence than for the conduction band (Fig. 6.12), injection of electrons dominates. On the other hand, holes are more easily trapped in the oxide than electrons, because the mobility of holes in SiO_2 is orders of magnitude smaller than that of electrons ($\sim 2\cdot 10^{-5}$ cm^2/Vs versus 20 to 40 cm^2/Vs).

Let us first consider the classical injection case. Electrons are heated up by the electric field in the channel.[1332,1075,1333] Some of them do not lose their energy to the phonons and can enter the oxide over the barrier (**lucky electrons**). But if the drain voltage is low, it alone cannot provide enough energy. Therefore, lower voltages should reduce the hot-electron damage.[1328] Yet electrons may not only lose but also gain energy from the lattice, whereby the energetic electrons are more likely to interact with high-energy phonons than thermal electrons. This may help the channel electrons to collect the energy needed to overcome the classical barrier.[1334]

However, even in oxides as thick as 12 nm it is **tunneling** that dominates the leakage[1244] if the applied voltage is as low as during the device operation. This is because the energy difference between the conduction bands of Si and SiO_2 is large and it is difficult for carriers to collect enough energy to surmount it. Defects in the oxide (Fig. 6.13) facilitate the tunneling:[272] the carrier tunnels first from one electrode to the defect site, and from there to the other electrode. A recent study[1284] has concluded that the charge traps due to boron atoms segregated to the gate oxide (Sect. 6.5.5c) play an important role in this process. High temperatures and charging during plasma processing in CMOS manufacturing cycles make the oxides more suspectible to degradation caused by leakage currents.[1284]

After a **direct tunneling** event (DET in Fig. 6.12) the electron emerges from the dielectric without causing much trouble. If the tunneling is in the **Fowler-Nordheim** regime (FNT in Fig. 6.12), the electron is injected into the oxide and there it gains energy, possibly inflicting damage. It is usually assumed that hot electrons in the oxide can create holes by impact ionization.[1335] This can happen, for example, by kicking valence-band electrons into defect sites with electron states in the forbidden gap (Fig. 6.13). The mobility of holes in SiO_2 is low and they are likely to be captured in the film (particularly in the presence of hydrogen, Sect. 6.5.5d).

Recent studies[1336] show that holes excited from the substrate or from

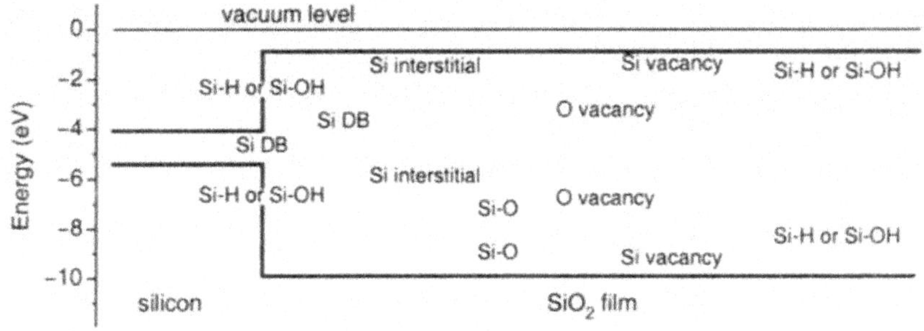

Fig. 6.13. Schematic energy diagram of interfacial and SiO_2 defects. From Ref. 268.

the gate by hot electrons emerging from the oxide are the dominating source of the damage, at least in pMOSFETs (see page 10 for the nomenclature). This **Anode Hole Injection** (AHI, Fig. 6.12, right) is caused by an Auger-like effect. A ballistic electron recombines with a hole in the channel of a pMOS device. The released energy is transferred to another hole. The excited hole may have enough energy to enter the oxide to become trapped there. Alternatively, the hot hole injected into the oxide may recombine with another electron, releasing ~9 eV, which is more than enough to break a Si–O bond (Table 1.4, page 57). The same process is possible in an operating nMOSFET if the gate is made of p-type material.

Typically, nMOSFET gates are made of n-type polysilicon. There are no holes available for AHI in this case, apart from the holes created in the course of impact electron-hole generation. But the energy of such holes is small. The most damage is then caused directly by **hot electrons**. The electrons are the most destructive where they have the highest energy, that is, near to the anode terminal. In nMOSFETs this is the interface with the polysilicon gate (Fig. 6.12, left). The primary damage is thus localized away from the oxide/substrate interface. The cathode damage comes possibly from positively charged H atoms which are cracked from Si–H or Si–OH sites by the hot electrons (Sect. 6.5.5d), drift through the oxide to the cathode interface,[1245,1337] and produce positively charged electron traps there (page 427). When the bias is reversed (that is, when the substrate is the anode), the amount of interfacial traps generated at the SiO_2/substrate interface increases by two orders of magnitude.[1338]

When the oxide is thin or the electric field is strong, the carriers may tunnel away from the trap sites.[1312] In addition, positively trapped defects may be neutralized or converted into negatively charged defects by the electron leakage current.[1245] A trapped carrier may become permanently localized if the charged defect undergoes strong lattice relaxation (Sections 6.2.1 and 6.5.5d). In this case, there is an additional energy barrier in the space of atomic coordinates, preventing the charge from being removed.

Since the damaged sites are mostly positively charged, they attract more leakage current, which in turn creates more damage. In addition, positively charged traps are potential recombination centers for electrons. The energy released during recombination of hot electrons with strongly bonded holes may spread the damage. As a consequence, defect "trees" grow from the anode to the cathode,[1316] eventually short-cutting the dielectric.

Impact ionization may occur not only in the oxide but also in the channel, whereby **Drain Avalanche Hot Carriers** (DAHC) are produced. The electrons accelerated by the drain voltage may generate hot holes in this way. A fraction of these holes may be injected into the oxide, contributing to the damage. Since channel lengths are scaled down faster than power supply voltages, electric fields in channels are increasing to high values, above 400 kV/cm. The mechanism which dominates the hot carrier injection and trapping depends on the drain and gate voltages. The spatial distribution of the damage along the channel has been recently discussed in Ref. 1339.

Another factor which influences the dielectric strength of the oxide is the **interface roughness** (Sect. 6.3). The most natural way in which a rough interface facilitates a shortcut is through thickness variation. Most of the current flows through the thinnest parts of the dielectric and inflicts a concentrated damage.[1081,991]

An important parameter is not only the roughness itself but first of all the spatial frequency spectrum of the roughness amplitude, $r(f)$. If the interface is undulated with a lateral wave length large compared to the film thickness t_{ox}, the oxide will easily conform even to a large roughness amplitude. Thickness variations will be minimal. But undulation wave lengths comparable or smaller that t_{ox} are critical, as their amplitude is likely to translate directly into thickness variation.[1166]

An oxide with a smooth interface[1079] may have a strength as high as 16 MV/cm. However, the interface roughness and interface defects are correlated. The mechanism responsible for this correlation is still not well understood. It is difficult to judge whether the high strength of such oxides is directly due to less roughness or is primarily caused by a fewer number of various charged and uncharged, atomic-size defects in a smooth interface.

Remarkably high reliability can be achieved[1110] by growing the oxide in electrically neutral *atomic* oxygen (**remote plasma oxidation**) rather than by conventional dry oxidation in O_2 ambient. Such oxides may have twice as large Q_{bd} as control oxides prepared in standard way. In addition, in contrast to dry oxidation, Q_{bd} of remote plasma oxides is hardly degraded by lowering the oxidation temperature from 900°C to 700°C.

Finally, let us note that reliability is a very active research field of significant practical importance. We attempted to present a consistent interpretation of the published data, focusing on recent advancements. However, given the multitude of oxide preparation methods, the rapid development of new technologies, and the difficult experimental access to the processes which control reliability of dielectric films, we could not avoid intended and unintended simplifications and omissions. Therefore, this brief presentation should not be viewed as the ultimate explanation of the breakdown mechanism, but rather as a somewhat subjective introduction to the subject.

b. *Drive current and high-frequency performance*

A key electrical parameter of a transistor is its **drive current**[21] (or saturation current) that is, the maximal current I_{Dsat} that can flow through it. Flaws in the gate oxide reduce the drive current density i_{Dsat}, which in turn increases the time needed to reload the capacitors controlled by the transistor. To maintain the speed of the circuit, the designer offsets the decrease in i_{Dsat} by drawing a wider transistor that passes more current but occupies more room on the die and dissipates more power.

Rough interfaces and defects in the oxide deteriorate I_{Dsat} by scattering carriers in the channel[1340] (see also Refs. 21, 1212, and references in Ref. 1079). The scattering occurs through irregular electric and strain fields as well as through random boundary conditions for the current flow. The effectiveness of the scattering depends on the wavelength of the interfacial roughness, in particular because electrons do not "see" a potential fluctuation which has a frequency higher than their own wavelength. A quantum mechanical correction is therefore needed to properly describe the influence of interfacial roughness on the mobility of carriers.[1341]

Mobility reduction comes also from increased dopant scattering. The dopant scattering increases because shorter channel lengths imply higher doping levels. To compensate the negative influence of higher doping, smaller devices require better oxides with smoother interfaces and fewer charge traps. The roughness is a decreasing function of oxidation temperature and it tends to be smaller for oxynitrides (Sect. 6.3.4) and steam oxides than for standard oxides (Sect. 6.3.2). But if the nitrogen is placed too close to the interface (in the first monolayer), the mobility may be degraded.[1131]

When the interface microroughness is comparable to the monatomic step height on Si(001), which is 1.35 Å, the mobility of electrons in the channel can reach[1071] 90% of its bulk value μ_0. Increasing the roughness to 5 Å rms decreases the mobility to $\sim 0.6\mu_0$. The mobility is thus optimal if the interface microroughness is reduced to steps of monatomic height.

Similarily, scattering degrades the **high-frequency performance** of the transistor. The frequency response is limited by two factors: by the time the carriers need to travel between the source and the drain (transient time τ) and by the RC time constant τ_{RC} of the device. Reduced carrier mobility worsens both factors: the transient time $\tau \sim 1/\mu$ and the RC time constant $\tau_{RC} \sim 1/g_m \sim 1/\mu$, where g_m is the transconductance $g_m = \partial J/\partial V$. Limits on the operating frequency are of particular concern in circuits built for communication purposes. In such circuits, transitors must be designed to process signals in the microwave regime (above 1 GHz). However, the major limitation in this case comes from interconnects (Sect. 1.2.2).

c. *Instabilities and noise*

Noise[1342] sources are typically quantified by their Power Spectral Density (PSD). The two-sided PDS is the Fourier transform of the voltage autocorrelation function. In electrical engineering the one-sided PSD is more common. It is defined for positive frequencies only and equals twice the two-sided PSD.

Three fundamental types of noise contribute to the noise power spectral density S of an electric circuit:[21] thermal noise, flicker noise, and shot noise. Both thermal and shot noise are characterized by a "white" Fourier spectrum, that is, their PSDs do not depend on the frequency f. In contrast, flicker noise has a characteristic $1/f$ frequency dependence. A typical plot of $\ln(S)$ of the noise power density $S(f)$ versus $\ln(f)$ of the frequency f looks like 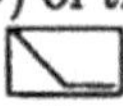, with low frequencies dominated by flicker noise (the line with the negative slope) and high frequences dominated by white noise (the horizontal line). The frequency at which the two lines meet is known as the **corner frequency** f_c. In MOS devices, the magnitude of f_c is usually between 100 kHz and 1 MHz.

Thermal noise is proportional to the thermal energy kT and to the resistance R. This is a natural relationship, because thermal noise and resistivity have the same source: random, thermal jumps of atoms. Thermal noise is usually the main source of noise in a circuit. It comes from passive elements; a real culprit may be inductivities with low quality Q. Therefore, in low-noise designs, certain passive elements are taken off the circuit. This helps to keep the noise contained. Thermal noise in a well-designed circuit can be reduced so much that other types of noise become important.

To be more specific, we write down the one-sided PDS of a thermal resistor noise. The voltage PDS can be expressed as

$$S_{\mathrm{V}}^{\mathrm{th}}(f) = 4kTR. \tag{6.34}$$

This is useful if the two ends of the resistor are open. If the resistor ends are short-circuited, the current PDS is more meaningful. It is given by

$$S_{\mathrm{I}}^{\mathrm{th}}(f) = \frac{4kT}{R}. \tag{6.35}$$

These so-called **Nyquist equations** are a special case of the fluctuation-dissipation theorem, which represents a general link between fluctuations (noise) and dissipation (loss, resistance) for systems in thermodynamic equilibrium. The Nyquist equations are sufficient for virtually all electrical applications. However, a quantum statistical correction needs to be included in the infrared region of the spectrum.

Of more concern to us is the **flicker** noise. It is caused by exchange of carriers between the gate channel and the interface traps. The mean-square noise current I (square root of current PDS) is directly proportional[1066] to the density of the interface traps N_{ST} and to the square of the oxide thickness, t_{ox}, and inversely proportional to the channel area LW:

$$I \sim \frac{t_{\mathrm{ox}}^2 N_{ST}}{LW}, \tag{6.36}$$

where L and W are the effective channel length and width. Therefore, the flicker noise increases when the device is scaled down, unless the oxide thickness is decreased to match the decrease of channel dimensions (see references in Ref. 1343). When t_{ox} is so small that the oxide is in the direct tunneling regime (page 355), the gate leakage current contributes to the noise.[1344] But the flicker noise decreases with t_{ox} even in this case.[1344]

The importance of flicker noise depends on the application. Since modern devices operate at frequencies well above a typical corner frequency, flicker noise can hardly influence the operation of the circuit within a single clock cycle. A potential difficulty is that lengths of signals sent by the clock are affected. If uncontrolled, the phase of the clock signal would randomize after many clock cycles (**timing jitter**[1345]), leading to synchronization problems. This is avoided if the phase of the clock is corrected periodically.

It is more difficult to deal with the 1/f noise in telecommunication. Here, the noise transfers into variations of the carrier frequency (the so-called **phase noise**). Phase noise is the frequency-domain equivalent of oscillator timing jitter describing the same physical effects.[1345] As a consequence of the phase noise, the minumum width of communication channels is broadened. A typical channel width is 1 MHz, which is close to a typical corner frequency f_c. The importance of flicker noise will increase with the tendency to reduce the bandwidths in order to compress information. This will require improvement of the atomic-scale quality of SiO_2/Si interfaces: the interfaces must contain as few electrically active traps as possible.

The need to reduce the flicker noise of MOS devices increases concurrently with the trend in telecommunication to replace bipolar circuits with CMOS circuits. This trend reflects the desire of designing low-cost system solutions on a single chip, known as System On a Chip (SOC).

Transconductance g and flicker noise power density $S(f)$ are correlated through their dependence on the interface trap density. Experiment yields[1343] an exponential relationship of the form $S(f) = A\exp(B/g)$. Low-field oxide degradation by hot carriers also increases the flicker noise level.[1343] This may be consistent with the observation that low-field hot carriers hardly change the interface charge density.[1346] The damaged sites can exchange charge with the channel, which produces noise and, at the same time, prevents the accumulation of charge.

Already in 0.35 μm technologies the noise power can be spread between **different devices** in a circuit by several orders of magnitude. This is because the noise of a transistor is governed by a few traps. An average transistor of area $A = 0.5{\times}2\ \mu m^2$ and with interface trap density $D = 5 \cdot 10^{10}/cm^2$ has $AD = 500$ defects. Only a fraction of them has electronic states sufficiently close to the Fermi level to exchange carriers with the channel within a reasonable time. The noise figure becomes thus sensitive to even small variations in technological process parameters.

Shot noise is of little interest to us. It comes from the quantization of electric charges: electrons and holes behave like particles, hence the current arrives in portions, like series of shots from a machine gun, but at random instants. For a *pn* junction, the current one-sided PSD of shot noise is

$$S_I^{sh}(f) = 2qI \tag{6.37}$$

(q is the elementary charge and I is the direct current through the diode). Shot noises are important in bipolar but not relevant in MOS devices.

In complex CMOS circuits one is faced with the problem of (white) **supply** and **substrate** noise.[1345] Power supply and the substrate may become the dominant noise sources in mixed-signal integrated circuits.

d. *Threshold voltage*

The influence of the **fixed charge** Q_f on the threshold voltage becomes obvious when one realizes that the presence of this charge causes band bending on the silicon side of the interface. Therefore, in order to flatten the energy bands one must apply a corrective gap voltage that induces a compensating charge in the gate. In practice this means that the threshold voltate is changed by this corrective "flatband shift" voltage V_{FB}. If we assume that Q_f resides at the interface, we can easily estimate the fixed charge contribution V_{FB}^f to V_{FB} as:

$$V_{FB}^f = -\frac{Q_f}{C_{ox}}, \text{ or} \tag{6.38}$$

$$V_{FB}^f = -4.5 \cdot 10^{-14} Q_f t_{ox}, \tag{6.39}$$

where the latter formula returns V_{FB}^f in volts, t_{ox} is the oxide thickness in nanometers, Q_f is given in electrons per cm^2, and the dielectric constant of $\epsilon_{ox} = 4$ is assumed for SiO_2; $C_{ox} = \epsilon_{ox}/t_{ox}$ is the capacitance of the oxide layer. The other factors which affect V_{FB} are the mobile charge and the work function difference between the gate and the substrate.[268]

With a 5 nm oxide and $5 \cdot 10^{10}$ charged defects per cm^2 (typical conditions nowadays[1222]), the absolute value of the flatband shift is less than 12 mV. Does it mean that there is no problem with threshold voltage?

Only to a certain extent. Leakage currents (Sect. 6.6.2a) degrade the oxide by producing charged defects; the critical defect density[1338] to cause breakdown is $\sim 10^{14}$ cm^{-2}. This corresponds to V_{FB} of more than 20 V! The charged traps may therefore render the transistor useless even before they manage to kill it by a regular discharge (Fig. 6.10, page 431).

Contaminants like sodium can be critical (page 358), producing a **mobile oxide** charge which changes its distribution in the oxide with time under the influence of the electric field. This causes dramatic instabilities[821] in the threshold voltage V_T. Fortunately, this is not of much concern today. Ultraclean processing technologies (Sect. 1.4.2c and page 365) have helped to reduce the amout of such ions down to insignificant traces.

Several emerging problems are associated with interactions between dopants and gate oxides. **Segregation** and pile-up of dopant atoms at the interface can affect the threshold voltage even by a few hundreds meV (Fig. 6.7, page 418). Segregation of boron from the polysilicon gate to the dioxide and **penetration** of the boron atoms through the oxide (Sect. 6.5.5c) influence the threshold voltage as well; in this case, the effect is enhanced as the oxide thickness is reduced.[1282,991]

Misplaced B atoms leave a depleted region in the gate,[1282] produce charged defects in the oxide,[1262] and change the dopant distribution in the channel.[1063] This can lead to V_T shifts of the order of a hundred mV. For example,[1282] annealing a 3 nm gate oxide at 1000°C for 40 seconds, as needed to activate the B acceptors in the gate, changed the V_T by 80 meV (the gate was 200 nm thick and implanted with $5 \cdot 10^{15}$ cm^2 BF_2^+ at 20 keV). For 4 nm oxides, the corresponding shift was negligible. Since boron diffusion in polysilicon proceeds preferentially along grain bondaries, this type of influence has stochastic character, reflecting the statistical distribution of the grains. As a consequence, B penetration effects can strongly degrade transistor matching performance in deep-submicron CMOS circuits.[1283]

The problems with boron penetration can be minimized by implantation of B^+ atoms instead of BF_2^+ molecules,[1282] by using nitridated oxides (Sect. 6.3.4), or by a special design of the gate. For instance, the gate can be implanted with nitrogen in order to slow down boron diffusion along the grain boundaries, or it can be horizontally separated by an extremely thin (~1 nm, or roughly three atomic layers) oxide membrane[1347] which acts as a diffusion barrier for fluorine from BF_2 implants.

A fool's brain digests philosophy into folly, science into superstition, and art into pedantry. Hence University education.

– G. B. Shaw

[Universities should be] nurseries of scientific research and mental cultivation, [free of] professional conceit or exclusivity.

– W. C. Röntgen

CHAPTER 7

Afterword

J. Dąbrowski

Philosophical remarks. The industry knows enough to produce better and better devices; who needs basic research? Conclusions and outlook.

7.1. Do We Need an Afterword?

In one of his lectures,[1348] Stephen Hawking recalled that there are two ways to get paid for what one enjoys; one of them is to become a scientist. Certainly, science has been so amazingly succesfull largely because science is the *passion* of scientsts. But the growing tendency to commercialize all research may change this passion into duty. The main purpose of this Afterword is to find out how curiosity and joy may be preserved as the main motivations for scientific work (Sections 7.2 – 7.4). We will do this by analysing the dangers and challenges posed to scientists by the commercialization of science and the globalization of economy. A concise summary of the whole book is then given in Section 7.5.

Our opinion is that science may remain enjoyable and encompassing a broad community, provided that many scientists will manage to harmoniously incorporate purely scientific, commercial, and educational aspects into their research. The commercial values can be recognized and exploited only by those who know industrial technologies well enough to estimate the impact of their work on issues of practical interest.[1349] Similarly, the educational values can be fully recognized and exploited only by those who appreciate problems brought to a man on the street by the changing world. Indeed, education means not only a professional, technical literacy. At least equally important is that persons considered as educated deal sensibly with contemporary social and even philosophical or religious dilemmas. We will collect arguments and examples which propose and illustrate a way in which science in general, and surface science in particular, can be integrated into the changing framework of our civilization without losing its identity.

7.2. A Glance Behind

We have described silicon surfaces and interfaces as seen from various perspectives: of experimentalists' labs, theorists' computers, engineers' desks, and technicians' clean rooms. But our discourse began with questions about the role of science in the world controlled by money (page 1). We now come back to these philosophically tinged remarks.

We asked why should one do science. And who profits from science? First, a scientific excercise can be valued for its own sake. Such pure science resembles alpinism: scientists feel like climbing snowy summits of knowledge and relish vast expanses of intellectual freedom. This certainly gives satisfaction to the participants: the scientific community applauds as new laws are snatched from Nature. But the game is expensive and few goverments and organizations sponsor it unconditionally.

Second, science may support commercial applications. This generates financial profit. Applied science is welcome by the industry, especially when salaries of the scientists are paid from governmental resources. Industrial pressure increases the practical value of research. The drawback is, however, that this pressure reduces the researcher's intellectual freedom by shifting emphasis from time for thinking to Time-To-Market (TTM).

Third, results of scientific investigations enhance our understanding how the world works. Clear, easy to grasp examples of basic rules in action have an indispensable educational value for students, newcomers to the field, and even beyond the academic community. In addition, taking part in scientific successes and failures allows one to appreciate that "how?" is the only type of question that may be answered credibly by science. Scientific "why?" must have no end and the answer can be proven unique only within a given theory. This experience may result in philosophical understanding, a milestone on the way to wisdom.

The profit from science goes thus to scientists, to industrial tycoons, and to the mankind. The only problem is that basic, applied, and educational forms of science, the three variations which produce the three different types of profit, are often perceived as mutually exclusive. Basic studies are time consuming, motivated by curiosity, and highly specialized. Succesfull applications need fast results and have roots in market analysis, whereby untimely ideas and too much curiosity are a hinderance. Educational message implies broad overview, generalization, and simplification.

Though investigations of Si surfaces are basic in their nature, we noted that they do have commercial and educational values (Section 1.1). We also noted that they can reasonably guarantee intellectual freedom of the scientist, but only if a balance between all these aspects of his or her research is established. We kept returning to technological and educational sides of silicon surface science. We now collect these thoughts, remarks, and observations in an attempt to analyse the current stand and possible future of fundamental science in our commercialized world.

7.3. Fundamental Science in a Commercial World

Science is often thought to be the driving force of technological progress. It is however money that makes the world go round and it is money that makes it accelerate. Money is the crank shaft which transfers motion from the engine powered by human ambitions, greed, diversity, and misfortunes. This engine may be built of human ingeniousness, yet ingeniousness does not necessarily mean science. Scientists are gear wheels in only one part of the machine. The tenders consider this part important but not crucial.

It may seem a trivial observation that progress, be it technological, cultural, or social, has been related to money since money was invented. The commercial aspect of progress used to be simple: in order to get more money, producers followed the market trends. But recent information revolution and its close follower, market globalization, have changed commercialism from such a passive opinion *observer* into the active opinion *maker*. No more does financial profit direct the progress to suit the needs of mankind. A new trend has established itself: the needs of the mankind are directed to sustain enough progress that profit is guaranteed. This is not so trivial, not necessarily nice, and is often forgotten or ignored.

Science and scientists are also affected by this transformation. Research costs money; it seems more and more natural that it should therefore return profit.[1] Science *can* be profitable. It is just the matter of research focus and management.[1350] Customers are waiting.[22] The key issue is: *should* science be profitable? Is a commercialized science a science at all[1351,1352]?

7.3.1. *The dangers of commercialization*

Scientific work is motivated chiefly by researcher's curiosity. Commercialization would inevitably phase out curiosity-driven studies.[1353] Can this destroy science? The answer depends on what we understand as "science".

Is someone who develops a cheaper method to clean wafers a scientist or an engineer? Since engineering is an "application of science and mathematics by which the properties of matter [...] are made useful to people",[1354] this researcher is certainly an engineer. Naturally, being an engineer does not exclude being a scientist. The objective of a scientific study is to collect "knowledge covering general truths".[1354] The notion of generality is broad enough to encompass wafer-clean research projects. Nevertheless, is it "truth" that is pursued?

The scientific definition of truth is operational: a statement can be accepted as true only if it is provable. This links the meaning of truth firmly to the reason why a particular question was raised. Truth reflects therefore a relation between the problem and the solution: something is true if it solves the problem. *No other definition of truth can be supported by scientific research.* A discovery of a cheaper clean does reveal a new truth which is at least as general as a discovery of a new surface reconstruction. If the latter is a scientific achievement, so is the former.

Commercialism is not a threat to science. It is a threat to *scientists*.

The post-academic science[1351] organized efficiently along the rules of market analysis would turn scientists into craftsmen hired to solve specific problems. The difference between an "academic" scientist and an "industrial" engineer is not in the research subject but in the attitude. The engineer derives satisfaction from customer's praise, while the key factor which makes research attractive to the scientist is the freedom to choose the subject. The danger commercialization constitutes to the scientist is that it will take this freedom away from him.

Even more: the only way to maintain a socially acceptable living standard may be to open a private busines, such as a consulting firm.[1] Owning a tiny consulting firm in Silicon Valley or an ice cream stand in Miami may be an excellent opportunity to make money. But a scientist, forced to secure his income in this way, will turn into a businessman.

Another important difference between a scientific and an engineering research is that a scientist looks chiefly for *mechanisms* which control the investigated effect, while an engineer focuses on the *application* of this effect. In an ideal world, both ways are complementary. But the world is not ideal. There is a growing tendency to speed up and commercialize scientific studies, whereby taxpayer's money is spent to support huge national and international concerns. In consequence, scientists are being transformed quite rapidly into underpaid engineers.

Thus science will survive. Only that scientists will have to go.

One might summarize these remarks by saying that engineers are instrumental for the progress, but scientists are not instrumental for science. A future world without academic scientists is imaginable. But what would happen to the progress if the academic tradition fell? We would evolve into specialists led by market experts, both groups inclined to consider as wasteful any thought beyond those needed to make money. And the most efficient way to boost profits is: *aim to satisfy the most primitive carvations of an average uneducated man, so that new cheap products can be sold expensively in large quantities.* Invention and competition would be substituted by vanity and brutality. If anyone has the antidote today, it is not censors and lawyers, but the whole academic community.

7.3.2. *Can science be saved?*

Can science remain academic and free, if only to give our civilization a chance to develop its human character? Maybe, but not without help of industrial partners. They control the money that science needs to survive. It would be, however, destructive to rely on them alone, because they care for profit rather than knowledge. So academic science must remain an independent entity selling to the industry in order to assure financial assistance and to the public to earn its support and maintain broad education.

A scientist must therefore know how does his or her expertise match current and anticipated industrial needs. Besides being prerequisite to practical relevance, this knowledge allows one to take the initiative in formulating technologically attractive *long-term* research goals. The latter provides protection from industrial pressure. Such a pressure is unavoidable in short-term studies: the customer waits impatiently for the results he paid for.

At the same time, cultivating educational and philosophical links may help to preserve curiosity as widely appraised penetration. Scientists must thus try to view and present their work from a broad perspective; otherwise the educational effect will diminish. They must also look for *their own* scientific problems to solve, so that their minds are trained not only in providing creative solutions but also in setting creative directions. Science and worldview must become mutually correlated[1355] in a transparent way that influences all inhabitants of our "global village".

Philosophical and educational activities can be maintained only in a system which is partially detached from market. It follows that fundamental science cannot be organized in a commercial way. Governmental support is imperative. Interaction with the non-industrial world outside science should be at best through the public opinion which influences national budgets. This is realistic only in a country in which the electorate views science at least as a kind of a cultural care program, as essential as health care programs. So scientific outcome must be presented to the public as clearly and attractively as possible. This is not an easy task,[1356] but we must popularize science in our own interest.

7.3.3. *Science for technology*

Many of the issues we discussed come close to industrial applications. But we must be aware that *direct* links between the basic knowledge (particularly this on the atomic level) and the problems bothering technologists are rare. The practical difficulties are solved in a pragmatic, phenomenological way. Engineers do not care for atoms and seldom ask scientists for suggestions. The industry builds its progress on its own experiences. It may use diagnostic tools developed in scientific labs but rarely adopts our solutions. The few prominent exceptions prove this rule general.

A scientific study invokes an industrial interest when an idea of a possibly profitable application emerges. Many companies are ready to sponsor university programs that are in this phase (though potential problems with the ownership of patents may be a hinderance[1350]). But this interest often ends as soon as a broad scientific interest begins. When the technology needed to manufacture the application is so established that chances for a profitable production are high enough, the industrial partners prefer – if only for security reasons – to continue the study alone. And they continue in a cost efficient, pragmatic rather than scientific way. The consequence is that long-term, in-depth studies are difficult to combine with a substantial support from the industry. The same is true for projects which are too far (more than a few years) ahead of the contemporary stand of technology.

Naturally, sometimes it is wise to sponsor a fundamentally oriented research even after a technical solution was found. We recall a few examples from the MOS technology (Sect. 1.2.2). Microelectronics reached a critical milestone when the reason for instabilities of gate oxides was found (it was the mobile charge due to metal contamination) and methods to deal with it were drafted (Sect. 6.3). Reliable MOS became then a matter of merely a technical development. But it was the formalization of basic interactions between oxygen, Si substrate, and SiO_2 that opened the way to practical simulations of oxidation (Sect. 6.4) so that new production steps could be designed with fewer trial-and-error experiments. Near future may bring an advent of nitridated gate oxides (Sect. 6.3.4). It was basic experiments with nitrogen-containing species (Sections 5.6 and 6.5.5d) that made this feasible. University search for CMOS-compatible SiO_2 substitutes in deep-submicron transistors brought a number of nitridation techniques, some of them very promising (Sect. 5.6). Investigations of the mechanism of hydrogen interaction with clean Si surfaces gave birth to the useful idea that the life time of transistors would increase significantly if deuterium rather than hydrogen were used in the passivation step (Sect. 6.5.5d).

Perhaps the most direct influence of the fundamental ideas on applications is seen in methodology. This is not a new development. When medieval scholars could not decide how many souls fit on a pin appex, one of these scholars formulated the economy principle, known today as the Occam's Razor: *theories should contain only necessary elements.* Recently, when surface scientists could not decide how atoms are arranged on the top of a crystal, a group of these scientists took a sharp pin and built a Scanning Tunneling Microscope (STM, page 30). Both tools, though conceived to deal with very basic problems, found their important practical applications. The economy principle is today a corner stone of the whole modern science, fundamental as well as applied. STM descendants, Atomic Force Microscopy and Scanning Capacitance Microscopy, are paving their way to become standard analytic instruments in technology labs.

To name one more example: Low Energy Electron Diffraction (LEED, page 34) helped to solve contamination problems (page 358). LEED patterns typical for clean surfaces are suppressed even by a fraction of a monolayer of a foreign substance deposited in an uncontrolled way, as in a "dirty" environment. LEED was the first monitor of chemical reactions on Ge and Si clean surfaces (Sect. 2.2.1). In fact, "the roots of all today's powerful experimental tools, techniques, procedures, and fundamental concepts at the disposal of surface science and engineering can be traced to that gigantic effort on the study of semiconductor surfaces" (Harry C. Gatos[3]).

In the long run it pays off to go deeper into details even in order to find out why does an established technique actually work. Such a "pursuit of truth" is sometimes viewed as wasteful, but in fact it builds a solid foundation for our practical knowledge. It assures that the castle of this knowledge will not fall when the conditions around it change. An exclusively pragmatic research needs a fast success measured by short-time profit and cannot afford this kind of approach. So there must be fundamental scientists broadening the paths cut by the engineers and checking for hidden passages and alternative ways. Otherwise, our technology may one day face a solid wall of impossibility and see a mess of ignorance in the rear.

How does the academic community respond to the growing demand for industrially-oriented research? Our impression is that those who avoid excessive stress or want intellectual independence tend to consider close links with industry as dangerous and troublesome. More scientists will be therefore interested in cooperation with the industry if:

- Their suspicions are proven false. This requires that
  - industrial pressure does not enter academic labs;
  - academia is not treated as a cheap labor force;
  - academic research has *no* imposed confidentiality.
- Their curiosity is awaken and sustained. This can be done by
  - exposing faculty to technological problems;
  - pouring industrial money into long-term projects;
  - strongly encouraging and supporting university patents.

We hope that co-operation between science and industry will evolve along these (or similar) lines.

7.3.4. *Science for all*

Let us now turn to the educational role of scientific studies. The purpose of modern education is training people to become members of a welfare, democratic society of self-knowing and self-realized individuals. Besides technical qualifications and the ability to deal creatively with new problems, a full education should therefore include skills needed to recognize and develop one's own capabilities and personality. Education does not imply a scientific background. But in order to be considered educated, a person lacking experience with scientific methods should at least understand them.

There are a few important points rarely raised in public discussions on the educational role of science. We want to highlight some of them. We will begin with remarks on the influence of specialized scientific publications on the professional education. Next, we will illustrate relations between "scientific experience" and problems common to everyone.

Clearly, education aims to bring up professionals. **Technical skills** and **creativity** are developed efficiently when students are given clear-cut examples and problems typical for contemporary science and technology. This puts certain duties on scientists, even those not involved in teaching.

To wit, few university professors are in the lucky position of being able to provide a full spectrum of didactic and timely examples and problems derived from their own research. The way from laboratories and production lines to lecture rooms is usually a long and winding one. It leads through patents, conferences, journal papers, review articles, and monographs. The latter are the easiest to comprehend, but also the most outdated. State-of-the-art knowledge would reach students and newcomers to the field much faster if specialized publications tried more often to make their key points understandable to people beyond small circles of hard-core experts. An introduction clearly stating the purpose of the work, explaining the main message and its implications, and placing them in the context of other studies enhances the educational value of the paper. It may also have a nice and measurable side-effect: an increase in the number of citations.

Science excites and motivates science students and scientists, this is clear. Yet we would postulate more than that. We would say that science should be valued also beyond the scientific community. And even more: that this should be true for surface science as well. But is there anything about the knowledge on atomic structures of a surface that an **average man** would find exciting? What is attractive in solid state physics, or, for that matter, in physics or science at all?

There are two major sources of human mental activities: curiosity and eagerness for comfort. The latter is the origin of technology. The former emanates science, philosophy, and religion. Science has a key role in the quartet of technology, science, philosophy, and religion because it is the only one of them that rightfully affects the other three.

However, scientific contributions to everyday's gadgets invoke only a limited interest. This may be because they are often taken for granted. Best selling to the public is this kind of science that influences philosophy and religion. "Frontier" research, as cosmology, particle physics, molecular biology, or space travel stimulates our phantasy by giving us eyes to see our selves in new dimensions. It tackles the very core of our curiosity, the metaphysical mystery of the existence of the Universe, the Life, the Self.

Though surface science does not fall into this "philosophically gifted" category, its theories and methods are sufficiently advanced to represent the abstract character and sophistication of the modern scientific approach. At the same time, the combination of scanning tunneling microscopy and *ab initio* calculations made the quantum world of surface atoms almost tangible. Surfaces may not only be one of the few places where atomistic science and industrial technologies meet (Sect. 1.1), but also the unique place where objects from the counter-intuitive quantum physics world can be *observed* in the space and time that we call real.

In this context, we argued that the experiences of scientists who studied silicon surfaces should have a sobering effect on those who want to view our scientific knowledge about the Universe as fully objective (Sections 2.1.1 and 2.3.4). Such an idealistic view assumes that science is based on objective facts. But this objectivity may be far from perfect because the amount of data is often so huge and the resources to process them so limited that only a subset of the data can be studied carefully. The size of this subset depends on the intuition of the researcher (Sect. 2.1.1). Since scientists working in the same area are biased by comparable education and have similar expectations, the eventual consensus and internal consistency of conclusions do not guarantee that the solution is correct (Figure 2.1, page 85). And although some people believe that continued intensive studies may lead to "fully objective understanding" of the pivotal issues,[a] this hypothetical development has, naturally, no effect on the status of our knowledge *today*.

In brief, it follows that in practice scientific conclusions cannot be

[a] Note that every theory T builds on assumptions. Since these assumptions cannot be proven within T, the term "fully objective understanding" is circular, i.e., it is a moving target. In the text we note that problems with it begin even before this circularity.

free from subjectivity. The only statement a scientist can make with a full confidence is a technical prescription that given specified conditions a specified effect will occur. All beyond that is an educated speculation.

Consider the consequences of this simple fact. There is a direct relation between religious **beliefs** and exposure to science: more science means more atheism. This "scientific atheism" has two major roots. First, some people hope that science can free their worldviews from subjective components, so that unprovable statements (epistemologically equivalent to beliefs) become unnecessary. As we noted, this hope is futile: arbitrary assumptions are unavoidable, no matter how advanced our science is.[a]

Second, it happens frequently that people simply lose their interest in religion. They explain this saying that the notion of God or gods was conceived and maintained to fill gaps in knowledge and to provide a feeling of security in the world full of injustice and incurable illness. As our knowledge and living conditions improve, religion becomes obsolete.

While this explantation of the origin of religion may be correct, the rest of the argumentation misses an essential point. The motivation which compells a sane person to accept or reject a worldview can be traced back to psychological predispositions or to the influence of the environment. Building on the former may lead to self-realization and internal peace. Unconditional surrender to the latter is likely to cause disappointment and stress.

Is the scientific approach appropriate to find out if one follows one's own way, and not the way imposed by the media, school, and friends?

Yes and no. No, because the distinction between "predisposition" and "influence" is fuzzy. And yes, if the purpose is to approximate one's *most suitable* way of life. This resembles the classical problem of describing the shape and finding the minimum of a complicated function. The elements considered during the construction of one's worldview correspond to the basis set used to approximate this function. The analysis of the worldviews incorporating these elements represents the minimization algorithm.

So followers of the scientific path to self-realization must keep their eyes and minds widely open, remaining ready to consider various opinions. This is needed because too narrow a view, just like too small a basis set in a numerical calculation (Sections 1.3.1b and 2.3.4), may be responsible for chosing the wrong direction. Careless or too hasty optimization can end up in a local extremum. In addition, new experiences may gradually change one's attitude to certain ideas.

The bottom line is that an "internal search" is a must for anyone, atheist and theist alike, who claims using scientific criteria in his or her life. Some occasional reconsideration of one's opinions seems unavoidable. Although science and religion are mutually unrelated, an impenetrable wall around one's atheism or religion is hardly a scientific construct.

The fundamental rule, without which a reasoning is not truly scientific, is: *try to be aware of your assumptions.* We have just seen that one should obey this rule when constructing and maintaining one's worldview. But assumptions play a key role not only in natural science and philosophy.

It is easy to see that the scientific attitude assists us in dealing with **daily** problems. Being aware of assumptions works against all nationalisms, rascisms, and religious or sexual intollerance that plague our planet. Not that it could erase them automatically. The world is still full of crimes and wars. Traumatic experiences can condition anyone: emotions can be stronger than reason. But the scientific way of thinking would help to get rid of preposessions. A consequent scientist – atheist and theist alike – should try to recognize and reject them anyway.

Precisely the same rule can help one to confront even petty controversies. After the first passions calm down, a sincere analysis of the situation is likely to show that the core of the conflict is in different assumptions made by the parties rather than in someone's ill will. Since assumptions always indicate these points where real knowledge is missing, identifying them is a good start to settle the dispute.

As next we consider an example how a simple observation of the way in which science works can bring us to **epistemological** conclusions. There is an intriguing correlation between the phenomenal world of physics and the logical world of mathematics (Sect. 1.3). The ability of abstract mathematical models to approximate real events progressively is remarkable. Can this correlation between phenomens and logic be accidental? Do mathematical concepts exist in the same sense as stones or planets do?

Note that mathematical ideas represent formalized patterns of efficient thinking. On the other hand, physical facts correspond to the data which feed our thoughts. The correlation between these two different activities of the brain (thinking and data acquisition) may indicate that the evolution protected each of its creatures with a cognition filter. This filter assures that the aspects of the data which are fundamentally unprocessable to the system do not pass to the conscious. The world of a sane animal would be therefore sane, understandable, orderly on any level of logical sophistication. Indeed, while the ability to predict boosts the chances of survival, all beyond that is a confusing noise.[b]

In other words, the observation that the world around us can be described mathematically and controlled technically apparently without limits may merely disclose an internal limitation of our cognition system.

Does such a viewpoint violate the economy principle by introducing an unspecified surrounding of the observable world? Not necessarily. What we did is that we interpreted two facts which must be interpreted anyway: that physics has rules and that these rules are mathematically consistent. We did it by invoking intrinsic cognition limits. This approach, akin to *inventionism*, is at least as efficient as other possibilities.[1357] For example, as the seemingly popular *platonic* interpretation which says that mathematical concepts and physical rules posess a specific kind of existence (being a form of "objective truth") and that they are mutually related.[1357]

Last but not least, the modern world imposes some **social duties** on science. First of all, the scientific, rational attitude must be taught to people of all professions. This education is needed to increase chances that our children will live in a human world in which people will not be persuaded to exchange their unique personalities for the standards suiting the needs of computer game producents, film and news makers, religious fanatics, or national leaders.

Second, it may turn out soon that the prominent side-effect imposed on mankind by the globalization of economy is an extremely high, permanent unemployment. Though enough material goods will be available for nearly all, the commercially oriented industry will offer few reasonable jobs. Art and research, practiced on any level of sophistication (from naive through amateur up to world class) will then be the only productive activities left to an average human. The options will be tourism, sport, internet surfing, computer games, pornography, drugs, crime, and riots.

How to summarize this Section from the point of view of *surface* scientists? We would say that although results of our studies are not automatically predestinated for first pages of popular magazines, our daily work provides us with enough first-hand material to support a serious analysis of ideas that concern everyone (Sect. 7.3.4).

---

[b] Its echo may sound in the subconscious, ocassionally causing unexplainable experiences.

Fig. 7.1. An atomic-size logo of the DIMES (Delft Institute for Microelectronics and Submicron Technology), drawn on Si(001) by Sven Rogge. From Ref. 1358. The background dark spots are hydrogenated (chemically passive) surface atoms, the bright foreground is made of unsaturated (chemically active) dangling bonds of Si. The ability to control the chemical activity of arbitrarily shaped arrays constructed from atoms by STM may help one to build and study extremely small microelectronic and optoelectronic devices before the ascent of technologies to assemble such structures in production lines.

7.4. How Can Silicon Surfaces Contribute?

Combining practical relevance with scientific excellence has always been a major objective of semiconductor surface studies. Semiconductor surface science was launched by technological needs.[3] Later, it managed to absorb this initial stimulus and began a wide survey on its own (Chapters 2-4), nevertheless cultivating the technological links (Chapters 5-6). Some of the results were crucial to the industrial development. Transistors were invented and silicon was introduced. Progress in SiO_2/Si(001) films and interfaces (Chapter 6) gave birth to MOS technologies (Sect. 1.2.2) and boosted microelectronics with its mobile phones and internet. MBE techniques made growth of III–V heterostructures possible and opened the way to optoelectronics with its CD players and LED traffic lights. Some of the results were interesting to everyone. Maybe the most spectacular of these popularizing outcomes is that we can venture into the nano-world to "see" and manipulate individual atoms with truly atomic precission (Fig. 7.1)!

Can this succesfull trend be continued? Below we will consider sample challenges which surface scientists may want to accept. They are related to possible commercial and educational demands.

An important **pragmatic** challenge is to assist in building a bridge between atomistic understanding of physical mechanisms and practical control and design of microchip technology. Surface studies can be helpful here not only because many technological steps depend on surface processes, but also because observations of surface structures and dynamics may help to gain insight into less accessible bulk processes (Sections 1.4.2a and 4.3.1a).

In practice, this means that such physical processes as diffusion and chemical reactions must be understood and approximately described in the language of easily solvable differential equations.[28] Thus, starting from an analysis of atomic structures and processes, one would formulate mesoscopic models which can be treated, for example, by Monte Carlo calculations, and complete the chain by providing a predictable and transferable model for a standard industrial process simulator. Such a task must begin with studies of the relevant materials (silicon matrix and individual atoms, oxygen, halogens, transition metals). Atomic-scale studies of silicon surfaces and interfaces are just a part of this program, but an important one.

A difficult but essential detail is that no consistent theoretical and experimental apparatus exists which would allow for a predictive description of atomic-scale processes and for a conclusive, direct verification of these predictions. Again, surface science is a perfect vehicle for a project which would attempt to construct such an apparatus. Given all the marvellous surface-sensitive techniques (Sect. 1.3.2) as STM, standing waves, photoemission, even holography, one can develop much more comprehensive and comprehendible methods to access surface atomic-sized structures than those existing for atomic-sized structures in the bulk. To succor this we have the constantly improving, *ab initio* numerical methods (Sect. 1.3.1). Their independence on arbitrary parametrization makes it easier to formulate unbiased theoretical models. Intellectually unbiased means more reliable (Sect. 2.1.1); hence less frequent failures of models and faster progress.

But *ab initio* calculations, though formally "parameter-free", must necessarily contain a number of (well defined) physical and numerical approximations (Sect. 1.3.1a), so their applicability to particular systems must be experimentally verified. Here again the surfaces of silicon provide the essential factors: accessibility of the atomic-scale details to experimental methods, and pragmatic relevance of the constituting materials.

Surface science has also a number of **educational** advantages. Lectured to non-physicists, it can show the "true face of science":

1. The lecture can be simple and illustrative on any level of sophistication;
2. One can show the interplay between knowledge and speculation.

Lectured to physicists, it can illustrate "atomic-scale physics in action":

1. One can venture into the atomic world of quantum mechanics;
2. One can discuss pros and cons of solid-state theoretical methods.

Finally, surface science labs may help students in their future work:

1. Interactions between surfaces and the ambient will become apparent;
2. Surface-sensitive measurements and UHV techniques will be mastered.

But is there any **philosophical** challenge for surface science? Surface scientists can hardly contribute to the knowledge about the way the Universe was born. Still, it is *not only* strings, baby universes, cloning techniques, or DNA gymnastics themselves that should influence our philosophy and religion. What should really matter is the astonishing fact that we, mere humans, are able to proceed towards a self-consistent description of the world in which we live. Solid state physics and particularly its frontier, the surface science, provide paramount examples of this ability. If not taken for granted but carefully analysed, the experience gained by a surface scientist in everyday work may help him or her on the way to become an intellectually independent, self-realized person.

7.5. Overview and Summary

When we started the work on this monography four years ago, in 1995, we had in mind a book that would be helpful to students as well as to experienced researchers. It should present the flavor and scope of advanced semiconductor surface science to the former, while to the latter it should provide an assistance in the writing of publications and project proposals. This is reflected in the choice and coverage of the material. It remains the task of the reader to decide whether this formula, which implies a mixture of general and detailed information, was successful or not.

The scientific Chapters which are now behind us started from the very basic ideas of semiconductor surface physics and led through atomistic details of the clean Si(001) surface and adsorption of foreign atoms on

Si(001) down to some practical issues in a key step during the manufacture of microelectronic chips: Si(001) oxidation. On the way, we browsed through the properties and usage of Si, we introduced the basics of Metal-Oxide-Semiconductor (MOS) technology (Sect. 1.2), we analysed the role of various theoretical and experimental methods of modern surface science (Sect. 1.3), we overviewed the pivotal concepts and problems related to studies and applications of Si surfaces (Sect. 1.4), and we described the major structural properties of their primary orientations (Sect. 1.5): Si(001), Si(111), and several Si(11n), in particular Si(113) and Si(110).

The discussion that followed the introductory Chapter 1 was focused chiefly to the technologically relevant Si(001). But our interest was not restricted to technological problems. On the contrary, we spent most of the time on such basic physical details as the atomic and electronic structures of the clean and perfect surface, surface defects, and various metastable objects (Chapter 3), and the time evolution of these structures in processes related to vibration and diffusion of surface atoms (Chapter 4).

The needs of microelectronics were mostly in the background, but often acting as a beacon. Thus the treatment of the chemical adsorption (Chapter 5) was more fundamental than applied. But because its scope was limited mostly to the elements interacting with Si surfaces during CMOS technological processes, we were able to highlight practical implications of these fundamental studies. Chapter 6 was devoted to fundamental as well as practical sides of a very applied issue of silicon oxidation. On one hand we considered, for example, atomic-scale structures and processes at the interface (Sect. 6.5). On the other hand, we described the role of oxide films in the modern silicon technology (Sect. 6.2.2), the methods to prepare such films (Sect. 6.3), and the practical problems related to SiO_2/Si films and interfaces (Sect. 6.6). When talking about oxidation models (Sect. 6.4), we paid attention to the actual and potential role of these models in practical process simulation tools. In all cases and in all Chapters we tried to combine experimental data with their interpretation in the light of various theoretical results.

In order to stay on a course balanced between the applied and the academic sides of science, we made also remarks on methodological, philosophical, and psychological aspects of scientific research. In particular, Chapter 2 used the history of astonishingly long studies leading to the discovery of the atomic structure of the clean Si(001) surface as an illustration of psychologically conditioned interaction between expectations, knowledge, conclusions, and accuracy of investigations. We also devoted most of the Afterword to the problems and challenges that may await science in the years to come.

We hope that this monograph will assist the reader in everyday work and provide at least some ideas for the future.

Warszawa, 24 maja 1999

References

Each entry is followed by a list of pages on which it is cited in the text.

[1] C. Duke, *Physics World* (August 1997) p. 17. *(448, 449)*
[2] D. J. Eaglesham, *Physics World* **11**, 41 (1995). *(3)*
[3] H.C. Gatos, *Surf. Sci.* **299**, 1 (1994). *(3, 345, 452, 458)*
[4] D. R. Lide (editor) *CRC Handbook of Chemistry and Physics*, 73rd edition (CRC Press, 1993). *(3, 352, 369)*
[5] S. D. Collins, *J. Electrochem. Soc.* **144**, 2242 (1997). *(4, 60)*
[6] R. Maboudian, *Surf. Sci. Rep.* **30**, 209 (1998). *(4)*
[7] M. Mehregany, S. D. Senturia, J. H. Lang, P. Nagarkar, *IEEE Transactions on Electron Devices* **39**, 2060 (1992). *(4)*
[8] N. Yazdi, F. Ayazi, K. Najafi, *Proceedings of the IEEE* **86**, 1640 (1998). *(4)*
[9] A. A. G. Tomlinson, *Materials Science Foundations* **3**, 1 (1998). *(4)*
[10] J. R. Chelikowsky, M. L. Cohen, *Phys. Rev. B* **10**, 5095 (1974). *(5, 6)*
[11] A. Garcia, C. Elsässer, J. Zhu, S. G. Louie, L. M. Cohen, *Phys. Rev. B* **46**, 9829 (1992). *(6)*
[12] M. L. Cohen, T. K. Bergstresser, *Phys. Rev.* **141**, 789 (1966). *(6)*
[13] O. Madelung (editor), , in *Landolt-Börnstein, New Series* Vol. 22a (Springer 1987) p. 14. *(6)*
[14] R. Del Sole, L. Reining, R. W. Godby, *Phys. Rev. B* **49**, 8024 (1994). *(6)*
[15] A. Fleszar, W. Hanke, *Phys. Rev. B* **56**, 10228 (1997). *(6)*
[16] M. M. Rieger, L. Steinbeck, I. D. White, H. N. Rojas, R. W. Godby, *Computer Physics Communications* **117**, 211 (1999). *(6)*
[17] D.H. Rich, T. Miller, G. E. Franklin, T.C. Chiang, *Phys. Rev. B* **39**, 1438 (1989). *(6)*
[18] T. P. Ma, *IEEE Transactions on Electron Devices* **45**, 680 (1998). *(7, 308, 309, 310, 319, 320, 321, 342)*
[19] H. H. Tseng, *IEDM Technical Digest*, 647 (1997). *(7, 320)*
[20] K. Stokbro, K. W. Jacobsen, J.K. Nørskov, D. M. Deaven, C.-Z. Wang, K. M. Ho, *Surf. Sci.* **360**, 221 (1996). *(7, 238, 251)*
[21] S. M. Sze, 2nd edition (John Wiley & Sons, Inc., 1981). *(7, 67, 353, 439, 440)*
[22] *The National Technology Roadmap for Semiconductors*, (Semiconductor Industry Association, San Jose, CA, 1997); International Technology Roadmap for Semiconductors, 1998 Update. *(7, 11, 14, 61, 65, 66, 156, 329, 341, 356, 377, 430, 432, 448)*
[23] S. Mori, Y. Yamaguchi, M. Sato, H. Meguro, H. Tsunoda, E. Kamiya, K. Yoshikawa, N. Ari, E. Sakagami, *IEEE Transactions on Electron Devices* **43**, 47 (1996). *(10, 367)*
[24] S. Verdonckt-Vandebroek, E. F. Crabbé, B.S. Meyerson, D.L. Harame, P. J. Restle, J.M.C. Stork, J. B. Johnson, *IEEE Transactions on Electron Devices* **41**, 90 (1994). *(10)*
[25] K. Liu, S. K. Ray, S. K. Oswal, S. Banerjee, *IEEE Electron Device Letters* **19**, 13 (1998). *(10)*
[26] C. A. Warwick, R.-H. Yanm, Y. O. Kim, A. Ourmazd, *AT&T Technical Journal* **93**, 50 (September/October 1993). *(11)*
[27] C. Fiegna, H. Iwai, T. Wada, T. Saito, E. Sangiorgi, B. Riccò, *IEEE Transac-*

tions on Electron Devices **41**, 941 (1994). *(11)*

[28] J. Dąbrowski, H.-J. Müssig, M. Duane, S. T. Dunham, R. Goossens, H.-H. Vuong, , *Advances in Solid State Physics* **38**, 565 (1999); posted at `http://www.ihp-ffo.de/chipps.html`. *(11, 12, 15, 18, 33, 54, 65, 123, 129, 241, 309, 326, 346, 347, 359, 363, 422, 425, 430, 459)*

[29] A.-H. Lo, D. A. Buchanan, Y. Taur, W. Wang, *IEEE Electron Device Letters* **18**, 209 (1997). *(11)*

[30] R. Wilson, *EE Times - Headline News*, Internet page posted at `http://techweb.cmp.com/eet/news/97/982news/bell.html` (1997). *(11, 365, 432)*

[31] D. J. Foster, in *Electronic Materials. From Silicon to Organics*, edited by L. S. Miller and J. B. Mullin (Plenum Press, New York 1991) p. 173. *(12, 235, 236)*

[32] E. Chason, S.T. Picraux, J.M. Poate, J. O. Borland, M.I. Current, T. Diaz de la R ubia, D. J. Eaglesham, O. W. Holland, M. E. Law, C. W. Magee, J. W. Mayer, J. Melngailis, A. F. Tasch, *J. Appl. Phys.* **R81**, 6513 (1997). *(12, 60)*

[33] M. A. Foad, D. Jennings, *Solid State Technology* **41**, 43 (December 1998). *(12, 13, 347, 365)*

[34] P. K. Chu, S. B. Felch, P. Kellerman, F. Sinclair, L. Larson, B. Mizuno, *Solid State Technology* **42**, 55 (September 1999). *(12, 347)*

[35] IBM pages, `http://www.ibm.com/News/1997/09/ls9709222.html` (1997). *(13)*

[36] P. C. Andricacos, *Interface* **8**, 32 (Spring 1999). *(13)*

[37] C. T. Gabriel, J. P. McVittie, *Solid State Technology* **34**, 81 (June 1992). *(14, 354, 429, 434)*

[38] C. T. Gabriel, *Solid State Technology* **40**, 49 (March 1997). *(14, 354, 366, 429, 434)*

[39] P. Burggraaf, *Solid State Technology* **42**, 22 (March 1999). *(14)*

[40] J. A. McClay, A. S. L. McIntyre, *Solid State Technology* **42**, 57 (June 1999). *(14)*

[41] L. R. Harriott, *Materials Science in Semiconductor Processing* **1**, 93 (1998). *(14)*

[42] P. Hohenberg, W. Kohn, *Physical Review* **136B**, 864 (1964). *(16)*

[43] W. Kohn, L. J. Sham, *Phys. Rev.* **140**, A 1133 (1965). *(16, 20, 21, 111)*

[44] L. Hedin, *Phys. Rev.* **139**, A796 (1965). *(16, 22)*

[45] P. N. Keating, *Phys. Rev.* **145**, 637 (1966). *(16)*

[46] W. A. Harrison, S. Ciraci, *Phys. Rev. B* **10**, 1516 (1974). *(16)*

[47] D.J. Chadi, *Phys. Rev.* **B19**, 2074 (1979). *(16, 28)*

[48] F. Stillinger, T. Weber, *Phys. Rev. B* **31**, 5262 (1985). *(16)*

[49] J. Tersoff, *Phys. Rev. Lett.* **56**, 632 (1986). *(16)*

[50] B. W. Dodson, *Phys. Rev. B* **33**, 7361 (1986). *(16)*

[51] F. Stillinger, T. Weber, *Phys. Rev. B* **36**, 1208 (1987). *(16)*

[52] B. W. Dodson, *Phys. Rev. B* **35**, 2795 (1987). *(16)*

[53] K. E. Khor, S. Das Sarma, *Phys. Rev. B* **36**, 7733 (1987). *(16)*

[54] M. S. Hybertsen, S. G. Louie, *Phys. Rev. Lett.* **58**, 1551 (1987). *(16, 111)*

[55] P. Krüger, J. Pollmann, *Phys. Rev. B* **38**, 10578 (1988). *(16)*

[56] O. L. Alerhand, D. Vanderbilt, R. D. Meade, J.D. Joannopoulos, *Phys. Rev. Lett.* **61**, 1973 (1988). *(16, 113, 114, 139, 141, 249, 253, 255)*

[57] F. Sankey, D. J. Niklewski, *Phys. Rev. B* **40**, 3979 (1989). *(16, 28)*

[58] R. O. Jones, O. Gunnarsson, *Rev. Mod. Phys.* **61**, 689 (1989). *(16, 20, 111)*

[59] J. Olsen, P. Joergenson, J. Simons, *Chem. Phys. Lett.* **169**, 463 (1990). *(16)*

[60] J.H. Wilson, J.D. Todd, A. P. Sutton, *J. Phys.: Cond. Mat.* **2**, 10259 (1990). *(16)*

[61] Y.-T. Lu, H Metiu, *Surf. Sci.* **245**, 150 (1991). *(16, 18)*
[62] K.A. Fichthorn, W.H. Weinberg, *J. Chem. Phys.* **95**, 1090 (1991). *(16, 18)*
[63] J. Tersoff, *Phys. Rev. B* **43**, 9377 (1991). *(16, 340)*
[64] J. P. Perdew, *Physica B* **172**, 1 (1991). *(16, 22)*
[65] G. Ortiz, *Phys. Rev.* **B45**, 11328 (1992). *(16)*
[66] J. P. LaFemina, , *Surf. Sci. Reports* **16**,133 (1992). *(16, 28)*
[67] J. P. Perdew, J. A. Chevary, S. H. Vosko, K. A. Jackson, M. A. Pederson, D. J. Singh, C. Fiolhais, *Phys. Rev. B* **46**, 6671 (1992). *(16, 20, 22)*
[68] R. W. Godby, in *Topics in Applied Physics*, Vol. 69, edited by J. C. Fuggle and J. E. Inglesfield, (Springer, Berlin, 1992), p. 51. *(16)*
[69] C. J. Umrigar, X. Gonze, , in *High Performance Computing and its Application to the Physical Sciences*, proceedings of the Mardi Gras '93 Conference (World Scientific, Singapore, 1993). pp. 43-59 *(16, 22)*
[70] C. Filippi, C. J. Umrigar, M. Taut, *J. Chem. Phys.* **100**, 1290 (1994). *(16)*
[71] J. E. Inglesfield, in *Cohesion and Structure of Surfaces*, Vol. 4 edited by K. Binder, M. Bowker, J. E. Inglesfield, and P. J. Rous, (Elsevier Science, 1995), p. 63. *(16)*
[72] M. Rohlfing, P. Krüger, J. Pollmann, *Phys. Rev. B* **52**, 1905 (1995). *(16, 23, 121, 143, 144, 147)*
[73] H. Rücker and M. Methfessel, *Phys. Rev. B* **52**, 11059 (1995). *(16)*
[74] P.C.L. Stephenson, M.W. Radny, P.V. Smith, *Surf. Sci.* **366**, 177 (1996). *(16)*
[75] L.E. Carter, E. A. Carter, *Surf. Sci.* **360**, 200 (1996). *(16, 278)*
[76] B. J. Spencer, J. Tersoff, *Phys. Rev. Lett.* **79**, 4858 (1997). *(16, 19)*
[77] R. Q. Hood, M. Y. Chou, A. J. Williamson, G. Rajagopal, R. J. Needs, W. M. C. Foulkes, *Phys. Rev. Lett.* **78**, 3350 (1997). *(16, 21)*
[78] G. E. Engel, *Phys. Rev. Lett.* **78**, 3515 (1997). *(16, 22)*
[79] M. R. Radeke, E. A. Carter, *Annu. Rev. Phys. Chem.* **48**, 243 (1997). *(16, 20, 265, 270)*
[80] A. Groß, *Surf. Sci. Rep.* **32**, 291 (1998). *(16, 20, 265, 270, 272, 274)*
[81] M. Fuchs, M. Bockstedte, E. Pehlke, M. Scheffler, *Phys. Rev. B* **57**, 2134 (1998). *(16, 22, 264)*
[82] N. Gidopoulos, *Phys. Rev. B* **57**, 2146 (1998). *(16, 19)*
[83] F. Aryasetiawan, O. Gunnarson, *Rep. Prog. Phys.* **61**, 237 (1998). *(16)*
[84] M. Jaraiz, L. Pelaz, J. Barbolla, G.H. Gilmer, C. S. Rafferty, *E-MRS 1999 Spring Meeting, Book of Abstracts*, L-3 (1999); to be published in *Materials Science in Semiconductor Processing* (1999). *(18)*
[85] P. M. Fahey, P.B. Griffin, J.D. Plummer, *Rev. Mod. Phys.* **61**, 289 (1989). *(19, 308, 353, 415)*
[86] D. M. Ceperley, *Advances in Chemical Physics* **93**, 1 (1996). *(20)*
[87] G. Rajagopal, R. J. Needs, S. D. Kenny, W. M. C. Foulkes, A. J. James, *Phys. Rev. Lett.* **73**, 1959 (1994). *(20)*
[88] D. M. Ceperley, B. J. Alder, *Phys. Rev. Lett.* **45**, 566 (1980). *(20)*
[89] J. P. Perdew, A. Zunger, *Phys. Rev. B* **23**, 5048 (1981). *(20)*
[90] R. M. Dreizler, E. K. U. Gross, *Density Functional Theory* (Springer, Berlin 1990) *(20)*
[91] M.C. Payne, M. P. Teter, D. C. Allan, T.A. Arias, J.D. Joannopoulos, *Rev. Mod. Phys.* **64**, 1045 (1992). *(20)*
[92] E. K. U. Gross, R. M. Dreizler (editors), *Density Functional Theory* (Plenum, New York 1995); *(21)*
[93] A. Khein, N. W. Ashcroft, *Phys. Rev. Lett.* **78**, 3346 (1997). *(21)*
[94] N. D. Mermin, *Physical Review* **137**, A1441 (1965). *(21, 55)*

[95] J. P. Perdew, K. Burke, M. Enzerhoff, *Phys. Rev. Lett.* **77**, 3865 (1996). *(22)*
[96] P. Nachtigall, K. D. Jordan, A. Smith, H. Jónsson, *J. Chem. Phys.* **104**, 148 (1996). *(22, 265)*
[97] D. M. Bylander, L. Kleinman, "Requested!" *Phys. Rev. B* **41**, 7868 (1990). *(22)*
[98] A. Seidl, A. Görling, P. Vogl, J. A. Majewski, M. Levy, *Phys. Rev. B* **53**, 3764 (1996). *(22)*
[99] P. Bendt, A. Zunger, *Phys. Rev. Lett.* **50**, 1684 (1983). *(23)*
[100] R. Car, M. Parrinello, *Phys. Rev. Lett.* **55**, 2471 (1985). *(23, 114)*
[101] M. Bockstedte, A. Kley, J. Neugebauer, M. Scheffler, *Computer Physics Communications* **107**, 187 (1997). *(23)*
[102] J. Neugebauer, C. G. Van de Walle, *Phys. Rev. B* **50**, 8067 (1994). *(25)*
[103] B.D. Yu, Y. Miyamoto, O. Sugino, T. Sasaki, T. Ohno, *Phys. Rev. B* **58**, 3549 (1998). *(25, 322, 339, 340)*
[104] S. G. Louie, S. Froyen, M. L. Cohen, *Phys. Rev. B* **B26**, 1738 (1982). *(25)*
[105] A. Baldereschi, *Phys. Rev. B* **7**, 5212 (1973). *(26)*
[106] H. J. Monkhorst, J. D. Pack, *Phys. Rev. B* **13**, 5188 (1976). *(26)*
[107] W. R. Fehlner, S. H. Vosko, *Canadian Journal of Physics* **55**, 2041 (1978). *(26)*
[108] M. J. Puska, S. Pöykkö, M. Pesola, R. M. Nieminen, *Phys. Rev. B* **58**, 1318 (1998). *(26)*
[109] X. P. Li, W. Nunes, D. Vanderbilt, *Phys. Rev. B* **47**, 10891 (1993). *(27, 28)*
[110] W. Kohn, *Phys. Rev. Lett.* **76**, 3168 (1996). *(27)*
[111] E. L. Briggs, D. Sullivan, J. Bernholc, *Phys. Rev. B* **54**, 14362 (1996). *(27)*
[112] C.M. Goringe, E. Hernández, M.J. Gillan, I. J. Bush, *Computer Physics Communications* **102**, 1 (1997). *(27)*
[113] E. Hernández, M.J. Gillan, C.M. Goringe, *Phys. Rev. B* **53**, 7147 (1996). *(27)*
[114] P. Ordejón, E. Artacho, J. M. Soler, *Phys. Rev. B* **53**, 10441 (1996). *(27)*
[115] A.P. Smith, J.K. Wiggs, H. Jónsson, H. Yan, L.R. Corrales, P. Nachtigall, K. D. Jordan, *J. Chem. Phys.* **102**, 1044 (1995). *(27, 28)*
[116] C. Yang, S. Y. Lee, H. C. Kang, *J. Chem. Phys.* **107**, 3295 (1997). *(27, 126)*
[117] E. Penev, P. Kratzer, M. Scheffler, *J. Chem. Phys.* **110**, 3986 (1999). *(27)*
[118] J. C. Wu, I. V. Ionova, E. A. Carter, *Phys. Rev. B* **49**, 13488 (1994). *(27, 266)*
[119] A. Groß, M. Scheffler, M. J. Mehl, D. A. Papaconstantopulos, *Phys. Rev. Lett.* **82**, 1209 (1999). *(28)*
[120] E. Artacho, F. Yndurian, *Phys. Rev. B* **44**, 6169 (1991). *(28)*
[121] C.M. Goringe, D.R. Bowler, E. Hernández, *Reports on Progress in Physics* **60**, 1447 (1997). *(28)*
[122] W. A. Harrison, *Surf. Sci.* **55**, 1 (1976). *(28, 97, 100)*
[123] D.R. Bowler, M. G. Bowler, *Phys. Rev. B* **57**, 15385 (1998). *(28, 178, 184)*
[124] *Analysis of Microelectronic Materials and Devices*, edited by M. Grasserbauer and H. W. Werner (Wiley, New York, 1991). *(28)*
[125] J.M. Walls, R. Smith, *Vacuum* **45**, 649 (1994). See also other article in this Special Issue (numbers 6 and 7) devoted to surface science techniques. *(28)*
[126] M. Sarikaya, *Ultramicroscopy* **47**, 1 (1992). See also other papers in this Special Issue devoted to resolution in microscopies. *(28)*
[127] H. I. Smith, *J. Vac. Sci. Technol. B* **13**, 2323 (1995). *(28)*
[128] K. Nakayama, J. H. Weaver, *Phys. Rev. Lett.* **82**, 980 (1999). *(28, 194, 195, 199)*
[129] Ch. Linsmeier, *Vacuum* **45**, 673 (1994). *(29)*
[130] G.V. Hansson, R.I.G. Uhrberg, *Surf. Sci. Rep.* **9**, 197 (1988). *(29)*
[131] H.-P. Steinrück, *Vacuum* **45**, 715 (1994). *(29, 143)*

[132] H. Ibach, D. L. Mills, *Electron Energy Loss Spectroscopy and Surface Vibrations* (Academic, New York, 1982). *(29, 43)*

[133] J. C. Fuggle, J. E. Inglesfield, in *Topics in Applied Physics*, Vol. 69 ("Unoccupied Electronic States") edited by J. C. Fuggle, J. E. Inglesfield, (Springer, Berlin, 1992), p. 1. *(29)*

[134] J. E. Griffith, G.P. Kochanski, *Critical Reviews in Solid State and Materials Sciences* **16**, 255 (1990). *(29, 165, 166, 178, 185, 188, 249, 253)*

[135] E. Bauer, *Rep. Prog. Phys.* **57**, 895 (1994). *(29)*

[136] M.P. Murrell, *Vacuum* **45**, 773 (1994). *(29)*

[137] L.G. Earwaker, *Vacuum* **45**, 783 (1994). *(29, 42)*

[138] J.M. Bermond, J. J. Metois, J.C. Heyraud, C. Alfonso, *Surf. Sci.* **331**, 855 (1995). *(29, 257, 261)*

[139] F.J. Himpsel, *Surf. Sci.* **299**, 525 (1994). *(29, 37)*

[140] M.D. Crapper, *Vacuum* **45**, 691 (1994). *(29)*

[141] U. Höfer, *Applied Physics A* **63**, 533 (1996). *(29, 264, 265, 270)*

[142] P. C. Zalm, *Vacuum* **45**, 753 (1994). *(29, 41)*

[143] J.A. Kubby, J. J. Boland, *Surf. Sci. Rep.* **26**, 61 (1996). *(29)*

[144] H. Neddermeyer, *Reports on Progress in Physics* **59**, 701 (1996). *(29, 79, 265, 297)*

[145] J.F. Watts, *Vacuum* **45**, 653 (1994). *(29)*

[146] G. Binnig, H. Rohrer, Ch. Gerber, E. Weibel, *Phys. Rev. Lett.* **49**, 57 (1982). *(30, 107)*

[147] R.J. Hamers, in *Springer Series in Surface Sciences*, Vol. 20 ("The Structure of Surfaces") edited by M. A. Van Hove, S. Y. Tong, (Springer, Berlin, 1992), p. 83. *(30)*

[148] J. Tersoff, D. R. Hamann, *Phys. Rev. B* **31**, 805 (1985). *(31, 32, 78)*

[149] B.S. Swartzentruber, *Phys. Rev. Lett.* **76**, 459 (1996). *(31, 205, 213, 234)*

[150] M. Tsukada, S. Watanabe, *Jap. J. Appl. Phys.* **32**, 1352 (1993). *(31)*

[151] R. M. Feenstra, J.A. Stroscio, A. P. Fein, *Surf. Sci.* **181**, 295 (1987). *(31)*

[152] S. Heinze, S. Blügel, R. Pascal, M. Bode, R. Wiesendanger, *Phys. Rev. B* **58**, 16432 (1998). *(32)*

[153] G.A.D. Briggs, A. J. Fisher, *Surf. Sci. Rep.* **33**, 1 (1999). *(32, 159, 265, 303)*

[154] J. Bardeen, *Phys. Rev. Lett.* **6**, 57 (1961). *(32)*

[155] M. Tsukada, K. Kobayashi, S. Ohnishi, "First-principles theory of the STM simulation" *J. Vac. Sci. Technol. A* **8**, 160 (1990). *(32)*

[156] J. M. Bass, C. C. Matthai, *Phys. Rev. B* **50**, 11212 (1994). *(32)*

[157] J. Dąbrowski, H.-J. Müssig, G. Wolff, *Surf. Sci.* **331**, 1022 (1995). *(32, 77, 159)*

[158] G. Binnig, C.F. Quate, C. Gerber, *Phys. Rev. Lett.* **56**, 930 (1986). *(32)*

[159] M. Komiyama, K. Tsujimichi, K. Tazawa, A. Hirotani, H. Yamano, M. Kubo, E. Broclawik, A. Miyamoto, *Surf. Sci.* **357**, 222 (1996). *(32)*

[160] M. Komiyama, K. Tazawa, K. Tsujimichi, A. Hirotani, M. Kubo, A. Miyamoto, *Thin Solid Films* **281**, 580 (1996). *(32)*

[161] A. S. Foster, L. N. Kantorovich, A. L. Shluger, A. Livshits, *E-MRS Spring Meeting, Book of Abstracts*, L-4 (1999); to be published in *Materials Science in Semiconductor Processing* (1999). *(32)*

[162] C.C. Williams, W. P. Hough, S. A. Rishton, *Appl. Phys. Lett.* **55**, 203 (1989). *(33)*

[163] G. M. Yu, P.B. Griffin, J. D. Plummer, *IEDM Technical Digest*, 717 (1998). *(33)*

[164] C. J. Kang, C. K. Kim, J. D. Lera, Y. Kuk, K. M. Mang, J. G. Lee, K. S. Suh, C.C. Williams, *Appl. Phys. Lett.* **71**, 1546 (1997). *(33)*

[165] H. Edwards, R. McGlothlin, R. San Martin, E. U, M. Gribelyuk, R. Mahaffy, C. K. Shih, R. S. List, V. A. Ukraintsev, *Appl. Phys. Lett.* **72**, 698 (1998). *(33)*
[166] C. J. Davisson, L. H. Germer, *Phys. Rev.* **30**, 705 (1927). *(34)*
[167] J. B. Pendry, *Low Energy Electron Diffraction* (Academic Press, New York, 1974). *(34)*
[168] M. A. Van Hove, W.H. Weinberg, C. M. Chan, *Low Energy Electron Diffraction* (Springer, Berlin, 1986). *(34)*
[169] J. J. Lander, *Phys. Rev.* **91**, 1382 (1953). *(35)*
[170] D. Briggs, M. P. Seah (editors), *Practical Surface Analysis, Vol. I: Auger and X-ray Photoelectron Spectroscopy* (Wiley, Chichester 1990). *(35, 37)*
[171] *Auger Spectroscopy and Electronic Structure, Springer Series in Surface Sciences*, Vol. 18, edited by G. Cubiotti, G. Mondio, K. Wandelt (Springer, Berlin, 1989). *(35)*
[172] E. H. S. Burhop, *The Auger Effect and Other Radiationless Transitions* (Cambridge University Press, 1952). *(35)*
[173] L. E. Davis, N. C. MacDonald, P. W. Palmberg, G. E. Riach, R. E. Weber, *Handbook of Auger Electron Spectroscopy* (Physical Electronics Industries, Inc., 1976). *(35)*
[174] L. A. Harris, *J. Appl. Phys.* **39**, 1419 (1968). *(36)*
[175] L. A. Harris, *J. Appl. Phys.* **39**, 1428 (1968). *(36)*
[176] H. Robinson, W. J. Rawlinson, *Philosophical Magazine* **28**, 277 (1914). *(37)*
[177] H. Robinson, *Philosophical Magazine* **50**, 241 (1925). *(37)*
[178] K. Siegbah, C. Nordling, A. Fahlman, K. Nordberg, K. Hamrin, J. Hedman, G. Johansson, T. Bergmark, S. E. Karlsson, I. Lindgren, B. Lindberg, *ESCA: Atomic, Molecular, and Solid State Structure Studied by Means of Electron Spectroscopy* (Almquist and Wilsells, Uppsala, 1967). *(37)*
[179] F.J. Himpsel, B.S. Meyerson, F.R. McFeely, J. F. Morar, A. Taleb-Ibrahimi, J. A. Yarmoff, *Proceedings of the Enrico Fermi School on Photoemission and Absorption Spectroscopy of Solids and Interfaces with Synchrotron Radiation*, ed. by M. Campagna and R. Rosei (North Holland, Amsterdam 1990) p. 203. *(37, 83, 118)*
[180] E. Pehlke, M. Scheffler, *Phys. Rev. Lett.* **71**, 2338 (1993). *(37, 120, 134, 143, 153, 154)*
[181] H.J. Osten, M. Methfessel, G. Lippert, H. Rücker, *Phys. Rev. B* **52**, 12179 (1995). *(37)*
[182] M. Methfessel, D. Hennig, M. Scheffler, *Surface Review and Letters* **2**, 197 (1995). *(37)*
[183] M. P. Seah, in *Practical Surface Analysis, Vol. I: Auger and X-ray Photoelectron Spectroscopy*, ed. by D. Briggs and M. P. Seah (Wiley, Chichester 1990) p. 201. *(38)*
[184] W. E. Spicer, *Phys. Rev.* **112**, 114 (1958). *(39)*
[185] C. N. Berglund, W. E. Spicer, *Physical Review* **136**, 1044A (1964). *(39)*
[186] D. E. Eastman, J. K. Cashion, *Phys. Rev. Lett.* **27**, 1520 (1971). *(39)*
[187] B. E. Warren, *X-Ray Diffraction* (Dover, New York, 1990). *(40)*
[188] B. W. Batterman, *Phys. Rev.* **133**, A 759 (1964). *(40)*
[189] Y. Qian, P.F. Lyman, M. J. Bedzyk, *Scanning Microscopy* **9**, 969 (1995). *(40)*
[190] D. P. Woodruff, *Surface and Interface Analysis* **11**, 25 (1988). *(41)*
[191] P. R. Watson, *J. Chem. Ref. Data* **21**, 123 (1992). *(41)*
[192] A. Benninghoven, F. G. Rüdenauer, H.W. Werner, *Secondary Ion Mass Spectrometry* (Wiley, New York, 1987). *(41)*
[193] D. Briggs, A. Brown, J. C. Vickerman, *Handbook of Static Secondary Ion Mass*

Spectrometry (Wiley, New York, 1989). *(42)*
[194] S. V. Teplov, V. V. Zastavnjuk, V. Bykov, J. W. Rablais, *Surf. Sci.* **310**, 436 (1994). *(42)*
[195] I. Kamiya, M. Katayama, E. Noura, M. Aono, *Surf. Sci.* **242**, 404 (1991). *(42)*
[196] L.C. Feldman, J. W. Mayer, *Fundamentals of Surface and Thin Film Analysis* (North Holland, Amsterdam, 1986). *(43)*
[197] E. Bauer, *Surf. Sci.* **299**, 102 (1994). *(44)*
[198] J. M. Cowley, *Ultramicroscopy* **47**, 187 (1992). *(45)*
[199] K. Tsuno, *Ultramicroscopy* **50**, 245 (1993). *(45)*
[200] W. D. Rau, F.H. Baumann, H.-H. Vuong, B. Heinemann, W. Höppner, C. S. Rafferty, P. Schwander, A. Ourmazd, *IEDM Technical Digest*, 713 (1998). *(45)*
[201] R. L. Park, H. H. Madden, *Surf. Sci.* **11**, 188 (1968). *(47)*
[202] E. A. Wood, *J. Appl. Phys.* **35**, 1306 (1964). *(47)*
[203] C. Kittel, *Introduction to Solid State Physics* (Wiley, New York, 1976). *(49, 57)*
[204] X.-J. Kong, C. T. Chan, K. M. Ho, Y. Y. Ye, *Phys. Rev. B* **42**, 9357 (1990). *(49)*
[205] G. Wulff, *Z. Kristallogr. Mineral.* **34**, 449 (1901). *(50)*
[206] C. Jayaprakash, W. F. Saam, *Phys. Rev. B* **30**, 3916 (1984). *(50)*
[207] M. A. Schlüter, in *The Chemical Physics of Solid Surfaces and Heterogeneous Catalysis*, ed. by D. A. King and D. P. Woodruff (Elsevier 1988) p. 37. *(50)*
[208] D. J. Eaglesham, A.E. White, L.C. Feldman, N. Moriya, D.C. Jacobson, *Phys. Rev. Lett.* **70**, 1643 (1993). *(52, 74)*
[209] J. M. Bermond, J. J. Metois, X. Egea, F. Floret, *Surf. Sci.* **330**, 48 (1995). *(52)*
[210] S. V. Khare, N.C. Bartelt, T.L. Einstein, *Phys. Rev. Lett.* **75**, 2148 (1995). *(52)*
[211] T. Doi, M. Ichikawa, S. Hosoki, K. Ninomiya, *Surf. Sci.* **351**, 135 (1996). *(52)*
[212] S. Song, S. G. J. Mochrie, *Phys. Rev. B* **51**, 10068 (1995). *(53, 73)*
[213] A.A. Baski, L.J. Whitman, *Phys. Rev. Lett.* **74**, 956 (1995). *(53, 74)*
[214] S. Takeda, T. Kamino, *Phys. Rev. B* **51**, 2148 (1995). *(54, 76)*
[215] J. Dąbrowski, H.-J. Müssig, G. Wolff, S. Hinrich, *Surf. Sci.* **411**, 54 (1998). *(54, 76, 77, 262, 422)*
[216] G. R. Parr, R. A. Donnelly, R. A. Levy, W. E. Palke, *J. Chem. Phys.* **68**, 381 (1978). *(55)*
[217] R. G. Pearson, *Science* **151**, 172 (1966). *(55)*
[218] W. Yang, G. R. Parr, *Proc. Natl. Acad. Sci. USA* **82**, 6723 (1985). *(55)*
[219] A. Filippetti, D. Vanderbilt, W. Zhong, Y. Cai, G. B. Bachelet, *Phys. Rev. B* **52**, 11793 (1995). *(55)*
[220] K. D. Brommer, M. Galván, A. Dal Pino Jr., J.D. Joannopoulos, *Surf. Sci.* **314**, 57 (1994). *(55)*
[221] P. Polizer, *J. Chem. Phys.* **86**, 1072 (1987). *(56)*
[222] J. B. Rem, J. Holleman, J. F. Verweij, *J. Electrochem. Soc.* **144**, 2101 (1997). *(56)*
[223] J.E. Northrup, *Phys. Rev. Lett.* **62**, 2487 (1989). *(56)*
[224] *Landolt-Börnstein, I. Band: Atom- und Molekularphysik*, Teil 2 "Molekeln I", sechste Auflage (Springer 1951). *(57)*
[225] L. Pauling, *The Nature of Chemical Bond*, 3rd edition (Cornell University, Ithaca Press, NY, 1960) p. 85. *(57, 289)*
[226] W. K. Burton, N. Cabrera, F. C. Frank, *Phil. Trans. Roy. Soc. A* **243**, 299 (1951). *(58, 236, 257)*
[227] R. L. Schwoebel, E. J. Shipsey, *J. Appl. Phys.* **37**, 3682 (1966). *(58, 224)*
[228] Y.W. Mo, M. G. Lagally, *Surf. Sci.* **248**, 313 (1991). *(58, 224, 225, 257, 260)*

[229] B.S. Swartzentruber, *Surf. Sci.* **374**, 277 (1997). *(58, 223, 229, 234)*
[230] S. Tanaka, N.C. Bartelt, C.C. Umbach, R. M. Tromp, J.M. Blakely, *Phys. Rev. Lett.* **78**, 3342 (1997). *(58, 223, 237, 257)*
[231] M.A. Lutz, R. M. Feenstra, P.M. Mooney, J. Tersoff, J.O. Chu, *Surf. Sci.* **316**, L1075 (1994). *(59)*
[232] R. A. A. Kubiak, E.H.C. Parker, in *The Chemical Physics of Solid Surfaces and Heterogeneous Catalysis*, Vol. 5 (*Surface properties of electronic materials*) ed. by D. A. King, D. P. Woodruff, (Elsevier Science, 1988), p. 309 *(59)*
[233] D. J. Eaglesham, *J. Appl. Phys.* **77**, 3597 (1995). *(59, 256, 257)*
[234] J. Musil, *Vacuum* **47**, 145 (1996). *(60)*
[235] L. F. Tz. Kwakman, M. Omini, D. Levy, D. Malgouyres, *Solid State Phenomena* **65**, 31 (1999). *(61, 62, 279, 361)*
[236] G. J. Norga, M. Platero, K. A. Black, A. J. Reddy, J. Michel, L. C. Kimerling, *J. Electrochem. Soc.* **144**, 2801 (1997). *(61, 62)*
[237] W. Fyen, L. Mouche, M. Meuris, M. M. Heyns, J. Zahka, *J. Electrochem. Soc.* **144**, 2189 (1997). *(61, 62, 195)*
[238] M. Heyns, M. Meuris, P. Mertens (editors), *Proceedings of the Fourth International Symposium on Ultra Clean Processing of Silicon Surfaces*, Ostend, Belgium, Sept. 1998, *Solid State Phenomena* **65** (1999). *(61)*
[239] K. Saga, T. Hattori, *Appl. Phys. Lett.* **71**, 3670 (1997). *(61, 63, 64, 358)*
[240] W. B. Henley, L. Jastrzebski, N. F. Haddad, *Material Research Society Symposia Proceedings* **262**, 993 (1992). *(61)*
[241] E.R. Weber, *Applied Physics A* **30**, 1 (1983). *(61)*
[242] C.J. Sofield, A. M. Stoneham, *Semiconductor Science and Technology* **10**, 215 (1995). *(61, 66, 345, 381, 408, 429)*
[243] J. Ruzyllo, E. Röhr, M. Baeyens, T. Bearda, P. W. Mertens, M. M. Heyns, *Solid State Phenomena* **65**, 85 (1999). *(61)*
[244] M. Meuris, P. W. Mertens, A. Opdebeeck, H. F. Schmidt, M. Depas, G. Vereecke, M. M. Heyns, A. Philipossian, *Solid State Technology* **38**, 109 (July 1995). *(62, 361)*
[245] H. F. Winters, J. W. Coburn, *Surf. Sci. Rep.* **14**, 161 (1992). *(62, 277, 278, 279)*
[246] Y. B. Kim, M.R. Baklanov, T. Conrad, S. Vanhaelemeersch, W. Vandervorst, *Solid State Phenomena* **65**, 291 (1999). *(62)*
[247] L. Mouche, F. Tardif, J. Derrien, *J. Electrochem. Soc.* **142**, 2395 (1995). *(62)*
[248] F. De Smedt, C. Vinckier, I. Cornelissen, S. De Gendt, M. Meuris, G. Gilis, G. Heyns, *Solid State Phenomena* **65**, 81 (1999). *(62)*
[249] F. Tardif, T. Lardin, Y. Abolafia, A. Daniel, P. Boelen, C. Cowache, I. Kashkoush, R. Novak, *Solid State Phenomena* **65**, 19 (1999). *(62, 63, 361)*
[250] M. Baeyens, W. Hub, B. O. Kolbesen, A. R. Martin, P. W. Mertens, *Solid State Phenomena* **65**, 23 (1999). *(62, 63)*
[251] T. Ohmi, M. Miyashita, M. Itano, T. Imaoka, I. Kawanabe, *IEEE Transactions on Electron Devices* **39**, 537 (1992). *(63)*
[252] T. Takahagi, I. Nagai, A. Ishitani, H. Kuroda, Y. Nagasawa, *J. Appl. Phys.* **64**, 3516 (1988). *(63, 64, 65, 278, 280, 286)*
[253] S. R. Kasi, M. Liehr, P.A. Thiry, H. Dallaporta, M. Offenberg, *Appl. Phys. Lett.* **59**, 108 (1991). *(63, 64, 280, 358)*
[254] H. Okumura, T. Akane, Y. Tsubo, S. Matsumoto, *J. Electrochem. Soc.* **144**, 3765 (1997). *(63, 282)*
[255] M. Tabe, *Appl. Phys. Lett.* **45**, 1073 (1984). *(63, 64, 280)*
[256] M. Hirose, T. Yasaka, M. Hiroshima, M. Takakura, S. Mizayaki, *Appl. Phys.*

Lett. **59**, 38 (1991). *(63)*
[257] Y. Morita, H. Tokumoto, *J. Vac. Sci. Technol. A* **14**, 854 (1996). *(63, 268, 269)*
[258] F.J. Grunthaner, P. J. Grunthaner, *Materials Science Reports* **1**, 65 (1986). *(63)*
[259] B. Garrido, J. Montserrat, J. R. Morante, *J. Electrochem. Soc.* **143**, 4059 (1996). *(63)*
[260] M. Izawa, Y. Fujisaki, *Appl. Surf. Sci.* **115**, 149 (1997). *(63)*
[261] J. A. Venables, in *Chemistry and Physics of Solid Surcaces*, Vol. IV (Springer, Heidelberg, 1982). *(64)*
[262] G. Lippert, *Thesis* Universität der Bundeswehr, München, Germany (1995). *(64)*
[263] K. Saga, T. Hattori, *J. Electrochem. Soc.* **L144**, 250 (1997). *(65)*
[264] J. Ramm, E. Beck, A. Zueger, A. Dommann, R. E. Pixley, *Thin Solid Films* **228**, 23 (1993). *(65)*
[265] H. Froitzheim, in *The Chemical Physics of Solid Surfaces and Heterogeneous Catalysis*, Vol. 5 (*Surface properties of electronic materials*) ed. by D. A. King, D. P. Woodruff, (Elsevier Science, 1988), p. 183 *(65, 264, 265, 286, 345)*
[266] M.A. Lutz, R. M. Feenstra, J.O. Chu, *Surf. Sci.* **328**, 215 (1995). *(65, 79)*
[267] F.M. D'Heurle, *Journal de Physique IV C* **6**, 29 (1996). *(65, 322, 326)*
[268] S. Wolf, *Silicon Processing for the VLSI Era*, vol. 3 ("The Submicron MOSFET") (Lattice Press, Sunset Beach, 1995). *(65, 345, 358, 376, 410, 433, 434, 437, 443)*
[269] R. H. Fowler, L. W. Nordheim, *Proc. Roy. Soc. London A* **119**, 173 (1928). *(67)*
[270] A. Gupta, P. Fang, M. Song, M.-R. Lin, D. Wollesen, K. Chen, C. Hu, *IEEE Electron Device Letters* **18**, 580 (1997). *(67, 356, 430)*
[271] M. Depas, B. Vermeire, P. W. Mertens, R. L. van Meirhaeghe, M. M. Heyns, *Solid State Electronics* **38**, 1465 (1995). *(67)*
[272] B. Ricco, G. Gozzi, M. Lanzoni, *IEEE Transactions on Electron Devices* **45**, 1554 (1998). *(67, 436)*
[273] T.M. Jung, R. Kaplan, S. M. Prokes, *Surf. Sci.* **289**, L577 (1993). *(67, 74)*
[274] L. Fotiadis, R. Kaplan, *Thin Solid Films* **184**, 415 (1990). *(67, 74)*
[275] P.M. Mooney, F.K. LeGoues, J. Tersoff, J.O. Chu, *J. Appl. Phys.* **75**, 3968 (1994). *(67)*
[276] K. Sudoh, T. Yoshinobu, H. Iwasaki, *Jap. J. Appl. Phys.* **37**, 5870 (1998); *Surf. Sci.* **419**, 128 (1999). *(68)*
[277] E. Schröder-Bergen, W. Ranke, *Surf. Sci.* **259**, 323 (1991). *(68, 73, 249, 250)*
[278] J. J. Lander, G. W. Gobeli, J. Morrison, *J. Appl. Phys.* **34**, 2298 (1963). *(69, 70)*
[279] K.C. Pandey, *Phys. Rev. Lett.* **47**, 1913 (1981). *(69, 70)*
[280] J.E. Northrup, M. L. Cohen, *Phys. Rev. Lett.* **49**, 1349 (1982). *(69, 71)*
[281] R.I.G. Uhrberg, G.V. Hansson, J. M. Nicholls, S.A. Flodström, *Phys. Rev. Lett.* **48**, 1032 (1982). *(69, 71)*
[282] F.J. Himpsel, P. M. Marcus, R. M. Tromp, I. P. Batra, M. R. Cook, F. Jona, H. Liu, *Phys. Rev. B* **30**, 2257 (1984). *(69)*
[283] H. Sakama, A. Kawazu, K. Ueda, *Phys. Rev. B* **34**, 1367 (1986). *(69)*
[284] R. E. Schlier, H. E. Farnsworth, *J. Chem. Phys.* **30**, 917 (1959). *(69, 71)*
[285] D. Haneman, *Phys. Rev.* **121**, 1093 (1961). *(69, 90, 100)*
[286] F.J. Himpsel, P. Heimann, T.C. Chiang, D. E. Eastman, *Phys. Rev. Lett.* **45**, 1112 (1980). *(69, 101, 119, 134, 143)*
[287] S. Brennan, J. Stühr, R. Jaeger, J. E. Rowe, *Phys. Rev. Lett.* **45**, 1414 (1980).

(69)
[288] J.C. Woicik, P. Pianetta, T. Kendelewicz, *Phys. Rev. B* **40**, 12463 (1989). *(69)*
[289] R. Seiwatz, *Surf. Sci.* **2**, 473 (1964). *(70, 92)*
[290] A. Selloni, E. Tosatti, *Solid State Commun.* **17**, 387 (1975). *(70)*
[291] W. A. Harrison, *J. Vac. Sci. Technol.* **14**, 883 (1977). *(70)*
[292] R. M. Feenstra, A. P. Fein, W. A. Thompson, *Phys. Rev. Lett.* **56**, 608 (1986). *(71)*
[293] J.A. Stroscio, R. M. Feenstra, A. P. Fein, *Phys. Rev. Lett.* **57**, 2579 (1986). *(71)*
[294] R. Feder, W. Mönch, *Solid State Commun.* **50**, 311 (1984). *(71)*
[295] R.S. Becker, G.S. Higashi, Y. J. Chabal, A. J. Becker, *Phys. Rev. Lett.* **65**, 1917 (1990). *(71)*
[296] G. Binnig, H. Rohrer, Ch. Gerber, E. Weibel, *Phys. Rev. Lett.* **50**, 120 (1983). *(71, 107)*
[297] K. Takayanagi, Y. Tanishiro, M. Takahashi, S. Takahashi, *Surf. Sci.* **164**, 367 (1985). *(71)*
[298] K. Takayanagi, Y. Tanishiro, M. Takahashi, S. Takahashi, *J. Vac. Sci. Technol. A* **3**, 1502 (1985). *(71)*
[299] S. Y. Tong, H. Huanq, C. M. Wei, W. E. Packard, F. K. Men, G. S. Glander, M.B. Webb, *J. Vac. Sci. Technol. A* **6**, 615 (1988). *(71, 72)*
[300] M. Fujita, H. Nagayoshi, *Surf. Sci.* **242**, 229 (1991). *(71, 72)*
[301] J. Dąbrowski, H.-J. Müssig, S. Hinrich, unpublished. *(72, 135, 150, 151, 168, 169, 180, 181, 182, 186, 233, 285, 315, 422, 423)*
[302] R.L. Headrick, I. K. Robinson, E. Vlieg, L.C. Feldman, *Phys. Rev. Lett.* **63**, 1253 (1989). *(72)*
[303] P. J. Bedrossian, R. D. Meade, K. Mortensen, D.M. Chen, J. A. Golovchenko, D. Vanderbilt, *Phys. Rev. Lett.* **63**, 1257 (1989). *(72)*
[304] I.-W. Lyo, E. Kaxiras, Ph. Avouris, *Phys. Rev. Lett.* **63**, 1261 (1989). *(72)*
[305] I. Stich, M.C. Payne, R. D. King-Smith, J.-S. Lin, L. J. Clarke, *Phys. Rev. Lett.* **68**, 1351 (1992). *(72)*
[306] K. D. Brommer, M. Needels, B. E. Larson, J.D. Joannopoulos, *Phys. Rev. Lett.* **68**, 1355 (1992). *(72)*
[307] I. Stich, K. Terakura, B. E. Larson, *Phys. Rev. Lett.* **74**, 4491 (1995). *(72)*
[308] R. M. Feenstra, M.A. Lutz, *J. Vac. Sci. Technol. B* **9**, 716 (1991). *(72)*
[309] D. Vanderbilt, *Phys. Rev. B* **36**, 6209 (1987). *(72)*
[310] R.I.G. Uhrberg, E. Landemark, L. S. O. Johansson, *Phys. Rev. B* **39**, 13525 (1989). *(73)*
[311] B. Garni, D.E. Savage, M. G. Lagally, *Surf. Sci.* **235**, L324 (1990). *(73)*
[312] R. M. Feenstra, M.A. Lutz, *Phys. Rev. B* **42**, 5391 (1990). *(73)*
[313] N. Osakabe, Y. Tanishiro, K. Yagi, G. Honjo, *Surf. Sci.* **102**, 424 (1981). *(73)*
[314] A. V. Latishev, A.L. Aseev, A.B. Krasilnikov, S.I. Stenin, *Surf. Sci.* **227**, 24 (1990). *(73)*
[315] S. Song, S. G. J. Mochrie, *Phys. Rev. Lett.* **73**, 995 (1994). *(73)*
[316] S. Song, M. Yoon, S. G. J. Mochrie, *Surf. Sci.* **334**, 153 (1995). *(73)*
[317] S. C. Erwin, A.A. Baski, L.J. Whitman, *Phys. Rev. Lett.* **77**, 687 (1996). *(74)*
[318] A. Laracuente, S. C. Erwin, L.J. Whitman, *Appl. Phys. Lett.* **74**, 1397 (1999). *(74, 265)*
[319] J. Wie, X.-S. Wang, J. L. Goldberg, N.C. Bartelt, E. D. Williams, *Phys. Rev. Lett.* **68**, 3885 (1992). *(74)*
[320] B. Li, *Thesis*, University of Maryland (1993). *(74)*
[321] X.-S. Wang, W.H. Weinberg, "Surface structure of Si(112)" *Surf. Sci.* **314**, 71 (1994). *(74)*

[322] A.A. Baski, L.J. Whitman, *J. Vac. Sci. Technol. A* **13**, 1469 (1969). *(74)*
[323] A.A. Baski, S. C. Erwin, L.J. Whitman, *Science* **269**, 1556 (1995). *(74)*
[324] T. Suzuki, H. Minoda, Y. Tanishiro, K. Yagi, T. Sueyoshi, T. Sato, M . Iwatsuki, *Surf. Sci.* **357**, 522 (1996). *(74)*
[325] J.M. Gibson, M.L. McDonald, F.C. Unterwald, *Phys. Rev. Lett.* **55**, 1765 (1985). *(74)*
[326] J. Dąbrowski, H.-J. Müssig, in *Proceedings of the 23rd International Conference on The Physics of Semiconductors*, ed. by M. Scheffler and R. Zimmermann (World Scientific, Singapore 1996) p. 795. *(74, 76, 265)*
[327] B.Z. Olshanetsky, V. I. Mashanov, *Surf. Sci.* **111**, 414 (1981). *(75)*
[328] J. Knall, J.B. Pethica, J.D. Todd, J.H. Wilson, *Phys. Rev. Lett.* **66**, 1733 (1991). *(75)*
[329] J. Dąbrowski, H.-J. Müssig, G. Wolff, *Phys. Rev. Lett.* **73**, 1660 (1994). *(75, 76, 78)*
[330] J. Dąbrowski, H.-J. Müssig, G. Wolff, *J. Vac. Sci. Technol. B* **13**, 1597 (1995). *(75)*
[331] J. Dąbrowski, H.-J. Müssig, G. Wolff, H. Rücker, *Proc. 22th Int. Conf. Phys. Semicond.*,Vancouver 1994, edited by D. J. Lockwood (World Scientific, Singapore 1995) p. 427. *(75)*
[332] H.-J. Müssig, J. Dąbrowski, S. Hinrich, invited talk at the MRS 1999 Spring Meeting, San Francisco, to be published in the Proceedings of Symposium R. *(76, 295)*
[333] S. Takeda, M. Kohayama, K. Ibe, *Philosophical Magazine A* **70**, 287 (1994). *(76)*
[334] W. Ranke, *Phys. Rev. B* **41**, 5243 (1990). *(77)*
[335] K. Jacobi, U. Myler, P. Althainz, *Phys. Rev. B* **41**, 10721 (1990). *(77)*
[336] T. Suzuki, H. Minoda, Y. Tanishiro, K. Yagi, *Preprint* **1996**, 18 (1996). *(77)*
[337] S. M. Scholz, K. Jacobi, *Surf. Sci.* **369**, 117 (1996). *(77, 286)*
[338] W. Ranke, Y.R. Xing, *Surf. Sci.* **381**, 1 (1997). *(77, 296, 297, 301)*
[339] T. Suzuki, H. Minoda, Y. Tanishiro, K. Yagi, submitted to *Surf. Sci.* (1999). *(77)*
[340] E. T. Yu, K. Barmak, P. Ronsheim, M.B. Johnson, P. McFarland, J.-M. Halbout, *J. Appl. Phys.* **79**, 2115 (1996). *(78)*
[341] L. Butz, H. Lüth, *Surf. Sci.* **365**, 807 (1996). *(78)*
[342] Y. Yamamoto, *Surf. Sci.* **313**, 155 (1994). *(78, 79, 339)*
[343] T. Ichinokawa, H. Ampo, S. Miura, A. Tamura, *Phys. Rev. B* **31**, 5183 (1985). *(79, 81, 335)*
[344] E.J. van Loenen, D. Dijkkamp, A.J. Hoeven, *Journal of Microscopy* **152**, 487 (1988). *(79)*
[345] H. Ampo, S. Miura, K. Kato, Y. Ohkawa, A. Tamura, *Phys. Rev. B* **34**, 2329 (1986). *(79, 265)*
[346] Y. Yamamoto, S.-i. Kitamura, M . Iwatsuki, *Jap. J. Appl. Phys.* **31**, L635 (1992). *(79, 81)*
[347] H. Kim, N. Taylor, T. Spila, G. Glass, S. Y. Park, J.E. Greene, J. R. Abelson, *Surf. Sci.* **380**, L496 (1997). *(79)*
[348] A. I. Shkrebtii, C. M. Bertoni, R. Del Sole, B. Nesterenko, *Surf. Sci.* **239**, 227 (1990). *(79)*
[349] B.Z. Olshanetsky, A. A. Shklyaev, *Surf. Sci.* **67**, 581 (1977). *(79)*
[350] Y. Yamamoto, T. Sueyoshi, T. Sato, M . Iwatsuki, *Jap. J. Appl. Phys.* **32**, 1808 (1993). *(79)*
[351] A.J. Hoeven, D. Dijkkamp, E.J. van Loenen, P. J. G. M. van Hooft, *Surf. Sci.*

211, 165 (1989). *(81)*

[352] F. Jona, *IBM Journal of Research and Development* **9**, 375 (1965). *(81)*

[353] T. Sakurai, H. D. Hagstrum, *J. Vac. Sci. Technol.* **13**, 807 (1976). *(81)*

[354] P. Martensson, G.V. Hansson, P. Chiaradia, *Phys. Rev. B* **31**, 2581 (1985). *(81)*

[355] S. Becker, B.S. Swartzentruber, J.S. Vickers, *J. Vac. Sci. Technol. A* **6**, 472 (1988). *(81)*

[356] H. Neddermeyer, St. Tosch, *Phys. Rev. B* **38**, 5784 (1988). *(81)*

[357] R. E. Schlier, H. E. Farnsworth, in *Semiconductor Surface Physics* (University of Pennsylvania Press, 1957), p. 3. *(83, 134)*

[358] G. A. Somorjai, H. H. Farrell, *Advan. Chem. Phys.* **20**, 215 (1971). *(83)*

[359] F.J. Himpsel, E. D. Eastman, *J. Vac. Sci. Technol.* **16**, 1297 (1979). *(83, 86, 99, 100, 110, 125, 143)*

[360] D. J. Chadi, *Phys. Rev. Lett.* **43**, 43 (1979). *(83, 100, 134, 136, 143, 238)*

[361] M. T. Yin, M. L. Cohen, *Phys. Rev. B* **24**, 2303 (1981). *(83, 102, 105, 134)*

[362] F. F. Abraham, I. P. Batra, *Surf. Sci.* **163**, L752 (1985). *(83)*

[363] R. M. Tromp, R.J. Hamers, J.E. Demuth, *Phys. Rev. Lett.* **55**, 1303 (1985). *(83, 85, 86, 87, 107, 108, 120, 134, 149, 157, 238)*

[364] D. Haneman, *Rep. Prog. Phys.* **50**, 1045 (1987). *(83, 107)*

[365] R.J. Hamers, R. M. Tromp, J.E. Demuth, *Surf. Sci.* **181**, 346 (1987). *(83, 110, 113, 134)*

[366] F. Bechstedt, D. Reichardt, *Surf. Sci.* **202**, 83 (1988). *(83, 114)*

[367] F. K. Men, W. P. Packard, M.B. Webb, *Phys. Rev. Lett.* **61**, 2469 (1988). *(83, 113, 139, 140, 249, 255)*

[368] Z. Zhu, N. Shima, M. Tsukada, *Phys. Rev. B* **40**, 11868 (1989). *(83, 114, 134)*

[369] L. S. O. Johansson, R.I.G. Uhrberg, P. Mårtensson, G.V. Hansson, *Phys. Rev. B* **42**, 1305 (1990). *(83, 121, 125, 143, 145, 147, 148)*

[370] S. Tang, A.J. Freeman, B. Delley, *Phys. Rev. B* **45**, 1776 (1992). *(83)*

[371] L. S. O. Johansson, B. Reihl, *Surf. Sci.* **269**, 810 (1992). *(83, 143, 147, 148)*

[372] J. Dąbrowski, M. Scheffler, *Appl. Surf. Sci.* **56**, 15 (1992). *(83, 119, 120, 134, 143, 149, 238)*

[373] R. A. Wolkow, *Phys. Rev. Lett.* **68**, 2636 (1992). *(83, 85, 87, 119, 120, 134, 150)*

[374] H. Tochihara, T. Amakusa, M . Iwatsuki, *Phys. Rev. B* **50**, 12262 (1994). *(83, 87, 120, 134, 150, 165, 238)*

[375] J. Dąbrowski, E. Pehlke, M. Scheffler, *J. Vac. Sci. Technol. B* **12**, 2675 (1994). *(83, 134, 139, 140)*

[376] M. Takahasi, S. Nakatani, Y. Ito, T. Takahashi, X. W. Zhang, M. Ando, *Surf. Sci.* **338**, L846 (1995). *(83, 121, 134, 135, 138)*

[377] A. Ramstad, G. Brocks, P. J. Kelly, *Phys. Rev. B* **51**, 14504 (1995). *(83, 121, 122, 134, 137, 138)*

[378] H. Tamura, K. Nishii, H. Minoda, Y. Tanishiro, K. Yagi, *Surf. Sci.* **357**, 576 (1996). *(83, 134, 236)*

[379] E. P. Stoll, A. Baratoff, A. Mangili, D. Maric, *Mater. Sci. Eng. B* **37**, 155 (1996). *(83, 134, 149)*

[380] S. Kitamura, M . Iwatsuki, *Jap. J. Appl. Phys.* **35**, L668 (1996). *(83, 133, 134, 135)*

[381] R. Felici, I. K. Robinson, C. Ottaviani, P. Imperatori, P. J. Eng, P. Perfetti, *Surf. Sci.* **375**, 55 (1997). *(83, 121, 134, 138)*

[382] H. Shigekawa, K. Hata, K. Miyake, M. Ishida, S. Ozawa, *Phys. Rev. B* **55**, 15448 (1997). *(83, 120, 140, 145, 149, 238, 243)*

[383] H. E. Farnsworth, R. E. Schlier, T. H. George, R. M. Burger, *J. Appl. Phys.*

29, 1150 (1958). *(86, 88, 98)*
[384] J.E. Northrup, *Phys. Rev. Lett.* **54**, 815 (1985). *(86, 106)*
[385] H. E. Farnsworth, R. E. Schlier, T. H. George, R. M. Burger, *J. Appl. Phys.* **26**, 252 (1955). *(89)*
[386] M. Green, R. Seiwatz, *J. Chem. Phys.* **37**, 458 (1962). *(89)*
[387] W. A. Harrison, *Electronic Structure and the Properties of Solids* (W. H. Freeman and Co., 1980). *(90, 97, 133)*
[388] J. J. Lander, J. Morrison, *J. Appl. Phys.* **33**, 2089 (1962). *(92, 245, 294)*
[389] J. J. Lander, J. Morrison, *J. Chem. Phys.* **37**, 729 (1962). *(92, 238)*
[390] S. E. Trullinger, S. L. Cunningham, *Phys. Rev. Lett.* **30**, 913 (1973). *(93)*
[391] H. Ibach, J. E. Rowe, *Surf. Sci.* **43**, 481 (1974). *(94, 95)*
[392] P. Masri, P. W. Tasker, *Surf. Sci.* **149**, 209 (1985). *(95)*
[393] P. Masri, P. W. Tasker, J. P. Hoare, J. Harding, *Surf. Sci.* **173**, 439 (1986). *(95)*
[394] P. Masri, *Surf. Sci. Rep.* **9**, 293 (1988). *(95)*
[395] J. C. Phillips, *Surf. Sci.* **40**, 459 (1973). *(95)*
[396] J. D. Levine, *Surf. Sci.* **34**, 90 (1973). *(95)*
[397] J. E. Rowe, H. Ibach, *Phys. Rev. Lett.* **31**, 102 (1973). *(95, 143)*
[398] S. J. White, D. P. Woodruff, *Journal of Physics C, Letters* **9**, 451 (1976). *(95)*
[399] J. E. Rowe, H. Ibach, *Phys. Rev. Lett.* **32**, 421 (1974). *(96, 143)*
[400] J. A. Appelbaum, G. A. Baraff, D. R. Hamann, *Phys. Rev. B* **11**, 3822 (1975). *(96, 133)*
[401] J. A. Appelbaum, G. A. Baraff, D. R. Hamann, *Phys. Rev. B* **12**, 5749 (1975). *(96)*
[402] J. A. Appelbaum, G. A. Baraff, D. R. Hamann, *Phys. Rev. B* **14**, 588 (1976). *(96, 143)*
[403] F. Jona, H. D. Shih, A. Ignatiev, D. W. Jepsen, P. M. Marcus, *Journal of Physics C, Letters* **10**, 67 (1977). *(96, 97)*
[404] J. A. Appelbaum, G. A. Baraff, D. R. Hamann, *Phys. Rev. B* **15**, 2408 (1977). *(96)*
[405] J. A. Strozier Jr., D. W. Jepsen, F. Jona, in *Surface Physics of Materials*, ed. by J.M. Blakely, (Academic Press, New York, 1975), p. 1. *(97)*
[406] G. P. Kerker, S. G. Louie, M. L. Cohen, *Phys. Rev. B* **17**, 706 (1978). *(97)*
[407] K.A.R. Mitchell, *Surf. Sci.* **64**, 797 (1977). *(98)*
[408] S. J. White, D. P. Woodruff, *Surf. Sci.* **64**, 131 (1977). *(98)*
[409] S. J. White, D. P. Woodruff, *Surf. Sci.* **63**, 254 (1977). *(98)*
[410] T. D. Poppendieck, T. C. Ngoc, M.B. Webb, *Surf. Sci.* **75**, 287 (1978). *(98)*
[411] J. A. Appelbaum, D. R. Hamann, *Surf. Sci.* **74**, 21 (1978). *(98, 134)*
[412] S. Y. Tong, A. L. Maldondo, *Surf. Sci.* **78**, 459 (1978). *(99, 134)*
[413] F. Jona, H. D. Shih, D. W. Jepsen, P. M. Marcus, *Bulletin APS* **24**, 467 (1979). *(99)*
[414] M. J. Cardillo, G. F. Becker, *Phys. Rev. Lett.* **40**, 1148 (1978). *(99)*
[415] E. K. Grimmelmann, J. C. Tully, *Bulletin APS* **24**, 467 (1979). *(99)*
[416] I. Estermann and O. Stern, *Z. Phys.* **61**, 95 (1930). *(99)*
[417] F. Jona, H. D. Shih, D. W. Jepsen, P. M. Marcus, *J. Phys. C* **12**, 455 (1979). *(100)*
[418] R. M. Tromp, R. G. Smeenk, F. W. Saris, *Solid State Commun.* **39**, 755 (1981). *(101)*
[419] R. M. Tromp, R. G. Smeenk, F. W. Saris, D. J. Chadi, *Surf. Sci.* **133**, 137 (1983). *(101, 105, 109)*
[420] J. Ihm, M. L. Cohen, D.J. Chadi, *Phys. Rev. B* **21**, 4592 (1980). *(101, 143)*
[421] A. Mazur, J. Pollmann, *Phys. Rev. B* **26**, 7086 (1982). *(101, 102)*

[422] J. Pollmann, S. T. Pantelides, *Phys. Rev. B* **18**, 5524 (1978). *(101)*

[423] J. Pollmann, R. Kalla, P. Krüger, A. Mazur, G. Wolfgarten, *Applied Physics A* **41**, 21 (1986). *(102, 111, 149)*

[424] M. A. Bowen, J. D. Dow, R. E. Allen, *Phys. Rev. B* **26**, 7083 (1982). *(102)*

[425] W. S. Yang, F. Jona, P. M. Marcus, *Solid State Commun.* **43**, 847 (1982). *(102, 104, 134, 242)*

[426] M. Aono, Y. Hou, C. Oshima, Y. Ishizawa, *Phys. Rev. Lett.* **49**, 567 (1982). *(102)*

[427] W. S. Verwoerd, *Surf. Sci.* **99**, 581 (1980). *(102)*

[428] W. S. Verwoerd, *Surf. Sci.* **103**, 404 (1981). *(102)*

[429] D. Haneman, *Advances in Physics* **31**, 165 (1982). *(102)*

[430] A. Redondo, W. A. Goddard III, *J. Vac. Sci. Technol.* **21**, 344 (1982). *(102)*

[431] W. A. Goddard III, A. Redondo, *J. Vac. Sci. Technol.* **16**, 1308 (1979). *(102)*

[432] W. S. Verwoerd, *Surf. Sci.* **129**, 419 (1983). *(104)*

[433] W. Mönch, P. Koke, S. Krueger, *J. Vac. Sci. Technol.* **19**, 313 (1981). *(104)*

[434] W. S. Verwoerd, *Surf. Sci.* **108**, 153 (1981). *(104)*

[435] W. S. Yang, F. Jona, P. M. Marcus, *Phys. Rev. B* **28**, 2049 (1983). *(104)*

[436] C. B. Duke, A. Paton, B. W. Holland, *Proc. 17th Int. Conf. Phys. Semicond.*, ed. by D. J. Chadi and W. A. Harrison (Springer, Berlin 1984) p. 59. *(105)*

[437] Y. S. Shu, W. S. Yang, F. Jona, P. M. Marcus, in *Springer Series in Surface Sciences*, Vol. 2 (*The Structure of Surfaces*) ed. by M. A. Van Hove, S. Y. Tong, (Springer, Berlin, 1985), p. 293. *(105)*

[438] H. H. Farrell, F. Stucki, J. Anderson, D. J. Frankel, G. J. Lapeyre, M. Levinson, *Phys. Rev. B* **30**, 721 (1984). *(105)*

[439] K.C. Pandey, *Proc. 17th Int. Conf. Phys. Semicond.*, ed. by D. J. Chadi and W. A. Harrison (Springer, Berlin 1984) p. 55. *(105, 157, 195, 199)*

[440] K.C. Pandey, *Phys. Rev. Lett.* **49**, 223 (1982). *(105)*

[441] R.J. Hamers, R. M. Tromp, J.E. Demuth, *Phys. Rev. B* **34**, 5343 (1986). *(108, 158, 163, 178)*

[442] R. S. Brink, W. S. Verwoerd, *Surf. Sci.* **154**, L203 (1985). *(109)*

[443] B.Z. Olshanetsky, A. A. Shklyaev, *Surf. Sci.* **82**, 445 (1979). *(109)*

[444] R. Kaplan, *Surf. Sci.* **93**, 145 (1980). *(109)*

[445] R. D. Bringans, R.I.G. Uhrberg, M. A. Olmstead, R. Z. Bachrach, *Phys. Rev. B* **34**, 7447 (1986). *(110)*

[446] P. Mårtensson, A. Circenti, G.V. Hansson, *Phys. Rev. B* **33**, 8855 (1986). *(110)*

[447] R.J. Hamers, Ph. Avouris, F. Bozso, *Phys. Rev. Lett.* **59**, 2071 (1987). *(110, 309)*

[448] F.J. Himpsel, Th. Fauster, *J. Vac. Sci. Technol. A* **2**, 815 (1984). *(110)*

[449] J. Pollmann, P. Krüger, A. Mazur, *J. Vac. Sci. Technol. B* **5**, 945 (1987). *(111)*

[450] M. S. Hybertsen, S. G. Louie, *Phys. Rev. Lett.* **55**, 1418 (1985). *(111)*

[451] T. Tabata, T. Aruga, Y. Murata, *Surf. Sci.* **179**, L63 (1987). *(112, 140, 238, 246)*

[452] R. McGrath, R. Cimino, W. Braun, G. Thorton, I.T. McGovern, *Vacuum* **38**, 251 (1988). *(112)*

[453] D.H. Rich, A. Samsavar, T. Miller, H. F. Lin, T.C. Chiang, J.-E. Sundgren, J. E. Green, *Phys. Rev. Lett.* **58**, 579 (1987). *(112, 117, 118)*

[454] H. D. Rich, T. Miller, T.C. Chiang, *Phys. Rev. B* **37**, 3124 (1988). *(112, 117)*

[455] M.C. Payne, N. Roberts, R. J. Needs, M. Needels, J.D. Joannopoulos, *Surf. Sci.* **211**, 1 (1989). *(114, 115, 121)*

[456] E. Artacho, F. Ynduráin, *Phys. Rev. Lett.* **62**, 2491 (1989). *(114)*

[457] B.I. Craig, P.V. Smith, *Surf. Sci.* **218**, 569 (1989). *(114)*

[458] S. Ihara, S. L. Ho, T. Uda, M. Hirao, *Phys. Rev. Lett.* **65**, 1909 (1990). *(115, 157, 159, 216)*

[459] I. P. Batra, *Phys. Rev. B* **41**, 5048 (1990). *(115)*

[460] N. Roberts, R. J. Needs, *Surf. Sci.* **236**, 112 (1990). *(115, 157, 161, 162)*

[461] K. Kobayashi, K. Terakura, S. Blügel, *Technical Report of ISSP A* **2366**, 1 (1990). *(115)*

[462] K. Kobayashi, Y. Morikawa, K. Terakura, S. Blügel, *Phys. Rev. B* **45**, 3469 (1992). *(115)*

[463] P. Badziag, W. S. Verwoerd, A. M. Van Hove, *Phys. Rev. B* **43**, 2058 (1991). *(115)*

[464] M. Tsuda, T. Hoshino, S. Oikawa, I. Ohdomari, *Phys. Rev. B* **44**, 11241 (1991). *(115)*

[465] G.P. Kochanski, J. E. Griffith, *Surf. Sci.* **249**, L293 (1991). *(115)*

[466] S. Hawking, "My position", in *Black Holes and Baby Universies*, (Bantam Books, New York, 1994), p. 41. *(116)*

[467] F.J. Himpsel, *Phys. Rev. Lett.* **69**, 551 (1992). *(118, 119)*

[468] G. K. Wertheim, D. M. Riffe, J. E. Rowe, P. H. Citrin, *Phys. Rev. Lett.* **67**, 120 (1991). *(118, 153)*

[469] S. Doniach, M. Šunjić, *J. Phys. C* **3**, 285 (1970). *(118)*

[470] E. Landemark, C.J. Karlsson, Y.-C. Chao, R.I.G. Uhrberg, *Phys. Rev. Lett.* **69**, 1588 (1992). *(118, 120, 143, 152, 153)*

[471] D.S. Lin, T. Miller, T.C. Chiang, *Phys. Rev. Lett.* **67**, 2187 (1991). *(118, 153)*

[472] J. E. Rowe, G. K. Wertheim, T.C. Chiang, *Phys. Rev. Lett.* **69**, 550 (1992). *(118)*

[473] X. Yang, R. Cao, J. Terry, P. Pianetta, *J. Vac. Sci. Technol. B* **10**, 2013 (1992). *(119)*

[474] D.S. Lin, J. A. Carlise, T. Miller, T.C. Chiang, *Phys. Rev. Lett.* **69**, 552 (1992). *(119)*

[475] R. H. Miwa, *Surf. Sci.* **418**, 55 (1998). *(119)*

[476] J. Dąbrowski, E. Pehlke, M. Scheffler, *Proc. 21th Int. Conf. Phys. Semicond.*, Beijing 1992, ed. by P. Jiang and H.-Z. Zheng (World Scientific, Singapore 1992) p. 389. *(120, 143, 153, 154)*

[477] D. Band, H. Wengelnik, H. Neddermeyer, *J. Vac. Sci. Technol. B* **12**, 2015 (1994). *(120)*

[478] A. W. Munz, Ch. Ziegler, W. Gopel, *Phys. Rev. Lett.* **74**, 2244 (1995). *(120)*

[479] L. Kipp, D.K. Biegelsen, J.E. Northrup, L.-E. Swartz, R. D. Bringans, *Phys. Rev. Lett.* **76**, 2810 (1996). *(120, 143)*

[480] M. Uchikawa, M. Ishida, K. Miyake, K. Hata, R. Yoshizaki, H. Shigekawa, *Surf. Sci.* **357**, 468 (1996). *(120, 150)*

[481] J.E. Northrup, *Phys. Rev. B* **47**, 10032 (1993). *(121, 122, 138, 143, 147, 149, 243)*

[482] C. Kress, M. Fiedler, F. Bechstedt, in Proceedings of the *4'th International Conference on the Formation of Semiconductor Interfaces* ed. by B. Lengeler, H. Lüth, W. Mönch, and J. Pollmann (World Scientific, Singapore 1994) p. 19. *(121, 143, 147)*

[483] M. Rohlfing, P. Krüger, J. Pollmann, *Phys. Rev. B* **52**, 13753 (1995). *(121, 136, 143, 144, 145)*

[484] Y.C. Wang, T.A. Arias, J.D. Joannopoulos, *Phys. Rev. B* **47**, 10497 (1993). *(121, 157, 159, 160, 161, 162, 163, 164, 195, 196, 215, 216, 217)*

[485] P. Boguslawski, Q.-M. Zhang, Z. Zhang, J. Bernholc, *Phys. Rev. Lett.* **72**, 3694 (1994). *(121, 173, 177, 178, 179, 182, 187, 232)*

[486] A. Garcia, J.E. Northrup, *Phys. Rev. B* **48**, 17350 (1993). *(121, 140)*
[487] J. Dąbrowski, E. Pehlke, M. Scheffler, *Phys. Rev. B* **49**, 4790 (1994). *(121, 139, 140)*
[488] A. I. Shkrebtii, R. Del Sole, *Phys. Rev. Lett.* **70**, 2645 (1993). *(121)*
[489] P. Krüger, J. Pollmann, *Applied Physics A* **59**, 487 (1994). *(121)*
[490] P. Krüger, J. Pollmann, *Phys. Rev. Lett.* **74**, 1155 (1995). *(121)*
[491] J. Fritsch, P. Pavone, *Surf. Sci.* **344**, 159 (1995). *(121, 122, 143, 144, 147, 148, 238, 241, 242)*
[492] K. Cho, J.D. Joannopoulos, *Phys. Rev. B* **53**, 4553 (1996). *(121)*
[493] V. Milman, S. J. Pennycook, D. E. Jesson, *Thin Solid Films* **272**, 375 (1996). *(121, 241)*
[494] M.B. Webb, F. K. Men, B.S. Swartzentruber, R. Kariotis, M. G. Lagally, *Surf. Sci.* **242**, 23 (1991). *(121, 140, 141, 142, 233, 249, 254)*
[495] J. Tersoff, *Phys. Rev. B* **45**, 8833 (1992). *(121, 141, 157, 197)*
[496] K. Cho, J.D. Joannopoulos, *Phys. Rev. Lett.* **71**, 1387 (1993). *(121, 134)*
[497] E. Fontes, J.R. Patel, F. Comin, *Phys. Rev. Lett.* **70**, 2790 (1993). *(121)*
[498] E. L. Bullock, R. Gunnella, L. Patthey, T. Abukawa, S. Kono, C. R. Natoli, L. S. O. Johansson, *Phys. Rev. Lett.* **74**, 2756 (1995). *(121, 134, 138, 153)*
[499] I. H. Hong, S. C. Shyu, Y. C. Chou, C. M. Wei, *Phys. Rev. B* **52**, 16884 (1995). *(122, 134, 135)*
[500] A. I. Shkrebtii, R. Di Felice, C. M. Bertoni, R. Del Sole, *Phys. Rev. B* **51**, 11201 (1995). *(122, 141, 238, 242)*
[501] S. J. O'Shea, R. M. Atta, M. P. Murrell, and M. E. Welland, *J. Vac. Sci. Technol. B 13*, 1945 (1995). *(124, 430)*
[502] J. Xu, S.H. Overbury, J. Wendelken, *Phys. Rev. B* **53**, 4245 (1996). *(124, 156)*
[503] S. D. Kevan, N.G. Stoffel, *Phys. Rev. Lett.* **53**, 702 (1984). *(125)*
[504] S. D. Kevan, *Phys. Rev. B* **32**, 2344 (1985). *(125, 240)*
[505] H.J.W. Zandvliet, A. van Silfhout, M. J. Sparnaay, *Phys. Rev. B* **39**, 5576 (1989). *(125)*
[506] H.J.W. Zandvliet, B. S. Swartzentruber, W. Wulfhekel, B. J. Hattink, B. Poelsema, *Phys. Rev. B* **57**, 6803 (1998). *(125)*
[507] N. Jedrecy, M. Sauvage-Simkin, R. Pinchaux, J. Massies, N. Greiser, V. H. Etgens, *Surf. Sci.* **230**, 197 (1990). *(134)*
[508] G. Jayaram, P. Xu, L. D. Marks, *Phys. Rev. Lett.* **71**, 3489 (1993). *(134)*
[509] M.B. Webb, *Surf. Sci.* **299**, 454 (1994). *(134, 249)*
[510] T. Kawamura, *Phase Transitions* **53**, 215 (1995). *(134, 135, 209)*
[511] R.I.G. Uhrberg, E. Landemark, Y.-C. Chao, *Journal of Electron Spectroscopy and Related Phenomena* **75**, 197 (1995). *(134, 143)*
[512] T. Aruga, Y. Murata, *Phys. Rev. B* **34**, 5654 (1986). *(135, 195)*
[513] R. Englman, *The Jahn-Teller Effect in Molecules and Crystals* (Wiley, New York, 1972). *(137)*
[514] K. Inoue, Y. Morikawa, K. Terakura, M. Nakayama, *Phys. Rev. B* **49**, 14774 (1994). *(137, 238, 241, 242, 243, 245)*
[515] V. I. Marchenko, *Sov. Phys. JEPT Lett.* **33**, 381 (1981). *(139, 255)*
[516] M. Kubota, Y. Murata, *Phys. Rev. B* **49**, 4810 (1994). *(140, 238, 247)*
[517] T. W. Poon, S. Yip, P. S. Ho, F. F. Abraham, *Phys. Rev. B* **45**, 3521 (1992). *(141, 249, 254, 255)*
[518] B.S. Swartzentruber, N. Kitamura, M. G. Lagally, M.B. Webb, *Phys. Rev. B* **47**, 13432 (1993). *(141, 184, 249, 254)*
[519] H.J.W. Zandvliet, H.B. Elswijk, *Phys. Rev. B* **48**, 14269 (1993). *(141, 249, 254, 255)*

[520] E. Bauer, M. Mundschau, W. Swiech, W. Telieps, *J. Vac. Sci. Technol. A* **9**, 1007 (1991). *(142, 249, 253, 257, 325)*
[521] R. G. Pearson, *Inorganic Chemistry* **27**, 734 (1988). *(142)*
[522] M. Galván, A. Dal Pinto Jr., J.D. Joannopoulos, *Phys. Rev. Lett.* **70**, 21 (1993). *(142, 143)*
[523] R.I.G. Uhrberg, G.V. Hansson, J. M. Nicholls, S.A. Flodström, *Phys. Rev. B* **24**, 4684 (1981). *(143)*
[524] Y. Enta, S. Suzuki, S. Kono, *Phys. Rev. Lett.* **65**, 2704 (1990). *(143, 147, 148)*
[525] R.I.G. Uhrberg, G.V. Hansson, *Critical Reviews in Solid State and Materials Sciences* **17**, 133 (1991). *(143, 145, 265)*
[526] L. Gavioli, M. G. Betti, A. Cricenti, C. Mariani, *Journal of Electron Spectroscopy and Related Phenomena* **76**, 541 (1995). *(147)*
[527] R. M. Feenstra, *Phys. Rev. B* **44**, 13791 (1991). *(152)*
[528] M. Mundschau, E. Bauer, W. Telieps, W. Swiech, *Philosophical Magazine A* **61**, 257 (1990). *(155)*
[529] C.S. Chang, Y. M. Huang, C. C. Chen, T. T. Tsong, *Surf. Sci.* **367**, L8 (1996). *(156, 194)*
[530] Ph. Avouris, D. G. Cahill, *Ultramicroscopy* **42**, 838 (1992). *(156, 166, 177, 178, 187, 286, 287, 293)*
[531] M. K.-J. Johansson, S. M. Gray, L. S. O. Johansson, *Phys. Rev. B* **53**, 1362 (1996). *(156, 166)*
[532] T. Sugano, *Appl. Surf. Sci.* **60**, 698 (1992). *(156)*
[533] P. J. Bedrossian, *Phys. Rev. B* **44**, 13783 (1991). *(156)*
[534] J. Xu, S.H. Overbury, J. F. Wendelken, *J. Vac. Sci. Technol. B* **14**, 918 (1996). *(156)*
[535] N. Roberts, R. J. Needs, *J. Phys.: Cond. Mat.* **1**, 3139 (1989). *(157)*
[536] Z.-Y. Zhang, H. Metiu, *Phys. Rev. B* **48**, 8166 (1993). *(157, 165, 215, 216, 217, 222)*
[537] M. Chander, Y.Z. Li, J.C. Patrin, J. H. Weaver, *Phys. Rev. B* **48**, 2493 (1993) *(157, 165, 297, 303, 305)*
[538] L. Andersohn, U. Köhler, *Surf. Sci.* **284**, 77 (1993). *(157, 165, 297, 301, 302, 303, 305)*
[539] F. K. Men, A. R. Smith, K.-J. Chao, Z. Zhang, C. K. Shih, *Phys. Rev. B* **52**, 8650 (1995). *(157, 165, 194, 195, 197)*
[540] J.-Y. Koo, J.-Y. Yi, C. Hwang, D.Y. Kim, S. Lee, D. Shin, *Phys. Rev. B* **52**, 17269 (1995). *(157, 158, 163, 165, 195)*
[541] V. A. Ukraintsev, J. T. Yates, *Surf. Sci.* **346**, 31 (1996). *(157, 164, 194, 195, 336, 337)*
[542] K. Terakura, T. Uda, T. Yamasaki, T. Miyazaki, *Surf. Sci.* **357**, 394 (1996). *(157, 161, 162, 165, 166, 167, 169, 217, 220, 221, 265)*
[543] A. R. Smith, F. K. Men, K.-J. Chao, Z. Zhang, C. K. Shih, *J. Vac. Sci. Technol. B* **14**, 909 (1996). *(158, 194, 195, 196)*
[544] H.J.W. Zandvliet, H. K. Louwsma, P. E. Hegeman, B. Poelsema, *Phys. Rev. Lett.* **75**, 3890 (1995). *(158, 338)*
[545] P. J. Bedrossian, *Surf. Sci.* **301**, 223 (1994). *(162, 195)*
[546] R.J. Hamers, U. Köhler, *J. Vac. Sci. Technol. A* **7**, 2854 (1989). *(165, 167)*
[547] A. Vittadini, A. Selloni, M. Casarin, *Phys. Rev. B* **49**, 11191 (1994). *(165, 271, 272, 285, 304)*
[548] A. Kobayashi, F. Grey, E. Snyder, M. Aono, *Phys. Rev. B* **49**, 8067 (1994). *(165, 170)*
[549] J. E. Vasek, Z.-Y. Zhang, C. T. Salling, M. G. Lagally, *Phys. Rev. B* **51**, 17207

(1995). *(165, 275, 276, 304)*
[550] T. Miyazaki, T. Uda, K. Terakura, *Mater. Sci. Eng. B* **37**, 168 (1996). *(165, 166, 167)*
[551] T. Uda, K. Terakura, *Phys. Rev. B* **53**, 6999 (1996). *(165, 166, 167, 168, 169)*
[552] Z. Zhang, M.A. Kulakov, B. Bullemer, *Surf. Sci.* **369**, L131 (1996). *(165, 170, 303)*
[553] V. A. Ukraintsev, Z. Dohnalek, J.T. Yates Jr., *Surf. Sci.* **388**, 132 (1997). *(165, 170)*
[554] G. A. Baraff, E. O. Kane, M. A. Schlüter, *Phys. Rev. B* **21**, 5662 (1980). *(168, 169, 351)*
[555] O. Sugino, A. Oshiyama, *Phys. Rev. Lett.* **68**, 1858 (1992). *(168, 169)*
[556] J. Dąbrowski, M. Scheffler, *Materials Science Forum* **83**, 735 (1992). *(169, 351)*
[557] E.D. Williams, *Surf. Sci.* **299**, 502 (1994). *(171, 249, 257)*
[558] X. Tong, P. A. Bennet, *Phys. Rev. Lett.* **67**, 101 (1991). *(172, 249)*
[559] E. Pehlke, J. Tersoff, *Phys. Rev. Lett.* **67**, 465 (1991). *(172, 249, 252, 253, 254)*
[560] E. Pehlke, J. Tersoff, *Phys. Rev. Lett.* **67**, 1290 (1991). *(172)*
[561] J. J. de Miguel, C.E. Aumann, R. Kariotis, M. G. Lagally, *Phys. Rev. Lett.* **67**, 2830 (1991). *(172, 249, 258)*
[562] J. Tersoff, E. Pehlke, *Phys. Rev. Lett.* **68**, 816 (1992). *(173, 249, 251, 252)*
[563] R. M. Tromp, M. C. Reuter, *Phys. Rev. Lett.* **68**, 820 (1992). *(173, 185, 201, 204, 249, 251)*
[564] D.J. Chadi, *Phys. Rev. Lett.* **59**, 1691 (1987). *(173, 178, 188, 190)*
[565] B.S. Swartzentruber, Y.W. Mo, R. Kariotis, M. G. Lagally, M.B. Webb, *Phys. Rev. Lett.* **65**, 1913 (1990). *(176, 178, 183, 184)*
[566] F. Iwawaki, M. Tomitori, O. Nishikawa, *J. Vac. Sci. Technol. B* **9**, 711 (1991). *(177, 178, 188, 192, 193)*
[567] R.J. Hamers, E.S. Hirschorn, J.E. Demuth, *J. Vac. Sci. Technol. A* **8**, 195 (1990). *(178, 201, 202)*
[568] T. W. Poon, S. Yip, P. S. Ho, F. F. Abraham, *Phys. Rev. Lett.* **65**, 2161 (1990). *(178, 190, 249, 251, 253)*
[569] B.S. Swartzentruber, Y.W. Mo, M.B. Webb, M. G. Lagally, *J. Vac. Sci. Technol. A* **8**, 210 (1990). *(178, 255)*
[570] B.S. Swartzentruber, Y.W. Mo, M. G. Lagally, *Appl. Phys. Lett.* **58**, 822 (1991). *(178, 185, 201, 204, 257)*
[571] T. Yamaguchi, N. Fujima, *Journal of the Physical Society of Japan* **60**, 1028 (1991). *(178, 186, 187)*
[572] H.J.W. Zandvliet, H.B. Elswijk, E.J. van Loenen, I.S.T. Tsong, *Phys. Rev. B* **46**, 7581 (1992). *(178, 194, 198)*
[573] D. G. Cahill, R.J. Hamers, *Scanning Microscopy* **6**, 931 (1992). *(178, 183, 187, 227, 257)*
[574] N. Kitamura, B.S. Swartzentruber, M. G. Lagally, M.B. Webb, *Phys. Rev. B* **48**, 5704 (1993). *(178, 229, 230)*
[575] N.C. Bartelt, R. M. Tromp, E. D. Williams, *Phys. Rev. Lett.* **73**, 1656 (1994). *(178, 184, 231, 232, 250, 251, 261)*
[576] A. Oshiyama, *Phys. Rev. Lett.* **74**, 130 (1995). *(178, 188, 190, 223, 224, 227, 228, 257, 258, 259, 261)*
[577] Q.-M. Zhang, C. M. Roland, P. Boguslawski, J. Bernholc, *Phys. Rev. Lett.* **75**, 101 (1995). *(178, 181, 182, 207, 209, 210, 211, 217, 223, 224, 225, 226, 227, 232)*
[578] C. Pearson, B. Borovsky, M. Krueger, R. Curtis, E. Ganz, *Phys. Rev. Lett.* **74**, 2710 (1995). *(178, 184, 229, 230, 257)*

[579] T. Komura, M. Yoshimura, T. Yao, *J. Vac. Sci. Technol. B* **14**, 906 (1996). *(178, 181)*

[580] K. Kakitani, R. Sakagami, T. Kawamura, Y. Yagi, A. Yoshimori, *Surf. Sci.* **346**, 55 (1996). *(178, 238, 240, 248)*

[581] N. Akutsu, Y. Akutsu, *Surf. Sci.* **376**, 92 (1997). *(178, 184)*

[582] E. Pehlke, P. Kratzer, *Phys. Rev. B* **59**, 2790 (1999). *(178, 188, 265, 266, 272, 275)*

[583] A. R. Smith, F. K. Men, K.-J. Chao, C. K. Shih, *J. Vac. Sci. Technol. B* **14**, 914 (1996). *(180, 238, 248)*

[584] H. Tochihara, T. Sato, T. Sueyoshi, T. Amakusa, M . Iwatsuki, *Phys. Rev. B* **53**, 7863 (1996). *(182)*

[585] B.S. Swartzentruber, M. Schacht, *Surf. Sci.* **322**, 83 (1995). *(182, 229, 230)*

[586] H.J.W. Zandvliet, H.B. Elswijk, E.J. van Loenen, D. Dijkkamp, *Phys. Rev. B* **45**, 5965 (1992). *(183, 184)*

[587] D. E. Aspnes, J. Ihm, *Phys. Rev. Lett.* **57**, 3054 (1986). *(188)*

[588] P. E. Wierenga, J.A. Kubby, J. E. Griffith, *Phys. Rev. Lett.* **59**, 2169 (1987). *(188, 193)*

[589] J. E. Griffith, J.A. Kubby, P. E. Wierenga, R.S. Becker, J.S. Vickers, *J. Vac. Sci. Technol. A* **6**, 493 (1988). *(188, 193)*

[590] P. J. Bedrossian, T. Klitsner, *Phys. Rev. Lett.* **68**, 646 (1992). *(188, 189, 195, 207, 215, 229, 235, 257, 258)*

[591] D. J. Eaglesham, A.E. White, L.C. Feldman, N. Moriya, D.C. Jacobson, *Phys. Rev. Lett.* **72**, 2975 (1994). *(188, 190)*

[592] H.J.W. Zandvliet, S. Vandijken, B. Poelsema, *Phys. Rev. B* **53**, 15429 (1996). *(188, 189, 190, 191, 192)*

[593] H. Kroemer, in *Heteroepitaxy on Silicon*, ed. by J. C. C. Fan and J. M. Poate, MRS Symposia Proceedings Vol. 67 (MRS, Pittsburgh, 1986) p. 3. *(188)*

[594] K. Müller, E. Lang, L. Hammer, W. Grimm, P. Heilman, K. Heinz, in *Determination of Surface Structure by LEED*, ed. by P. M. Marcus and F. Jona, (Plenum, New York, 1984), p. 483. *(194, 195)*

[595] H. Niehus, U. Köhler, M. Copel, J.E. Demuth, *Journal of Microscopy* **152**, 735 (1988). *(194, 195, 336, 337)*

[596] H. Feil, H.J.W. Zandvliet, M.-H. Tsai, J. D. Dow, I.S.T. Tsong, *Phys. Rev. Lett.* **69**, 3076 (1992). *(194, 195, 196, 198)*

[597] K.E. Johnson, P.K. Wu, M. Sander, T. Engel, *Surf. Sci.* **290**, 213 (1993). *(194, 195, 199)*

[598] J. A. Martin, D.E. Savage, W. Moritz, M. G. Lagally, *Phys. Rev. Lett.* **56**, 1936 (1986). *(194, 195)*

[599] Y. Wei, L. Li, I.S.T. Tsong, *Nucl. Instr. Meth. Phys. Res. B* **115**, 572 (1996). *(194, 195)*

[600] D. Rioux, R. J. Pechman, M. Chander, J. H. Weaver, *Phys. Rev. B* **50**, 4430 (1994). *(194, 196, 278)*

[601] D. Rioux, F. Stepniak, R. J. Pechman, J. H. Weaver, *Phys. Rev. B* **51**, 10981 (1995). *(194, 198, 278, 284, 285, 286)*

[602] E. G. McRae, R. A. Malic, D. A. Kapilow, *Review of Scientific Instrument* **56**, 2077 (1985). *(195, 336)*

[603] Z.-Y. Zhang, H. Chen, B. Bolding, M. G. Lagally, *Phys. Rev. Lett.* **71**, 3677 (1993). *(195, 215, 216)*

[604] P. C. Weakliem, Z. Zhang, H. Metiu, *Surf. Sci.* **336**, 303 (1995). *(195)*

[605] B.S. Swartzentruber, C.M. Matzke, D.L. Kendall, J.E. Houston, *Surf. Sci.* **329**, 83 (1995). *(195)*

[606] Y. Wei, L. Li, I.S.T. Tsong, *Appl. Phys. Lett.* **66**, 1818 (1995). *(195, 268, 275)*
[607] H.J.W. Zandvliet, *Surf. Sci.* **377**, 1 (1997). *(195, 198, 338)*
[608] M. Chander, D. A. Goetsch, C. M. Aldao, J. H. Weaver, *Phys. Rev. B* **52**, 8288 (1995). *(198, 283)*
[609] H. Kahata, K. Yagi, *Surf. Sci.* **220**, 131 (1989). *(198)*
[610] R. N. Thomas, M. H. Francombe, *Appl. Phys. Lett.* **11**, 108 (1967). *(199)*
[611] H. Wang, R. Lin, X. Wang, *Phys. Rev. B* **36**, 7712 (1987). *(199)*
[612] K. Kato, T. Ide, T. Nishimori, T. Ichinokawa, *Surf. Sci.* **207**, 177 (1988). *(199)*
[613] S.M. Mokler, W.-K. Liu, N. Ohtani, B.A. Joyce, *Appl. Phys. Lett.* **59**, 3419 (1991). *(199)*
[614] T. Ide, T. Mizutani, *Phys. Rev. B* **45**, 1447 (1992). *(199)*
[615] R.I.G. Uhrberg, J.E. Northrup, D.K. Biegelsen, R. D. Bringans, L.-E. Swartz, *Phys. Rev. B* **46**, 10251 (1992). *(199, 200)*
[616] T. Sakamoto, T. Takahashi, E. Suzuki, A. Shoji, H. Kawanami, Y. Komiya, Y. Tarui, *Surf. Sci.* **86**, 102 (1979). *(199)*
[617] W.-K. Liu, S.M. Mokler, N. Ohtani, C. Roberts, B.A. Joyce, *Surf. Sci.* **264**, 301 (1992). *(199)*
[618] P. J. Bedrossian, E. Kaxiras, *Phys. Rev. Lett.* **70**, 2589 (1993). *(201, 203, 204)*
[619] M.J. Bronikowski, Y. Wang, R.J. Hamers, *Phys. Rev. B* **48**, 12361 (1993). *(201)*
[620] A. Rockett, *Surf. Sci.* **312**, 201 (1994). *(201, 209, 212, 229, 235)*
[621] D.R. Bowler, C.M. Goringe, *Phys. Rev.* **B58**, 3937 (1998). *(201)*
[622] T. Yamasaki, T. Uda, K. Terakura, *Phys. Rev. Lett.* **76**, 2949 (1996). *(207, 209, 210, 213, 214, 217, 218, 219, 220, 227, 232)*
[623] J.A. Venables, *Surf. Sci.* **299**, 798 (1994). *(208, 217, 257)*
[624] P. E. Blöchl, C. G. Van de Walle, S. T. Pantelides, *Phys. Rev. Lett.* **64**, 1401 (1990). *(208)*
[625] Y.W. Mo, J. Kleiner, M.B. Webb, M. G. Lagally, *Phys. Rev. Lett.* **66**, 1998 (1991). *(208, 209, 212, 215, 217, 222)*
[626] Zhenyu Zhang, Fang Wu, H.J.W. Zandvliet, B. Poelsema, H. Metiu, M.G. Lagally, *Phys. Rev. Lett.* **74**, 3644 (1995). *(208, 213, 214, 217)*
[627] T. Miyazaki, H. Hiramoto, M. Okazaki, *Jap. J. Appl. Phys.* **29**, L1165 (1990). *(209, 212)*
[628] Z.-Y. Zhang, Y.-T. Lu, H. Metiu, *Surf. Sci.* **255**, L543 (1991). *(209, 212)*
[629] G. Brocks, P. J. Kelly, R. Car, *Phys. Rev. Lett.* **66**, 1729 (1991). *(209, 210, 212)*
[630] Y.W. Mo, J. Kleiner, M.B. Webb, M. G. Lagally, *Surf. Sci.* **268**, 275 (1992). *(209, 224, 260)*
[631] R. A. Wolkow, *Phys. Rev. Lett.* **74**, 4448 (1995). *(209, 210, 213, 217, 218)*
[632] P. Boguslawski, Q.-M. Zhang, Z. Zhang, C. M. Roland, J. Bernholc, *Mater. Sci. Eng. B* **30**, 167 (1995). *(209, 211, 223, 224)*
[633] J. Wang, D. A. Drabold, A. Rockett, *Appl. Phys. Lett.* **66**, 1954 (1995). *(209, 210, 211, 223, 224, 225)*
[634] T. Doi, M. Ichikawa, S. Hosoki, K. Ninomiya, *Surf. Sci.* **343**, 24 (1995). *(212, 224, 236)*
[635] C. Pearson, M. Krueger, E. Ganz, *Phys. Rev. Lett.* **76**, 2306 (1996). *(212, 223, 224, 225, 257)*
[636] H.B. Elswijk, A.J. Hoeven, E.J. van Loenen, D. Dijkkamp, *J. Vac. Sci. Technol. B* **9**, 451 (1991). *(213, 218, 257)*
[637] P. J. Bedrossian, *Phys. Rev. Lett.* **74**, 3648 (1995). *(213, 214, 217)*
[638] G. Brocks, P. J. Kelly, *Phys. Rev. Lett.* **76**, 2362 (1996). *(213, 214, 217, 218, 220, 221)*

[639] A. van Dam, J. van Wingerden, M. J. Haye, P.M.L.O. Scholte, F. Tuinstra, *Phys. Rev. B* **54**, 1557 (1996). *(213, 217, 218, 219, 220)*

[640] P. J. Bedrossian, J.E. Houston, E. Chason, J.Y. Tsao, S.T. Picraux, *Phys. Rev. Lett.* **67**, 124 (1991). *(215, 217)*

[641] N. Kitamura, M. G. Lagally, M.B. Webb, *Phys. Rev. Lett.* **71**, 2082 (1993). *(215, 217)*

[642] Y.W. Mo, R. Kariotis, B.S. Swartzentruber, M.B. Webb, M. G. Lagally, *J. Vac. Sci. Technol. A* **8**, 201 (1990). *(217, 218, 221)*

[643] C.P. Toh, C.K. Ong, *Phys. Rev. B* **45**, 11120 (1992). *(217, 227)*

[644] G. Brocks, P. J. Kelly, R. Car, *Surf. Sci.* **269**, 860 (1992). *(217)*

[645] G. Brocks, P. J. Kelly, R. Car, *Phys. Rev. Lett.* **70**, 2786 (1993). *(217, 219)*

[646] T. Kawai, K. Watanabe, *Surf. Sci.* **357**, 830 (1996). *(217)*

[647] J. van Wingerden, M. J. Haye, P.M.L.O. Scholte, F. Tuinstra, *Surf. Sci.* **377**, 1006 (1997). *(217, 220, 221)*

[648] B.S. Swartzentruber, *Phys. Rev. B* **55**, 1322 (1997). *(217, 225)*

[649] C. Roland, G.H. Gilmer, *Phys. Rev. Lett.* **67**, 3188 (1991). *(224, 227)*

[650] M. Tsuda, S. Oikawa, S. Furukawa, *J. Cryst. Growth* **115**, 556 (1991). *(224)*

[651] C. Roland, G.H. Gilmer, *Phys. Rev. B* **46**, 13437 (1992). *(224, 227, 228, 257)*

[652] D. Srivastava, B.J. Garrison, *Phys. Rev. B* **47**, 4464 (1993). *(224)*

[653] S. Stoyanov, *J. Cryst. Growth* **94**, 751 (1989). *(224)*

[654] M. Ichikawa, T. Doi, *Appl. Phys. Lett.* **60**, 1082 (1992). *(224, 233, 236, 237)*

[655] C.P. Toh, S.H. Seow, C.K. Ong, *Surf. Sci.* **292**, 114 (1993). *(227, 229)*

[656] J. P. van der Eerden, G. Z. Liu, F. de Jong, M. J. Anders, *J. Cryst. Growth* **99**, 106 (1990). *(227)*

[657] H.J.W. Zandvliet, H.B. Elswijk, E.J. Loenen van, *Surf. Sci.* **272**, 264 (1992). *(228, 229, 230, 235)*

[658] H.J.W. Zandvliet, B. Poelsema, H.B. Elswijk, *Phys. Rev. B* **51**, 5465 (1995). *(228, 230, 231, 232)*

[659] H. Tokumoto, M . Iwatsuki, *Jap. J. Appl. Phys.* **32**, 1368 (1993). *(229, 230, 236)*

[660] A. Pimpinelli, J. Villain, D. E. Wolf, J. J. Metois, J.C. Heyraud, L. Elkinani, G. Uimin, *Surf. Sci.* **295**, 143 (1993). *(229, 231)*

[661] C. Ebner, J. V. Seiple, J. P. Pelz, *Phys. Rev. B* **52**, 16651 (1995). *(229, 290)*

[662] J. D. Erlebacher, M. J. Aziz, *Surf. Sci.* **374**, 427 (1997). *(229)*

[663] A.J. Hoeven, J.M. Lenssinck, D. Dijkkamp, E.J. van Loenen, J. Dieleman, *Phys. Rev. Lett.* **63**, 1830 (1989). *(230, 257, 258)*

[664] D. B. Abraham, P. J. Upton, *Phys. Rev. B* **39**, 736 (1989). *(231)*

[665] L. Kuipers, M. S. Hoogeman, J.W.M. Frenken, *Phys. Rev. Lett.* **71**, 3517 (1993). *(232)*

[666] S. Ma, G. F. Mazenko, *Phys. Rev. B* **11**, 4077 (1977). *(232)*

[667] A.V. Latyshev, A.B. Krasilnikov, A.L. Aseev, S.I. Stenin, *JETP Letters* **48**, 529 (1988). *(236)*

[668] S. Stoyanov, *Jap. J. Appl. Phys.* **30**, 1 (1991). *(236)*

[669] A. Natori, H. Fujimura, M. Fukuda, *Appl. Surf. Sci.* **60**, 85 (1992). *(236)*

[670] T. Yamaguchi, *Journal of the Physical Society of Japan* **61**, 3658 (1992). *(236, 237)*

[671] D. Kandel, E. Kaxiras, *Phys. Rev. Lett.* **76**, 1114 (1996). *(236, 237)*

[672] Y. Imry, S. Ma, *Phys. Rev. Lett.* **35**, 1399 (1975). *(238)*

[673] A. Ihm, D. H. Lee, J.D. Joannopoulos, J. J. Xiong, *Phys. Rev. Lett.* **51**, 1872 (1983). *(238, 240)*

[674] A. Saxena, E. T. Gawlinski, J. D. Gunton, *Surf. Sci.* **160**, 618 (1985). *(238)*

[675] D. S. Fisher, G. M. Grinstein, A. Khurana, *Physics Today* **41**, 56 (1988). *(238, 246)*

[676] V. E. Zubkus, E. E. Tornau, *Surf. Sci.* **216**, 23 (1989). *(238, 240)*

[677] J. R. Duchter, S. Lee, B. Hillebrands, G. J. McLaughlin, B. G. Nickel, G. I. Stegeman, *Phys. Rev. Lett.* **68**, 2464 (1992). *(238)*

[678] J. J. Métois, D. E. Wolf, *Surf. Sci.* **298**, 71 (1993). *(238, 249, 251)*

[679] Y. Murata, M. Kubota, *Phase Transitions* **53**, 125 (1995). *(238)*

[680] K. Terakura, T. Yamasaki, Y. Morikawa, *Phase Transitions* **53**, 143 (1995). *(238, 244, 245, 246)*

[681] J. Fritsch, U. Schröder, *Physics Reports* **309**, 209 (1999). *(238)*

[682] H. J. Leamy, G.H. Gilmer, K. A. Jackson, in *Surface Physics of Materials*, (Academic Press, New York, 1975) p. 121. *(249)*

[683] V. I. Marchenko, A. Ya. Parshin, *Sov. Phys. JETP* **52**, 129 (1980). *(249, 254)*

[684] O. L. Alerhand, A. Nihat Berker, J.D. Joannopoulos, D. Vanderbilt, R.J. Hamers, J.E. Demuth, *Phys. Rev. Lett.* **64**, 2406 (1990). *(249, 251)*

[685] C.C. Umbach, M.E. Keeffe, J.M. Blakely, *J. Vac. Sci. Technol. B* **9**, 721 (1991). *(249, 252)*

[686] J. J. de Miguel, C.E. Aumann, S.G. Jaloviar, R. Kariotis, M. G. Lagally, *Phys. Rev. B* **46**, 10257 (1992). *(249, 252)*

[687] J. Wasserfall, W. Ranke, *Surf. Sci.* **315**, 227 (1994). *(249, 252)*

[688] J. Wasserfall, W. Ranke, *Surf. Sci.* **331**, 1099 (1995). *(249)*

[689] R. M. Tromp, M. C. Reuter, *Phys. Rev. B* **47**, 7598 (1993). *(251)*

[690] O. L. Alerhand, A. Nihat Berker, J.D. Joannopoulos, D. Vanderbilt, R.J. Hamers, J.E. Demuth, *Phys. Rev. Lett.* **66**, 962 (1991). *(252)*

[691] N.C. Bartelt, T.L. Einstein, C. Rottman, *Phys. Rev. Lett.* **66**, 961 (1991). *(252)*

[692] N.-E. Lee, D. G. Cahill, J.E. Greene, *Phys. Rev. B* **53**, 7876 (1996). *(256, 257)*

[693] A.J. Hoeven, D. Dijkkamp, E.J. Loenen van, J.M. Lenssinck, J. Dieleman, *J. Vac. Sci. Technol. A* **8**, 207 (1990). *(256)*

[694] W. Ranke, Y.R. Xing, *Phys. Rev. B* **31**, 2246 (1985). *(257, 286, 288)*

[695] S. Clarke, M. Wilby, D. D. Vvedensky, T. Kawamura, T. Sakamoto, *Appl. Phys. Lett.* **54**, 2417 (1989). *(257, 258)*

[696] M. Mundschau, E. Bauer, W. Telieps, W. Święch, *Surf. Sci.* **223**, 413 (1989). *(257, 261)*

[697] A. J. Pidduck, D.J. Robbins, D. J. Young, G. Patel, *Thin Solid Films* **183**, 255 (1989). *(257, 259)*

[698] G. S. Bales, A. Zangwill, *Phys. Rev. B* **41**, 5500 (1990). *(257, 259, 261)*

[699] H. Hirayama, M. Hiroi, T. Ide, *Phys. Rev. B* **48**, 17331 (1993). *(257, 261)*

[700] N.C. Bartelt, T.L. Einstein, E. D. Williams, *Surf. Sci.* **312**, 411 (1994). *(257)*

[701] S. Harris, *Surf. Sci.* **311**, L712 (1994). *(257)*

[702] A. Pimpinelli, J.-J. Métois, *Phys. Rev. Lett.* **72**, 3566 (1994). *(257, 261)*

[703] D. Kandel, J. D. Weeks, *Phys. Rev. Lett.* **74**, 3632 (1995). *(257)*

[704] N.C. Bartelt, W. Theis, R. M. Tromp, *Phys. Rev. B* **54**, 11741 (1996). *(257)*

[705] J. van Wingerden, E.C. van Halen, K. Werner, P.M.L.O. Scholte, F. Tuinstra, *Surf. Sci.* **352**, 641 (1996). *(257, 259)*

[706] W. Theis, R. M. Tromp, *Phys. Rev. Lett.* **76**, 2770 (1996). *(260)*

[707] T. Morimoto, T. Ohguro, H. S. Momose, T. Iinuma, I. Kunishima, K. Suguro, I. Katakabe, H. Nakajima, M. Tsuchiaki, M. Ono, Y. Katsumata, H. Iwai, *IEEE Transactions on Electron Devices* **42**, 915 (1995). *(262, 325, 332)*

[708] V.G. Lifshits, A.A. Saranin, A.V. Zotov, *Surface Phases on Silicon* (Willey, Chichester, 1994). *(262, 265, 278, 286, 309, 336)*

[709] S. C. A. Gay, S.J. Jenkins, G. P. Srivastava, *J. Phys.: Cond. Mat.* **10**, 7751

(1998). *(262)*
[710] J. A. Kittl, W.-T. Shiau, D. Miles, K. E. Violette, J. C. Hu, Q.-Z. Hong, *Solid State Technology* **42**, 81 (June 1999). *(263, 325, 329)*
[711] J. N. Smith Jr., W. L. Fite, *J. Chem. Phys.* **37**, 898 (1962). *(264)*
[712] A. Vittadini, A. Selloni, M. Casarin, *Phys. Rev. B* **52**, 5885 (1995). *(264, 278, 297, 302, 304)*
[713] T. Sakurai, H. D. Hagstrum, *Phys. Rev. B* **14**, 1593 (1976). *(265)*
[714] J. A. Appelbaum, G. A. Baraff, D. R. Hamann, H. D. Hagstrum, T. Sakurai, *Surf. Sci.* **70**, 654 (1978). *(265)*
[715] K. Sinnah, M. G. Sherman, L. B. Lewis, W.H. Weinberg, J.T. Yates Jr., K. C. Janda, *Phys. Rev. Lett.* **62**, 567 (1989). *(265, 269, 271, 275)*
[716] H. Tokumoto, H. Miki, H. Bando, M. Ono, K. Kajumura, *J. Vac. Sci. Technol. A* **8**, 255 (1990). *(265, 286)*
[717] J. J. Boland, *Phys. Rev. Lett.* **65**, 3325 (1990). *(265, 267)*
[718] C.C. Cheng, J.T. Yates Jr., *Phys. Rev. B* **43**, 4041 (1991). *(265)*
[719] K. Fujita, N. Hirashita, *Jap. J. Appl. Phys.* **33**, 399 (1994). *(265)*
[720] K. W. Kołasiński, *International Journal of Modern Physics B* **9**, 2753 (1995). F *(265, 274)*
[721] K. W. Kołasiński, W. Nessler, K.-H. Bornscheuer, E. Hasselbrink, *Surf. Sci.* **331**, 485 (1995). *(265, 266)*
[722] Y. J. Chabal, M. A. Hines, D. Feijoo, *J. Vac. Sci. Technol. A* **13**, 1719 (1995). *(265)*
[723] P. Kratzer, B. Hammer, J. K. Nørskov, *Phys. Rev. B* **51**, 13432 (1995). *(265, 269, 270, 274)*
[724] E. Pehlke, M. Scheffler, *Phys. Rev. Lett.* **74**, 952 (1995). *(265, 270, 274)*
[725] A. Hoffman, M. Petravic, *Phys. Rev. B* **53**, 6996 (1996). *(265)*
[726] D. P. Adams, T. M. Mayer, B.S. Swartzentruber, *J. Vac. Sci. Technol. B* **14**, 1642 (1996). *(265)*
[727] S. Watanabe, Y. Sugita, *Surf. Sci.* **357**, 536 (1996). *(265)*
[728] A. Pasquarello, M. S. Hybertsen, R. Car, *J. Vac. Sci. Technol. B* **14**, 2809 (1996). *(265, 297)*
[729] P. Bratu, W. Brenig, A. Groß, M. Hartmann, U. Höfer, P. Kratzer, R. Russ, *Phys. Rev. B* **54**, 5978 (1996). *(265)*
[730] M. B. Raschke, U. Höfer, *Phys. Rev. B* **59**, 2783 (1999). *(265, 266)*
[731] A. Groß, M. Bockstedte, M. Scheffler, *Phys. Rev. Lett.* **79**, 701 (1997). *(265, 270, 272)*
[732] K. W. Kołasiński, W. Nessler, K.-H. Bornscheuer, E. Hasselbrink, *J. Chem. Phys.* **101**, 7082 (1994). *(266, 270, 273)*
[733] D. A. Hansen, M. R. Halbach, E.G. Seebauer, *J. Chem. Phys.* **104**, 7338 (1996). *(266)*
[734] J.H.G. Owen, D.R. Bowler, C.M. Goringe, K. Miki, G.A.D. Briggs, *Phys. Rev. B* **54**, 14153 (1996). *(266)*
[735] A. Vittadini, A. Selloni, M. Casarin, *Surf. Sci.* **289**, L625 (1993). *(266)*
[736] K. Oura, J. Yamane, K. Umezawa, M. Naitoh, F. Shoji, T. Hanawa, *Phys. Rev. B* **41**, 1200 (1990). *(267)*
[737] S. M. Gates, R. R. Kunz, C. M. Greenlief, *Surf. Sci.* **207**, 364 (1989). *(267, 268)*
[738] C. G. Van de Walle, P. J. H. Denteneer, Y. Bar-Yam, S. T. Pantelides, *Phys. Rev. B* **39**, 10791 (1989). *(267, 298)*
[739] J. J. Boland, *Surf. Sci.* **261**, 17 (1992). *(267)*
[740] D. R. Olander, M. Balooch, J. Abrefah, W. J. Siekhaus, *J. Vac. Sci. Technol. B5*, 1404 (1987). *(267)*

[741] S.K. Jo, B. Gong, G. Hess, J.M. White, J.G. Ekerdt, *Surf. Sci.* **394**, L162 (1997). *(267, 268)*

[742] A. Vittadini, A. Selloni, R. Car, M. Casarin, *Phys. Rev. B* **46**, 4348 (1992). *(267, 268)*

[743] Copel, R. M. Tromp, *Phys. Rev. Lett.* **72**, 1236 (1994). *(267, 275, 276)*

[744] J.E. Northrup, *Phys. Rev. B* **44**, 1419 (1991). *(268)*

[745] D. Gräf, M. Grundner, R. Schulz, *J. Vac. Sci. Technol. A* **7**, 808 (1989). *(268, 281, 306, 307)*

[746] G.J. Pietsch, U. Köhler, M. Henzler, *J. Appl. Phys.* **73**, 4797 (1993). *(268, 281)*

[747] S. Watanabe, N. Nakayama, T. Ito, *Appl. Phys. Lett.* **59**, 1458 (1991). *(268)*

[748] Y. J. Chabal, G.S. Higashi, K. Raghavachari, V. A. Burrows, *J. Vac. Sci. Technol. A* **7**, 2104 (1989). *(268)*

[749] M. Niwano, J.-i. Kageyama, K. Kurita, K. Kinashi, I. Takahashi, N. Miyamoto, *J. Appl. Phys.* **76**, 2157 (1994). *(269, 280, 297)*

[750] E. Kondoh, M.R. Baklanov, F. Jonckx, K. Maex, *Materials Science in Semiconductor Processing* **1**, 107 (1998). *(269, 279, 280)*

[751] Y. Nakagawa, A. Ishitani, T. Takahagi, H. Kuroda, H. Tukumoto, M. .Ono, K. Kajimura, *J. Vac. Sci. Technol. A* **8**, 262 (1990). *(269)*

[752] M.L. Wise, B. G. Koehler, P. Gupta, P. A. Coon, S.M. George, *Surf. Sci.* **258**, 166 (1991). *(269, 271, 272)*

[753] U. Höfer, L. Li, T. F. Heinz, *Phys. Rev. B* **45**, 9485 (1992). *(269, 270, 271, 272)*

[754] J.W. Sharp, G. Eres, *Surf. Sci.* **320**, 169 (1994). *(269)*

[755] K. W. Kołasiński, W. Nessler, A. de Meijere, E. Hasselbrink, *Phys. Rev. Lett.* **72**, 1356 (1994). *(269, 270)*

[756] W. Brenig, A. Groß, R. Russ, *Z. Phys. B* **96**, 231 (1994). *(270)*

[757] J. J. Boland, *Phys. Rev. Lett.* **67**, 1539 (1991). *(270, 271, 272)*

[758] M. R. Radeke, E. A. Carter, *Phys. Rev. B* **54**, 11803 (1996). *(270, 273, 274)*

[759] M.P. D'Evelyn, Y. L. Lang, L. F. Sutcu, *J. Chem. Phys.* **96**, 852 (1992). *(271)*

[760] C. J. Wu, E. A. Carter, *Chem. Phys. Lett.* **185**, 172 (1991). *(271)*

[761] P. Nachtigall, K. D. Jordan, K. C. Janda, *J. Chem. Phys.* **95**, 8652 (1991). *(271)*

[762] P. Nachtigall, K. D. Jordan, C. Sosa, *Journal of Physical Chemistry* **97**, 11666 (1993). *(271)*

[763] J.E. Northrup, *Phys. Rev. B* **51**, 2218 (1995). *(271)*

[764] Z. Jing, G. Lucovsky, J. L. Whitten, *Surf. Sci.* **296**, L33 (1993). *(273)*

[765] P. Nachtigall, K. D. Jordan, C. Sosa, *J. Chem. Phys.* **101**, 8073 (1994). *(273)*

[766] M. R. Radeke, E. A. Carter, *Phys. Rev. B* **55**, 4649 (1997). *(273)*

[767] W. Widdra, S. I. Yi, R. Maboudian, G.A.D. Briggs, W.H. Weinberg, *Phys. Rev. Lett.* **74**, 2074 (1995). *(275)*

[768] S.A. Buntin, *J. Chem. Phys.* **108**, 1601 (1998). *(275)*

[769] P. Kratzer, E. Pehlke, M. Scheffler, M. B. Raschke, U. Höfer, *Phys. Rev. Lett.* **81**, 5596 (1998). *(275)*

[770] P. Kratzer, *Chem. Phys. Lett.* **288**, 396 (1998). *(275)*

[771] P. Kratzer, *J. Chem. Phys.* **106**, 6752 (1997). *(275)*

[772] M. Tao, L. P. Hunt, *J. Electrochem. Soc.* **144**, 2221 (1997). *(275, 276, 287)*

[773] B.S. Meyerson, F.J. Himpsel, K. J. Uram, *Appl. Phys. Lett.* **57**, 1034 (1990). *(275)*

[774] J. Nara, T. Sasaki, T. Ohno, *Phys. Rev. Lett.* **79**, 4421 (1997). *(276, 305)*

[775] S. Jeong, A. Oshiyama, *Phys. Rev. Lett.* **79**, 4425 (1997). *(276)*

[776] J. Spitzmüller, M. Fehrenbacher, H. Rauscher, R. J. Behm, *Surf. Sci.* **377**, 1001 (1997). *(276)*

[777] J. Spitzmüller, M. Fehrenbacher, H. Rauscher, R. J. Behm, *Phys. Rev. B* **55**, 4659 (1997). *(276)*
[778] J.H.G. Owen, D.R. Bowler, C.M. Goringe, K. Miki, G.A.D. Briggs, *Surf. Sci.* **382**, L687 (1997). *(276)*
[779] S.M. Gates, S. K. Kulkarni, *Appl. Phys. Lett.* **58**, 2963 (1991). *(276)*
[780] M.J. Bronikowski, Y. Wang, M. T. McEllistrem, D. Chen, R.J. Hamers, *Surf. Sci.* **298**, 50 (1993). *(276)*
[781] Y. Tezuka, N. Kitano, N. Nakano, "Requested!" *J. Electrochem. Soc.* **142**, 3569 (1995). *(277)*
[782] M.R. Baklanov, S. Vanhaelemeersch, W. Storm, Y. B. Kim, W. Vandervorst, K. Maex, *J. Vac. Sci. Technol. A* **15**, 3005 (1997). *(277)*
[783] C. T. Gabriel, *J. Vac. Sci. Technol. B* **9**, 370 (1991). *(277)*
[784] W. M. Greene, J. B. Kruger, G. Kooi, *J. Vac. Sci. Technol. B* **9**, 366 (1991). *(277)*
[785] M. Chander, D. Rioux, J. H. Weaver, *Phys. Rev. Lett.* **71**, 4154 (1993). *(277, 278, 284)*
[786] Y.-H. Lee, M.-M. Chen, *J. Vac. Sci. Technol. B* **4**, 468 (1986). *(278)*
[787] R.B. Jackman, H. Ebert, J.S. Foord, *Surf. Sci.* **176**, 183 (1986). *(278)*
[788] C. G. Van de Walle, F.R. McFeely, S. T. Pantelides, *Phys. Rev. Lett.* **61**, 1867 (1988). *(278)*
[789] T. M. Mayer, M. S. Ameen, D. J. Vitkavage, in *The Chemical Physics of Solid Surfaces and Heterogeneous Catalysis*, Vol. 5 (*Surface properties of electronic materials*) ed. by D. A. King, D. P. Woodruff, (Elsevier Science, 1988), p. 427 *(278)*
[790] J.R. Engstrom, M. M. Nelson, T. Engel, *Surf. Sci.* **215**, 437 (1989). *(278, 279)*
[791] R.B. Jackman, R. J. Price, J.S. Foord, *Appl. Surf. Sci.* **36**, 296 (1989). *(278, 283, 284)*
[792] M. Hirose, T. Yasaka, M. Takakura, S. Mizayaki, *Solid State Technology* **34**, 43 (December 1991). *(278, 280, 281)*
[793] M. Chander, Y.Z. Li, J.C. Patrin, J. H. Weaver, *Phys. Rev. B* **47**, 13035 (1993). *(278)*
[794] J. J. Boland, *Science* **262**, 1703 (1993). *(278)*
[795] D. Rioux, M. Chander, Y.Z. Li, J. H. Weaver, *Phys. Rev. B* **49**, 11071 (1994). *(278, 284, 285)*
[796] A. Szabo, T. Engel, *J. Vac. Sci. Technol. A* **12**, 648 (1994). *(278, 282, 283)*
[797] M. Chander, D. A. Goetsch, C. M. Aldao, J. H. Weaver, *Phys. Rev. Lett.* **74**, 2014 (1995). *(278, 283)*
[798] J. R. Sanchez, C. M. Aldao, J. H. Weaver, *J. Vac. Sci. Technol. B* **13**, 2230 (1995). *(278)*
[799] W. C. Simpson, J. A. Yarmoff, *Surf. Sci.* **359**, 135 (1996). *(278, 279)*
[800] J. Lohmüller, H.H. Bertschat, H. Granzer, H. Haas, G. C. Schatz, W.-D. Zeitz, ISOLDE Collaboration, *Surf. Sci.* **360**, 213 (1996). *(278)*
[801] T. Kanashima, Y. Kurioka, T. Imai, H. Yamamoto, M. Okuyama, *Jap. J. Appl. Phys.* **36**, 2460 (1997). *(278, 413)*
[802] I.-M. Lee, A. Jansons, C. G. Takoudis, *J. Vac. Sci. Technol. B* **15**, 880 (1997). *(278, 282, 296)*
[803] M. Hierlemann, C. Werner, A. Spitzer, *J. Vac. Sci. Technol. B* **15**, 935 (1997). *(278, 282)*
[804] G. A. de Wijs, A. De Vita, A. Selloni, *Phys. Rev. Lett.* **78**, 4877 (1997). *(278, 282)*
[805] G. S. Mathad, D. W. Hess, M. Meyyappan, *Interface* **8**, 34 (Summer 1999).

(278)
[806] G. S. Khoo, C.K. Ong, *Phys. Rev. B* **52**, 2574 (1995). *(278)*
[807] P.-L. Cao, R.-H. Zhou, *J. Phys.: Cond. Mat.* **5**, 2897 (1993). *(278)*
[808] M. Seel, M. Bagus, *Phys. Rev. B* **28**, 2023 (1983). *(279)*
[809] L. E. Carter, E. A. Carter, *Journal of Physical Chemistry* **100**, 873 (1996). *(279)*
[810] F.A. Houle, *Phys. Rev. Lett.* **61**, 1871 (1988). *(279)*
[811] G.S. Higashi, R.S. Becker, Y. J. Chabal, A. J. Becker, *Appl. Phys. Lett.* **58**, 1656 (1991). *(280)*
[812] G.J. Pietsch, U. Köhler, O. Jusko, M. Henzler, P. O. Hahn, *Appl. Phys. Lett.* **60**, 1321 (1992). *(280)*
[813] S. Kasi, M. Liehr, S. Cohen, *Appl. Phys. Lett.* **58**, 2975 (1991). *(280)*
[814] H. Ubara, T. Imura, A. Hiraki, *Solid State Commun.* **50**, 673 (1984). *(280)*
[815] J.-N. Chazalviel, F. Ozanam, *J. Appl. Phys.* **81**, 7684 (1997). *(281)*
[816] B. Ren, F. M. Liu, J. Xie, B. W. Mao, Y. B. Zu, Z. Q. Tian, *Appl. Phys. Lett.* **72**, 933 (1998). *(281)*
[817] G. W. Trucks, K. Raghavachari, G.S. Higashi, Y. L. Chabal, *Phys. Rev. Lett.* **65**, 504 (1990). *(281)*
[818] G.S. Higashi, Y. L. Chabal, G. W. Trucks, K. Raghavachari, *Appl. Phys. Lett.* **56**, 656 (1990). *(281)*
[819] Y. Nishioka, Y. Ohji, K. Mukai, T. Sugano, Y. Wang, T. P. Ma, *Appl. Phys. Lett.* **54**, 1127 (1989). *(281, 364, 417)*
[820] A. C. Dillon, M. L. Wilse, M. B. Robinson, S.M. George, *J. Vac. Sci. Technol. A* **13**, 1 (1995). *(282)*
[821] A. S. Grove, *Physics and Technology of Semiconductor Devices* (Wiley, New York, 1967). *(282, 345, 348, 358, 430, 444)*
[822] P. Krüger, J. Pollmann, *Phys. Rev. B* **47**, 1898 (1993). *(282)*
[823] A. Poredda, K. Jug, *Surf. Sci.* **313**, 168 (1994). *(282)*
[824] K. Nakayama, C. M. Aldao, J. H. Weaver, *Phys. Rev. Lett.* **82**, 568 (1998). *(282, 283, 284, 285)*
[825] W. Yang, Z. Dohnálek, W.J. Choyke, J.T. Yates Jr., *Surf. Sci.* **392**, 8 (1997). *(283)*
[826] Z. Dohnálek, W. Yang, V. A. Ukraintsev, W.J. Choyke, J.T. Yates Jr., *Surf. Sci.* **392**, 17 (1997). *(283)*
[827] J. Schaefer, W. Göpel, *Surf. Sci.* **155**, 535 (1985). *(286)*
[828] M.P. D'Evelyn, M. M. Nelson, T. Engel, *Surf. Sci.* **186**, 75 (1987). *(286, 289, 290)*
[829] F.J. Himpsel, F.R. McFeely, A. Taleb-Ibrahimi, J. A. Yarmoff, G. Hollinger, *Phys. Rev. B* **38**, 6084 (1988). *(286, 398, 399, 406)*
[830] Y. Miyamoto, A. Oshiyama, *Phys. Rev. B* **41**, 12680 (1990). *(286, 289, 291)*
[831] J.R. Engstrom, D.J. Bonser, M. M. Nelson, T. Engel, *Surf. Sci.* **256**, 317 (1991). *(286, 290, 345)*
[832] R. Kliese, B. Röttger, D. Badt, H. Neddermeyer, *Ultramicroscopy* **42**, 824 (1992). *(286, 291, 296)*
[833] Y. Miyamoto, *Phys. Rev. B* **46**, 12473 (1992). *(286, 292)*
[834] Y.-K. Sun, D.J. Bonser, T. Engel, *J. Vac. Sci. Technol. A* **10**, 2314 (1992). *(286, 290, 295, 296, 304, 358, 372)*
[835] E. H. Poindexter, *Semiconductor Interfaces, Microstructures and Devices: Properties and Applications* ed. by Z. C. Feng (Institute of Physics Publishing, London 1993) p. 229. *(286, 351, 392, 407)*
[836] T. Engel, *Surf. Sci. Rep.* **18**, 91 (1993). *(286, 289, 294, 372, 398, 399)*

[837] J. V. Seiple, J. P. Pelz, *Phys. Rev. Lett.* **73**, 999 (1994). *(286, 294)*
[838] M. Udagawa, M. Niwa, I. Sumita, *Jap. J. Appl. Phys.* **33**, 375 (1994). *(286, 293, 393)*
[839] V. D. Borman, E. P. Gusev, Yu. Yu. Lebedinski, V. I. Troyan, *Phys. Rev. B* **49**, 5415 (1994). *(286, 294)*
[840] J. V. Seiple, J. P. Pelz, *J. Vac. Sci. Technol. A* **13**, 772 (1995). *(286, 294, 296)*
[841] A. Munkholm, S. Brennan, F. Comin, L. Ortega, *Phys. Rev. Lett.* **75**, 4254 (1995). *(286, 392)*
[842] M. Niwa, K. Okada, R. Sinclair, *Appl. Surf. Sci.* **100**, 425 (1996). *(286, 353, 358, 392)*
[843] T. Uchiyama, M. Tsukada, *Phys. Rev. B* **53**, 7917 (1996). *(286, 289, 291)*
[844] M. Niwa, T. Kouzaki, K. Okada, M. Udagawa, R. Sinclair, *Jap. J. Appl. Phys.* **33**, 388 (1994). *(287, 357, 358, 392)*
[845] D. G. Cahill, Ph. Avouris, *Appl. Phys. Lett.* **60**, 326 (1992). *(287, 291)*
[846] A.J. Schell-Sorokin, J.E. Demuth, *Surf. Sci.* **157**, 273 (1985). *(289)*
[847] A. Oshiyama, Y. Miyamoto, *Proc. 20th Int. Conf. Phys. Semicond.*, Thessaloniki 1990, ed. by E. M. Anastassakis and J. M. Joannopoulos (World Scientific, Singapore 1990) p. 87. *(291)*
[848] H. Kageshima, K. Shiraishi, *Phys. Rev. Lett.* **81**, 5936 (1998). *(291, 410)*
[849] P.V. Smith, A. Wander, *Surf. Sci.* **219**, 77 (1989). *(291)*
[850] T. L. Cottrell, *The Strength of Chemical Bonds* (Academic Press, New York, 1958), p. 270. *(292)*
[851] A. Mascaraque, C. Ottaviani, M. Capozi, M. Pedio, E.G. Michel, *Surf. Sci.* **377**, 650 (1997). *(292)*
[852] F. W. Smith, G. Ghidini, *J. Electrochem. Soc.* **129**, 1300 (1982). *(292, 294)*
[853] T. Yasuda, L. Mantese, U. Rossow, D. E. Aspnes, *Phys. Rev. Lett.* **74**, 3431 (1995). *(293)*
[854] F. Rochet, M. Froment, C. D'Anterroches, H. Roulet, G. Dufour, *Philosophical Magazine B* **59**, 339 (1989). *(293)*
[855] H. Watanabe, M. Ichikawa, *Surf. Sci.* **408**, 95 (1998). *(295)*
[856] K. Wurm, R. Kliese, Y. Hong, B. Röttger, Y. Wei, H. Neddermeyer, I.S.T. Tsong, *Phys. Rev. B* **50**, 1567 (1994). *(295, 296)*
[857] M. Offenberg, M. Liehr, G. W. Rubloff, *J. Vac. Sci. Technol. A* **9**, 1058 (1991). *(295)*
[858] P. A. Thiel, T. E. Madey, *Surf. Sci. Rep.* **7**, 211 (1987). *(296, 297, 298, 299)*
[859] W. Ranke, D. Schmeisser, *Surf. Sci.* **149**, 485 (1985). *(296)*
[860] F. Meyer, *Surf. Sci.* **27**, 107 (1971). *(297)*
[861] K. Fujiwara, M. Nishijima, *Physics Letters A* **55**, 211 (1975). *(297)*
[862] H. Ibach, H. Wagner, D. Bruchmann, *Solid State Commun.* **42**, 457 (1982). *(297)*
[863] S. Ciraci, H. Wagner, *Phys. Rev. B* **27**, 5180 (1983). *(297, 301)*
[864] D. Schmeisser, F.J. Himpsel, G. Hollinger, *Phys. Rev. B* **27**, 7813 (1983). *(297)*
[865] Y. J. Chabal, S. B. Christman, *Surf. Sci.* **29**, 6974 (1984). *(297)*
[866] W. Ranke, Y.R. Xing, *Surf. Sci.* **157**, 339 (1985). *(297)*
[867] M. Morita, T. Ohmi, E. Hasegawa, M. Kawakami, K. Suma, *Appl. Phys. Lett.* **55**, 562 (1989). *(297, 360, 392, 395)*
[868] Ph. Avouris, I.-W. Lyo, *Surf. Sci.* **242**, 1 (1991). *(297)*
[869] C. Poncey, F. Rochet, G. Dufour, H. Roulet, F. Sirotti, G. Panaccione, *Surf. Sci.* **338**, 143 (1995). *(297)*
[870] R. Konečný, D. J. Doren, *J. Chem. Phys.* **106**, 2426 (1997). *(297, 301, 302, 303)*

[871] N. Franco, J. Chrost, J. Avila, M.C. Asensio, C. Müller, E. Dudzik, A. J. Patchett, I.T. McGovern, T. Giebel, R. Lindsay, V. Fritzsche, A.M. Bradshaw, D. P. Woodruff, *Appl. Surf. Sci.* **123**, 219 (1998). *(297)*
[872] M. Niwano, M. Terashi, M. Shinohara, D. Shoji, N. Miyamoto, *Surf. Sci.* **401**, 364 (1998). *(297, 304, 305, 306)*
[873] P. W. Atkins, *Physical Chemistry*, fifth edition (Oxford, New York, 1994). *(299, 387, 389)*
[874] J. A. Odutola, T. R. Dyke, *J. Chem. Phys.* **72**, 5062 (1980). *(300)*
[875] F. H. Stillinger, *Science* **209**, 451 (1980). *(300)*
[876] Q. Y. Tong, T.-H. Lee, U. M. Gösele, M. Reiche, J. Ramm, E. Beck, *J. Electrochem. Soc.* **144**, 384 (1997). *(300)*
[877] E. Kaxiras, J.D. Joannopoulos, *Phys. Rev. B* **37**, 8842 (1988). *(300)*
[878] R. J. Baierle, M. J. Caldas, private communication; work submitted to *Phys. Rev. B* (1999). *(300)*
[879] D. Marx, M. E. Tuckerman, J. Hutter, M. Parrinello, *Nature* **397**, 601 (1999). *(300)*
[880] W. Ranke, *Surf. Sci.* **369**, 137 (1996). *(301)*
[881] Y. J. Chabal, *Phys. Rev. B* **29**, 3677 (1984). *(305)*
[882] M. M. Moslehi, K. C. Saraswat, *IEEE Transactions on Electron Devices* **32**, 106 (1985). *(308, 309, 310, 318)*
[883] P. Fahey, G. Barbuscia, M. M. Moslehi, R. W. Dutton, *Appl. Phys. Lett.* **46**, 784 (1985). *(308, 309)*
[884] A. G. Schrott, Q. X. Su, S. C. Fain Jr., *Surf. Sci.* **123**, 223 (1982). *(309, 317, 319)*
[885] T. Sugi, T. Ito, H. Ishikawa, *Appl. Phys. Lett.* **45**, 966 (1984). *(309, 318)*
[886] M. Nishijima, H. Kobayashi, K. Edamoto, M. Onchi, *Surf. Sci.* **137**, 473 (1984). *(309, 413)*
[887] A. Galachant, D. Saidi, *J. Vac. Sci. Technol. B* **3**, 985 (1985). *(309, 318)*
[888] K. Edamoto, S. Tanaka, M. Onchi, M. Nishijima, *Surf. Sci.* **167**, 285 (1986). *(309)*
[889] Ph. Avouris, F. Bozso, R.J. Hamers, *J. Vac. Sci. Technol. B* **5**, 1387 (1987). *(309, 314, 317, 318, 369)*
[890] S. Tanaka, M. Onchi, M. Nishijima, *Surf. Sci.* **191**, L576 (1987). *(309)*
[891] E. K. Hlil, L. Kubler, J.L. Bischoff, D. Bolmont, *Phys. Rev. B* **35**, 5913 (1987). *(309, 314, 318)*
[892] L. Kubler, J.L. Bischoff, D. Bolmont, *Phys. Rev. B* **38**, 13113 (1988). *(309, 314, 318, 319)*
[893] M. J. Dresser, P. A. Taylor, R. M. Wallace, W.J. Choyke, J.T. Yates Jr., *Surf. Sci.* **218**, 75 (1989). *(309)*
[894] Ph. Avouris, R. Wolkow, *Phys. Rev. B* **39**, 5091 (1989). *(309)*
[895] J.L. Bischoff, L. Kubler, D. Bolmont, *Surf. Sci.* **209**, 115 (1989). *(309)*
[896] Y. Taguchi, M. Fujisawa, Y. Kuwahara, M. Onchi, M. Nishijima, *Surf. Sci.* **217**, L413 (1989). *(309, 317)*
[897] G. Rangelov, J. Stober, B. Eisenhut, Th. Fauster, *Phys. Rev. B* **44**, 1954 (1991). *(309, 317, 318)*
[898] R.-H. Zhou, P.-L. Cao, S.-B. Fu, *Surf. Sci.* **249**, 129 (1991). *(309, 315)*
[899] E. A. Slaughter, J. L. Lnand, *J. Vac. Sci. Technol. A* **10**, 66 (1992). *(309, 315)*
[900] Z. Jing, G. Lucovsky, J. L. Whitten, *Phys. Rev. B* **49**, 14003 (1994). *(309)*
[901] G. Dufor, F. Rochet, H. Roulet, F. Sirotti, *Surf. Sci.* **304**, 33 (1994). *(309, 314, 318)*
[902] J. Stober, B. Eisenhut, G. Rangelov, Th. Fauster, *Surf. Sci.* **321**, 111 (1994).

(309, 314, 316, 317, 318)
[903] T.L. Bush, D.O. Hayward, T.S. Jones, *Surf. Sci.* **331**, 306 (1995). *(309, 317, 318)*
[904] T. Fukuda, K. Saito, M. Ohue, K. Shima, N. Momma, *Jap. J. Appl. Phys.* **34**, L937 (1995). *(309, 321)*
[905] K. Nakamae, K. Kurosawa, Y. Takigawa, W. Sasaki, Y. Izawa, M. Okuda, *Jap. J. Appl. Phys.* **34**, L1482 (1995). *(309)*
[906] A.A. Saranin, O.L. Tarasova, V.G. Kotljar, E.A. Khramtsova, V.G. Lifshits, *Surf. Sci.* **331**, 458 (1995). *(309)*
[907] I.J.R. Baumvol, F.C. Stedile, *phys. stat. sol. (b)* **192**, 253 (1995). *(309, 318, 381, 391, 408)*
[908] M. Mukhopadhyay, S. K. Ray, C. K. Maiti, *J. Vac. Sci. Technol. B* **14**, 1682 (1996). *(309, 310)*
[909] N. Franco, J. Avila, M. E. Davila, M.C. Asensio, D. P. Woodruff, O. Schaff, V . Fernandez, K.-M. Schindler, V. Fritzsche, A.M. Bradshaw, *Phys. Rev. Lett.* **79**, 673 (1997). *(309, 314, 315)*
[910] E. Fattal, M.R. Radeke, G. Reynolds, E. A. Carter, *J. Phys. C* **101**, 8658 (1997). *(309, 315)*
[911] A. Izumi, H. Matsumura, *Appl. Phys. Lett.* **71**, 1371 (1997). *(309, 310, 318, 319)*
[912] T. Kubo, T. Ems, A. Atli, T. Aruga, N. Takagi, M. Nishijima, *Surf. Sci.* **382**, 214 (1997). *(309, 317)*
[913] C. Tindall, L. Li, O. Takaoka, Y. Hasegawa, T. Sakurai, *Surf. Sci.* **380**, 481 (1997). *(309, 315)*
[914] J.R. Elmiger, R. Schieck, M. Kunst, *J. Vac. Sci. Technol. A* **15**, 2418 (1997). *(310, 320)*
[915] J. J. Schmitt, B. L. Halpern, *Method and apparatus for the deposition of solid films of material from a jet stream entraining the gaseous phase of said material US Patent* 4 788 082 (1988); *Method for microwave plasma assisted supersonic gas jet deposition of thin films US Patent* 5 356 672 (1994). *(310, 320)*
[916] H.-J. Müssig, J. Dąbrowski, *German Patent* Nr. 195 22 054 (28.11.1996). *(311)*
[917] K. Ando, A. Ishitani, K. Hamano, *Appl. Phys. Lett.* **59**, 1081 (1991). *(311, 319)*
[918] M. V. Coleman, D. J. D. Thomas, *phys. stat. sol.* **25**, 241 (1968). *(312)*
[919] F. L. Edelmann, *Struktur von Festkörpern der Mikroelektronik* (Akademie-Verlag, Berlin 1985) p. 88. *(312, 347)*
[920] B.-R. Zhang, Z. Yu, G. J. Collins, T. Hwang, W. H. Richie, *J. Vac. Sci. Technol. A* **7**, 176 (1989). *(312, 320, 371)*
[921] I. Idrenstedt, C. Brosset, *Acta Chem. Scand.* **18**, 1879 (1964). *(313)*
[922] S. R. Srinivasa, L. Cartz, J. D. Jorgensen, J. D. Wolrton, R. A. Beyerlein, M. Billy, *J. Appl. Cryst.* **10**, 167 (1977). *(313)*
[923] S. Deleonibus, F. Martin, M. Heitzmann, J. Ch. Guibert, A. M. Papon, *J. Electrochem. Soc.* **144**, 164 (1997). *(313, 378)*
[924] A. Karimi, O. R. Shojaei, T. Kruml, J. L. Martin, *Thin Solid Films* **308**, 334 (1997). *(313)*
[925] M.D. Naeem, C. Parks, K. Wangemann, H. Glawischnig, *Thin Solid Films* **290**, 485 (1996). *(314)*
[926] P. J. Chen, M. L. Colaianni, J.T. Yates Jr., *Surf. Sci.* **274**, L605 (1992). *(315)*
[927] Y. Bu, M.C. Lin, *Surf. Sci.* **311**, 385 (1994). *(315)*
[928] H. Wormeester, A. van Silfhout, E. G. Keim, A. G. B. M. Sasse, *Surf. Sci.* **233**, L249 (1990). *(316)*

[929] A. G. B. M. Sasse, A. van Silfhout, *Phys. Rev. B* **40**, 1773 (1989). *(316, 317)*
[930] Z. Ying, W. Ho, *Phys. Rev. Lett.* **60**, 57 (1988). *(316)*
[931] E. Ekwelundu, A. Ignatiev, *Surf. Sci.* **179**, 119 (1987). *(316)*
[932] Y. Taguchi, M. Fujisawa, M. Nishijima, *Surf. Sci.* **233**, L251 (1990). *(316)*
[933] H. Kato, J. Lee, K. Sawabe, Y. Matsumoto, *Surf. Sci.* **386**, 93 (1997). *(317)*
[934] E. G. Keim, L. Wolterbeek, A. van Silfhout, *Surf. Sci.* **180**, 565 (1987). *(317)*
[935] A. G. Schrott, S. C. Fain, *Surf. Sci.* **111**, 39 (1981). *(317)*
[936] M. Fujisawa, Y. Taguchi, Y. Kuwahara, M. Onchi, M. Nishijima, *Phys. Rev. B* **39**, 12918 (1989). *(318)*
[937] M. Copel, R. M. Tromp, H. J. Timme, K. Penner, T. Nakao, *J. Vac. Sci. Technol. A* **14**, 462 (1996). *(319)*
[938] Y. Manabe, T. Yu, *J. Appl. Phys.* **66**, 2475 (1989). *(319)*
[939] B. Y. Kim, H. F. Luan, D. J. Kwong, *IEDM Technical Digest*, 463 (1997). *(320, 368)*
[940] S. Song, H. F. Luan, Y. Y. Chen, M. Gardner, J. Fulford, M. Allen, D. J. Kwong, *IEDM Technical Digest*, 373 (1998). *(320, 368)*
[941] S.-T. Chang, N. M. Johnson, S. A. Lyon, *Appl. Phys. Lett.* **44**, 316 (1984). *(320, 367, 368)*
[942] T. Y. Chu, W. Ting, J. H. Ahn, S. Lin, D.L. Kwong, *Appl. Phys. Lett.* **59**, 1412 (1991). *(320, 368, 370)*
[943] Y. Ma, T. Yasuda, G. Lucovsky, *J. Vac. Sci. Technol. B* **11**, 1533 (1993). *(320)*
[944] C.-K. Wang, T.-L. Ying, C.-S. Wei, L.-M. Liu, H.-C. Cheng, M.-S. Lin, *Jap. J. Appl. Phys.* **34**, 4736 (1995). *(321)*
[945] P. I. L. Smeys, P.B. Griffin, K. C. Saraswat, *J. Appl. Phys.* **78**, 2837 (1995). *(321)*
[946] A. Korkin, J. V. Cole, A. A. Bagaturyants, A. A. Safonov, K. P. Novoselov, D. Sengupta, J. Adams, *E-MRS Spring Meeting, Book of Abstracts*, L-2 (1999); *(321)*
[947] A. A. Bagaturyants, K. P. Novoselov, A. A. Safonov, L. L. Savchenko, J. V. Cole, A. Korkin, to be published in *Materials Science in Semiconductor Processing* (1999). *(321)*
[948] J. S. Pan, A.T.S. Wee, C.H.A. Huan, H. S. Tan, K.L. Tan, *Vacuum* **47**, 1495 (1996). *(321)*
[949] Y. Y. Chen, I. M. Liu, M. Gardner, J. Fulford, D.L. Kwong, *IEDM Technical Digest*, 639 (1997). *(321)*
[950] L.-K. Han, S. W. Crowder, M. Hargrove, E. Wu, S. H. Lo, E. Guarin, E.F. Crabbé, L. Su, *IEDM Technical Digest*, 643 (1997). *(321)*
[951] R. L. Opila, J. P. Chang, M. Du, J. Bevk, Y. Ma, M. Weldon, Y. J. Chabal, A. Gurevich, *Solid State Phenomena* **65**, 257 (1999). *(321)*
[952] A. Markwitz, H. Baumann, W. Grill, A. Knop, E.F. Krimmel, K. Bethge, *Nucl. Instr. Meth. Phys. Res. B* **89**, 362 (1994). *(321)*
[953] J. S. Pan, A.T.S. Wee, C.H.A. Huan, H. S. Tan, K.L. Tan, *Appl. Surf. Sci.* **115**, 166 (1997). *(321)*
[954] R. Nieh, W.-J. Qi, Y. Jeon, B. H. Lee, A. Lucas, L. Kang, J. C. Lee, M. Gardner, M. C. Gilmer, in *Ultrathin SiO_2 and High-K Materials for ULSI Gate Dielectrics*, ed. by H. R. Huff, C. A. Richter, M. L. Green, G. Lucovsky, and T. Hattori MRS Symposia Proceedings Vol. 567 (MRS, Warrendale, 1999) p. 521. *(321)*
[955] B. Doyle, H. R. Soleimani, A. Philipossian, *IEEE Electron Device Letters* **16**, 301 (1995). *(321, 358)*

[956] C. T. Liu, Y. Ma, M. Oh, P. W. Diodato, K. R. Stiles, J. R. McMacken, F. Li, C. P. Chang, K. P. Cheung, J. I. Colonell, W. Y. C. Lai, R. Liu, E.J. Lloyd, J. J. Miner, C. S. Pai, H. Vaidya, J. Frackoviak, A. Timko, F. Klemens, H. Maynard, J. C. Clemens, *IEDM Technical Digest*, 589 (1998). *(321, 425, 431)*
[957] U. M. Gösele, K. N. Tu, *J. Appl. Phys.* **53**, 3252 (1982). *(322)*
[958] J. Tersoff, *Phys. Rev. Lett.* **52**, 465 (1984). *(322, 325)*
[959] S. Kono, Y. Enta, T. Abukawa, N. Nakamura, K. Anno, S. Suzuki, *Physica Scripta Volume T* **31**, 96 (1990). *(322)*
[960] K. Maex, *Physics World* (November 1995) p. 35. *(322)*
[961] E.G. Colgan, J.P. Gambino, Q.Z. Hong, *Materials Science and Engineering R* **16**, 43 (1996). *(322, 325, 330)*
[962] R. Pretorius, *Thin Solid Films* **290**, 477 (1996). *(322)*
[963] C. Chaneliere, J. L. Autran, R. A. B. Devine, B. Balland, *Materials Science and Engineering R* **22**, 269 (1998). *(322, 341, 342)*
[964] P. P. Apte, S. Potla, D. A. Prinslow, G. Pollack, D. Scott, K. Varahramyan, *IEDM Technical Digest*, 729 (1998). *(322)*
[965] J.R. Tucker, C. Wang, P. S. Carney, *Appl. Phys. Lett.* **65**, 618 (1994). *(324)*
[966] M.-K. Ieong, P. M. Solomon, S. E. Laux, H.-S. P. Wong, D. Chidambarro, *IEDM Technical Digest*, 733 (1998). *(324)*
[967] R.T. Tung, J.M. Poate, J.C. Bean, J.M. Gibson, D.C. Jacobson, *Thin Solid Films* **93**, 77 (1982). *(325)*
[968] H. Föll, P. S. Ho, K. N. Tu, *Philosophical Magazine A* **45**, 31 (1982). *(325, 331)*
[969] R.T. Tung, J.M. Gibson, J.M. Poate, *Phys. Rev. Lett.* **50**, 429 (1983). *(325, 331, 334, 335)*
[970] F. Comin, J. E. Rowe, P. H. Citrin, *Phys. Rev. Lett.* **51**, 2402 (1983). *(325)*
[971] R.T. Tung, *Phys. Rev. Lett.* **52**, 461 (1984). *(325)*
[972] C. Calandra, O. Bisi, G. Ottaviani, *Surf. Sci. Rep.* **4**, 271 (1984). *(325)*
[973] D. R. Hamann, L. F. Mattheiss, *Phys. Rev. Lett.* **54**, 2517 (1985) *(325, 335)*
[974] W. R. L. Lambrecht, N. E. Christensen, P. Blöchl, *Phys. Rev. B* **36**, 2493 (1987). *(325, 332, 334, 337)*
[975] L. Braicovich, in *The Chemical Physics of Solid Surfaces and Heterogeneous Catalysis*, Vol. 5 (*Surface properties of electronic materials*) ed. by D. A. King, D. P. Woodruff, (Elsevier Science, 1988), p. 235 *(325)*
[976] D. Loretto, J.M. Gibson, S. M. Yalisove, *Phys. Rev. Lett.* **63**, 298 (1989). *(325)*
[977] H. L. Meyerheim, U. Döbler, A. Puschman, *Phys. Rev. B* **44**, 5738 (1991). *(325)*
[978] H. von Känel, *Materials Science Reports* **8**, 193 (1992). *(325)*
[979] C. Schwarz, U. Schrärer, P. Sutter, R. Stalder, N. Onda, H. von Känel, *J. Cryst. Growth* **127**, 659 (1993). *(325)*
[980] Y. Khang, S. J. Kahng, K. M. Mang, D. Jeon, J. H. Lee, Y. N. Kim, Y. Kuk, *J. Vac. Sci. Technol. B* **12**, 2094 (1994). *(325)*
[981] Z. Ma, L.H. Allen, *Phys. Rev. B* **49**, 13501 (1994). *(325, 328)*
[982] F. Hong, G.A. Rozgonyi, *J. Electrochem. Soc.* **141**, 3480 (1994). *(325)*
[983] J.A. Kittl, D. A. Prinslow, P. P. Apte, M. F. Pas, *Appl. Phys. Lett.* **67**, 2308 (1995). *(325)*
[984] H. von Känel, C. Schwarz, S. Goncalves-Conto, E. Müller, L. Miglio, F. Tavazza, G. Malegori, *Phys. Rev. Lett.* **74**, 1163 (1995). *(325)*
[985] S. Komiya, H. Tomita, K. Ikeda, Y. Horii, T. Nakamura, *Jap. J. Appl. Phys.* **35**, 242 (1996). *(325)*
[986] J.A. Kittl, Q.Z. Hong, *Thin Solid Films* **290**, 473 (1996). *(325)*
[987] J. Cardenas, S.-L. Zhang, B. G. Svensson, C. S. Petersson, *J. Appl. Phys.* **80**,

762 (1996). *(325)*

[988] R.T. Tung, *Appl. Phys. Lett.* **68**, 1933 (1996). *(325, 329)*

[989] B.D. Yu, Y. Miyamoto, O. Sugino, T. Sasaki, T. Ohno, *Appl. Phys. Lett.* **72**, 1176 (1998). *(325, 329)*

[990] M. E. Alperin, T. C. Hollaway, R. A. Haken, C. D. Gosmeyer, R. V. Karnaugh, W. D. Parmantie, *IEEE Transactions on Electron Devices* **32**, 141 (1985). *(327)*

[991] H. S. Momose, S.-i. Nakamura, T. Ohguro, T. Yoshitomi, R. Morifuji, T. Morimoto, Y. Katsumara, H. Iwai, *IEEE Transactions on Electron Devices* **45**, 691 (1998). *(329, 347, 356, 420, 430, 432, 438, 444)*

[992] J.-Y. Tsai, J. Sun, K. F. Yee, C. M. Osburn, *IEEE Electron Device Letters* **17**, 331 (1996). *(329)*

[993] P.-H. Bricout, E. Dubois, *IEEE Transactions on Electron Devices* **43**, 1251 (1996). *(329)*

[994] L. R. Zheng, L. S. Hung, S. Q. Feng, P. Revesz, J. W. Mayer, G. Miles, *Appl. Phys. Lett.* **48**, 767 (1986). *(330)*

[995] Q.Z. Hong, S.Q. Hong, F.M. D'Heurle, J.M.E. Harper, *Thin Solid Films* **253**, 479 (1994). *(330)*

[996] K. C. Saraswat, D. L. Brors, J. A. Fair, K. A. Monning, R. Beyers, *IEEE Transactions on Electron Devices* **30**, 1497 (1983). *(330)*

[997] A.E. White, K.T. Short, R. C. Dynes, J. P. Garno, J.M. Gibson, *Appl. Phys. Lett.* **50**, 95 (1987). *(331)*

[998] R. W. Fathauer, T. George, W. T. Pike, *J. Appl. Phys.* **72**, 1874 (1992). *(331)*

[999] K. C. R. Chiu, J.M. Poate, J. E. Rowe, T.T. Sheng, A. G. Cullis, *Appl. Phys. Lett.* **38**, 988 (1981). *(331)*

[1000] R.T. Tung, J.C. Bean, J.M. Gibson, J.M. Poate, D.C. Jacobson, *Appl. Phys. Lett.* **40**, 684 (1982). *(331)*

[1001] B. M. Clemens, R. Sinclair, *Materials Research Bulletin* **15**, 19 (February 1990). *(331)*

[1002] J.L. Batstone, J.M. Gibson, R.T. Tung, A.F.J. Levi, *Appl. Phys. Lett.* **52**, 828 (1988). *(331)*

[1003] M. A. L. Dass, D. B. Fraser, C.-S. Wei, *Appl. Phys. Lett.* **58**, 1308 (1991). *(332)*

[1004] J. S. Byun, J. J. Kim, W. S. Kim, H.J. Kim, *J. Electrochem. Soc.* **142**, 2805 (1995). *(332)*

[1005] F. Hong, G.A. Rozgonyi, B.K. Patnaik, *Appl. Phys. Lett.* **64**, 2241 (1994). *(332)*

[1006] S.L. Hsia, T. Y. Tan, P. Smith, G.E. McGuire, *J. Appl. Phys.* **72**, 1864 (1992). *(332)*

[1007] G.A. Rozgonyi, J. H. Lee, D. Knoesen, D. P. Adams, B.K. Patnaik, N.R. Parikh, A. S. M. Salih, P. Balducci, *Appl. Phys. Lett.* **58**, 729 (1991). *(332)*

[1008] G. Rangelov, Th. Fauster, *Surf. Sci.* **365**, 403 (1996). *(334)*

[1009] R.T. Tung, K. K. Wg, J.M. Gibson, A.F.J. Levi, *Phys. Rev. B* **33**, 7077 (1986). *(335)*

[1010] A. E. Dolbak, B.Z. Olshanetsky, S.I. Stenin, S. A. Teys, T. A. Gavrilova, *Surf. Sci.* **218**, 37 (1989). *(336)*

[1011] D. Dijkkamp, E.J. Loenen, A.J. Hoeven, J. Dieleman, *J. Vac. Sci. Technol. A* **8**, 218 (1990). *(336, 339)*

[1012] M. Y. Lee, P. A. Bennett, *Phys. Rev. Lett.* **75**, 4460 (1995). *(336)*

[1013] M. Yoshida, K. Furusho, *Jap. J. Appl. Phys.* **3**, 521 (1964). *(336, 337)*

[1014] Y.-J. Chang, J. L. Erskine, *Phys. Rev. B* **28**, 5766 (1983). *(337)*

[1015] R.J. Wilson, S. Chiang, *Phys. Rev. Lett.* **58**, 2575 (1987). *(337)*

[1016] T. Ichinokawa, T. Tani, A. Sayama, *Surf. Sci.* **219**, 395 (1989). *(337)*

[1017] Y. J. Chabal, R.J. Culbertson, L.C. Feldman, J. E. Rowe, *J. Vac. Sci. Technol.* **18**, 880 (1981). *(337)*
[1018] Y. Yamamoto, S. Ino, T. Ichikawa, *Jap. J. Appl. Phys.* **25**, L331 (1986). *(339)*
[1019] U. Myler, K. Jacobi, *Surf. Sci.* **220**, 353 (1989). *(339)*
[1020] K. Ishiyama, Y. Taga, A. Ichimiya, *Phys. Rev. B* **51**, 2380 (1995). *(339, 340)*
[1021] R. Singh, R. K. Ulrich, *Interface* **8**, 26 (Summer 1999). *(341, 377)*
[1022] P. N. Dunn, *Solid State Technology* **41**, 44 (June 1998). *(341)*
[1023] H.-S. Kim, D. C. Gilmer, S. A. Campbell, D. L. Polla, *Appl. Phys. Lett.* **69**, 3860 (1996). *(341, 342)*
[1024] X. Guo, T. P. Ma, T. Tamagawa, B. L. Halpern, *IEDM Technical Digest*, 377 (1998). *(341, 342, 357)*
[1025] M. Saitoh, T. Mori, H. Tamura, *IEDM Technical Digest*, 680 (1986). *(341)*
[1026] H. F. Luan, B. Z. Wu, L. G. Kang, B. Y, Kim, R. Vrtis, D. Roberts, D.L. Kwong, *IEDM Technical Digest*, 609 (1998). *(341, 342)*
[1027] M. C. Gilmer, T.-Y. Luo, H. R. Huff, M. D. Jackson, S. Kim, G. Bersuker, P. Zeitzoff, L. Vishnubhotla, G. A. Brown, R. Amos, D. Brady, V. H. C. Watt, G. Gale, J. Guan, B. Nguyen, G. Williamson, P. Lysaght, K. Torres, F. Geyling, C. F. H. Gondran, J. A. Fair, M. T. Schulberg, T. Tamagawa, in *Ultrathin SiO_2 and High-K Materials for ULSI Gate Dielectrics*, ed. by H. R. Huff, C. A. Richter, M. L. Green, G. Lucovsky, and T. Hattori MRS Symposia Proceedings Vol. 567 (MRS, Warrendale, 1999) p. 323. *(341)*
[1028] Y. Numasawa, S. Kamiyama, M. Zenke, M. Sakamoto, *IEDM Technical Digest*, 44 (1989). *(341)*
[1029] S.-O. Kim, J. S. Byun, H. J. Kim, *Thin Solid Films* **206**, 102 (1991). *(341)*
[1030] G. Q. Lo, D.L. Kwong, S. Lee, *Appl. Phys. Lett.* **60**, 3286 (1992). *(341)*
[1031] W. R. Hitchens, W. C. Krusell, D. M. Dobkin, *J. Electrochem. Soc.* **140**, 2615 (1993). *(341)*
[1032] D. Laviale, J. C. Oberlin, R. A. B. Devine, *Appl. Phys. Lett.* **65**, 2021 (1994). *(341, 342)*
[1033] D. Park, Q. Lu, T.-J. King, C. Hu, A. Kalnitsky, S.-P. Tay, C.-C. Cheng, *IEDM Technical Digest*, 381 (1998). *(341, 342)*
[1034] Y. Nishioka, S. Kumura, H. Shinriki, K. Mukai, *J. Electrochem. Soc.* **134**, 410 (1987). *(342)*
[1035] S. Banerjee, B. Shen, I. Chen, J. Bohlman, G. Brown, R. Doering, *J. Appl. Phys.* **65**, 1140 (1989). *(342)*
[1036] S. K. Zhang, Z. W. Fu, L. Ke, F. Lu, Q. Z. Qin, X. Wang, *J. Appl. Phys.* **84**, 335 (1998). *(342)*
[1037] R. A. B. Devine, L. Vallier, J. L. Autran, P. Paillet, J. L. Leray, *Appl. Phys. Lett.* **68**, 1775 (1996). *(342)*
[1038] C. A. Billman, P. H. Tan, K. J. Hubbard, D. G. Schlom, in *Ultrathin SiO_2 and High-K Materials for ULSI Gate Dielectrics*, ed. by H. R. Huff, C. A. Richter, M. L. Green, G. Lucovsky, and T. Hattori MRS Symposia Proceedings Vol. 567 (MRS, Warrendale, 1999) p. 409. *(342)*
[1039] J. P. Chang, M. L. Steigerwald, R.M. Fleming, R. L. Opila, G. B. Alers, *Appl. Phys. Lett.* **74**, 3705 (1999). *(342)*
[1040] J. P. Chang, R. L. Opila, G. B. Alers, M. L. Steigerwald, H. C. Lu, E. Garfunkel, T. Gustafsson, *Solid State Technology* **42**, 43 (February 1999). *(343)*
[1041] W. S. Lau, K. K. Khaw, P. W. Qian, N. P. Sandler, P.K. Chu, *J. Appl. Phys.* **79**, 8841 (1996). *(343)*
[1042] K. J. Hubbard, D. G. Schlom, *J. Mater. Res.* **11**, 2575 (1996). *(343)*
[1043] S. T. Pantelides, (Pergamon, New York, 1978). *(345, 358, 392)*

[1044] E. H. Nicollian, J. R. Brews, *MOS Physics and Technology* (Wiley-Interscience, New York 1982). *(345, 358)*

[1045] F. P. Fehlner, *Low Temperature Oxidation* (Wiley, New York 1986). *(345)*

[1046] C. R. Helms, B. E. Deal (editors) *The Physics and Chemistry of* SiO_2 *and the Si-*SiO_2 *Interface* (Plenum Press, New York 1988). *(345, 360)*

[1047] P. Balk (editor), *The Si-SiO2 System* (Elsevier, Amsterdam, 1988). *(345)*

[1048] E. A. Irene, *CRC Crit. Rev. Solid State Mater. Sci.* **14**, 175 (1988). *(345)*

[1049] G.S. Higashi, Y. J. Chabal, in *Handbook of Semiconducting Wafer Cleaning Technology*, ed. by W. Kern, (Park Ridge: Noyes Publ., 1993), p. 433. *(345, 358, 361, 362, 392, 430)*

[1050] H. Z. Massoud, E. H. Poindexter, C. R. Helms (editors), *The Physics and Chemistry of SiO2 and the Si-SiO2 Interface* (The Electrochemical Proceedings Series, Pennington, New York 1996). *(345, 392)*

[1051] S. Yokoyama, T. Nagata, Y. Kuroda, T. Doi, T. Namba, K. Miyake, T. Miyamoto, S. Mizayaki, M. Koyanagi, M. Hirose, *J. Vac. Sci. Technol. A* **13**, 629 (1995). *(347)*

[1052] P. Packan, S. Thompson, E. Andideh, S. Yu, T. Ghani, J. Dandford, M. T. Bohr, *IEDM Technical Digest*, 505 (1998). *(347)*

[1053] R. W. G. Wyckoff, in *Crystal Structures*, Vol. 1 (Willey, New York, 1982), p. 239. *(348)*

[1054] F. R. Boyd, J. L. England, *Journal of Geophysical Research* **65**, 749 (1960). *(349)*

[1055] S. Tsuneyuki, M. Tsukada, H. Aoki, Y. Matsui, *Phys. Rev. Lett.* **61**, 869 (1988). *(350, 352)*

[1056] S. T. Pantelides, W. A. Harrison, *Phys. Rev. B* **13**, 2667 (1976). *(351)*

[1057] J. P. Rino, I. Ebbsjö, R. K. Kalia, A. Nakano, P. Vashishta, *Phys. Rev. B* **47**, 3503 (1993). *(351)*

[1058] A. H. Edwards, W. B. Fowler, *Phys. Rev. B* **26**, 6649 (1982). *(351)*

[1059] R. B. Fair, *J. Electrochem. Soc.* **144**, 708 (1997). *(351, 413, 420)*

[1060] G. D. Watkins, J. R. Troxell, *Phys. Rev. Lett.* **44**, 593 (1980). *(351)*

[1061] D.J. Chadi, K. J. Chang, *Phys. Rev. B* **39**, 10063 (1989). *(351)*

[1062] P. Burggraaf, *Solid State Technology* **41**, 38 (June 1998). *(352)*

[1063] L. Manchanda, in *Semiconductor Characterization: Present Status and Future Needs* ed. by W. M. Bullis, D. G. Seiler, and A. C. Diebold (Woodbury, New York 1996) p. 123. *(352, 420, 432, 444)*

[1064] P. O. Hahn, M. Henzler, *J. Vac. Sci. Technol. A* **2**, 574 (1984). *(352, 391, 392, 396, 429, 430)*

[1065] H. H. Vuong, C. S. Rafferty, S. A. Eshraghi, J. L. Lentz, P. M. Zeitzoff, M. R. Pinto, S.J. Hillenius, *IEEE Transactions on Electron Devices* **43**, 1144 (1996). *(353)*

[1066] H. Wong, Y.-C. Cheng, *IEEE Transactions on Electron Devices* **38**, 1883 (1991). *(353, 392, 441)*

[1067] P.B. Griffin, S. W. Crowder, J. M. Knight, *Appl. Phys. Lett.* **67**, 482 (1995). *(353, 392, 413, 419, 421, 422)*

[1068] Y.-L. Shen, S. Suresh, I. A. Blech, *J. Appl. Phys.* **80**, 1388 (1996). *(353)*

[1069] H. A. Rueda, S. Cea, M. E. Law, *SISPAD'97*, 53 (1997). *(353)*

[1070] N. E. B. Cowern, P. C. Zalm, P. van der Sluis, D. J. Gravesteijn, W.B. Boer, *Phys. Rev. Lett.* **72**, 2585 (1994). *(353)*

[1071] T. Ohmi, K. Kotani, A. Teramoto, M. Miyashita, *IEEE Electron Device Letters* **12**, 652 (1991). *(353, 392, 393, 440)*

[1072] W. J. Taylor, T.Y. Tan, U. Gösele, *Appl. Phys. Lett.* **59**, 2007 (1991). *(353)*

[1073] S. P. Ashburn, S. Krishnan, G. A. Dixit, M. Rodder, K. Taylor, T. Breedijk, I.-C. Chen, M. W. Goodwin, A. L. Esquivel, *IEDM Technical Digest*, 449 (1997). *(354)*
[1074] J. E. Vinson, J. J. Liou, *Proceedings of the IEEE* **86**, 399 (1998). *(354, 429, 434)*
[1075] T. H. Ning, P. W. Cook, R. H. Dennard, C. M. Osburn, S. E. Schuster, *IEEE Transactions on Electron Devices* **26**, 346 (1979). *(354, 433, 436)*
[1076] J. J. Ellis-Monaghan, R. B. Hulfachor, K. W. Kim, M. A. Littlejohn, *IEEE Transactions on Electron Devices* **43**, 1123 (1996). *(354)*
[1077] H.-H. Vuong, C. S. Rafferty, J. R. McMacken, J. Ning, S. Chaudhry, *SISPAD'97*, 85 (1997). *(354, 413, 418)*
[1078] S. Wolf, *Silicon Processing for the VLSI Era* (Lattice Press, Sunset Beach, 1993), p. 119. *(355)*
[1079] S. L. Wu, C. L. Lee, T. F. Lei, M. S. Liang, *J. Appl. Phys.* **72**, 1378 (1992). *(355, 357, 358, 438, 439)*
[1080] R. P. S. Thakur, Y. Chen, E. H. Poindexter, R. Singh, *Interface* **8**, 20 (Summer 1999). *(356)*
[1081] H. S. Momose, M. Ono, T. Yoshitomi, T. Ohguro, S.-i. Nakamura, M. Sato, H. Iwai, *IEDM Technical Digest*, 1233 (1996). *(356, 430, 438)*
[1082] D. A. Muller, T. Sorsch, S. Moccio, F. H. Baumann, K. Evans-Lutterodt, G. Timp, *Nature* **399**, 758 (1999). *(356)*
[1083] L. Manchanda, H. W. Lee, J. E. Bower, F.H. Baumann, W.L. Brown, C. J. Case, R. C. Keller, Y. O. Kim, E. J. Laskowski, M. D. Morris, M. D. Morris, R. L. Opila, P.J. Silverman, T. W. Sorsch, G. R. Weber, *IEDM Technical Digest*, 605 (1998). *(356)*
[1084] B. E. Deal, A. S. Grove, *J. Appl. Phys.* **36**, 3770 (1965). *(358, 363, 381, 382)*
[1085] A. M. Lin, R. W. Dutton, D. A. Antoniadis, W. A. Tiller, *J. Electrochem. Soc.* **128**, 1121 (1981). *(358)*
[1086] H. Z. Massoud, J.D. Plummer, E. A. Irene, *J. Electrochem. Soc.* **132**, 2685 (1985). *(358, 381)*
[1087] R. M. Tromp, G. W. Rubloff, P. Balk, F.K. LeGoues, E.J. van Loenen, *Phys. Rev. Lett.* **55**, 2332 (1985). *(358, 372)*
[1088] E. A. Lewis, E. A. Irene, *J. Electrochem. Soc.* **134**, 2332 (1987). *(358, 381)*
[1089] A. Ourmazd, D. W. Taylor, J. A. Rentschler, J. Bevk, *Phys. Rev. Lett.* **59**, 213 (1987). *(358, 392, 393, 398, 400, 401)*
[1090] R. Singh, *J. Appl. Phys.* **R63**, 59 (1988). *(358, 365)*
[1091] J. P. Colinge, *Silicon-on-Insulator Technology* (Kluwer Press, 1991). *(358, 380)*
[1092] R. W. Dutton, Z. Yu, *Technology CAD: Computer Simulation of IC Processes and Devices* (Kluwer Academics Publishers, Boston, 1993). *(358, 366, 374, 381, 391)*
[1093] S. Wolf, R. N. Tauber, *Silicon Processing for the VLSI Era*, Vol. 1–3 (Lattice Press, Sunset Beach, 1993). *(358)*
[1094] M. Morita, T. Ohmi, *Jap. J. Appl. Phys.* **33**, 370 (1994). *(358, 359, 392)*
[1095] G. Franco, V. Raineri, F. Frisina, E. Rimini, *Nucl. Instr. Meth. Phys. Res. B* **96**, 99 (1995). *(358)*
[1096] A. Chin, W. J. Chen, T. Chang, R. H. Kao, B. C. Lin, C. Tsai, J. C.-M. Huang, *IEEE Electron Device Letters* **18**, 417 (1997). *(358, 368, 413)*
[1097] R. Sharangpani, R. P. S. Thakur, N. Shah, S. P. Tay, *Solid State Technology* **41**, 91 (October 1998). *(358, 364, 425)*
[1098] H. Iwai, H. S. Momose, *IEDM Technical Digest*, 163 (1998). *(358, 365, 432)*
[1099] J. H. Stathis, D. J. DiMaria, *IEDM Technical Digest*, 167 (1998). *(358, 432)*

[1100] M. Yamin, *IEEE Transactions on Electron Devices* **12**, 88 (1965). *(358, 430)*
[1101] E. H. Snow, A. S. Grove, B. E. Deal, C. T. Sah, *J. Appl. Phys.* **36**, 1664 (1965). *(358, 430)*
[1102] E. Yon, W. H. Ko, A. B. Kuper, *IEEE Transactions on Electron Devices* **13**, 276 (1966). *(358, 430)*
[1103] H. Kobayashi, Y. Yamashita, T. Mori, Y. Nakato, K. H. Park, Y. Nishioka, *Surf. Sci.* **326**, 124 (1995). *(360, 406)*
[1104] H. Kobayashi, Y. Yamashita, T. Mori, Y. Nakato, T. Komeda, Y. Nishioka, *Jap. J. Appl. Phys.* **34**, 959 (1995). *(360)*
[1105] M. Morita, T. Ohmi, E. Hasegawa, A. Teramoto, *Jap. J. Appl. Phys.* **29**, L2392 (1990). *(360)*
[1106] W. Kern, D. A. Puotinen, *RCA Rev.* **31**, 187 (1970). *(360)*
[1107] T. Hattori, *J. Vac. Sci. Technol. B* **11**, 1528 (1993). *(361)*
[1108] H. Kobayashi, Y. Yamashita, Y. Nakato, T. Komeda, Y. Nishioka, *Appl. Phys. Lett.* **69**, 2276 (1996). *(361, 406, 429)*
[1109] A. Terrasi, J. Almeida, C. Coluzza, G. Margaritondo, *Nucl. Instr. Meth. Phys. Res. B* **116**, 416 (1996). *(362)*
[1110] M. Nagamine, H. Itoh, H. Satake, A. Toriumi, *IEDM Technical Digest*, 593 (1998). *(362, 381, 382, 439)*
[1111] R. R. Razouk, L. N. Lie, B. E. Deal, *J. Electrochem. Soc.* **128**, 2214 (1981). *(363, 383)*
[1112] P. Münzinger, W. Lerch, R. Mader, N. Kobayashi, *Solid State Technology* **41**, 121 (June 1998). *(363, 365)*
[1113] J. Kim, S. T. Ahn, *IEEE Electron Device Letters* **18**, 385 (1997). *(363, 372)*
[1114] M. M. Moslehi, S. C. Shatas, K. C. Saraswat, *Appl. Phys. Lett.* **47**, 1353 (1985). *(365)*
[1115] M. M. Moslehi, K. C. Saraswat, S. C. Shatas, *Appl. Phys. Lett.* **47**, 1113 (1985). *(365)*
[1116] R. J. Kriegler, Y.-C. Cheng, D. R. Colton, *J. Electrochem. Soc.* **119**, 388 (1972). *(365, 366)*
[1117] D. W. Hess, B. E. Deal, *J. Electrochem. Soc.* **124**, 735 (1977). *(366, 383)*
[1118] S. B. Desu, C. H. Peng, T. Shi, P. A. Agaskar, *J. Electrochem. Soc.* **139**, 2682 (1992). *(366)*
[1119] K. Kwok, E. Yieh, S. Robles, B. Nguyen, *J. Electrochem. Soc.* **141**, 2172 (1994). *(366)*
[1120] J. W. Klaus, O. Sneh, S.M. George, *Science* **278**, 1934 (1998). *(366)*
[1121] D.M. Knotter, *Appl. Surf. Sci.* **99**, 99 (1996). *(367)*
[1122] A. J. Bauer, W. R. Aderhold, E. P. Burte, *Microelectronics Journal* **24**, 415 (1993). *(367, 372)*
[1123] C. H. Bjorkman, T. Yasuda, C. E. Shearon, Y. Ma Jr., G. Lucovsky, U. Emmerichs, C. Meyer, K. Leo, H. Kurz, *J. Vac. Sci. Technol. B* **11**, 1521 (1993). *(367)*
[1124] W.Y. Yang, R. Jayaraman, C.G. Sodini, *IEEE Transactions on Electron Devices* **35**, 935 (1988). *(368)*
[1125] G. J. Dunn, S. A. Scott, *IEEE Transactions on Electron Devices* **37**, 1719 (1990). *(368)*
[1126] F. A. Sewell Jr., H. A. R. Wegener, E. T. Lewis, *Appl. Phys. Lett.* **14**, 45 (1969). *(368)*
[1127] H. Hwang, W. Ting, B. Maiti, D.L. Kwong, J. Lee, *Appl. Phys. Lett.* **57**, 1010 (1990). *(368)*
[1128] M.L. Green, D. Brasen, K. W. Evans-Lutterodt, L.C. Feldman, K. Krisch, W.N. Lennard, L. Manchanda, M.-T. Tang, *Appl. Phys. Lett.* **65**, 848 (1994). *(368,*

369, 370)

[1129] B. Maiti, P. J. Tobin, V. Misra, R. I. Hedge, K. G. Reid, C. Gelatos, *IEDM Technical Digest*, 651 (1997). *(368, 370, 428)*

[1130] T. Ogata, M. Inoue, T. Nakamura, N. Tsuji, K. Kobayashi, K. Kawase, H. Kurokawa, T. Kaneoka, Y. Ohno, H. Miyoshi, *IEDM Technical Digest*, 597 (1998). *(368)*

[1131] M. Takayanagi-Takagi, Y. Toyoshima, *IEDM Technical Digest*, 575 (1998). *(368, 369, 440)*

[1132] G. Timp, K. K. Bourdelle, J. E. Bower, F. H. Baumann, T. Boone, R. Cirelli, K. Evans-Lutterodt, J. Garno, A. Ghetti, H. Gossmann, M. Green, D. Jacobson, Y. Kim, R. Kleinman, F. Klemens, A. Kornblit, C. Lochstampfor, W. Mansfield, S. Moccio, D. A. Muller, L. E. Ocola, M. L. O'Malley, J. Rosamilia, J. Sapjeta, P. Silverman, T. Sorsch, D. M. Tennant, W. Timp, B. F. Weir, *IEDM Technical Digest*, 615 (1998). *(368)*

[1133] H. Kukuda, T. Arakawa, S. Ohno, *Electronics Letters* **26**, 1505 (1990). *(368)*

[1134] D. G. J. Sutherland, H. Akatsu, M. Copel, F.J. Himpsel, T. A. Callcott, J. A. Carlisle, D. L. Ederer, J. J. Jia, I. Jimenez, R. Perera, D. K. Shuh, L. J. Terminello, W. M. Tong, *J. Appl. Phys.* **78**, 6761 (1995). *(369)*

[1135] S. R. Kaluri, D. W. Hess, *Appl. Phys. Lett.* **69**, 1053 (1996). *(369)*

[1136] A. Kamath, D.L. Kwong, Y. M. Sun, P. M. Blass, S. Whaley, J.M. White, *Appl. Phys. Lett.* **70**, 63 (1997). *(369)*

[1137] G.-M. Rignanese, A. Pasquarello, J.-C. Charlier, X. Gonze, R. Car, *Phys. Rev. Lett.* **79**, 5174 (1997). *(369)*

[1138] R. I. Hedge, P. J. Tobin, K. G. Reid, B. Maiti, S. A. Ajuria, *Appl. Phys. Lett.* **66**, 2882 (1995). *(369, 371)*

[1139] J.-J. Ganem, S. Rigo, I. Trimaille, I.J.R. Baumvol, F.C. Stedile, *Appl. Phys. Lett.* **68**, 2366 (1996). *(369)*

[1140] M. L. Green, D. Brasen, L.C. Feldman, W.N. Lennard, H.T. Tang, *Appl. Phys. Lett.* **67**, 1600 (1995). *(370)*

[1141] W. Ting, H. Hwang, J. Lee, D.L. Kwong, *Appl. Phys. Lett.* **57**, 2808 (1990). *(370)*

[1142] K. Kim, Y.-H. Lee, M.-S. Suh, C.-J. Youn, K.-B. Lee, H. J. Lee, *J. Electrochem. Soc.* **143**, 3372 (1996). *(370, 383, 413)*

[1143] H.T. Tang, W.N. Lennard, L.C. Feldman, M.L. Green, D. Brasen, *Nucl. Instr. Meth. Phys. Res. B* **108**, 347 (1996). *(370)*

[1144] G. Weidner, D. Krüger, *Appl. Phys. Lett.* **62**, 294 (1993). *(370)*

[1145] N.-I. Lee, J.-W. Lee, S.-H. Hur, H.-S. Kim, C.-H. Han, *IEEE Electron Device Letters* **18**, 486 (1997). *(370)*

[1146] M. Bhat, J. Kim, J. Yan, G. W. Yoon, L.-K. Han, D.L. Kwong, *IEEE Electron Device Letters* **15**, 421 (1994). *(370)*

[1147] Y. Okada, P. J. Tobin, K. G. Reid, R. I. Hegde, B. Maiti, S. A. Ajuria, *Symposium on VLSI Technology Digest*, 105 (1994). *(370)*

[1148] E. Briner, Ch. Meiner, A. Rothen, *Journal de Chimie Physique* **23**, 609 (1926). *(371)*

[1149] P. Poveda, A. Glachant, *Surf. Sci.* **323**, 258 (1995). *(371)*

[1150] A. Glachant, P. Soukiassian, S. T. Kim, S. Kapoor, A. Papageorgopoulos, Y. Baros, *J. Appl. Phys.* **70**, 2387 (1991). *(371)*

[1151] A. Glachant, P. Soukiassian, P.S. Mangat, J., Peng, S. T. Kim, *Surf. Sci.* **56**, 802 (1991). *(371)*

[1152] J. P. Li, A. J. Steckl, *J. Electrochem. Soc.* **L140**, 89 (1993). *(372)*

[1153] S.M. Hu, *J. Appl. Phys.* **R 70**, 53 (1991). *(373)*

[1154] K. Eriguchi, Y. Harada, M. Niwa, *IEDM Technical Digest*, 175 (1998). *(373, 399)*
[1155] D. R. Wolters, A. T. A. Zegers-van Duynhoven, *Microelectronics Journal* **24**, 333 (1993). *(373, 378, 389, 390)*
[1156] A. Pasquarello, M. S. Hybertsen, R. Car, *Nature* **396**, 58 (1998). *(374, 392, 393, 405, 410, 411)*
[1157] S.-F. Huang, P.B. Griffin, J.D. Plummer, P. Rissman, *SISPAD'97*, 49 (1997). *(374, 379, 381, 385, 389)*
[1158] J. J. Wortman, J. R. Hauser, R. M. Burger, *J. Appl. Phys.* **35**, 2122 (1964). *(374)*
[1159] P. H. Fuoss, L. J. Norton, S. Brennan, A. Fischer-Colbrie, *Phys. Rev. Lett.* **60**, 600 (1988). *(374, 392, 398, 400)*
[1160] S. Nag, A. Chatterjee, *Solid State Technology* **40**, 129 (September 1997). *(375, 378)*
[1161] B. E. Deal, M. Sklar, A. S. Grove, E. H. Snow, *J. Electrochem. Soc.* **114**, 266 (1967). *(376, 392)*
[1162] K. Hofmann, D. R. Young, G. W. Rubloff, *J. Appl. Phys.* **62**, 925 (1987). *(376)*
[1163] R. R. Razouk, B. E. Deal, *J. Electrochem. Soc.* **126**, 1573 (1979). *(377)*
[1164] N. Lifshitz, G. Smolinsky, *IEEE Electron Device Letters* **12**, 140 (1991). *(377, 425)*
[1165] H. C. Mogul, L. Cong, R. M. Wallace, P. J. Chen, T. A. Rost, K. Harvey, *Appl. Phys. Lett.* **72**, 1721 (1998). *(377, 425, 426)*
[1166] X. D. Chen, J.M. Gibson, *Appl. Phys. Lett.* **70**, 1462 (1997). *(377, 392, 393, 396, 438)*
[1167] J. A. Appels, K. Kooi, M. M. Paffen, J. J. H. Schatorji, W. H. C. G. Verkuylen, *Philips Research Reports* **25**, 118 (1970). *(377)*
[1168] B. Davari, C. Koburger, T. Furukawa, Y. Taur, W. Noble, A. Megdanis, J. Warnock, J. Mauer, *IEDM Technical Digest*, 92 (1988). *(378)*
[1169] H. Landis, P. Burke, W. Cote, W. Hill, C. Hoffman, C. Kaanta, C. Koburger, W. Lange, M. Leach, S. Luce, *Thin Solid Films* **220**, 1 (1992). *(378)*
[1170] D. Chidambarrao, J. P. Peng, G. R. Srinivasan, *J. Appl. Phys.* **70**, 4816 (1991). *(378)*
[1171] P. Sallagoity, M. Ada-Hanifi, M. Paoli, M. Haond, *IEEE Transactions on Electron Devices* **43**, 1900 (1996). *(379)*
[1172] A. Chatterjee, D. Rogers, J. McKee, I. Ali, S. Nag, I.-C. Chen, *IEDM Technical Digest*, 829 (1996). *(379)*
[1173] H. Watanabe, K. Shimizu, Y. Takeuchi, S. Aritome, *IEDM Technical Digest*, 833 (1996). *(379)*
[1174] D. R. Wolters, A. T. A. Zegers-van Duynhoven, *J. Electrochem. Soc.* **139**, 241 (1992). *(379, 381, 387, 392)*
[1175] U. Schwalke, M. Kerber, K. Koller, H. J. Jacobs, *IEEE Transactions on Electron Devices* **44**, 2070 (1997). *(379)*
[1176] S. S. Roth, K. J. Cooper, W. Ray, H. C. Kirsch, C. Grove, *J. Electrochem. Soc.* **141**, 2178 (1994). *(380)*
[1177] H.-C. Tseng, C.-Y. Chang, F.-M. Pan, L.-P. Chen, *J. Electrochem. Soc.* **144**, 2226 (1997). *(380)*
[1178] U. M. Gösele, E. Schroer, J.-Y. Huh, *Appl. Phys. Lett.* **67**, 241 (1995). *(380)*
[1179] S. Cristoloveanu, A. Simionescu, T. Wetteroth, H. Shin, D. Munteanu, P. Gentil, S. Hong, S. R. Wilson, *J. Electrochem. Soc.* **144**, 1468 (1997). *(380, 413)*
[1180] G. G. Shahidi, C. A. Anderson, B. A. Chappell, T. I. Chappell, J. H. Comfort, B. Davari, R. H. Dennard, R. L. Franch, P. A. McFarland, J. S. Neely, T. H. Ning,

M. R. Polcari, J. D. Warnock, *IEEE Transactions on Electron Devices* **41**, 2405 (1994). *(380)*
[1181] G. G. Shahidi, A. Ajmera, F. Assederaghi, R. J. Bolam, H. Hovel, E. Leobandung, W. Rausch, D. Sadana, D. Scheips, L. F. Wagner, L. Wissel, K. Wu, B. Davari, *IEEE Custom Integrated Circuits Conference*, 339 (1998). *(380)*
[1182] J. Bardeen, W. H. Brattain, W. Shockley, *J. Chem. Phys.* **14**, 714 (1946). *(381)*
[1183] N. Cabrera, N. F. Mott, *Reports on Progress in Physics* **12**, 163 (1948). *(381)*
[1184] F. J. Norton, *Nature* **171**, 701 (1961). *(381, 385)*
[1185] J. Blanc, *Appl. Phys. Lett.* **33**, 424 (1978). *(381, 386)*
[1186] W. A. Tiller, *J. Electrochem. Soc.* **127**, 619 (1980). *(381)*
[1187] M. Hamasaki, *Solid-State Electronics* **25**, 205 (1982). *(381)*
[1188] Y. Z. Lu, Y.-C. Cheng, *J. Appl. Phys.* **56**, 1608 (1984). *(381, 384, 386)*
[1189] E. A. Irene, E. A. Lewis, *Appl. Phys. Lett.* **51**, 767 (1987). *(381)*
[1190] H. Wong, Y.-C. Cheng, *J. Appl. Phys.* **64**, 893 (1988). *(381)*
[1191] D. R. Wolters, A. T. A. Zegers-van Duynhoven, *J. Appl. Phys.* **65**, 5126 (1989). *(381, 387)*
[1192] A. Kazor, *J. Appl. Phys.* **77**, 1477 (1995). *(381, 389)*
[1193] L. Verdi, A. Miotello, *Phys. Rev. B* **51**, 5469 (1995). *(381)*
[1194] R. M. C. de Almeida, S. Gonçalves, I.J.R. Baumvol, F.C. Stedile, paper submitted to *Nature* (1999), e-print available from `http://xxx.laln.gov.cond-mat/9901335` *(381, 391, 429)*
[1195] I. Fränz, W. Langheinrich, *Solid-State Electronics* **14**, 499 (1971). *(383)*
[1196] R. Ghez, Y. I. van der Meulen, *J. Electrochem. Soc.* **119**, 1100 (1972). *(383)*
[1197] D.-B. Kao, J. P. McVittie, W. D. Nix, K. C. Saraswat, *IEEE Transactions on Electron Devices* **35**, 25 (1988). *(383)*
[1198] N. E. McGruer, R. Singh, J. H. Weiss, K. Rajkanan, *J. Appl. Phys.* **62**, 3405 (1987). *(385)*
[1199] C.-J. Han, C. R. Helms, *J. Electrochem. Soc.* **135**, 1824 (1988). *(385)*
[1200] H. A. Schaeffer, *Journal of Non-Crystalline Solids* **38**, 545 (1980). *(386)*
[1201] G. Brebec, R. Seguin, C. Sella, J. Bevenot, J. C. Martin, *Acta Metall.* **28**, 327 (1980). *(386)*
[1202] G. K. Celler, L. E. Trimble, *Appl. Phys. Lett.* **54**, 1427 (1989). *(386)*
[1203] J. A. Costello, R. E. Tressler, *J. Electrochem. Soc.* **131**, 1944 (1984). *(386)*
[1204] J. C. Mikkelson, *Appl. Phys. Lett.* **45**, 1187 (1984). *(386)*
[1205] R. H. Doremus, *J. Electrochem. Soc.* **143**, 1992 (1996). *(386)*
[1206] C.-J. Han, C. R. Helms, *J. Appl. Phys.* **59**, 1767 (1988). *(386)*
[1207] P. J. Jorgensen, *J. Chem. Phys.* **37**, 874 (1962). *(386, 387)*
[1208] P. V. Dressendorfer, R. C. Baker, *Appl. Phys. Lett.* **36**, 933 (1980). *(388)*
[1209] R. Williams, *Phys.l Rev. A* **140**, 569 (1965). *(388)*
[1210] V. Senez, D. Collard, P. Ferreira, B. Baccus, *IEEE Transactions on Electron Devices* **43**, 720 (1996). *(389)*
[1211] D. Wolters, personal communication. *(390)*
[1212] J. R. Schrieffer, *The Physical Review* **97**, 641 (1955). *(392, 439)*
[1213] J. Blanc, C. J. Buiocchi, M. S. Abrahams, W. E. Ham, *Appl. Phys. Lett.* **30**, 120 (1977). *(392, 393)*
[1214] F.J. Grunthaner, P. J. Grunthaner, R. P. Vasquez, B. F. Lewis, J. Maserjian, A. Madhukar, *Phys. Rev. Lett.* **43**, 1683 (1979). *(392)*
[1215] R. Haight, L.C. Feldman, *J. Appl. Phys.* **53**, 4884 (1982). *(392)*
[1216] J. Robertson, M. J. Powell, *Appl. Phys. Lett.* **44**, 410 (1984). *(392)*
[1217] M.E. Welland, R. H. Koch, *Appl. Phys. Lett.* **48**, 724 (1986). *(392, 393, 429)*
[1218] I. Ohdomari, H. Akatsu, Y. Yukio, K. Kishimoto, *J. Appl. Phys.* **62**, 3751 (1987).

(392, 404)
[1219] K. Sakamoto, K. Yamanishi, F. Ichikawa, S. Ushio, *J. Appl. Phys.* **61**, 1553 (1987). *(392, 413)*
[1220] A. X. Chu, W. B. Fowler, *Phys. Rev. B* **41**, 5061 (1990). *(392, 393)*
[1221] M. Hane, Y. Miyamoto, A. Oshiyama, *Phys. Rev. B* **41**, 12637 (1990). *(392, 393, 402)*
[1222] D. M. Fleetwood, P. S. Winokur, R. A. Reber Jr., T. L. Meisenheimer, J. R. Schwank, M. R. Shaneyfelt, L .C. Riewe, *J. Appl. Phys.* **73**, 5058 (1993). *(392, 443)*
[1223] A. Pasquarello, M. S. Hybertsen, R. Car, *Phys. Rev. B* **53**, 10942 (1996). *(392, 393, 399)*
[1224] J. P. Peng, K. G. Lynn, P. Asoka-Kumar, D. P. Becker, D. R. Harshman, *Phys. Rev. Lett.* **76**, 2157 (1996). *(392, 407)*
[1225] H. Kageshima, K. Shiraishi, *Surf. Sci.* **380**, 61 (1997). *(392, 403, 404, 406)*
[1226] A. Pasquarello, M. S. Hybertsen, R. Car, *Appl. Phys. Lett.* **68**, 625 (1996). *(392, 410)*
[1227] B. Neri, P. Olivo, B. Ricco, *Appl. Phys. Lett.* **51**, 2167 (1987). *(392)*
[1228] T. A. Rabedeau, I. M. Tidswell, P.S. Pershan, J. Bevk, B.S. Freer, *Appl. Phys. Lett.* **59**, 706 (1991). *(393, 409, 410)*
[1229] A. Pasquarello, M. S. Hybertsen, R. Car, *Phys. Rev. Lett.* **74**, 1024 (1995). *(393)*
[1230] A. Tandia, G. Sarrabayrouse, A. Martinez, *Thin Solid Films* **296**, 122 (1997). *(393)*
[1231] R. B. Laughlin, J.D. Joannopoulos, D.J. Chadi, in *The Physics of SiO_2 and Its Interfaces*, ed. by S. T. Pantelides, (Pergamon, New York, 1978), p. 321. *(393)*
[1232] E. S. Daniel, J. T. Jones, O. J. Marsh, T. C. McGill, *J. Vac. Sci. Technol. B* **15**, 1089 (1997). *(393, 429, 431)*
[1233] G. Renaud, P. H. Fuoss, A. Ourmazd, J. Bevk, B.S. Freer, P. O. Hahn, *Appl. Phys. Lett.* **58**, 1044 (1991). *(393)*
[1234] T. A. Rabedeau, I. M. Tidswell, P.S. Pershan, J. Bevk, B.S. Freer, *Appl. Phys. Lett.* **59**, 3422 (1991). *(393, 409)*
[1235] C.L. Petersen, F. Grey, M. Aono, *Surf. Sci.* **377**, 676 (1997). *(393)*
[1236] P. O. Hahn, M. Henzler, *J. Appl. Phys.* **52**, 4122 (1981). *(396)*
[1237] S. T. Pantelides, M. Long, in *The Physics of SiO_2 and Its Interfaces*, ed. by S. T. Pantelides, (Pergamon, New York, 1978), p. 339. *(396, 398, 405)*
[1238] J.M. Gibson, M. Y. Lanzerotti, *Nature* **340**, 128 (1989). *(398, 409)*
[1239] Z. H. Lu, M. J. Graham, D. T. Jiang, K. H. Tan, *Appl. Phys. Lett.* **63**, 2941 (1993). *(399)*
[1240] K. Z. Zhang, M. M. Banaszak Holl, J. E. Bender IV, S. Lee, F.R. McFeely, *Phys. Rev. B* **54**, 7686 (1996). *(399)*
[1241] V. G. Zavodinsky, I. A. Kuyanov, *Surface Review and Letters* **3**, 1403 (1996). *(406)*
[1242] S. T. Dunham, *J. Appl. Phys.* **71**, 685 (1992). *(411, 413, 415)*
[1243] D. J. DiMaria, D. W. Dong, C. Falcony, T. N. Theis, J. R. Kirtley, J.C. Tsang, D. R. Young, F.L. Pesavento, S. D. Brorson, *J. Appl. Phys.* **54**, 5801 (1983). *(412)*
[1244] B. Fischer, A. Ghetti, L. Selmi, R. Bez, E. Sangiorgi, *IEEE Transactions on Electron Devices* **44**, 288 (1997). *(412, 436)*
[1245] D. J. DiMaria, J. W. Stasiak, *J. Appl. Phys.* **65**, 2342 (1989). *(412, 433, 437, 438)*
[1246] S. Mukherjee, P. E. Evans, *Thin Solid Films* **14**, 299 (1972). *(413)*
[1247] S.M. Hu, *J. Appl. Phys.* **45**, 1567 (1974). *(413, 415)*

[1248] S. I. Raider, R. A. Gdula, J. R. Petrak, *Appl. Phys. Lett.* **27**, 150 (1975). *(413)*
[1249] B. H. Vromen, *Appl. Phys. Lett.* **27**, 152 (1975). *(413)*
[1250] L. P. Hunter, *Handbook of Semiconductor Electronics*, third edition, ed. by L. P. Hunter (McGraw-Hill Book Company, 1976) p. 1. *(413, 420)*
[1251] S. P. Murarka, *Appl. Phys. Lett.* **34**, 587 (1979). *(413)*
[1252] R. B. Fair, in *Silicon Integrated Circuits, Part B*, ed. by Dawon Khang (Academic Press, 1981) p. 1. *(413, 416)*
[1253] G. A. Sai-Halasz, K.T. Short, J.S. Williams, *IEEE Electron Device Letters* **6**, 285 (1985). *(413)*
[1254] M. Orlowski, *Appl. Phys. Lett.* **55**, 1762 (1989). *(413, 418)*
[1255] G. Lucovsky, Z. Jing, D. R. Lee, *J. Vac. Sci. Technol. B* **14**, 2832 (1996). *(413, 425, 428)*
[1256] S. R. Kaluri, D. W. Hess, *J. Electrochem. Soc.* **144**, 2200 (1997). *(413)*
[1257] L. Perron, A. L. Lacaita, A. Pacelli, R. Bez, *IEEE Electron Device Letters* **18**, 235 (1997). *(413)*
[1258] W. K. Choi, F. W. Poon, F. C. Loh, K.L. Tan, *J. Appl. Phys.* **81**, 7386 (1997). *(413, 428)*
[1259] M.-y. Hao, D. Nayak, R. Rakkhit, *IEEE Electron Device Letters* **18**, 215 (1997). *(413)*
[1260] R. B. Fair, *IEEE Electron Device Letters* **18**, 244 (1997). *(413)*
[1261] C. T. Liu, Y. Ma, H.S. Luftman, S.J. Hillenius, *IEEE Electron Device Letters* **18**, 212 (1997). *(413)*
[1262] M. Itsumi, Y. Shigesato, S. Nakayama, S. Muramoto, *J. Appl. Phys.* **82**, 4629 (1997). *(413, 420, 430, 444)*
[1263] H. Kobayashi, S. Tachibana, K. Yamanaka, Y. Nakato, K. Yoneda, *J. Appl. Phys.* **81**, 7630 (1997). *(413, 414)*
[1264] J. Dąbrowski, H.-J. Müssig, *E-MRS 1999 Spring Meeting, Book of Abstracts*, L-5 (1999); `http://www.ihp-ffo.de/jd/muenster99.html`; to be published in *Materials Science in Semiconductor Processing* (1999). *(413, 417, 419, 422, 424)*
[1265] T. Y. Tan, U. M. Gösele, *Applied Physics A* **37**, 1 (1985). *(414)*
[1266] T. Y. Tan, B. J. Ginsberg, *Appl. Phys. Lett.* **42**, 448 (1983). *(414)*
[1267] C. P. Ho, J.D. Plummer, *J. Electrochem. Soc.* **126**, 1516 (1979). *(414)*
[1268] N. Jeng, S. T. Dunham, *J. Appl. Phys.* **72**, 2049 (1992). *(415)*
[1269] H. Bracht, N. A. Stolwijk, H. Mehrer, *Phys. Rev. B* **52**, 16542 (1995). *(415)*
[1270] H. Bracht, E. E. Haller, R. Clark-Phelps, *Phys. Rev. Lett.* **81**, 393 (1998). *(415)*
[1271] J. Zhu, T. Diaz de la Rubia, L. H. Yang, C. Mailhiot, G.H. Gilmer, *Phys. Rev. B* **54**, 4741 (1996). *(415)*
[1272] J. Dąbrowski, to be published in *Special Defects in Semi-Conducting Materials*, ed. by R. Agarwala *Solid State Phenomena* **71** (June/July 1999). *(415)*
[1273] S.M. Hu, *J. Appl. Phys.* **57**, 1069 (1985). *(415)*
[1274] S.M. Hu, *J. Appl. Phys.* **57**, 4527 (1985). *(415)*
[1275] S.M. Hu, *Appl. Phys. Lett.* **27**, 165 (1975). *(416)*
[1276] S. P. Murarka, *Phys. Rev. B* **16**, 2849 (1977). *(416)*
[1277] R. Kasnavi, P. Pianetta, Y. Sun, R. Mo, P.B. Griffin, J. D. Plummer, *IEDM Technical Digest*, 721 (1998). *(417)*
[1278] F. Lau, L. Mader, C. Mazure, Ch. Werner, M. Orlowski, *Applied Physics A* **49**, 671 (1989). *(417, 419, 421, 424)*
[1279] S. Mori, N. Ari, Y. Kaneko, K. Yoshikawa, *IEEE Transactions on Electron Devices* **38**, 270 (1991). *(417, 421)*
[1280] H. Sakamoto, S. Kumashiro, *SISPAD'97*, 81 (1997). *(418)*
[1281] H.-H. Vuong, C. S. Rafferty, J. Ning, J. R. McMacken, J. McKinley, F. A. Stevie,

SISPAD'98, 380 (1998). *(418, 419)*
[1282] T. Aoyama, K. Suzuki, H. Tashiro, Y. Tada, H. Arimoto, *IEDM Technical Digest*, 627 (1997). *(420, 444)*
[1283] H. P. Tuinhout, A. H. Montree, P.A. Stolk, *IEDM Technical Digest*, 631 (1997). *(420, 444)*
[1284] S. U. Kim, *IEEE Transactions on Electron Devices* **45**, 731 (1998). *(420, 429, 434, 436)*
[1285] T. Inoue, S. Horiuchi, S. Yoshii, Y. Tanuma, *Suppl. J. Jpn. Soc. Appl. Phys.* **44**, 329 (1975). *(420)*
[1286] R. B. Fair, *IEEE Electron Device Letters* **17**, 242 (1996). *(420)*
[1287] C. Y. Wong, F. S. Lai, *Appl. Phys. Lett.* **48**, 1658 (1986). *(420)*
[1288] T. Aoyama, K. Suzuki, Y. Toda, T. Yamazaki, K. Takashi, T. Ito, *J. Appl. Phys.* **77**, 417 (1995). *(420)*
[1289] S. I. Raider, in *The Physics and Chemistry of SiO2 and the Si-SiO2 Interface*, edited by C. R. Helms and B. E. Deal, (Plenum Press, New Your 1988) p. 35. *(420)*
[1290] V. Zubkov, S. Aronowitz, V. Sukharev, *E-MRS 1999 Spring Meeting, Book of Abstracts*, L-3 (1999). to be published in *Materials Science in Semiconductor Processing* (1999). *(420)*
[1291] P.A. Stolk, H.J. Gossmann, D. J. Eaglesham, D.C. Jacobson, C. S. Rafferty, G.H. Gilmer, M. Jaraiz, J.M. Poate, H.S. Luftman, T.E. Haynes, *J. Appl. Phys.* **81**, 6031 (1997). *(421)*
[1292] H.J. Gossmann, C. S. Rafferty, G. Hobler, H.-H. Vuong, D.C. Jacobson, M. Frei, *IEDM Technical Digest*, 725 (1998). *(421)*
[1293] H. R. Soleimani, *J. Electrochem. Soc.* **141**, 2182 (1994). *(421)*
[1294] K.C. Pandey, *Phys. Rev. Lett.* **57**, 2287 (1986). *(423)*
[1295] Y. Sato, M. Watanabe, K. Imai, *J. Electrochem. Soc.* **140**, 2679 (1993). *(424)*
[1296] R. J. Baierle, M. J. Caldas, J. Dąbrowski, H.-J. Müssig, W. Zavodinsky, paper accepted for presentation at International Conference of Defects in Semiconductors, Berkeley, USA, 1999; manuscript submitted. *(424)*
[1297] J. Mitsuhashi, S. Nakao, T. Matsukawa, *IEDM Technical Digest*, 386 (1986). *(425)*
[1298] I.J.R. Baumvol, E. P. Gusev, F.C. Stedile, F. L. Freire Jr., M.L. Green, D. Brasen, *Appl. Phys. Lett.* **72**, 450 (1998). *(426)*
[1299] K. L. Brower, S. M. Myers, *Appl. Phys. Lett.* **57**, 162 (1990). *(426)*
[1300] J. W. Lyding, T.C. Shen, J.S. Hubacek, J.R. Tucker, G.C. Abeln, *Appl. Phys. Lett.* **64**, 2010 (1994). *(426, 427)*
[1301] E. T. Foley, A. F. Kam, J. W. Lyding, Ph. Avouris, *Phys. Rev. Lett.* **80**, 1336 (1998). *(426)*
[1302] T.C. Shen, C. Wang, G.C. Abeln, J.R. Tucker, J. W. Lyding, Ph. Avouris, R. E. Walkup, *Science* **268**, 1590 (1995). *(427)*
[1303] C. G. Van de Walle, W. B. Jackson, *Appl. Phys. Lett.* **69**, 2441 (1996). *(427)*
[1304] C. G. Van de Walle, *Phys. Rev. B* **49**, 4579 (1994). *(427)*
[1305] C. G. Van de Walle, *J. Vac. Sci. Technol. A* **16**, 1767 (1998). *(427)*
[1306] K. Hess, I. C. Kizilyalli, J. W. Lyding, *IEEE Transactions on Electron Devices* **45**, 406 (1998). *(427)*
[1307] I. P. Ipatova, O. P. Chikalova-Luzina, K. Hess, *J. Appl. Phys.* **83**, 814 (1998). *(427)*
[1308] Z. Jing, G. Lucovsky, J. L. Whitten, *J. Vac. Sci. Technol. B* **13**, 1613 (1995). *(428)*
[1309] R. Degraeve, G. Groeseneken, I. De Wolf, H. E. Maes, *IEEE Transactions on*

Electron Devices **44**, 943 (1997). *(428)*
[1310] D. Park, C. Hu, *IEEE Electron Device Letters* **19**, 1 (1998). *(429, 433)*
[1311] S. C. Witczak, J. S. Suehle, M. Gaitan, *Solid-State Electronics* **35**, 345 (1992). *(429)*
[1312] K. P. Cheung, S. Martin, D. Misra, K. Steiner, J. I. Colonell, C. P. Chang, W. Y. C. Lai, C. T. Liu, R. Liu, C. S. Pai, *IEDM Technical Digest*, 437 (1997). *(429, 437)*
[1313] H. J. Wen, R. Ludeke, *J. Vac. Sci. Technol. B* **15**, 1080 (1997). *(429)*
[1314] D. M. McWhorter, P. S. Winokur, *Appl. Phys. Lett.* **48**, 133 (1986). *(429)*
[1315] R. DeJule, *Semiconductor International* **21**, 52 (May 1998). *(430)*
[1316] D. R. Wolters, A. T. A. Zegers-van Duynhoven, *Philosophical Transactions of the Royal Society A* **354**, 2327 (1996). *(430, 438)*
[1317] D. R. Wolters, J. F. Verwey, in *Instabilities in Silicon Devices*, edited by G. M. Barbottin, A. Vapaille, (Elsevier Science, 1986), p. 329. *(431)*
[1318] N. Klein, H. Gafni, *IEEE Transactions on Electron Devices* **13**, 281 (1966). *(431)*
[1319] D. J. DiMaria, E. Cartier, *J. Appl. Phys.* **78**, 3883 (1995). *(431)*
[1320] T. Sakura, H. Utsunomiya, Y. Kamakura, K. Taniguchi, *IEDM Technical Digest*, 183 (1998). *(431, 434)*
[1321] K. Okada, *Jap. J. Appl. Phys.* **36**, 1443 (1997). *(432)*
[1322] E. Wu, E. Nowak, J. Aitken, W. W. Abadeer, L.-K. Han, S. Lo, *IEDM Technical Digest*, 187 (1998). *(432, 433, 434)*
[1323] C. C. Chen, C.-Y. Chang, C.-H. Chien, T.-Y. Huang, H.-C. Lin, M. S. Liang, *Appl. Phys. Lett.* **74**, 3708 (1999). *(432)*
[1324] K. Noguchi, K. Tokashiki, T. Horiuchi, M. Miyamoto, *IEDM Technical Digest*, 441 (1997). *(433)*
[1325] S. Krishnan, A. Amerasekera, S. Rangan, S. Aur, *IEDM Technical Digest*, 601 (1998). *(433)*
[1326] J. McPherson, V. Reddy, K. Banerjee, H. Le, *IEDM Technical Digest*, 171 (1998). *(433)*
[1327] S. Ogura, P. J. Tseng, W. W. Walker, D. L. Critchlow, J. F. Shephard, *IEEE Transactions on Electron Devices* **27**, 1359 (1980). *(433)*
[1328] W. S. Whang, B. Davari, M. R. Wordeman, Y. Taur, C. C. H. Hsu, M. D. Rodrigues, *IEEE Transactions on Electron Devices* **39**, 959 (1992). *(433, 436)*
[1329] J. P. McVittie, *IEDM Technical Digest*, 433 (1997). *(434)*
[1330] S. Krishnan, S. Rangan, S. Hattangandy, G. Xing, K. Brennan, M. Rodder, S. Ashok, *IEDM Technical Digest*, 445 (1997). *(434)*
[1331] J. Suñé, E. Miranda, M. Nafria, X. Aymerich, *IEDM Technical Digest*, 191 (1998). *(434)*
[1332] W. Shockley, *Solid-State Electronics* **2**, 35 (1961). *(436)*
[1333] S. Tam, P.-K. Ko, C. Hu, *IEEE Transactions on Electron Devices* **31**, 1116 (1984). *(436)*
[1334] J Frey, *IEEE Circuits and Devices Magazine* **7**, 31 (November 1991). *(436)*
[1335] D. Arnold, E. Cartier, D. J. DiMaria, *Phys. Rev. B* **45**, 1477 (1992). *(436)*
[1336] J. D. Bude, B.E. Weir, P.J. Silverman, *IEDM Technical Digest*, 179 (1998). *(436)*
[1337] D. A. Buchanan, D. J. DiMaria, *J. Appl. Phys.* **67**, 7439 (1990). *(437)*
[1338] M. Inoue, A. Shimada, J. Shirafuji, *Jap. J. Appl. Phys.* **35**, 5921 (1996). *(437, 443)*
[1339] B. B. Jie, M. F. Li, C. L. Lou, W. K. Chim, D. S. H. Chan, K. F. Lo, *IEEE Electron Device Letters* **18**, 583 (1997). *(438)*

[1340] D. Vasileska, D. A. Ferry, *IEEE Transactions on Electron Devices* **44**, 577 (1997). *(439)*

[1341] C. S. Rafferty, B. Biegel, Z. Yu, M. G. Ancona, J. D. Bude, R. W. Dutton, *SISPAD'98*, 137 (1998). *(439)*

[1342] A. v. d. Ziel, *Noise in Solid State Devices and Circuits* (Wiley, New York, 1986). *(440)*

[1343] E. Simoen, P. Vasina, J. Sikula, C. Claeys, *IEEE Electron Device Letters* **18**, 480 (1997). *(441, 442)*

[1344] H. S. Momose, H. Kimijima, S.-i. Ishizuka, Y. Miyahara, T. Ohguro, T. Yoshitomi, E. Morifuji, S.-i. Nakamura, Y. Katsumara, T. Morimoto, H. Iwai, *IEDM Technical Digest*, 923 (1998). *(441)*

[1345] F. Herzel, B. Razavi, *IEEE Transactions on Circuit and Systems - II* **48**, 56 (1999). *(442, 443)*

[1346] J. S. Suehle, P. Chaparala, *IEEE Transactions on Electron Devices* **44**, 801 (1997). *(442)*

[1347] H. Ito, M. Sasaki, N. Kimizuka, K. Uwasawa, N. Nakamura, T. Ito, Y. Goto, A. Tsuboi, S. Watanuki, T. Ueda, T. Horiuchi, *IEDM Technical Digest*, 635 (1997). *(444)*

[1348] S. Hawking, "Oxford and Cambridge", in *Black Holes and Baby Universies*, (Bantam Books, New York, 1994), p. 13. *(445)*

[1349] P. F. Drucker, *Harvard Business Review, November-December*, 149 (1998). *(446)*

[1350] C. R. Helms, *Solid State Technology* **41**, 112 (August 1998). *(448, 450)*

[1351] J. Ziman, *Nature* **382**, 751 (1996). *(448)*

[1352] *Pure science in an age of attrition* (editorial) *Nature* **379**, 1 (1996). *(448)*

[1353] J. Woodall, *Interface* **8**, 11 (Summer 1999). *(448)*

[1354] *Webster's 9th New Collegiate Dictionary* (Merriam-Webster, 1988). *(448)*

[1355] W. Heisenberg, (R. Piper & Co. Verlag, München, 1969). *(450)*

[1356] S. Hawking, "Public attitudes toward science" in *Black Holes and Baby Universies*, (Bantam Books, New York, 1994), p. 27. *(450)*

[1357] J. D. Barrow, *Theories of Everything. The Quest for Ultimate Explanation* (Oxford University Press, Oxford, 1991). *(457)*

[1358] B. Geerlings, S. Radelaar, G. Janssen, *Vacuum Solutions* **7**, 13 (January/February 1999). *(458)*

Author Index

This index lists all author names cited, followed by the corresponding literature reference numbers. Since the index contains only family names, with all accents ignored, a single entry may combine contributions from a few persons.

Subject Index

www.ingramcontent.com/pod-product-compliance
Ingram Content Group UK Ltd.
Pitfield, Milton Keynes, MK11 3LW, UK
UKHW022109170926
8624IPUK00024B/349

9 789810 232863